Current Topics in
Developmental Biology
Volume 48

**Somitogenesis
Part 2**

Series Editors

Roger A. Pedersen and
Reproductive Genetics Division
Department of Obstetrics, Gynecology,
and Reproductive Sciences
University of California
San Francisco, California 94143

Gerald P. Schatten
Departments of Obstetrics–Gynecology
and Cell and Developmental Biology
Oregon Regional Primate Research Center
Oregon Health Sciences University
Beaverton, Oregon 97006-3499

Editorial Board

Peter Grüss
Max Planck Institute of Biophysical Chemistry
Göttingen, Germany

Philip Ingham
University of Sheffield, United Kingdom

Mary Lou King
University of Miami, Florida

Story C. Landis
National Institutes of Health/
National Institute of Neurological Disorders and Stroke
Bethesda, Maryland

David R. McClay
Duke University, Durham, North Carolina

Yoshitaka Nagahama
National Institute for Basic Biology, Okazaki, Japan

Susan Strome
Indiana University, Bloomington, Indiana

Virginia Walbot
Stanford University, Palo Alto, California

Founding Editors

A. A. Moscona
Alberto Monroy

Somitogenesis
Part 2

Edited by

Charles P. Ordahl
Department of Anatomy
Cardiovascular Research Institute
University of California
San Francisco, California

Academic Press
San Diego London Boston New York Sydney Tokyo Toronto

Cover Photo: A thoracic somite (stage VI) from a 2-day-old chicken embryo was iontophoretically microinjected, at mid-position in the dermomyotome dorso-medial lip, using diI, a fluorescent carbocyanine plasma membrane dye as described in Chapter 9. After 15 hours of embryo growth, a brightly labeled, epaxial myotome fiber that is fully elongated across the cranio-caudal somite axis appears. The myotome fiber, with a centrally located nucleus, spans the developing Von Ebner fissure and terminates at the cranial and caudal dermomyotome borders. The myotome fiber also exhibits a slight outward bow due to the presence of many juxtapositioned, unlabeled fibers located beneath it. Image courtesy of William F. Denetclaw, Jr.

This book is printed on acid-free paper.

Copyright © 2000 by ACADEMIC PRESS

All Rights Reserved.
No part of this publication may be reproduced or transmitted in any form or by any means, electronic or mechanical, including photocopy, recording, or any information storage and retrieval system, without permission in writing from the Publisher.

The appearance of the code at the bottom of the first page of a chapter in this book indicates the Publisher's consent that copies of the chapter may be made for personal or internal use of specific clients. This consent is given on the condition, however, that the copier pay the stated per copy fee through the Copyright Clearance Center, Inc. (222 Rosewood Drive, Danvers, Massachusetts 01923), for copying beyond that permitted by Sections 107 or 108 of the U.S. Copyright Law. This consent does not extend to other kinds of copying, such as copying for general distribution, for advertising or promotional purposes, for creating new collective works, or for resale. Copy fees for pre-2000 chapters are as shown on the title pages. If no fee code appears on the title page, the copy fee is the same as for current chapters.
0070-2153/00 $30.00

Explicit permission from Academic Press is not required to reproduce a maximum of two figures or tables from an Academic Press chapter in another scientific or research publication provided that the material has not been credited to another source and that full credit to the Academic Press chapter is given.

Academic Press
A Harcourt Science and Technology Company
525 B Street, Suite 1900, San Diego, California 92101-4495, U.S.A.
http://www.apnet.com

Academic Press
24-28 Oval Road, London NW1 7DX, UK
http://www.hbuk.co.uk/ap/

International Standard Book Number: 0-12-153148-1

PRINTED IN CANADA
99   00   01   02   03   04   05   FR   9   8   7   6   5   4   3   2   1

Contents

Contributors   ix
Preface   xi

1

Evolution and Development of Distinct Cell Lineages Derived from Somites
Beate Brand-Saberi and Bodo Christ

    I. Introduction   2
   II. Metamerism   2
  III. Somite Research Past and Present   6
  IV. Resegmentation   14
   V. The Muscle Lineage   26
  VI. Further Somitic Derivatives: Dermis, Smooth Muscle, and Angioblasts   32
 VII. Evolution of Metamerism in Vertebrates   34
      References   40

2

Duality of Molecular Signaling Involved in Vertebral Chondrogenesis
Anne-Hélène Monsoro-Burq and Nicole Le Douarin

    I. Origin of the Vertebrae and Intervertebral Disks   43
   II. Gene Expression in the Developing Somitic Cells   44
  III. The Dorsal Neural Tube Grafted Ectopically Promotes Dorsal Cartilage Formation in a Subectodermal Position   51
  IV. Ventral Axial Structures Prevent Subectodermal Cartilage Formation   54
   V. Molecular Pathways Leading to Chondrogenesis in the Vertebra   57
  VI. Conclusions: A Novel Model for Vertebra Development   65
      References   70

v

3

Sclerotome Induction and Differentiation
Jennifer L. Dockter

- I. Introduction 77
- II. Induction: Definitions 78
- III. Anatomical and Morphological Description of Sclerotome Development 79
- IV. Genetics 83
- V. Nonmolecular Approaches 85
- VI. The Molecular Approach 98
- VII. *Shh*, cAMP, and Sclerotome 106
- VIII. Issues Concerning *Shh* as the Sclerotome Inducer 106
- IX. Initiation of the Sclerotome 108
- X. *Shh, Noggin*, and Somite Chondrogenesis 109
- XI. Bone Morphogenetic Proteins and Sclerotome Development 110
- XII. Determination of the Sclerotome 110
- XIII. Epithelial–Mesenchymal Transitions and Sclerotome Formation 113
- XIV. Dorsal–Ventral Patterning of the Somite 113
- XV. Interactions between the Somitic Components 114
- XVI. Discussion 115
- XVII. Summary 117
- References 118

4

Genetics of Muscle Determination and Development
Hans-Henning Arnold and Thomas Braun

- I. Introduction 129
- II. The Origin of Skeletal Muscles in Vertebrate Organisms 130
- III. The MyoD Family of Muscle-Specific Transcription Factors 131
- IV. Development of the Limb Musculature 151
- V. Structures Surrounding the Somites and Signals Emanating from Them Affect Myogenesis 153
- VI. The Myogenic Potential of Mesoderm: Induction or Derepression? 155
- References 157

5

Multiple Tissue Interactions and Signal Transduction Pathways Control Somite Myogenesis
Anne-Gaëlle Borycki and Charles P. Emerson, Jr.

- I. Introduction 166
- II. Mesodermal Origins of Skeletal Muscle in Vertebrate Embryos 167
- III. Regulatory Genes That Control Somite Myogenesis 170

Contents

- IV. Tissue Interactions That Control Somite Myogenesis 174
- V. Signaling Molecules and Transduction Pathways Controlling Myogenesis in Somites 185
- VI. Overview and Future Perspectives 207
   References 212

6

The Birth of Muscle Progenitor Cells in the Mouse: Spatiotemporal Considerations

Shahragim Tajbakhsh and Margaret Buckingham

- I. Introduction 225
- II. Origins of Skeletal Muscle in Vertebrates 226
- III. Domains of the Dermomyotome 227
- IV. Differences between Somites at Different Axial Levels 235
- V. Myogenic Regulatory Factor Expression Patterns in the Somite and Myotome Heterogeneity 237
- VI. Extrinsic Factors Govern Somite Differentiation and the Activation of Myf5 and MyoD 242
- VII. Do Myf5 and MyoD Define Different Subpopulations of Muscle Cells? 246
- VIII. The Roles of Myf5 and MyoD in Muscle Progenitor Cell Determination 250
- IX. Conclusions 260
    References 261

7

Mouse–Chick Chimera: An Experimental System for Study of Somite Development

Josiane Fontaine-Pérus

- I. Introduction 269
- II. Technical Aspects 271
- III. Somitic Lineage into Somite Chimera 274
- IV. Epaxial Myogenic Lineage into Neural Tube Chimera 289
- V. What Is the Benefit of the Somite–Neural Tube Chimera? 292
- VI. Concluding Remarks 294
    References 296

8

Transcriptional Regulation during Somitogenesis

Dennis Summerbell and Peter W. J. Rigby

- I. Introduction 301
- I. Regulation of *Hox* Gene Expression 302
- III. Control of Myogenic Regulatory Factor Gene Expression 306

	IV.	Roles of Non-MRF Transcription Factors in Muscle Development	311
	V.	The Identification of Novel Somite-Specific Transcripts	312
		References	315

9

Determination and Morphogenesis in Myogenic Progenitor Cells: An Experimental Embryological Approach

Charles P. Ordahl, Brian A. Williams, and Wilfred Denetclaw

- I. Introduction 320
- II. The Diversity and Location of Muscle Precursor Cells in the Vertebrate Embryo 325
- III. When Do Myotome Precursor Cells Become Determined? 344
- IV. Gene Expression and Determination in Migratory Limb Muscle Precursor Cells 350
- V. A Cellular and Molecular Framework for Understanding Myogenic Tissue Development 353
- VI. Appendix: Key Word Guide 360
  References 361

Index 369
Contents of Previous Volumes 389

Contributors

Numbers in parentheses indicate the pages on which authors' contributions begin.

Hans Henning Arnold (129), Department of Cell and Molecular Biology, Institute of Biochemistry and Biotechnology, Technical University of Braunschweig, 38106 Braunschweig, Germany

Anne-Gaëlle Borycki (165), Department of Cell and Developmental Biology, University of Pennsylvania School of Medicine, Philadelphia, Pennsylvania 19104

Beate Brand-Saberi (1), Institute of Anatomy, University of Freiburg, D-79001 Freiburg, Germany

Thomas Braun (129), Department of Cell and Molecular Biology, Institute of Biochemistry and Biotechnology, Technical University of Braunschweig, 38106 Braunschweig, Germany

Margaret Buckingham (225), CNRS URA 1947, Department of Molecular Biology, Pasteur Institute, 75724 Paris Cedex 15, France

Bodo Christ (1), Institute of Anatomy, University of Freiburg, D-79001 Freiburg, Germany

Wilfred Denetclaw (319), Department of Anatomy, Cardiovascular Research Institute, University of California, San Francisco, San Francisco, California 94143

Jennifer Dockter (77), Department of Anatomy, Cardiovascular Research Institute, University of California, San Francisco, San Francisco, California 94143

Charles P. Emerson, Jr. (165), Department of Cell and Developmental Biology, University of Pennsylvania School of Medicine, Philadelphia, Pennsylvania 19104

Josiane Fontaine-Pérus (269), CNRS EP1593, Faculté des Sciences et des Techniques, 44322 Nantes Cedex 03, France

Nicole Le Douarin (43), Institut d'Embryologie Cellulaire at Moléculaire du CNRS et du Collège de France, 94736 Nogent-sur-Marne, France

Anne-Hélène Monsoro-Burq (43), Institut d'Embryologie Cellulaire at Moléculaire du CNRS et du Collège de France, 94736 Nogent-sur-Marne, France

Charles P. Ordahl (319), Department of Anatomy, Cardiovascular Research Institute, University of California, San Francisco, San Francisco, California 94143

Peter W. J. Rigby (301), Division of Eukaryotic Molecular Genetics, MRC National Institute for Medical Research, The Ridgeway, Mill Hill, London NW7 1AA, England

Dennis Summerbell (301), Division of Eukaryotic Molecular Genetics, MRC National Institute for Medical Research, The Ridgeway, Mill Hill, London NW7 1AA, England

Shahragim Tajbakhsh (225), CNRS URA 1947, Department of Molecular Biology, Pasteur Institute, 75724 Paris Cedex 15, France

Brian A. Williams (319), Department of Anatomy, Cardiovascular Research Institute, University of California, San Francisco, San Francisco, California 94143

Preface

Somitogenesis has fascinated embryologists for centuries. Somites were identified as the primordia of the segmented organization of vertebrates more than two centuries ago, but it was only at the end of the 19th century that embryologists deduced that the somites are embryonic tissue progenitor organs giving rise to cartilage, skeletal muscle, and dermis tissues. Today we recognize somites as intermediates in the overall process of somitogenesis, the development of the paraxial mesoderm from gastrulation and on into the fetal period. During the 20th century, the somite has become an increasingly attractive target for experimentalists in their investigations of the cellular and molecular underpinnings of fundamental processes in vertebrate development: segmentation, tissue morphogenesis, and cell fate determination. There are many reasons for the attractiveness of the somite as an experimental target, for example, its compact size and limited cell number. However, the special fascination of somitogenesis clearly derives from its very regularity—regularity in terms not only of the sequential, reiterative, and predictable nature of its development, but also of the relative constancy of somitogenesis in embryos from different branches of the vertebrate subphylum.

The 18 chapters in these volumes address a wide variety of developmental problems through the study of somitogenesis in the embryos of four vertebrate classes (fish, amphibians, birds, and mammals). Each also presents a different strategic approach to the study of somitogenesis encompassing genetics, molecular biology, cell biology, and experimental embryology, including xenotransplantation between vertebrate classes. That diversity, however, serves an underlying commonality of themes regarding fundamental problems in development, the mechanisms of segmentation, embryo growth and cellular morphogenesis, and the molecular/genetic control of development. Finally, because overlap between chapters has been kept to a minimum, it must be acknowledged that the chapters in these volumes cannot fully represent all of the new and exciting research currently being conducted on somitogenesis. The sequence of chapters is intended to proceed from the more general to the more specific; thus, early chapters deal with gastrulation and segmentation of the paraxial mesoderm, while later chapters deal with the subsequent elaboration of specific differentiated tissues from the somite.

Although the subject was originally intended to be covered in a single volume, two volumes became necessary because of the overall length of the 18 chapters. The first volume deals with the earliest phases of paraxial mesoderm development, including gastrulation and segmentation. The chapters in the second volume deal with the later development of somites, including the elaboration of specific differentiated tissues.

I particularly thank all of the authors, whose efforts have made this two-volume set possible and with whom it has been a pleasure to work. My involvement in organizing this project was only possible with the encouragement and support of my friend and colleague Roger Pedersen and with the patience and understanding of Craig Panner at Academic Press, who diplomatically and successfullly guided the project to completion. Finally, I acknowledge my debt of eternal gratitude to Professor Margaret Hess, who first introduced me to the study of embryology 36 years ago.

Charlie Ordahl

1 Evolution and Development of Distinct Cell Lineages Derived from Somites

Beate Brand-Saberi and Bodo Christ
Institute of Anatomy
University of Freiburg
D-79001 Freiburg, Germany

I. Introduction
II. Metamerism
III. Somite Research Past and Present
IV. Resegmentation
V. The Muscle Lineage
VI. Further Somitic Derivatives: Dermis, Smooth Muscle, and Angioblasts
VII. Evolution of Metamerism in Vertebrates
References

In the vertebrate embryo, the somites arise from the paraxial mesoderm as paired mesodermal units in a craniocaudal sequence. Segmentation is also the underlying principle of the body plan in annelids and arthropods. Genes controlling segmentation have been identified that are highly conserved in organisms belonging to different phyla. Segmentation facilitates movement and regionalization of the vertebrate body. Its traces in humans are, for example, vertebral bodies, intervertebral disks, ribs, and spinal nerves.

Somite research has a history of at least three centuries. Detailed morphological data have accumulated on the development of the avian somite. Especially in connection with the quail–chick interspecific marker system, progress was made toward an understanding of underlying mechanisms. At first each somite consists of an outer epithelium and a mesenchymal core. Later, the ventral portion of the somite undergoes de-epithelialization and gives rise to the sclerotome, whereas the dorsal portion forms the dermomyotome. The dermomyotome is the source of myotomal muscle cells and the dermis of the back. It also yields the hypaxial muscle buds at flank level and the myogenic cells invading the limb buds. The dorsal and ventral somitic domains express different sets of developmental control genes, for example, those of the *Pax* family. During later stages of development, the sclerotomes undergo a new arrangement called "resegmentation" leading to the fusion of the caudal half of one sclerotome with the cranial half of the following sclerotome. Further somitic derivatives include fibroblasts, smooth muscle, and endothelial cells. While sclerotome formation is controlled by the notochord, signals from the dorsal neural tube and ectoderm support the development of the dermomyotome. Myogenic precursor cells for the limb bud are recruited from the dermomyotome by the interaction of c-met with its ligand scatter factor (SF/HGF). In the evolution of metamerism in vertebrates, the first skeletal elements were primitive parts of neural arches, while axial elements developed only later in teleosts as pleurocentra and hypocentra.

I. Introduction

Somites are the earliest manifestation of body segmentation or metamerism in vertebrates. This feature is a principle realized in the body plans of invertebrate phyla such as worms and arthropods as well. In this review, we give an overview of segmentation and somite-derived cell lineages in a functional, phylogenetic, and developmental context. Moreover, we discuss some historic landmarks of research on this topic and show how these data have influenced our way of thinking. Special emphasis will be given to the historical development of the resegmentation concept of the sclerotome and to the myogenic lineage.

II. Metamerism

The fascinating pattern of metamerism underlies the body plans of annelid worms and insects as well as vertebrates (Fig. 1). Hence we may ask in what way segmentation of the body is useful for the individual and is therefore worth conserving during evolution. When we look at technology, we can understand what a construction of almost identical segments means: It enables these objects to bend. A chain and a train are composed of several to many individual entities arranged in a sequence. While the worm resembles more the chain, insects and vertebrates are more easily compared with a train where groups of railcar types correspond to tagmata or regionally grouped vertebrae.

The segmentally formed spine has a bearing on posture and locomotion. It

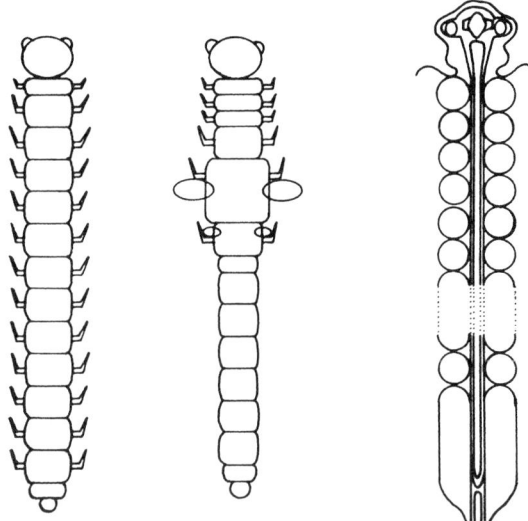

Figure 1 Metamerism in an idealized worm, arthropod, and vertebrate embryo.

1. Somitic Cell Lineages

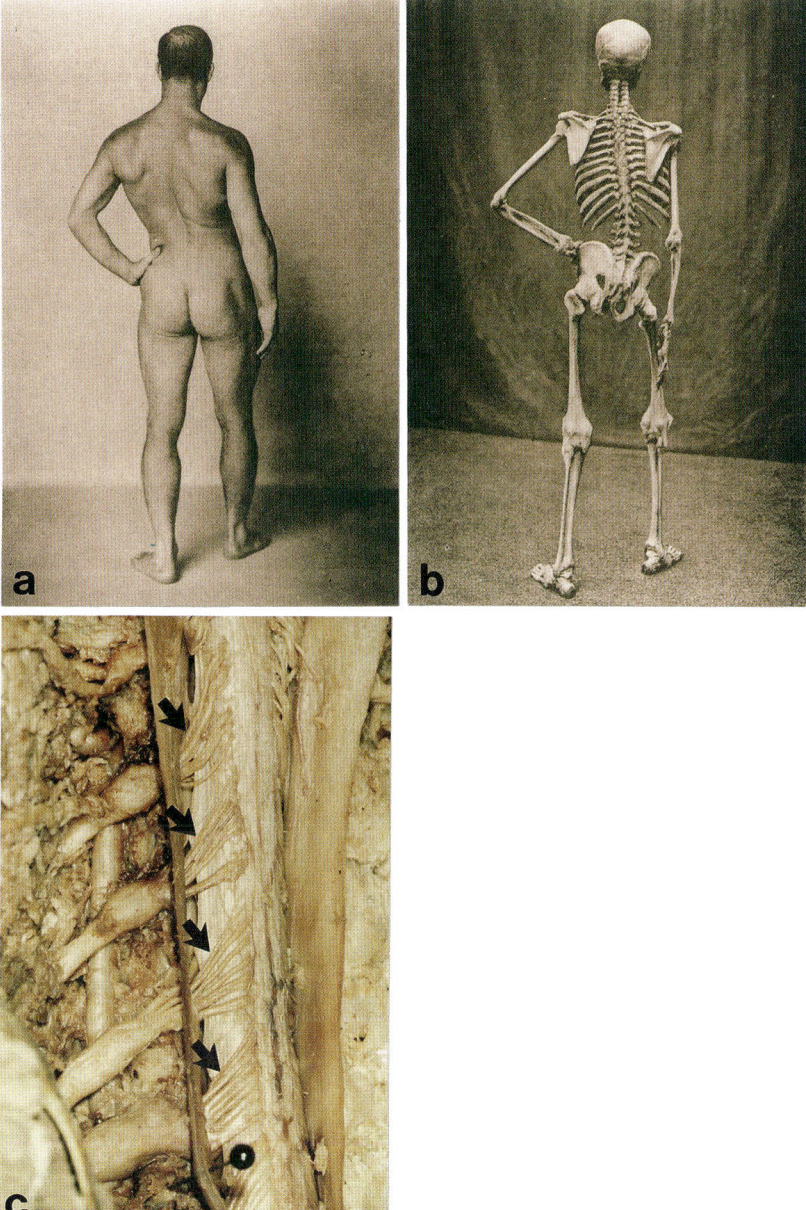

Figure 2 Metamerism of the human body. (a) The human body does not appear to be segmented. (b) Evidence for segmental structures: the vertebrae, ribs. (c) Segmental spinal nerves. Note the bundling of dorsal root fibers (arrows) (magnification: ×100).

allows us to change direction and to bend and rotate the body. The special freedom of movement of the neck and the head gives us the possibility to orientate our sensory organs such as eyes and ears. It allows bipeds in particular to look at the stars and to project angels and God into the sky. This means that segmentation of the human body, as an organizational prerequisite, has influenced the development of literature, philosophy, and religion.

Conditions leading to an inability to move the spine, like Bechterew's disease, result in very serious problems for the patients. Body posture is also meaningful in a social context for signaling information concerning age and status of an individual.

One striking difference between segmentation in insects such as *Drosophila* and vertebrates is the fact that in the fly, segmentation is a subdivision of the already existing body, whereas in vertebrates, segmentation starts in the head and proceeds caudally adding more segments by apposition. Moreover, looking at

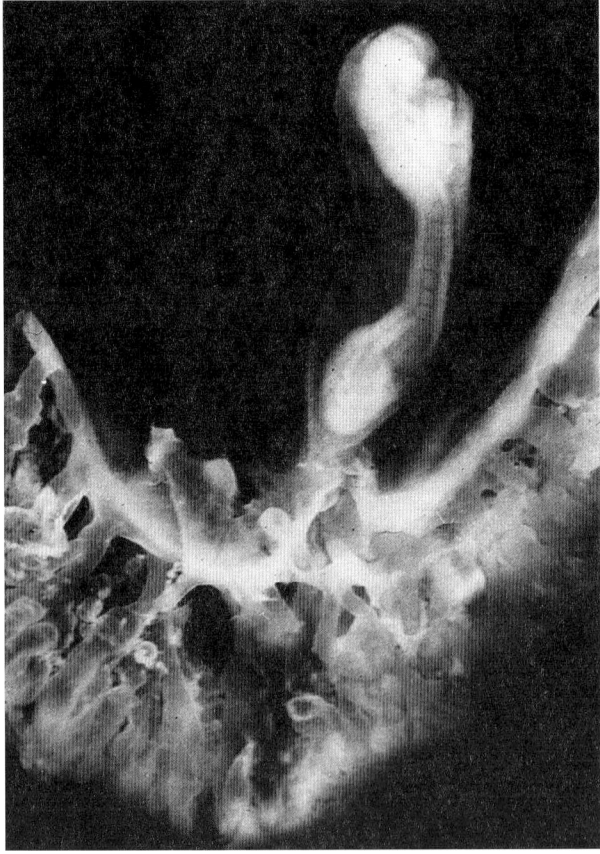

Figure 3 Human embryo with 14 somite pairs (magnification: ×20). (Original photograph by Erich Blechschmidt.)

1. Somitic Cell Lineages 5

the vertebrate body like that of a human, it is difficult to see segmental structures. However, when looking with the eyes of a human anatomist, you will become aware of the metamerism of vertebral bodies, intervertebral disks, ribs, spinal nerves etc. (Fig. 2).

Segmentation of the vertebrate body is especially evident during early embryonic development, where it becomes first manifest by the appearance of paired mesodermal structures, the somites (Fig. 3). Although this subject has attracted the attention of scientists for more than 300 years, it has only been in recent years that genes have been discovered that are involved in segmentation (Fig. 4; also see Chap. 6 by Rawls *et al.,* this volume).

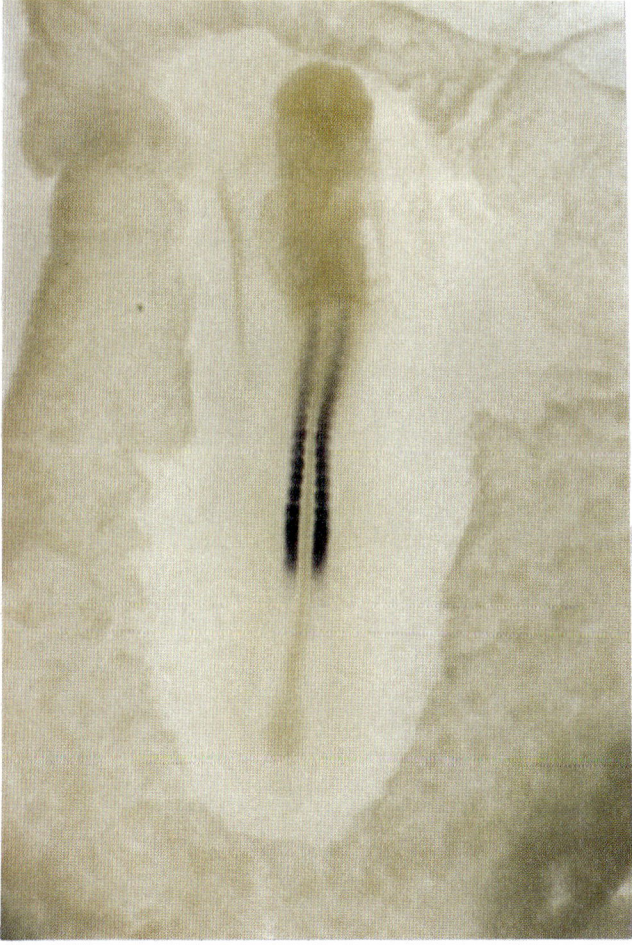

Figure 4 *Paraxis* expression in the 2-day chick embryo. Chick embryo of 2.5 days. *In situ* hybridization for *paraxis* (magnification: ×20). (Courtesy of Eric Olson, Dallas.)

III. Somite Research Past and Present

Let us now take a look at the somites as the primary elements of segmentation. In the chick embryo, a considerable body of information has accumulated concerning both morphological data and mechanisms.

First descriptions date back to 1689, when Malpighi published a paper on chick development and outlined embryos of different stages taking into account the segmentation of the body. The famous German embryologist Karl-Ernst von Baer in 1828 described chick somites which he called vertebra rudiments. Another very detailed description of the formation and differentiation of the somites goes back to His in 1886 (Fig. 5). His came to the conclusion that the somites form the wall of the aorta and do not contribute to the vertebral column. He supposed that the material for the vertebral column invades the embryo from extraembryonic sites. On the other hand, His was able to state that the dorsal root ganglia do not derive from the somites but from the neural tube.

During the following decades, many publications appeared dealing with the structure of somites in fishes, reptiles, amphibians, birds, and mammals. All of these observations have formed our view of somite development. In the chick, the early somite consists of an outer epithelium enveloping mesenchymal somitocoel cells (Fig. 6). Some hours later, this somite has changed its structure dramatically.

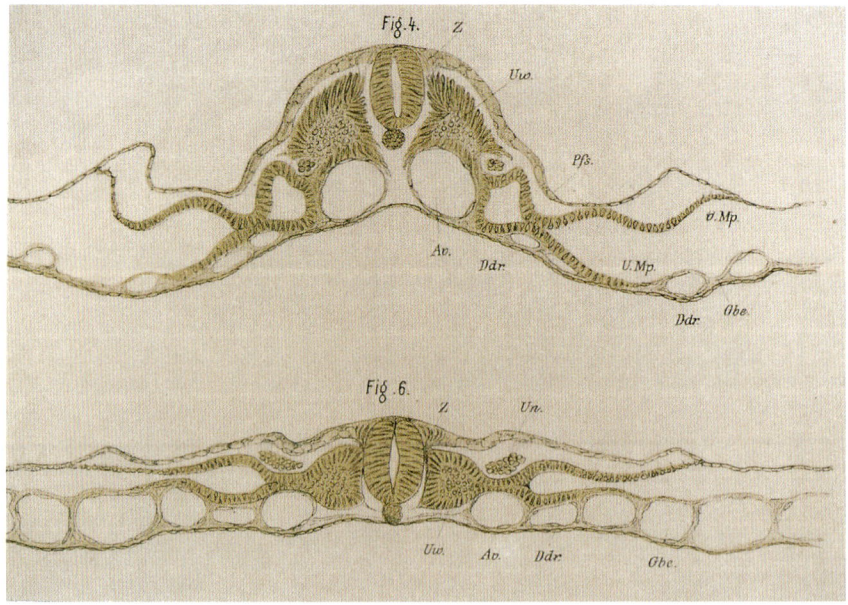

Figure 5 Original drawing of the chick embryo by Wilhelm His (1886).

1. Somitic Cell Lineages

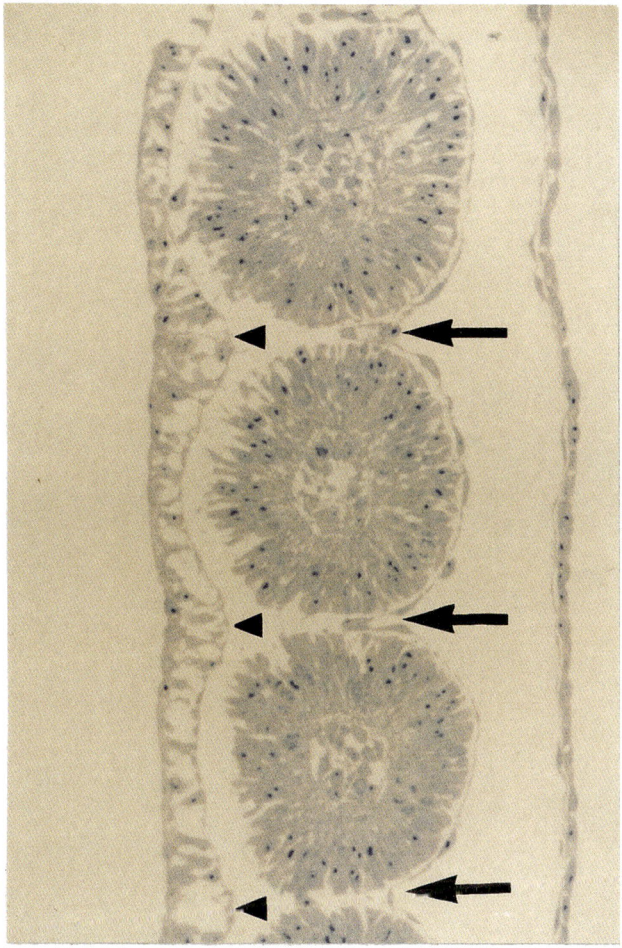

Figure 6 Semithin section of epithelial somite. In the center of each somite mesenchymal somitocoel cells are visible. Ectoderm on the left, endoderm on the right. Note the intersegmental vessels (arrows) and ectoderm wedges (arrowheads) (magnification: ×300).

The ventral part undergoes an epitheliomesenchymal transition forming the sclerotome (Fig. 7). This process is accompanied by a number of changes in the expression of developmental control genes, especially that of the paired-box (Pax) family. *Pax-3* which is initially expressed throughout the paraxial mesoderm is driven to the dorsal part of the somite (Goulding *et al.*, 1994), while *Pax-1* and *Pax-9* expression occupies the ventral part, which is destined to form the sclerotome (Fig. 8) (T. S. Müller *et al.*, 1996). The sclerotome yields the material for the vertebral column and ribs. The dorsal part has retained its epithelial structure and

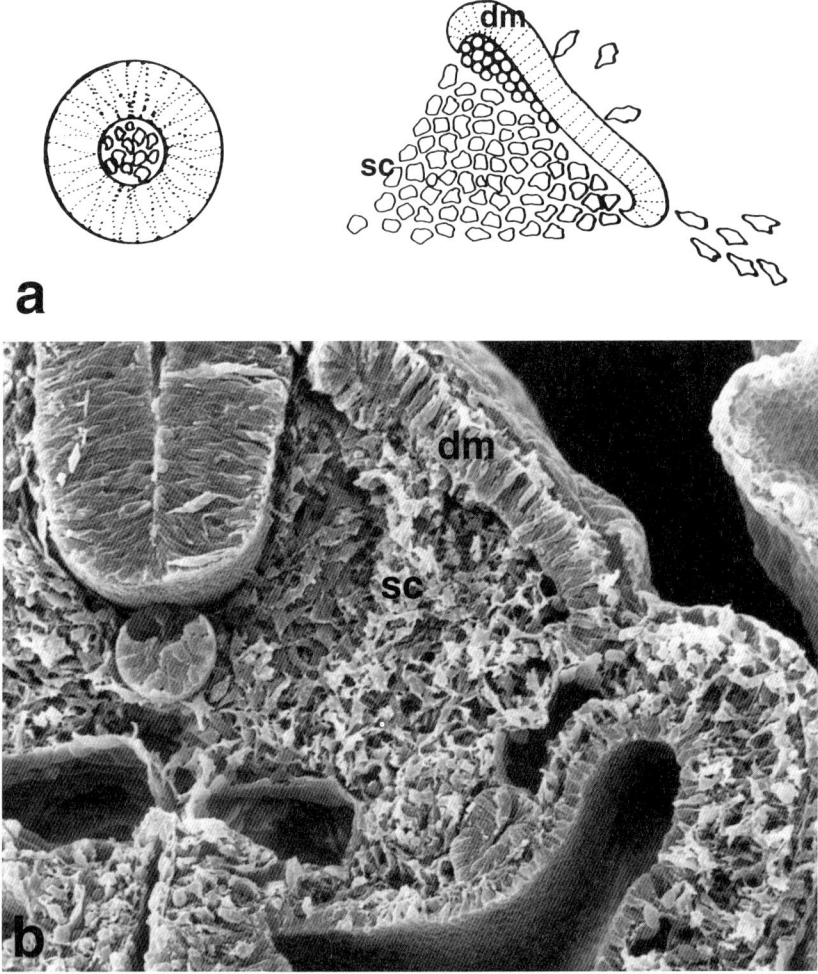

Figure 7 (a) Schematic drawing of sclerotome formation from the epithelial somite. (b) Scanning electron micrograph of a chick embryo. The ventral part of the somite has formed the mesenchymal sclerotome (magnification: ×550). dm, Dermomyotome; sc, sclerotome. (Courtesy of Heinz Jürgen Jacob, Bochum.)

represents the dermomyotome, which is the source of all trunk muscles and of the dermis of the back. From the dermomyotome, the myotome is formed by a complex rearrangement (Kaehn *et al.,* 1988; Denetclaw *et al.,* 1997). At the lateral edge of the dermomyotome, muscle precursor cells leave the epithelium as a result of a c-met/scatter factor cross-talk and migrate into the limb anlagen. Scatter factor/ hepatocyte growth factor (SF/HGF) is expressed in the limb bud mesenchyme and its receptor c-met by the cells of the lateral dermomyotomes. Although the phenomenon was described as early as 1895 by Fischel, later on rediscovered by Grim

1. Somitic Cell Lineages

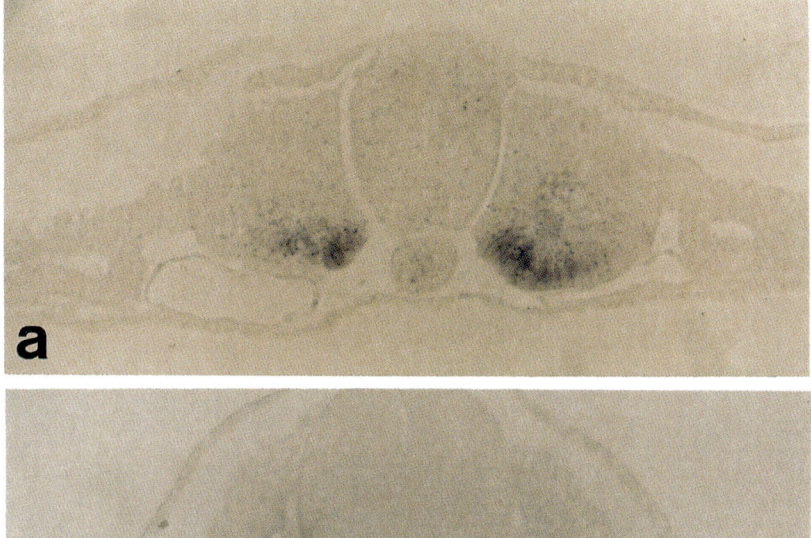

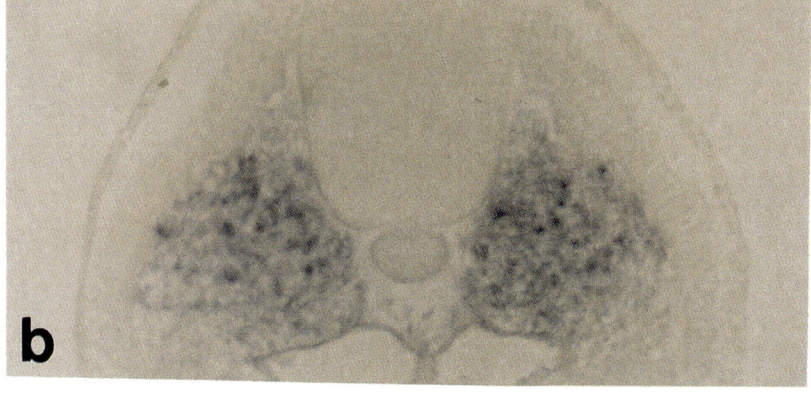

Figure 8 *Pax-1* expression in the epithelial somite (magnification: ×220) (a) and in the sclerotome (magnification: ×260) (b).

(1970), and experimentally shown in 1974 (Christ *et al.*, 1974; Christ *et al.*, 1977; Chevallier *et al.*, 1977), the molecular basis was found only in 1995 (Bladt *et al.*, 1995).

It is well known that an important morphological event preceding somitogenesis is the epithelialization of the paraxial mesoderm (Fig. 9). Olson and coworkers (Burgess *et al.*, 1996) have found that the basic helix–loop–helix (bHLH) transcription factor Paraxis is expressed in that part of the segmental plate in which the transition of the mesenchyme into an epithelium occurs (Fig. 4). Also at the transition zone, the expression of the Eph-related tyrosine kinase receptor cek-8 is located (Fig. 10). *Paraxis* expression depends on signals emanating from the ectoderm. After ectoderm removal, the epithelialization of the paraxial mesoderm and somite formation do not take place (Sosic *et al.*, 1997). This is also true of the formation of somite compartments that are important for the process of resegmentation (see below). *Paraxis* mutant mice do not develop a normal spine

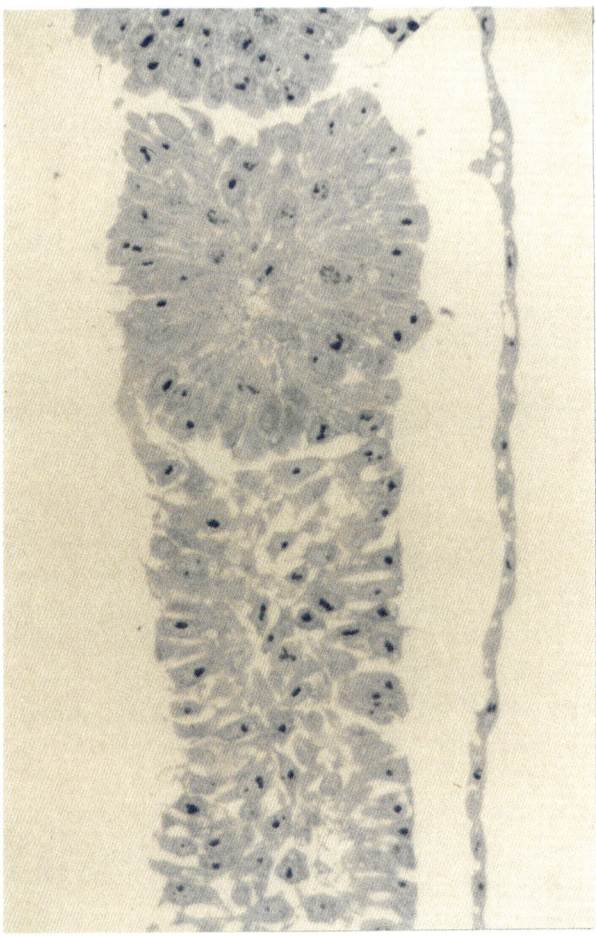

Figure 9 Semithin section through the segmental plate and most recently formed somites. Endoderm on the right. Epithelialization starts in the periphery of the segmental plate mesoderm (magnification: ×400).

(Burgess *et al.*, 1996). The segmental pattern of the epithelial somites becomes transmitted to the ectoderm and to the blood vessels that originate within the intersomitic clefts (Fig. 11). After formation of the sclerotome, the subdivision of the somite into a cranial and caudal part becomes visible (Fig. 12; Wilting *et al.*, 1994). mRNA of the basic helix–loop–helix protein Twist, for example, is strongly expressed in the caudal part of the sclerotome (Fig. 13; Füchtbauer, 1995).

Stern and colleagues (1986), as well as other authors, were able to show that neural crest cells and axons invade only the cranial half of the sclerotome, which differs from the caudal half with respect to several molecules. For instance, components of the extracellular matrix are differentially distributed in the cranial and caudal halves of the sclerotome and are implied in patterning the nerves (Fig. 14).

1. Somitic Cell Lineages

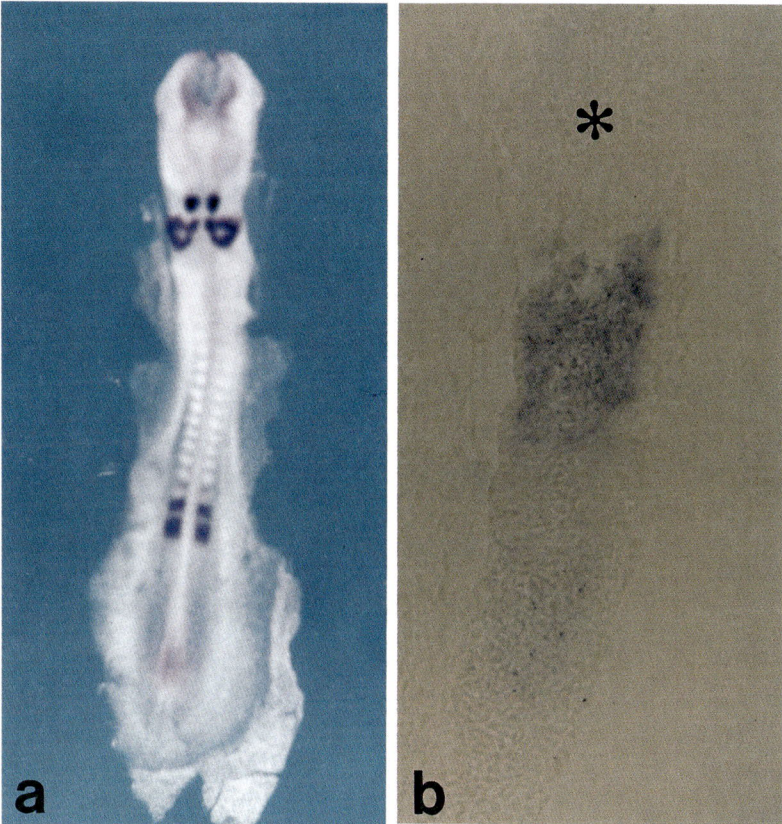

Figure 10 *Cek8* expression in the 2-day chick embryo. (a) Whole mount (magnification: ×16). Note the expression at the transition zone of segmental plate to somites. (Courtesy of Ketan Patel, London.) (b) Longitudinal section through the expression domain (magnification: ×170). Asterisk marks the last somite.

The caudal part of the sclerotome is characterized by a dense mesenchyme, and the cranial part contains the spinal nerve and thus can be called the neurotome (Fig. 15b). In addition to the craniocaudal polarity there is a gradient in a lateroaxial direction. At the level of still epithelial somites, the notochord is surrounded only by extracellular matrix material. Later on, this perinotochordal space becomes invaded by somite cells forming a perinotochordal sheath from which the vertebral bodies and intervertebral disks develop. The notochord induces *Pax-1* expression and the epitheliomesenchymal transition by secreting the signaling molecule Sonic hedgehog (Pourquié *et al.*, 1993; Brand-Saberi *et al.*, 1993; Johnson *et al.*, 1994). In transverse section, the caudal half of the sclerotome takes on a triangular shape. The edges of the triangle represent the rudiments of the neural arch, the rib, and the pedicle (Fig. 15a). Each half of a vertebra is made up of a lateral part giving rise to the neural arch, rib, and pedicle and a medial part giving

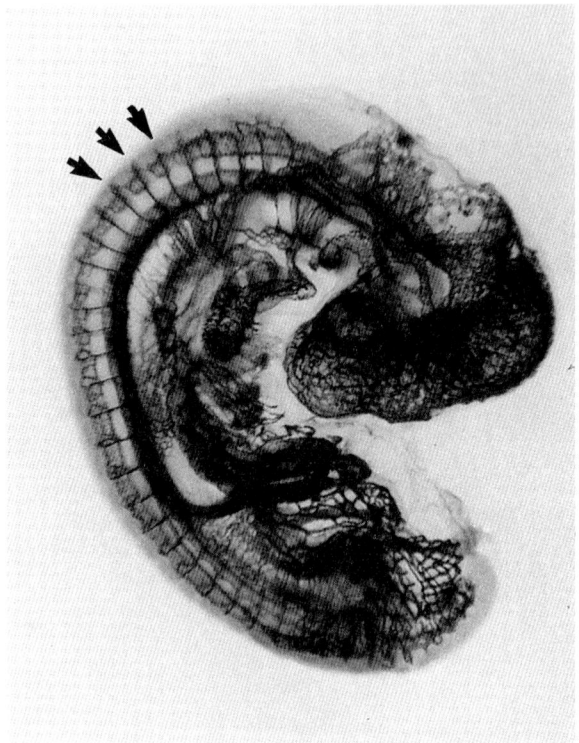

Figure 11 A 3-day chick embryo. Perfusion with India ink reveals the pattern of blood vessels. Note the segmental arrangement of arteries between somite border (arrows) (magnification: ×16).

rise to the central structures, the vertebral bodies, and intervertebral disks. Subsequently, both parts fuse at the level of the intervertebral disks and the cranial halves of the vertebral bodies. Proliferation markers such as proliferating cell nuclear antigen (PCNA) show that the vertebral bodies grow from the intervertebral disks (Fig. 15c). The cells of the vertebral bodies do not show S-phase nuclei. Differentiation, on the other hand, starts at the level of the vertebral bodies as shown by immunostaining for chondroitin-6-sulfate proteoglycan (Fig. 16). So initially, there is no sharp boundary between bodies and disks.

We have seen that the somites constitute segmentation. They represent primary segments, whereas segmental formation of blood vessels or nerves depends on somites. These structures can be considered as secondary segments. Strudel (1955) was able to show that the presence of segmentally arranged dorsal root ganglia is a prerequisite for the development of segmental neural arches. This means that secondary segmental structures impose their pattern on parts of the lateral sclerotome leading to segmental neural arches, which therefore represent tertiary segments (Fig. 17).

1. Somitic Cell Lineages

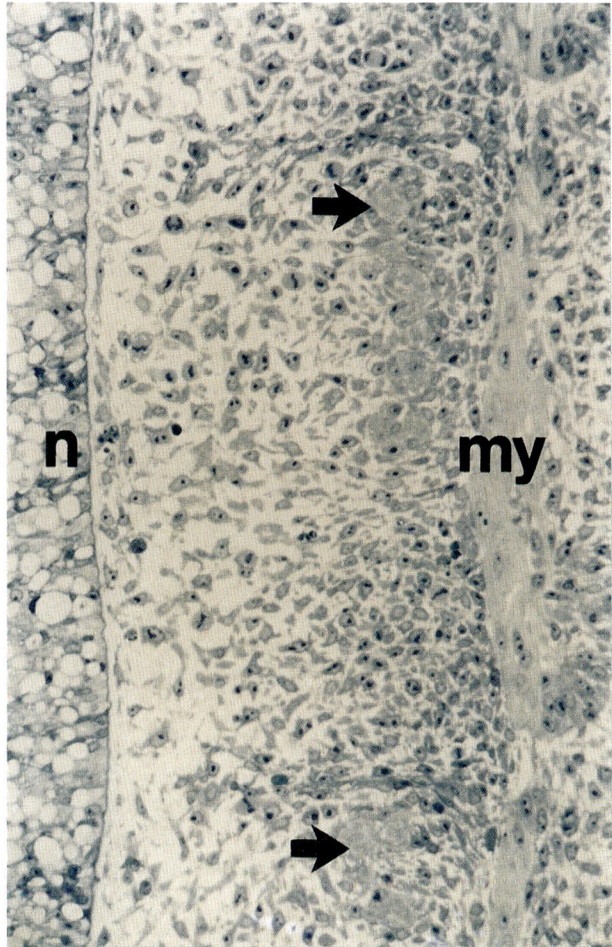

Figure 12 Semithin section through the sclerotome of a 4-day chick embryo. In the medial zone, cells are more loosely arranged than in the lateral zone. Cranial and caudal half can be distinguished by the position of axons (arrow) (magnification: ×420). n, notochord; my, myotome.

The dorsoventral polarity of the somite depends on signals from neighboring structures during early steps of compartment formation as well as during later development. If, for instance, the neural tube is very small, the neural arches also develop in a smaller size. If the axial organs are absent, the paraxial mesoderm of either side fuses in the midline and undergoes segmentation. Neither the vertebral column, nor the epaxial muscle develop without signals from surrounding tissues (Fig. 18).

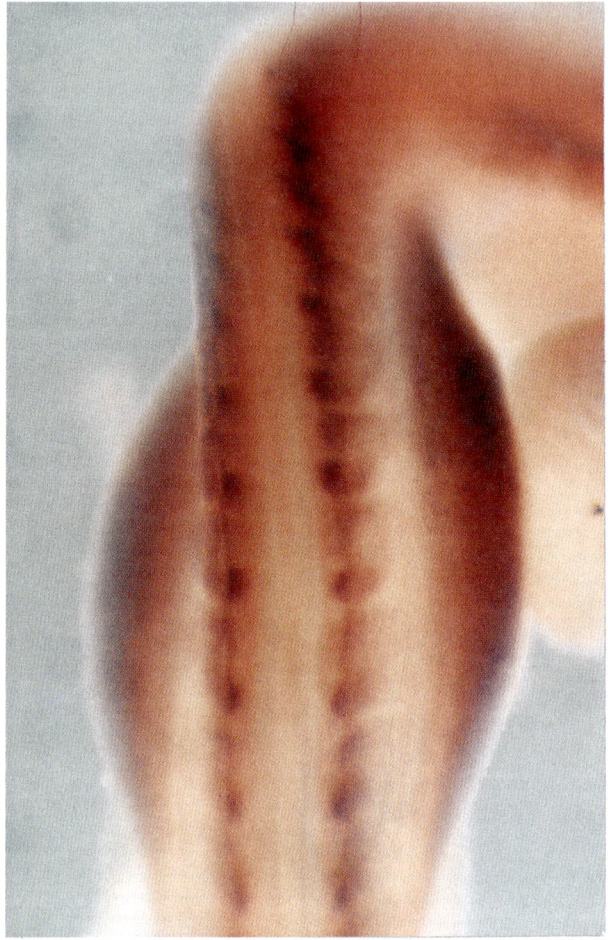

Figure 13 Expression of *twist* is stronger in the caudal half than in the cranial half of the sclerotome (magnification: ×65). (Courtesy of Ernst-Martin Füchtbauer, Freiburg.)

IV. Resegmentation

It was a great step forward in the research of segmentation when Robert Remak in 1850 published the first of his three volume work called *Untersuchungen über die Entwickelung der Wirbelthiere: Über die Entwickelung des Hühnchens im Ei* (Studies on the development of vertebrates. On the development of the chick embryo in the egg).

Remak called the somite "protovertebra" and distinguished between the dermomyotome he called "Rückentafel" or "Muskelplatte" (in translation: plate of the back or muscle plate) and the sclerotome he called "Wirbelkernmasse" (meaning:

1. Somitic Cell Lineages

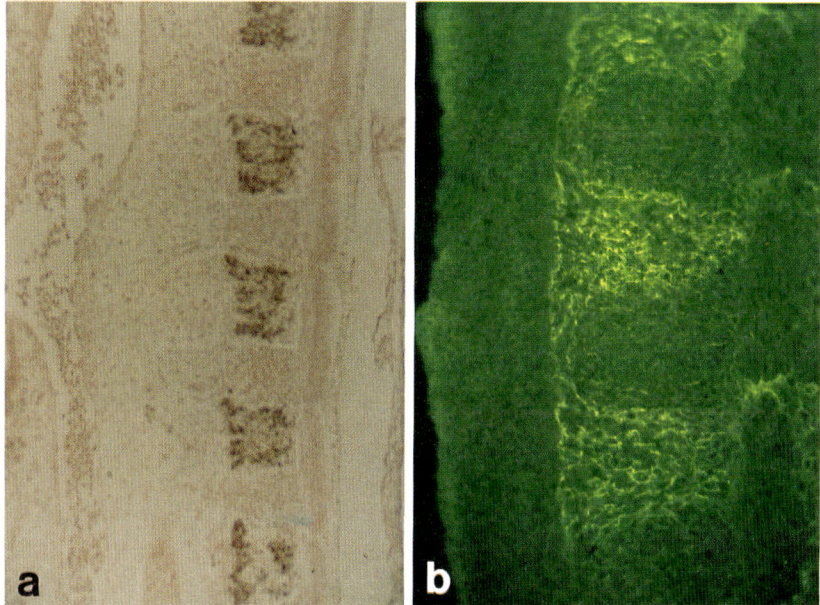

Figure 14 (a) HNK-1 immunostaining revealing position of axons in the cranial half of the sclerotome (magnification: ×170). (b) Distribution of tenascin in the sclerotome where it is concentrated in the cranial half. Immunofluorescent staining of a 2.5-day chick embryo (magnification: ×280).

mass of the vertebral centrals). He already described a difference between the cranial and caudal part of the sclerotome. The spinal nerve with its ganglion and roots appears in the cranial portion of the sclerotome from which Remak believed it to arise. The posterior part becomes condensed forming what Remak called the vertebral arch. During later stages Remak observed a striking change in the perinotochordal tissue and called this "Neue Gliederung" meaning "new segmentation" by which the definitive vertebral centra are formed. Remak's conclusions were based on the different positions of the dorsal root ganglia with respect to the somite and to the vertebral body (Fig. 19). It is interesting that according to Remak this new segmentation only concerns the central part of the vertebral column and not the arches, which were described to originate from the caudal sclerotomic halves. The term "new segmentation" expresses exactly what Remak observed and described. The term "resegmentation" is a bad translation giving this process a wrong meaning. The prefix *re* means "again, back to, received once more" and does not imply that the new segmentation is different from the original segmentation. However, because it is common to put false interpretations on poorly known events, we will not change this generally used term.

In 1881, Corning published a paper in which Remak's view was supported and an intrasegmentally located fissure was described that separated both halves of the sclerotome (Fig. 20). In 1888, von Ebner described the "intervertebral fissure"

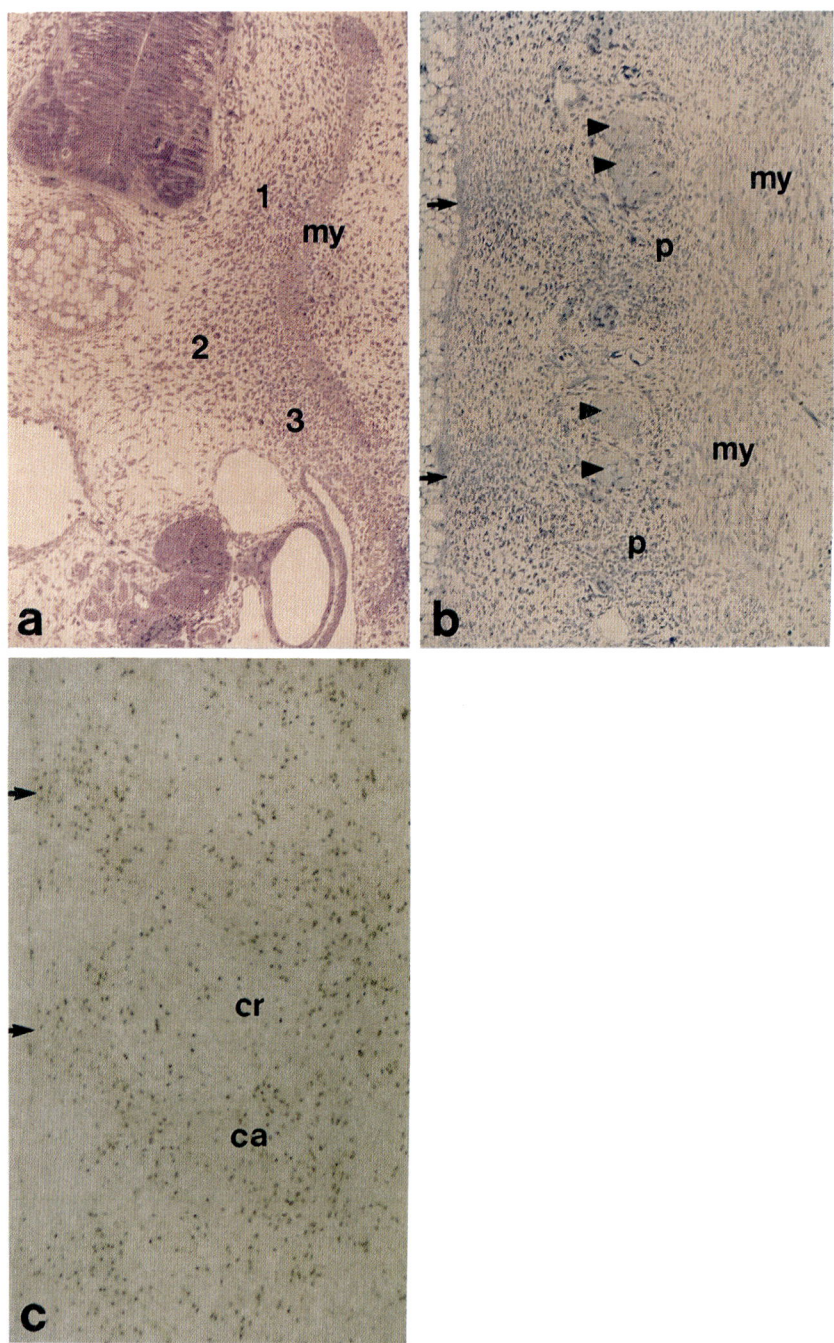

1. Somitic Cell Lineages

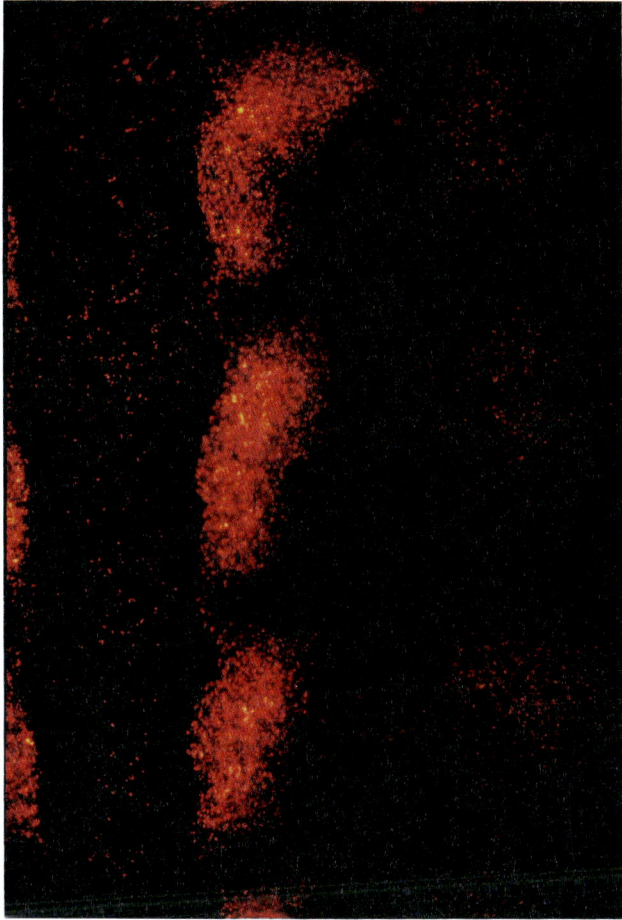

Figure 16 Chondroitin-6-sulfate proteoglycan as a marker of cartilage differentiation marks the site of the future vertebral bodies (magnification: ×160). Immunostaining of a 5-day chick embryo in frontal section.

(Fig. 21). Studying reptiles he concluded that this fissure represents a diverticulum of the somitocoel extending to the notochord. There was a long discussion on the meaning of resegmentation, and different models have been suggested during the twentieth century.

Figure 15 (a) Transverse section of a 4-day chick embryo at the level of the caudal sclerotome half (magnification: ×170). my, Myotome; 1, neural arch; 2, pedicle; 3, rib. (b) Frontal section of a 5.5-day chicken embryo (magnification: ×140). my, Myotome; p, pedicles; arrows, intervertebral disks; arrowhead, spinal nerve. (c) PCNA stained cells are more abundant in the caudal sclerotome half (magnification: ×210). Arrows, intervertebral disks; ca, caudal half; cr, cranial half.

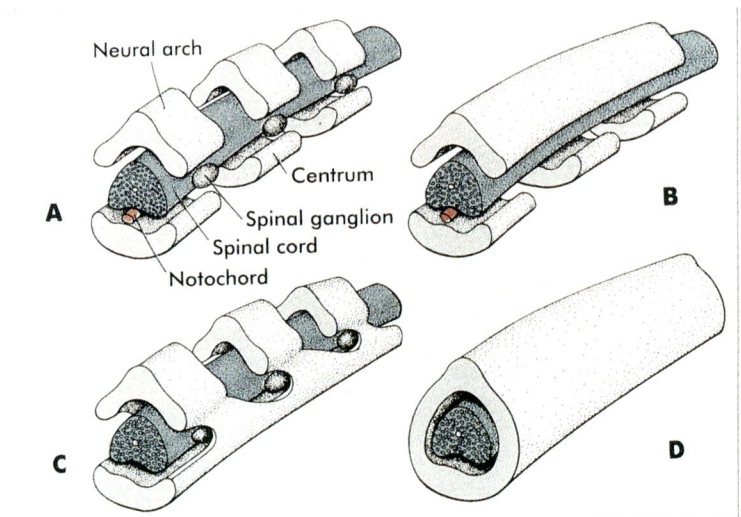

Figure 17 Diagram depicting the process of tertiary segmentation. When spinal ganglia are absent, the neural arches fuse to form a uniform cartilaginous plate above the neural tube (B, D). When the notochord is also lacking, the centra fuse (C, D). (A) Normal situation. (After B. K. Hall, 1978.)

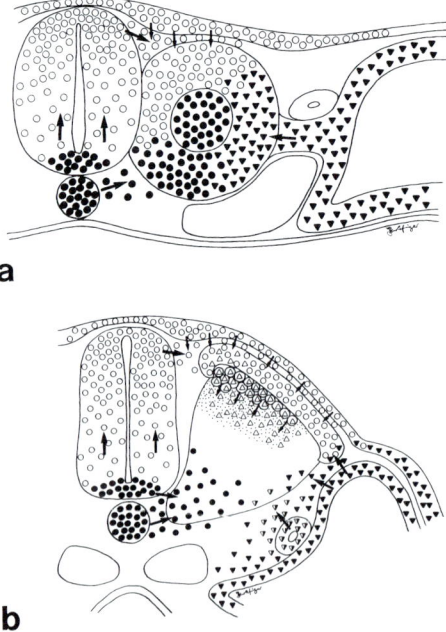

Figure 18 Local interactions between the paraxial mesoderm/somites and the neighboring organs during somitic compartment formation (a) and differentiation period (b).

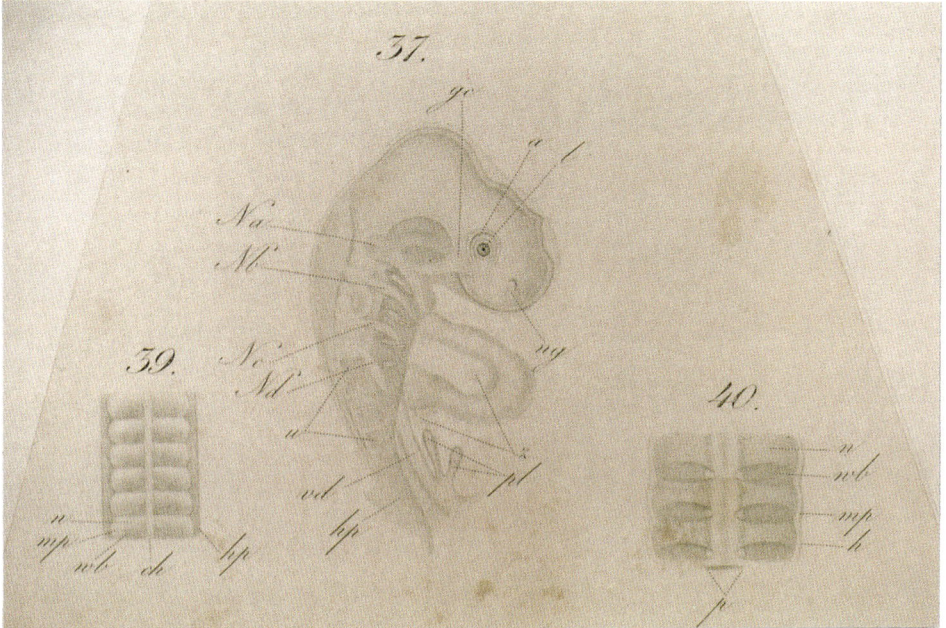

Figure 19 Resegmentation in the chick embryo according to Remak (1850). Na, trigeminal nerve; Nb, facial nerve; Nc, glossophasyngeal nerve; Nd, vapal nerve; u, somites; vd, foregut; hp, cervical plate; gc, ciliary ganglion; a, wall of eye vesicle; l, lens; ng, nasal pit; z, villi at venous end of heart; pl, primitive liver ducts; n, nerve; wb, vertabral arch; mp, muscle plate; h, primitive epidermis; and r, neural tube.)

Textbooks of embryology often show the process of resegmentation as seen in Fig. 22. In most of these drawings, the intrasegmental fissure subdivides the whole sclerotome up to the notochord to provide new segmental units that then form the definitive vertebra by intersegmental fusion. To get a better understanding of what really happens during the resegmentation process, the projection of somite boundaries and somite compartments on definitive segments has to be studied. This can be done by the application of fluorescent dyes or in avian embryos by the quail–chick marking system. After replacement of the fifth somite, for instance, the head–neck transitional zone is made up of quail cells (Wilting *et al.*, 1993). Another example of a somite derivative is the striated muscle of the esophagus, which derives from the second to the fifth somite (R. Huang, personal communication, 1998).

To address the question of resegmentation, we grafted single somites at the cervical, brachial, and thoracic level. After a reincubation period of 1 day, the somite derivatives sclerotome, myotome and dermatome are marked by quail cells. After a longer reincubation period, the caudal half of a vertebral body, the intervertebral disk, and the cranial half of the caudally adjacent vertebral body are formed by quail cells (Fig. 23a). This distribution also extends to the proximal part of the rib and the segmental muscle. If we look at the distal parts of the ribs, one somite

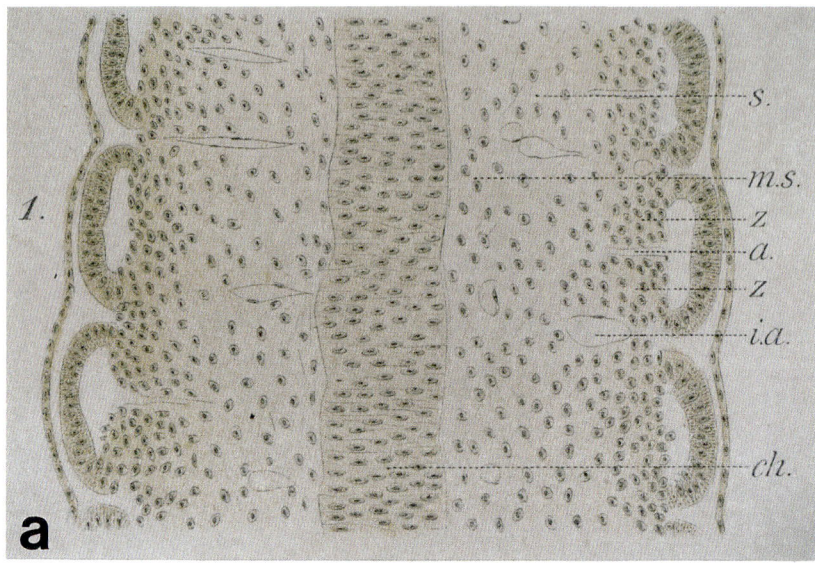
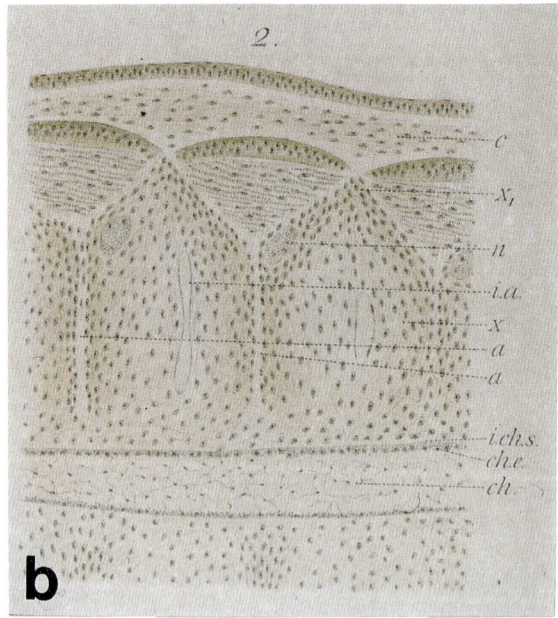

Figure 20 Resegmentation in the chick embryo according to Corning (1881). (a) Early phase in 3-day embryo. (b) Late phase in 5-day embryo. a, somitocoel; ch, notochord; a.ch.s, outer notochord sheath; i.ch.s, inner notochord sheath; ch.e, notochord epithelium; i.s, intervertebral fissure (cleft); i.a, intercostal artery; n, nerve; s, sclerotome; l.i, intermuscular ligament; c, vertebral centre; cu, cutis; x, primordia of transverse processes; and Tb, upper arch bases of transverse processes of outer notochord sheath.

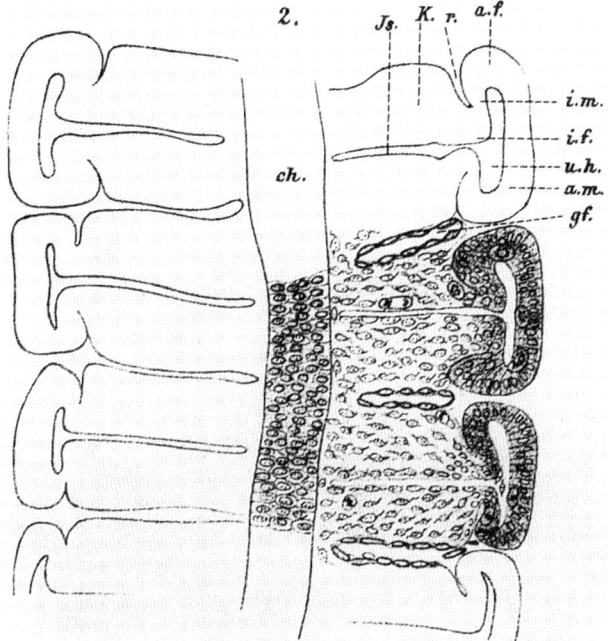

Figure 21 Drawing of sclerotome and dermomyotome in the reptilian embryo by von Ebner (1888). Note the "intervertebral fissure" subdividing the sclerotome. gf., intersegmental artery; ch., notochord; i.s., intrasegmental fissure; a.m., outer dermomyotome; i.m., inner dermomyotome; k, sclerotome; cf, anterior edge; and uh, somotocoel.

gives rise to the halves of two ribs that face each other and the intercostal muscle in between (Fig. 23b). It is interesting to look at the neural arches and their processes. Almost the whole arch develops from one somite. The caudal edge from the next cranial neural arch, where the muscle is attached, originates from the same somite, meaning that the muscle with its skeletal origin and its processes of insertion are formed by one somite. A real resegmentation of the neural arch does not take place (Huang *et al.*, 1996).

Huang has studied the projection of somite compartments on definitive structures. He grafted the mesenchymal somitocoel cells and found an interesting distribution of these cells (Fig. 24a). Most of them are located in the caudal half of the sclerotome close to the intrasegmental fissure (Fig. 24b). After a longer reincubation period, these cells can be seen in the intervertebral structure, the proximal part of the ribs, and in the surfaces of the intervertebral joints (Fig. 25). This means that moving structures and surfaces of the spinal joints are formed by somitocoel cells which are located within the center of the motion segment that represents the functional unit of the spine (Huang *et al.*, 1994).

In later stages of development, the segmental pattern of the skeletal elements and muscle can get lost. This applies, for instance, to the occipital and sacral re-

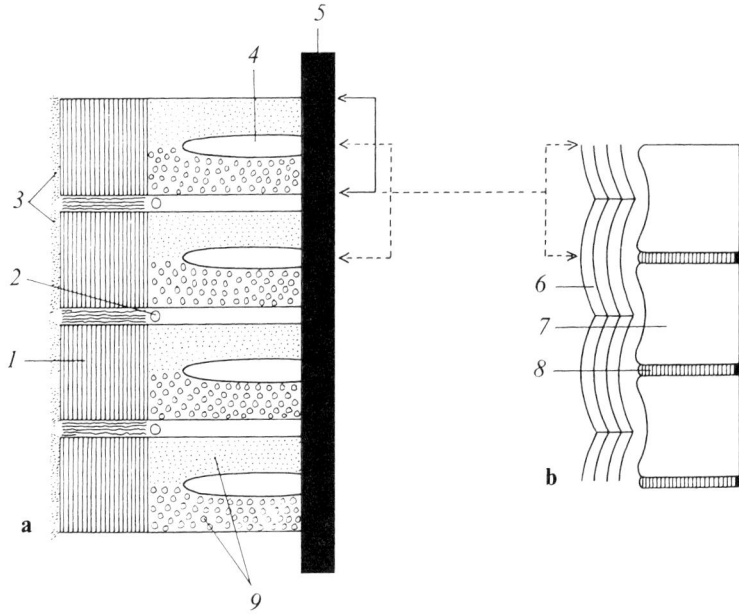

Figure 22 Modern view of the resegmentation, the intrasegmental fissure subdividing the whole sclerotome up to the notochord. 1, Myotome; 2, intersegmental artery; 3, dermatome; 4, intrasegmental fissure; 5, notochord; 6, segmental muscle; 7, vertebral body; 8, intervertebral disk; 9, sclerotome. [From Starck, D. (1979). "Embryologie: Ein Lehrbuch auf allgemein biologischer Grundlage."]

gion where the skeletal anlagen fuse to form the basioccipital and the sacral bone. This fusion or polymerization is accompanied by a downregulation of *Pax-1* in the intervertebral disks. The event of polymerization can also be seen in the superficial layer of the epaxial muscle, where cells migrate over four to seven segments. The molecular basis of this phenomenon is not yet clear. In connection with this finding, it is of special interest that *delta* knockout mice lose the segmental arrangement of myotomes.

Our present view of the complex process of resegmentation is summarized schematically in Fig. 26. Today, a new approach could be to compare details of resegmentation between flies and vertebrates. However, this is only possible when morphological data are taken into account.

Figure 23 Distribution of grafted quail cells after replacement of one somite, 6 days of reincubation shown by double-staining with antiquail (blue) and antidesmin (brown) in frontal section. (a) Quail cells form neighboring halves of two adjacent vertebral bodies (asterisk), intervertebral tissue, rib (r), connective tissue, and intercostal muscle (m). (b) Somite-derived cells form neighboring halves of two ribs (r), connective tissue, and muscle (m). (Courtesy of R. Huang, Freiburg.)

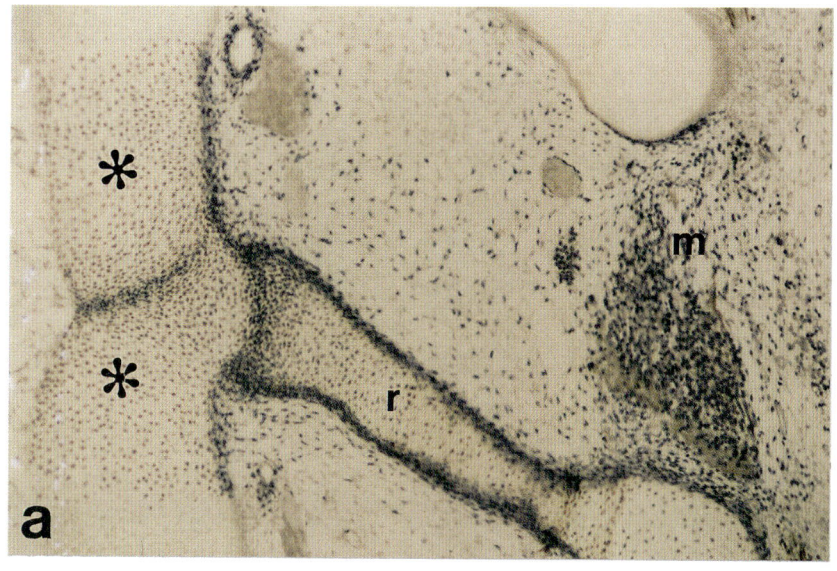
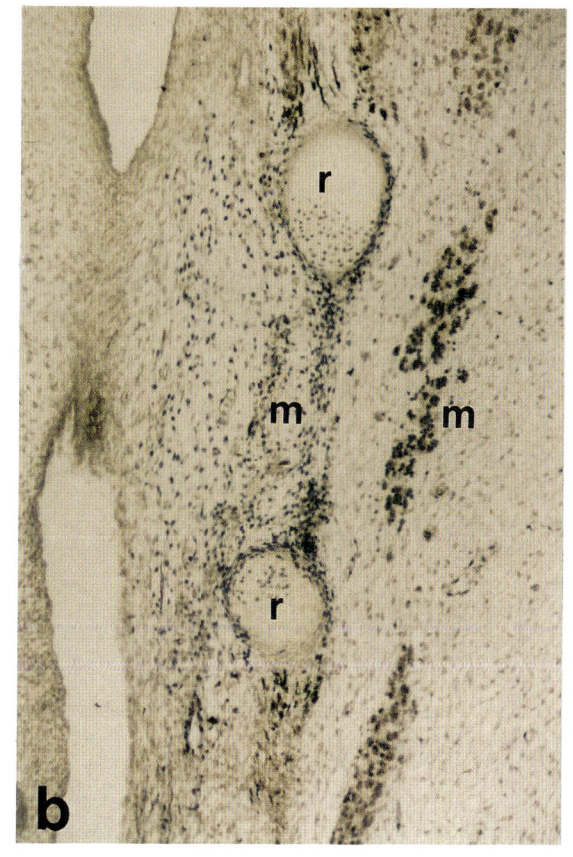

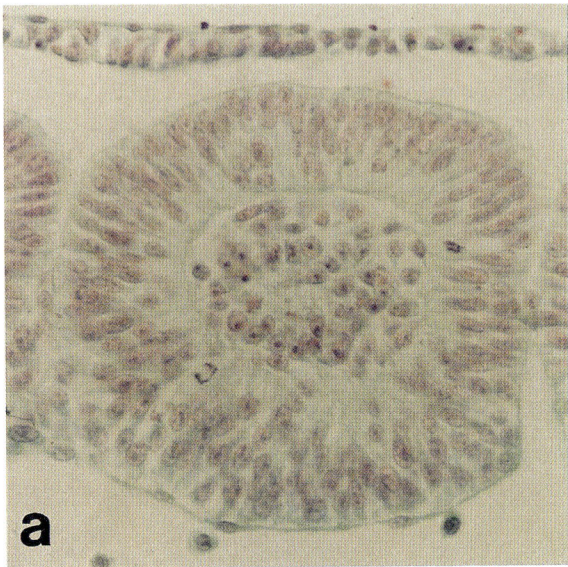
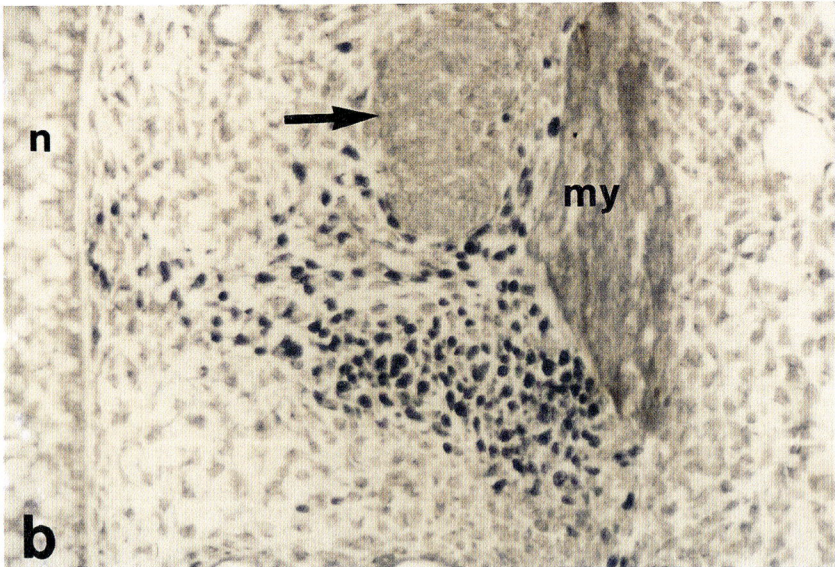

Figure 24 (a) Chick somite, 2.5 hr after replacement of somitocoel cells by quail cells. (b) After 2 days of reincubation, frontal sections show the distribution of quail cells in a triangular shaped area in the caudal sclerotome half and around the spinal nerve (arrow), my, Myotome; n, notochord. (Courtesy of R. Huang, Freiburg.)

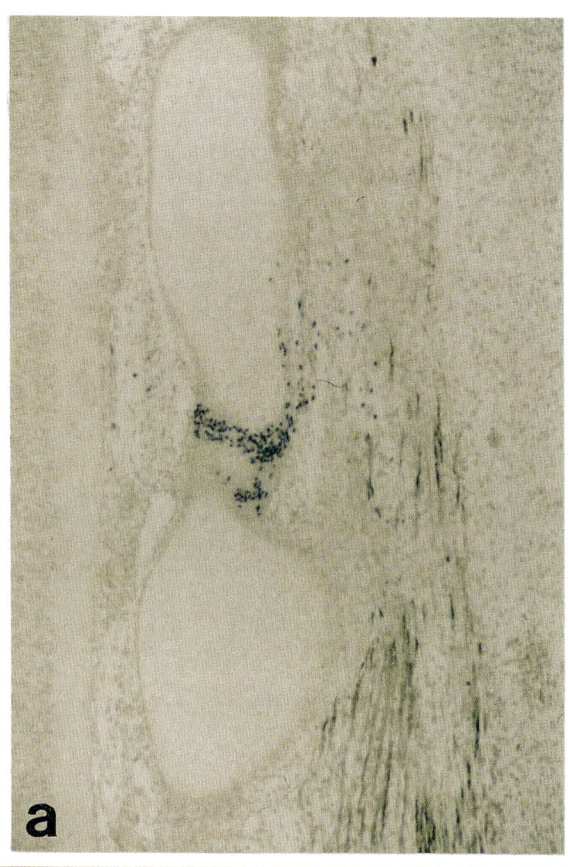
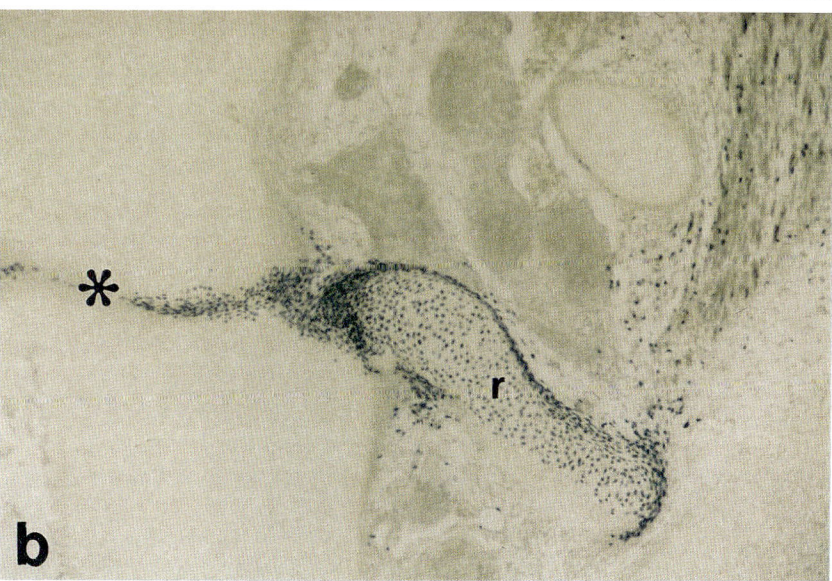

Figure 25 Six days after somitocoele cell grafting, quail cells are found in the intervertebral joint (a), intervertebral tissue (asterisk), and proximal rib (r) (b). (Courtesy of R. Huang, Freiburg.)

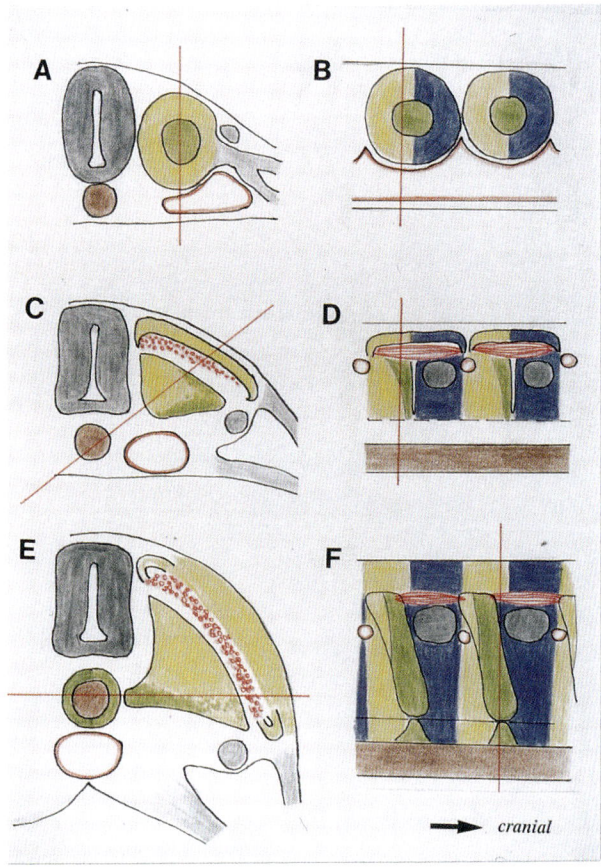

Figure 26 Schematic diagram of the resegmentation process summarizing current results. A, C, E, transverse sections; B, D, F, corresponding frontal sections.

V. The Muscle Lineage

As we have seen, the formation of the vertebral column involves a rearrangement of sclerotome cells that derive from the ventral half of the somite as a result of epitheliomesenchymal transformation. The total mass of skeletal muscle in the trunk, however, is derived from the dorsal half of the somite that forms the epithelial dermomyotome. There are two separate populations of muscle precursor cells; one is located in the medial dermomyotome and yields the intrinsic back muscle, and the other one arises from the lateral dermomyotome and yields the muscles of the ventrolateral body wall and the limbs (Ordahl and Le Douarin, 1992; Williams and Ordahl, 1994). The medial dermomyotome gives rise to the underlying myotome

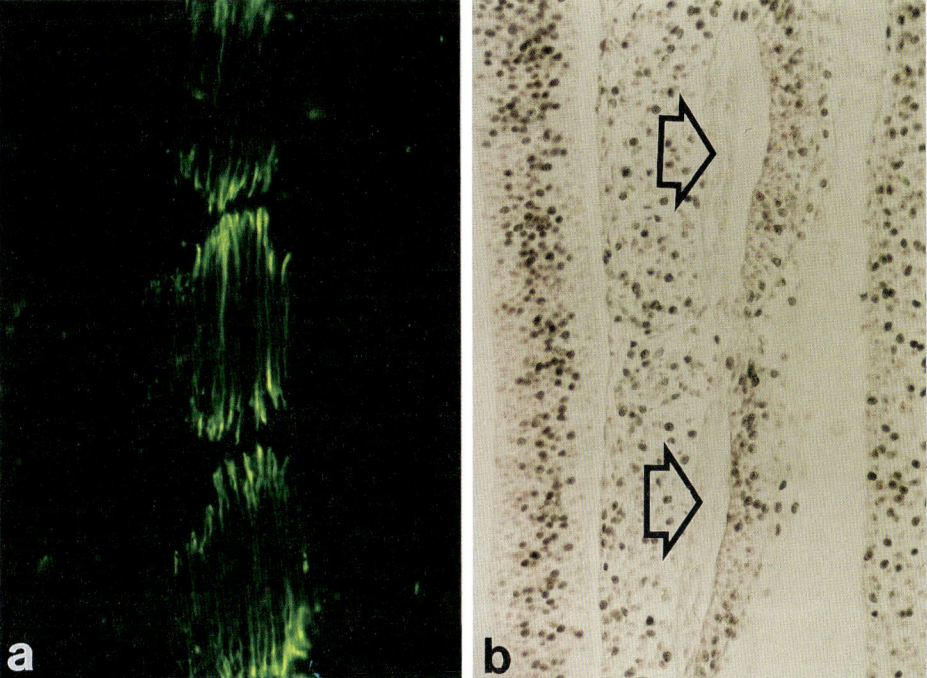

Figure 27 (a) Immunofluorescence staining of the murine myotome with antibodies against desmin to show the longitudinal myotome cells arranged in parallel. Sagittal section (magnification: ×140). (b) Anti-Bromodeoxyuridine (BrdU) staining after experimental exposure to BrdU *in ovo*. The myotomes (arrows) are completely devoid of S-phase-labeled cells. Frontal section (magnification: ×180).

of postmitotic longitudinally arranged myofibers (Fig. 27). The cells dislocate into the myotome from the edges of the dermomyotome and crawl along the ventromedial surface of the dermomyotome, initially spanning maximally one segment (Denetclaw *et al.*, 1997).

Later, myotome cells express markers of the MyoD family of bHLH transcription factors, the intermediate filament desmin, smooth muscle α-actinin, embryonic myosin, members of the myocyte enhancer factor (Mef) family, and others (Kaehn *et al.*, 1988; Pownall and Emerson, 1992; Olson *et al.*, 1995) (Fig. 28).

By means of interspecific grafting of single somites from quail to chick it can be shown that muscle precursor cells initially keep to the myotome of the same segment (Fig. 29a). After longer reincubation periods, quail nuclei are found also in segment-overbridging muscles of the intrinsic back musculature (Fig. 29b,c). The other muscle lineage arising from the lateral portion of the dermomyotome develops in dependence of its craniocaudal position either into muscle of the ventrolateral body wall or into limb muscle. At flank and neck level, double-layered

Figure 28 *In situ* hybridization for *MyoD* in a 4-day chick embryo. *MyoD* is expressed in the segmentally arranged myotomes, the branchial arches, and premuscle masses of the limb buds (magnification: ×20). (Courtesy of Ketan Patel, London.)

epithelial buds grow into the mesenchyme of the lateral plate and give rise to the ventral muscles of the neck and trunk (Fig. 30). At limb bud levels, the lateral dermomyotomes disintegrate under the influence of scatter factor/hepatocyte growth factor (SF/HGF) (Bladt *et al.*, 1995; Brand-Saberi *et al.*, 1996a). Myogenic precursor cells invade the limb buds as individual cells. During this process, they express *Pax-3*, one of the markers of the dorsal somite, as well as the cell adhesion molecule N-cadherin (Fig. 31). N-cadherin has been shown to be critically involved in the correct distribution of the myogenic cells in the limb bud mesenchyme (Brand-Saberi *et al.*, 1996b).

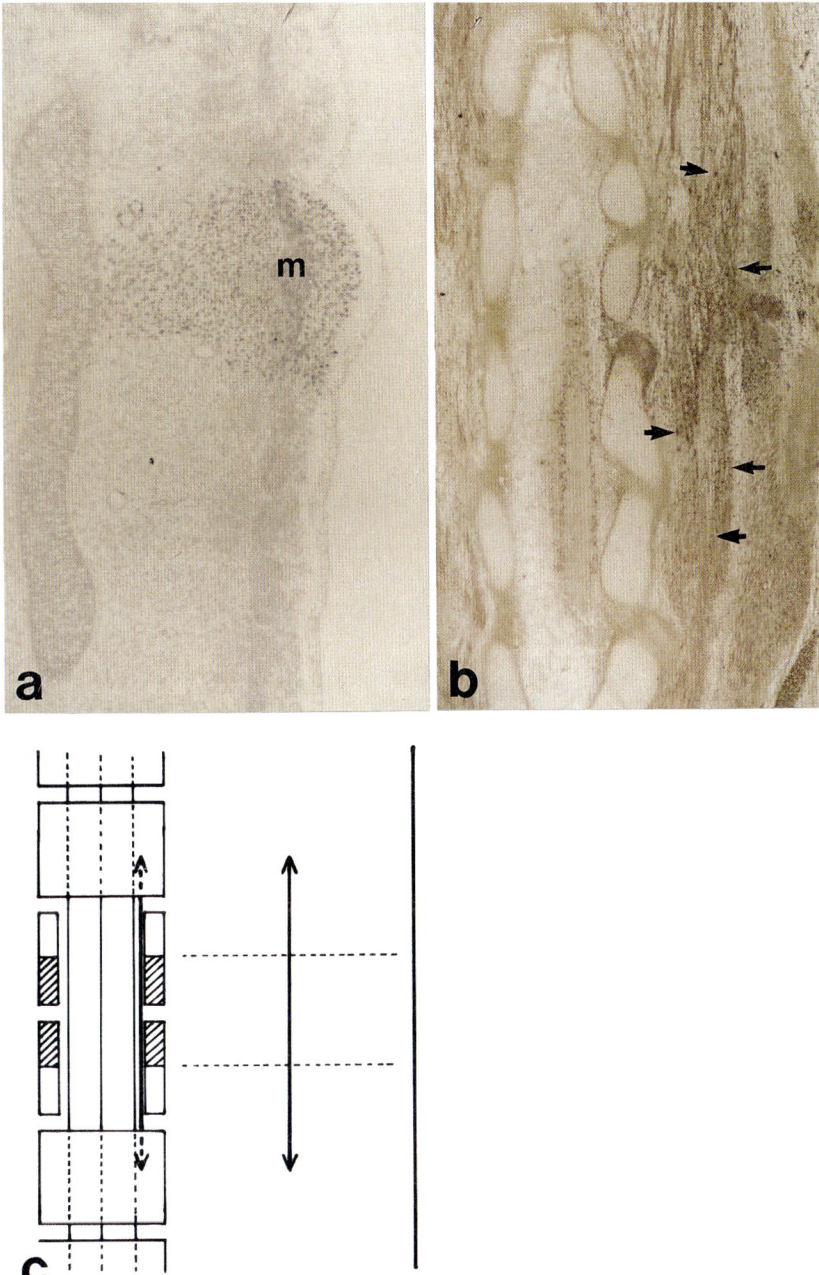

Figure 29 (a) Interspecific grafting of a single somite (magnification: ×160). After 2 days of reincubation, quail cells (blue) keep to the myotome (m) of their original segment. (b) Contributions to segment-overbridging intrinsic back muscle are seen after 6 days of reincubation (arrows). Frontal sections (magnification: ×40). (Courtesy of R. Huang, Freiburg.) (c) Schematic diagram of myocyte distribution from one segment to adjacent segments.

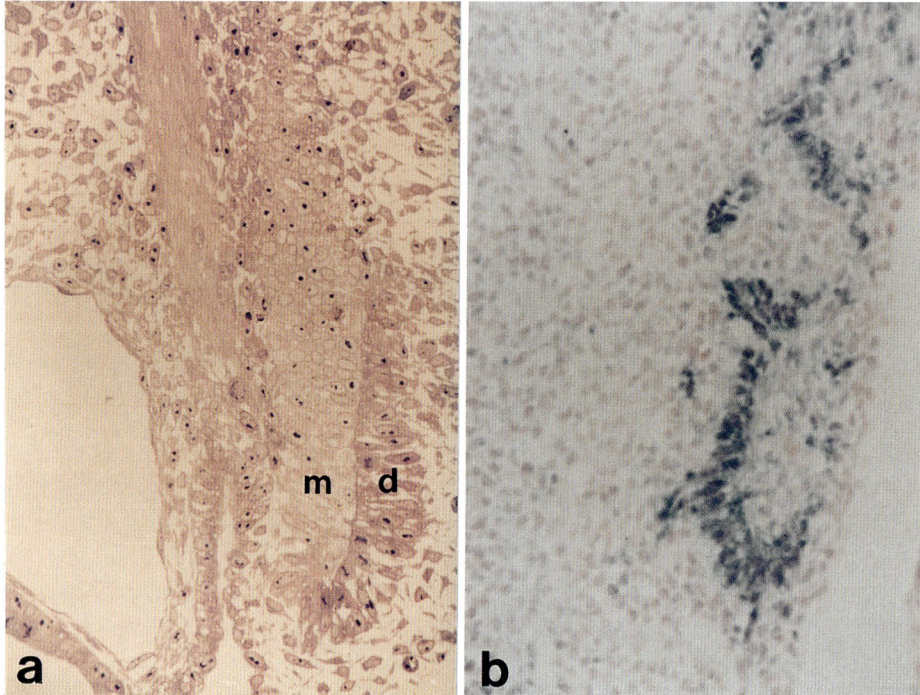

Figure 30 (a) Hypaxial muscle bud of a 4-day chick embryo, semithin section. Note that the bud consists of a ventral process of the myotome (m) and dermomyotome (d). (b) *Pax-3 in situ* hybridization of a 4-day chick embryo. Mixing and partial disintegration takes place (magnification: ×220).

The fate of myogenic cells of individual somites has been studied at the brachial level (Beresford, 1983; Zhi *et al.,* 1996). Somites 16–21 were found to give rise to wing muscle cells. The migration routes correspond to the position of the somites in relation to the limb bud (Fig. 32). The most cranial somite (16) takes part in the radial muscles and the most caudal somites (20, 21) in the ulnar muscles, reflecting their position with respect to the limb bud. The centrally located somites (17–19) are involved in all or most muscle primordia. This pattern of distribution is clearest in the forearm, whereas it is less distinct in the hand (Fig. 33). All brachial somites participate in extensor as well as flexor muscles. Moreover, each somite takes part in more than three muscle primordia in a reproducible fashion, and every muscle primordium is derived from at least three somites.

One of the most intriguing questions that has received a lot of attention is that of muscle specification. It has been known for a long time that the differentiation of the epaxial muscle depends on signals from the neural tube (Christ, 1970; Jacob *et al.,* 1975), whereas limb bud muscle can develop in the absence of axial organs (Rong *et al.,* 1992). Knockout technology and microsurgery in avian embryos have provided a wealth of information concerning the underlying mechanisms, which

1. Somitic Cell Lineages

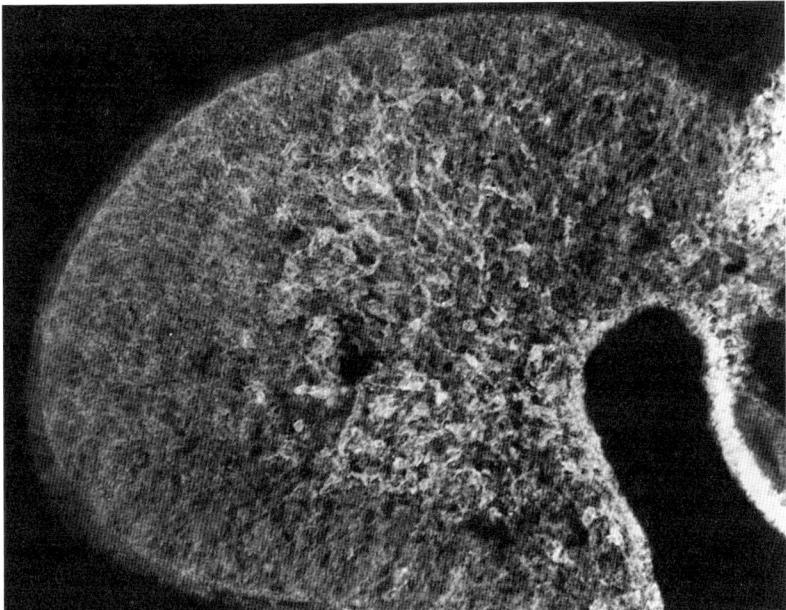

Figure 31 Immunofluorescence staining of myogenic cells in the limb bud by antibodies against the cell adhesion molecule N-cadherin (magnification: ×110).

we cannot discuss in detail here. The gist of what is currently known can be summarized as follows: While the dorsal neural tube and the ectoderm appear to provide signals that support muscle differentiation, the notochord seems to inhibit this process at early stages and to support it at later stages (Pourquié et al., 1993; Brand-Saberi et al., 1993; Buffinger and Stockdale, 1994, 1995; Münsterberg and Lassar, 1995; Münsterberg et al., 1995; Lassar and Münsterberg, 1996; Xue and Xue, 1996). The signaling molecules are believed to be Wnts from the dorsal neural tube and ectoderm possibly in combination with Sonic hedgehog from the notochord and floor plate (Stern et al., 1995). In the specification of the lateral myogenic lineage, signals from the lateral plate, such as BMP-4, also seem to play a role (Pourquié et al., 1995, 1996; Gamel et al., 1995). According to our own investigations, the maintenance or loss of N-cadherin also correlates well with the presence or absence of myogenic competence in somite-derived cells. N-cadherin has recently been shown to be involved in signal transduction in muscle precursor cells (Holt et al., 1994; Brand-Saberi et al., 1996b; George-Weinstein et al., 1997).

As we shall demonstrate in Section VI, muscle is the oldest derivative of the somites, since in fishes an osseous support of the body was initially unnecessary. In the evolution of fishes, a mesenchymal sclerotome develops late and is considerably smaller than the dermomyotome. In the chick embryo, this finds a parallel in the fact that initially the entire paraxial mesoderm expresses dorsal markers, until the sclerotome develops.

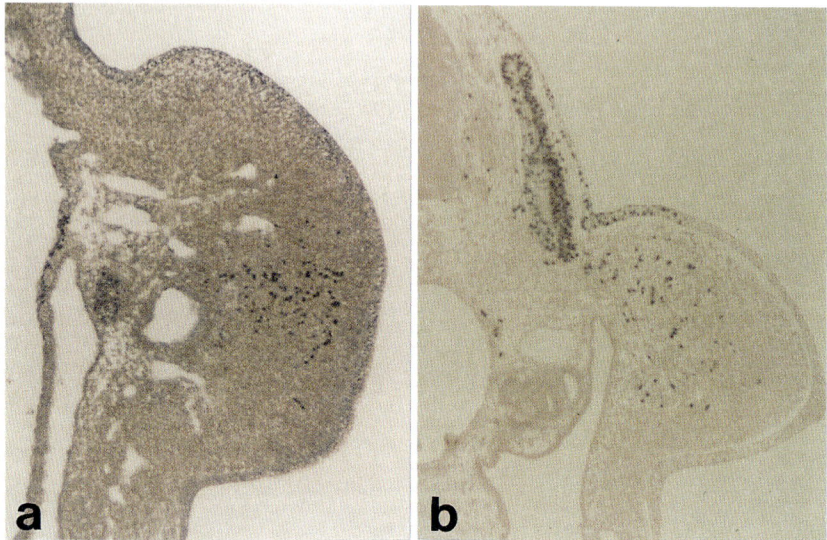

Figure 32 (a) Distribution of myogenic cells in the limb bud, 2 days after grafting somite 19. Quail cells (blue) are found in a central stripe of the limb bud. Frontal section. (Courtesy of Qixia Zhi, Freiburg.) (b) Myogenic cells from quail (blue) in the chick limb bud in transverse section after grafting of a brachial dermomyotome and overlying ectoderm (magnification: ×50).

VI. Further Somitic Derivatives: Dermis, Smooth Muscle, and Angioblasts

Muscle and cartilage are undoubtedly the most well-known derivatives of the somites. More recently, especially as a result of interspecific grafting using the quail–

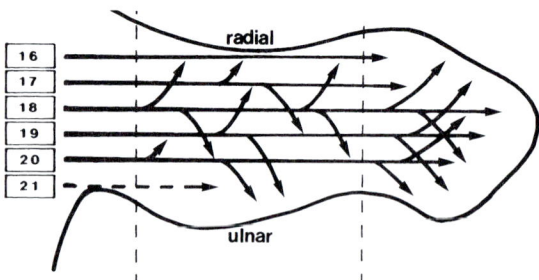

Figure 33 Schematic drawing of the hypothetical distribution of myogenic cells from somites 16–21 in the wing bud. Extensive mixing occurs, especially in the hand. In the forearm, somites 16 and 17 contribute mainly to radial muscle primordia, whereas caudal somites contribute mainly to the ulnar muscle primordia. (Courtesy of Qixia Zhi, Freiburg.)

chick marker system, the somitic origin of at least three other cell types has been established.

We know that the dermis of the back is derived from the dermomyotome (Christ et al., 1983; Christ and Ordahl, 1995). Dermal precursors migrate dorsally toward the epaxial skin ectoderm as a result of an epitheliomesenchymal transition. According to studies by Brill et al. (1995), this transformation process is controlled by neurotrophin-3 deriving from the neural tube. The somite-derived epaxial dermis forms a remarkably sharp lateral border toward the lateral plate-derived dermis of the hypaxial domain including the limb buds (Fig. 34).

Smooth muscle cells in the wall of blood vessels are formed from the local mesenchyme surrounding them. After somitic grafting, and a reincubation period of several days, the muscle layer of blood vessels in the tissue deriving from the graft was found to be made up of quail cells (Fig. 35). Operations were performed prior to the migration of neural crest cells through the somite so that smooth muscle has to be considered as another cell type that differentiates from somitic cells.

Finally, endothelial precursor cells (angioblasts) also reside in the somite (Fig. 36). It has been shown by Wilting et al. (1994) that all compartments of the epithelial somite including the somitocoel cells give rise to angioblasts. In this context, it is interesting to note that the receptor II for vascular endothelial growth

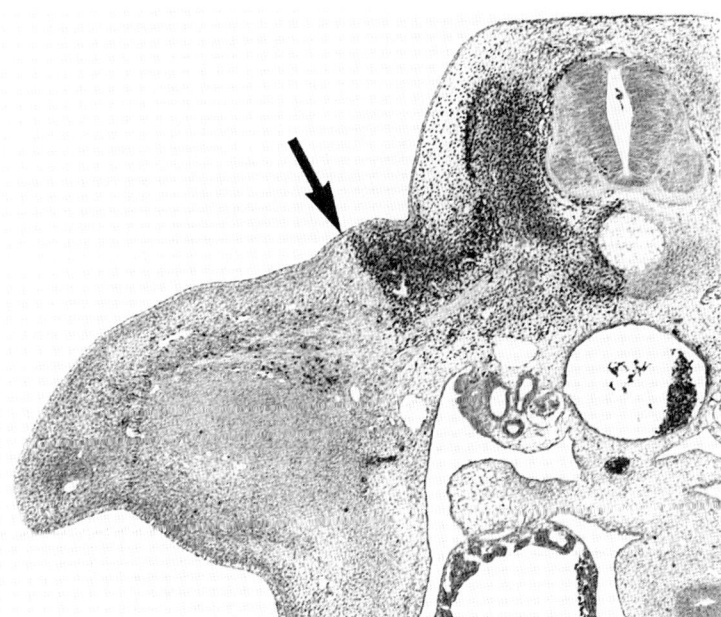

Figure 34 Chick embryo in transverse section after grafting of a single brachial somite. Quail derived epaxial dermis forms a sharp border at the proximal limb bud (arrow) (magnification: ×45). (Courtesy of Qixia Zhi, Freiburg.)

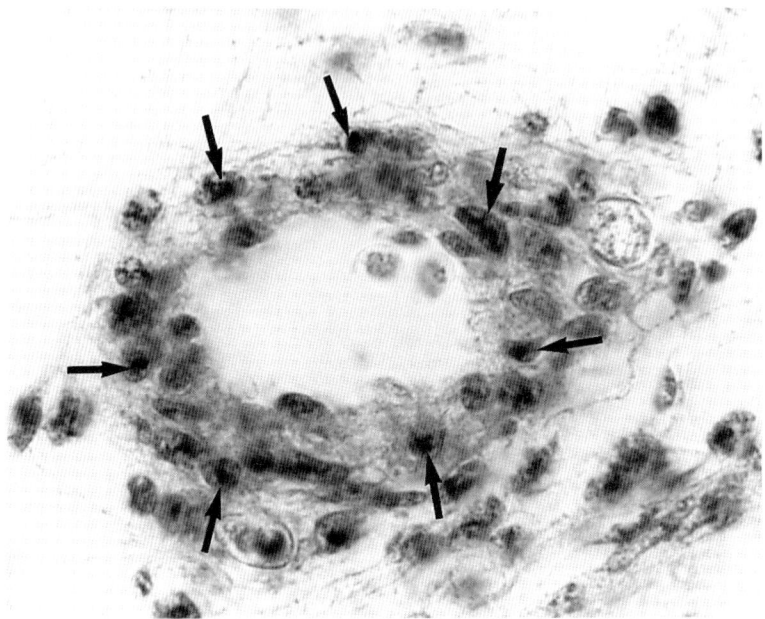

Figure 35 Smooth muscle cells (arrows) from quail forming the media layer of a chick blood vessel after grafting of several somites. Feulgen staining (magnification: ×750).

factor (VEGF) is expressed in the early somite, but only in its dorsolateral quadrant (Fig. 37) (Eichmann *et al.*, 1993).

Vessels tend to contain endothelial cells derived from the same somitic compartment that gives rise to the surroundings of the vessel (Wilting *et al.*, 1994). It is also less well known that not only the cartilage of the embryonic vertebral column is made up of somite-derived cells, but also the bone that replaces it during later development. Whereas osteoblasts derive from the paraxial mesoderm, osteoclasts take origin from blood-borne monocytes. The source of several fibroblast lineages, like that forming the meninx primitiva, are also the somites. However, it is still an unsolved problem where the connective tissue surrounding the epaxial muscle comes from.

VII. Evolution of Metamerism in Vertebrates

Highly conserved genes controlling segmentation in different phyla of the animal kingdom imply a long evolution of metamerism. Here, we shall give a brief outline thereof as far as the chordates are concerned. Moreover, some highly conserved genes will be discussed showing parallels in segmentation of arthropods and chordates.

In the lamprey, the notochord and its surrounding sheath serve as the main skeletal axis. However, there are certain vertebral elements present (Fig. 38a). The

1. Somitic Cell Lineages 35

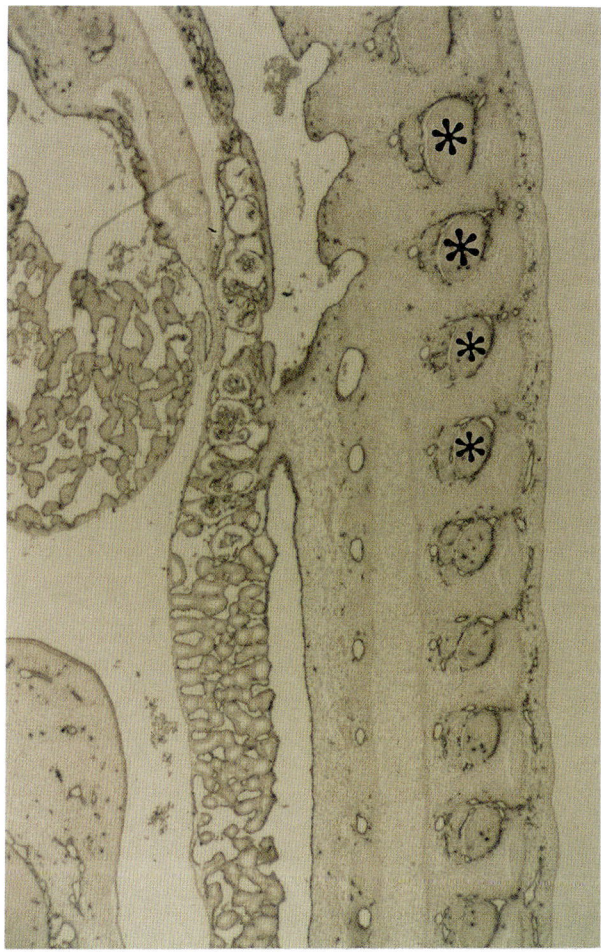

Figure 36 Sagittal section through a quail embryo, QH-1 antibody staining endothelial cells reveals segmental blood vessels surrounding spinal ganglia (asterisks) (magnification: ×45).

segmental boundaries are marked by the intersegmental blood vessels. Two skeletal elements in one segment represent primitive parts of the neural arches. The cranially located interdorsal can clearly be distinguished from the caudally located basidorsal. The afferent and efferent nerves have different positions within the segment. In contrast to gnathostome fishes, both vertebral elements are located intrasegmentally. The first trace of a primitive sclerotome can be found as epithelially structured diverticles ventrally facing the aorta as appendages of the dorsal somitic epithelium (Fig. 39a).

When looking at the shark embryo, the somite consists of two layers, the medial muscle layer and the lateral dermis layer. Both layers form the dorsomedial lip of the somite close to the neural tube. There is no sharp boundary between the

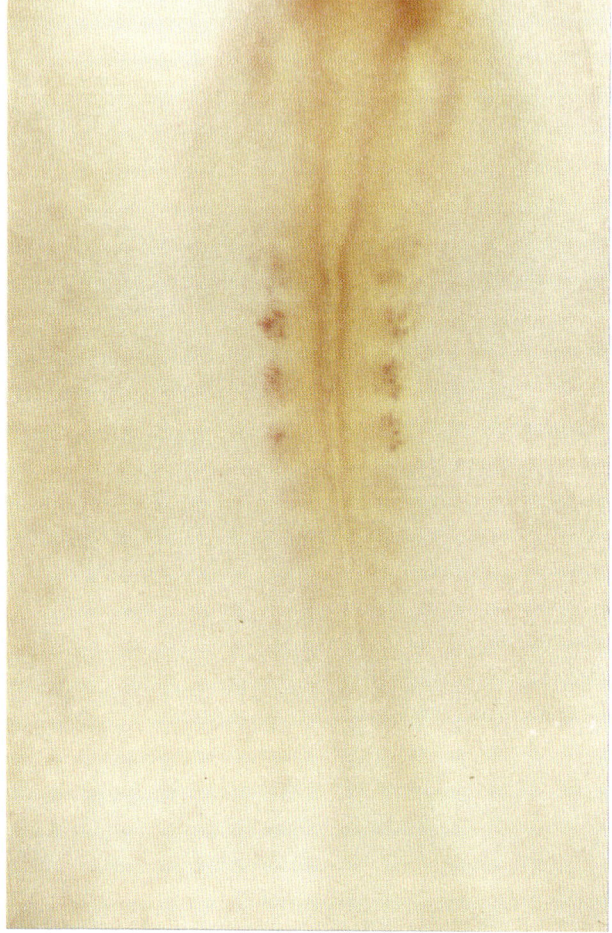

Figure 37 Whole mount *in situ* hybridization of a 2-day chick embryo for *VEGFR II (Quek 1)* (magnification: ×50). (Courtesy of Jörg Wilting, Freiburg.)

Figure 38 (a) Skeletal elements in relation to segment boundaries in the lamprey (*Petromyzon fluviatile*). 1, intersegmental blood vessel; 2, afferent nerve; 3, interdorsal; 4, efferent nerve; 5, basidorsal. (b) Vertebral column in the sturgeon (*Acipenser*). 1, spinal; 2, basidorsal; 3, sheath of connective tissue; 4, notochordal sheath; 5, notochord; 6, basiventral; 7, interdorsal; 8, interventral. Broken lines indicate segmental boundaries. (c) Phylogenetic development of the amniote vertebral body. *a*, Original situation; *b*, situation in *Seymouria*; *c*, situation in normal reptiles with intercentra. 1, Neural arch; 2, hypocentrum; 3, pleurocentrum; 4, intervertebral body. (d) Phylogenetic development of monospondyly in *Amia* (a) and in teleosts (b) from the hemi- and diplospondylic situation. 1, Hypocentrum; 2, pleurocentrum; 3, basidorsal, neural arch; 4, basiventral; 5, interdorsal; 6, interventral; 7, monospondylic vertebral body. [After Starck, D. (1978). "Vergleichende Anatomie der Wirbeltiere," Vol. 2., Springer-Verlag, Berlin, Heidelberg, and New York.]

1. Somitic Cell Lineages

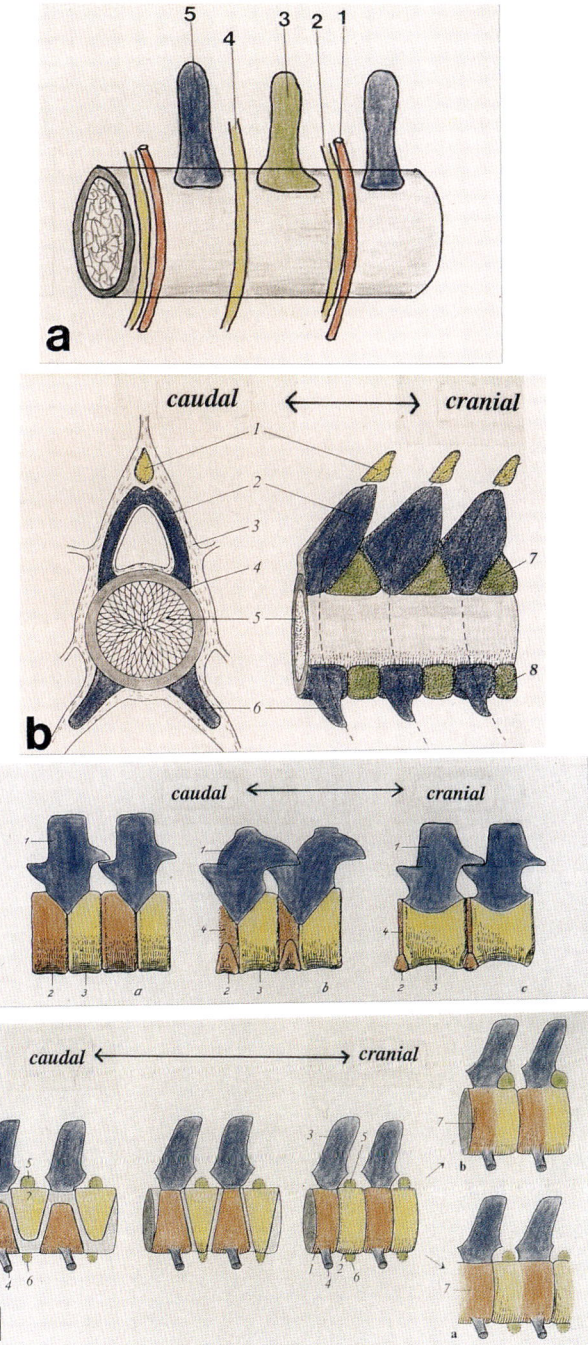

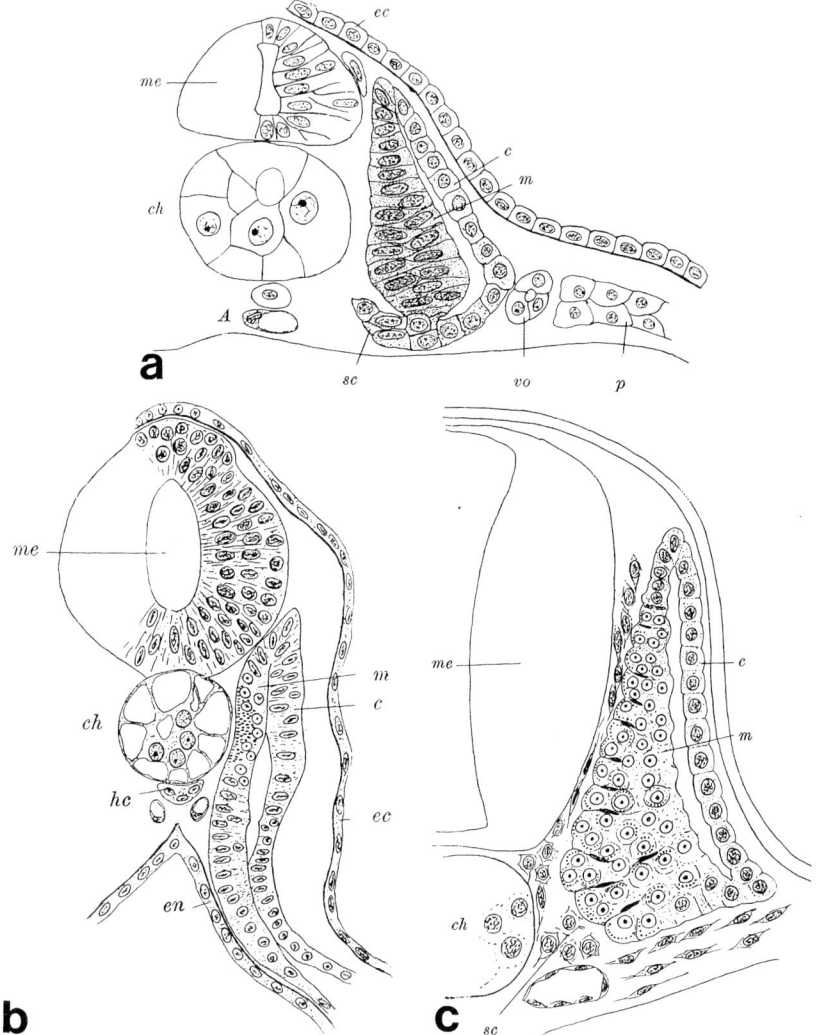

Figure 39 (a) Embryo of *Petromyzon fluviatile* (lamprey) in transverse section. (b) Shark embryo, transverse section. (c) Salmon embryo, transverse section. me, Neural tube; ch, notochord; m, myotome; c, dermatome; sc, sclerotome; vo, pronephric duct; ec, ectoderm; A, aorta; hc, hypochord; and en, endoderm. [From Hertwig, O. (1906). "Handbuch der Entwicklungslehre der Wirbeltiere."]

lateral plate mesoderm and the somite. The coelomic cavity is in connection with the somitocoel. There are only a few sclerotome cells (Fig. 39b). Their number increases in the salmon (teleost) embryo (Fig. 39c). In the sturgeon, an archaic teleost, the vertebral column still consists only of arch elements, the arcualia, but axial structures have not yet developed (Fig. 38b). We now have to distinguish between dorsal and ventral arch elements. The dorsal ones (interdorsal, basidorsal, and spinal) surround the spinal cord, whereas the ventral ones (interventral and basiventral) can be related to the aorta as hemal arches. According to the literature, the basidorsals and basiventrals are located intersegmentally, whereas the other elements lie within the segment.

The next step in evolution is the development of axial elements, the centra. We again find two central elements, the hypocentrum and the pleurocentrum. From this original situation, two different ways of development can be traced. The hypocentrum can be localized caudally or cranially within the definite central skeletal element (Fig. 38c). Nevertheless, Starck (1978) reported that during embryonic development as the sclerotome masses increase, each mass on each side of the notochord becomes divisible into a cranial area, in which the mesenchymal cells are less dense, and a caudal area, where the cells are closely aggregated. The less dense mesenchymal mass represents the interdorsal, whereas the caudal dense mass of mesenchyme is the basidorsal. The evolution of the amniote centrals can be seen in Fig. 38c. The pleurocentrum gives rise to the vertebral body and the hypocentrum to the intervertebral structure. The basidorsal develops into the neural arch. These comparative data show that neural arch elements can be found earlier than central ones during evolution and that both axial or central structures can be traced back to two skeletal elements, the pleurocentrum and the hypocentrum.

It has been shown that genes such as *hairy* and *engrailed* that are involved in segmentation of *Drosophila* are also expressed before and during somite formation in *Amphioxus* and zebrafish (M. Müller *et al.,* 1996; Holland *et al.,* 1997). Of special interest is the observation that the *engrailed* homologue *AmphiEn* is only expressed in the first eight somites that are formed by enterocoely (Holland *et al.,* 1995). This mode of segment formation may represent a more primitive mechanism. In vertebrates such as zebrafish, chick, and mouse that generate somites from mesodermal plates which are then subdivided, *engrailed* homologues are expressed in somite compartments after somites have been formed. This is similar to the more posterior somites of the chordate *Amphioxus,* which also derive from mesodermal blocks. Another phenomenon supporting the idea of segmentation homology in *Drosophila* and vertebrates is resegmentation. In vertebrates, vertebral body boundaries are shifted over half a segment as compared to somite boundaries (Remak, 1855). In *Drosophila,* the initial segment is the parasegment that gives rise to the posterior compartment of one definitive segment and the anterior compartment of the next caudal segment. It may be concluded from these observations that protostomes such as arthropods and deuterostomes such as chordates and vertebrates have a common ancestor that was already segmental (Holland *et al.,* 1997; De Robertis, 1997).

The construction principle of segmentation not only partitions the body into smaller units that facilitate local interactions, but also allows the formation of regional differences along the craniocaudal axis by minor variations of the same basic theme. This has resulted in the evolution of distinct body regions such as the thorax and abdomen in arthropods and the cervical, thoracic, lumbar, and sacral region in vertebrates. Thereby, new functions have arisen such as walking or flying in association with the development of paired body appendages. It will be interesting to see how the development of segments is established during evolution and ontogeny, as more signals and players in the process become characterized in such divergent organism as *Drosophila* and man.

Acknowledgments

Individual projects on which this article is based were supported by grants from the Deutsche Forschungsgemeinschaft. We thank Ellen Gimbel, Lidia Koschny, Ulrike Pein, Monika Schüttoff, and Günter Frank for excellent technical assistance. We are grateful to Birgit Strittmatter, Heike Bowe, and Ulrike Uhl for preparation of the manuscript.

References

Beresford, B. (1983). Brachial muscles in the chick embryo: The fate of individual somites. *J. Embryol. Exp. Morphol.* **77,** 99–116

Bladt, F., Riethmacher, D., Isenmann, S., Aguzzi, A., and Birchmeier, C. (1995). Essential role for the *c-met* receptor in the migration of myogenic precursor cells into the limb bud. *Nature* **376,** 768–771.

Brand-Saberi, B., Ebensperger, C., Wilting, J., Balling, R., and Christ, B. (1993). The ventralizing effect of the notochord on somite differentiation in chick embryos. *Anat. Embryol.* **188,** 239–245.

Brand-Saberi, B., Wilting, J., Ebensperger C., and Christ, B. (1996a). N-cadherin is involved in myoblast migration and muscle differentiation in the avian limb bud. *Dev. Biol.* **178,** 160–173.

Brand-Saberi, B., Müller T. S., Wilting, J., Christ, B., and Birchmeier, C. (1996b). Scatter factor/hepatocyte growth factor (SF/HGF) induces emigration of myogenic cells at interlimb level *in vivo*. *Dev. Biol.* **179,** 303–308.

Brill, G., Kahane, N., Carmeli, C., von Schack, D., Barde, Y. A., and Kalcheim, C. (1995). Epithelial–mesenchymal conversion of dermatome progenitors requires neural tube-derived signals: Characterization of the role of Neurotrophin-3. *Development* **121,** 2583–2594.

Buffinger, N., and Stockdale, F. E. (1994). Myogenic specification in somites: Induction by axial structures. *Development* **120,** 1443–1452.

Buffinger, N., and Stockdale, F. E. (1995). Myogenic specification of somites is mediated by diffusible factors. *Dev. Biol.* **169,** 96–108.

Burgess, R., Rawls, A., Brown, D., Bradley, A., and Olson, E. (1996). Requirement of the *paraxis* gene for somite formation and musculoskeletal patterning. *Nature* **384,** 570–573.

Chevallier, A., Kieny, M., and Mauger, A. (1977). Limb–somite relationship: Origin of the limb musculature. *J. Embryol. Exp. Morphol.* **41,** 245–258.

Christ, B. (1970). Experimente zur Lageentwicklung der Somiten. *Verh. Anat. Ges.* **64,** 555–564.

Christ, B., and Ordahl, C. P. (1995). Early stages of chick somite development. *Anat. Embryol.* **191,** 381–396.

Christ, B., Jacob, H. J., and Jacob, M. (1974). Ueber den Ursprung der Flugmuskulatur: Experimentelle Untersuchungen an Wachtel- und Huehnerembryonen. *Experientia* **30,** 1446–1448.

Christ, B., Jacob, H. J., and Jacob, M. (1977). Experimental analysis of the origin of the wing musculature in avian embryos. *Anat. Embryol.* **150,** 171–186.

Christ, B., Jacob, M., and Jacob, H. J. (1983). On the origin and development of the ventrolateral abdominal muscles in the avian embryo: An experiment and ultrastructural study. *Anat. Embryol.* **166,** 87–101.

Corning, H. K. (1881). Über die sogenannte Neugliederung der Wirbelsäule und über das Schicksal der Urwirbelhöhle bei Reptilien. *Morphol. Jb.* **17,** 611–622.

Denetclaw, W. F., Christ, B., and Ordahl, C. P. (1997). Location and growth of epaxial myotome precursor cells. *Development* **124**, 1601–1610.
De Robertis, E. M. (1997). The ancestry of segmentation. *Nature* **387**, 25–26.
Eichmann, A., Marcelle, C., Bréant, C., and Le Douarin, N. M. (1993). Two molecules related to the VEGF receptor are expressed in early endothelial cells during avian embryonic development. *Mech. Dev.* **42**, 33–48.
Fischel, A. (1895). Zur Entwicklung der ventralen Rumpf- und Extremitätenmuskulatur der Vögel und Säugetiere. *Morphol. Jb.* **23**, 544–561.
Füchtbauer, E.-M. (1995). Expression of m-twist during postimplantation development of the mouse. *Dev. Dyn.* **204**, 316–322.
Gamel, A. J., Brand-Saberi, B., and Christ, B. (1995). Halves of epithelial somites and segmental plate show distinct muscle differentiation behavior *in vitro* compared to entire somites and segmental plate. *Dev. Biol.* **172**, 625–639.
George-Weinstein, M., Gerhart, J., Blitz, J., Simak, E., and Knudsen, K. A. (1997). N-cadherin promotes the commitment and differentiation of skeletal muscle precursor cells. *Dev. Biol.* **185**, 14–24.
Goulding, M., Lumsden, A., and Paquette, A. J. (1994). Regulation of *Pax-3* expression in the dermomyotome and its role in muscle development. *Development* **120**, 957–971.
Grim, M. (1970). Differentiation of myoblasts and the relationship between somites and the wing bud of chick embryos. *Z. Anat. Entwicklungsgesch.* **132**, 260–271.
Hall, B. K. (1978). "Developmental and Cellular Skeletal Biology." Academic Press, New York and London.
Hertwig, O. (1906). "Handbuch der Entwicklungslehre der Wirbeltiere," Fischer Jena.
His, W. (1886). "Untersuchungen über die erste Anlage des Wirbeltierleibes. Die erste Entwicklung des Hühnchens im Ei." Vogel, Leipzig.
Holland, L. Z., Pace, D. A., Blink, M. L., Kene, M., and Holland, N. D. (1995). Sequence and expression of Amphioxus alkali myosin light chain (*AmphiMLC-alc*) throughout development: Implications for vertebrate myogenesis. *Dev. Biol.* **171**, 665–676.
Holland, L. Z., Kene, M., Williams, N. A., and Holland N. D. (1997). Sequence and embryonic expression of the Amphioxus *engrailed gene* (*AmphiEn*): The metameric pattern of transcription resembles that of its segment-polarity homolog in *Drosophila*. *Development* **124**, 1723–1732.
Holt, C. E., Lemaire, P., and Gurdon, J. B. (1994). Cadherin-mediated cell interactions are necessary for the activation of *MyoD* in *Xenopus* mesoderm. *Proc. Natl. Acad. Sci. U.S.A.* **91**, 10844–10848.
Huang, R., Zhi, Q., Wilting, J., and Christ, B. (1994). The fate of somitocoele cells in avian embryos. *Anat. Embryol.* **190**, 243–250.
Huang, R., Zhi, Q., Neubüser, A., Müller, T. S., Brand-Saberi, B., Christ, B., and Wilting, J. (1996). Function of somite and somitocoel cells in the formation of vertebral motion segment in avian embryos. *Acta Anat.* **155**, 231–241.
Jacob, M., Christ, B., and Jacob, H. J. (1975). Über die regionale Determination des paraxialen Mesoderms jünger Hühnerembryonen *Verh. Anat. Ges.* **69**, 263–269.
Johnson, R. L., Laufer, E., Riddle, R. D., and Tabin, C. (1994). Ectopic expression of *Sonic hedgehog* alters dorsal–ventral patterning of somites. *Cell* **79**, 1165–1173.
Kaehn, V., Jacob, H. J., Christ, B., Hinrichsen, K., and Poelmann, R. E. (1988). The onset of myotome formation in the chick. *Anat. Embryol.* **177**, 191–201.
Lassar, A. B., and Münsterberg, A. (1996). The role of positive and negative signals in somite patterning. *Curr. Opin. Neurobiol.* **6**, 57–63.
Malpighi, M. (1689). "Opera omnia" Lugduni, Batav.
Müller, M., Weizäcker, E., and Campos-Ortega, J. A. (1996). Expression domains of a zebrafish homologue of the *Drosophila* pair-rule gene *hairy* correspond to primordia of alternating somites. *Development* **122**, 2071–2078.
Müller, T. S., Ebensperger C., Neubüser, A., Koseki, H., Balling, R., Christ, B., and Wilting, J. (1996). Expression of avian *Pax-1* and *Pax-9* in the sclerotomes is controlled by axial and lateral tissues, but intrinsically regulated in pharyngeal endoderm. *Dev. Biol.* **178**, 403–417.

Münsterberg, A. E., and Lassar, A. B. (1995). Combinatorial signals from the neural tube, floor plate and notochord induce myogenic bHLH gene expression in the somite. *Development* **121,** 651–660.

Münsterberg, A. E., Kitajewski, J., Bumcrot, D. A., McMahon, A. P., and Lassar, A. B. (1995). Combinatorial signaling by Sonic hedgehog and Wnt family members induces myogenic bHLH gene expression in the somite. *Genes Dev.* **9,** 2911–2922.

Olson, E. N., Perry, W. M., and Schulz, R. A. (1995). Regulation of muscle differentiation by the MEF2 family of MADS box transcription factors. *Dev. Biol.* **172,** 2–14.

Ordahl, C. P., and Le Douarin, N. M. (1992). Two myogenic lineages within the developing somite. *Development* **114,** 339–353.

Pourquié, O., Coltey, M., Teillet, M. A., Ordahl, C., and Le Douarin, N. M. (1993). Control of dorsoventral patterning of somitic derivatives by notochord and floor plate. *Proc. Natl. Acad. Sci. U.S.A.* **90,** 5242–5246.

Pourquié, O., Coltey, M., Bréant, C., and Le Douarin, N. M. (1995). Control of somite patterning by signals from the lateral plate. *Proc. Natl. Acad. Sci. U.S.A.* **92,** 3219–3223.

Pourquié, O., Fan, C. M., Coltey, M., Hirsinger, E., Watanabe, Y., Bréant, C., Francis-West, P., Brickell, P., Tessier-Lavigne, M., and Le Douarin, N. M. (1996). Lateral and axial signals involved in avian and somite patterning: A role for BMP4. *Cell* **84,** 461–471.

Pownall, M. E., and Emerson, E. P. (1992). Sequential activation of myogenic regulatory genes during somite morphogenesis in quail embryos. *Dev. Biol.* **151,** 67–79.

Remak, R. (1850). "Untersuchungen über die Entwicklung der Wirbeltiere." Reimer, Berlin.

Rong, P. M., Teillet, M. A., Ziller, C., and Le Douarin, N. M. (1992). The neural tube/notochord complex is necessary for vertebral but not limb and body wall striated muscle differentiation. *Development* **115,** 657–672.

Sosic, D., Brand-Saberi, B., Schmidt, C., Christ, B., and Olson, E. N. (1997). Regulation of paraxis expression and somite formation by ectoderm—and neural tube—derived signals. *Dev. Biol.* **185,** 229–243.

Starck, D. (1978). Vergleichende Anatomie der Wirbeltiere," Vol. 2. Springer-Verlag, Berlin, Heidelberg, and New York.

Stern, C., Sisodaya, S., and Keynes, R. (1986). Interactions between neurites and somite cells: Inhibition and stimulation of nerve growth in the chick embryo. *J. Embryol. Exp. Morph.* **91,** 209–226.

Stern, H. M., Brown, A. M., and Hauschka, S. D. (1995). Myogenesis in paraxial mesoderm: Preferential induction by dorsal neural tube and by cells expressing Wnt-1. *Development* **121,** 3675–3686.

Strudel, G. (1955). L'action morphogène du tube nerveux et la chorde sur la différenciation des vertèbres et des muscles vertébraux chez l'embryon de poulet. *Arch. Anat. Microsc. Morphol. Exp.* **44,** 209–235.

von Baer, K. E. (1828). "Über die Entwicklungsgeschichte der Tiere," Vols. 1–3. Königsberg.

von Ebner, V. (1888). Urwirbel und Neugliederung der Wirbelsäule. *Sitzungsber. Akad. Wiss. Wien Math. Naturwiss. Kl. Abt. 3* **97,** 194–206.

Williams, B. A., and Ordahl, C. P. (1994). *Pax-3* expression in segmental mesoderm marks early stages in myogenic cell specification. *Development* **120,** 785–796.

Wilting, J., Kurz, H., Brand-Saberi, B., Steding, G., Yang, Y. X., Hasselhorn, M. M., Epperlein, H. H., and Christ, B. (1994). Kinetics and differentiation of somite cells forming the vertebral column: Studies on human and chick embryos. *Anat. Embryol.* **190,** 573–581.

Wilting, J., Brand-Saberi, B., Huang, R., Zhi, Q., Köntges, G., Kühlewein, M., Ordahl, C. P., and Christ, B. (1995a). The angiogenic potential of the avian somite. *Dev. Dyn.* **202,** 165–171.

Wilting, J., Ebensperger, C., Müller, T. S., Koseki, H., Wallin, J., and Christ, B. (1995b). *Pax-1* in the development of the cervico-occipital transitional zone. *Anat. Embryol.* **192,** 221–227.

Xue, X. J., and Xue, Z. (1996). Spatial and temporal effects of axial structures on myogenesis of developing somites. *Mech. Dev.* **60,** 73–82.

Zhi, Q., Huang, R., Christ, B., and Brand-Saberi, B. (1996). Participation of individual brachial somites in skeletal muscles of the avian distal wing. *Anat. Embryol.* **194,** 327–339.

2
Duality of Molecular Signaling Involved in Vertebral Chondrogenesis

Anne-Hélène Monsoro-Burq and Nicole Le Douarin
Institut d'Embryologie Cellulaire et Moléculaire du CNRS et du Collège de France
94736 Nogent-sur-Marne, France

I. Origin of the Vertebrae and Intervertebral Disks
II. Gene Expression in the Developing Somitic Cells
   A. *Pax* Gene Expression in the Sclerotome
   B. *Msx* Gene Expression in the Subectodermal Mesenchyme
   C. Origin and Fate of the Dorsal Mesenchyme
III. The Dorsal Neural Tube Grafted Ectopically Promotes Dorsal Cartilage Formation in a Subectodermal Position
IV. Ventral Axial Structures Prevent Subectodermal Cartilage Formation
   A. The Notochord
   B. The Floor Plate
V. Molecular Pathways Leading to Chondrogenesis in the Vertebra
   A. Sonic Hedgehog Protein Inhibits Subectodermal Chondrogenesis
   B. BMP4 Protein Promotes Dorsal Mesenchyme Development and Superficial Vertebral Cartilage Differentiation
VI. Conclusions: A Novel Model for Vertebra Development
References

I. Origin of the Vertebrae and Intervertebral Disks

The vertebral column, composed by the alternation of vertebrae and intervertebral disks, is the most conspicuous metameric structure of the vertebrate body. The vertebrae arise from the paraxial mesoderm previously segmented into somites. Somitic cells, initially lateral to the axial organs, eventually surround the spinal cord and notochord, to finally differentiate into cartilage and bone. Each vertebra (except for the atlas) comprises a ventral vertebral body (corpus vertebrae), forming around the notochord, and a pair of vertebral arches (arci vertebrae), which develop lateral and dorsal to the spinal cord, bear transverse processes (processi transversi), and are fused dorsally underneath the ectoderm by a more or less developed spinous process (processus spinosus). This general organization is found all along the anteroposterior (AP) axis. However, morphological features distinguish cervical from thoracic, lumbar, sacral, and coccygeal vertebrae. This AP pattern is fixed early in development, as shown by heterotopic transplantations of the unsegmented paraxial mesoderm along the AP axis (Kieny *et al.,* 1972) and is controlled by the expression of Hox genes (Kessel and Gruss, 1991; reviewed by Krumlauf, 1994; Burke *et al.,* 1995 and references therein). Intervertebral disks develop between the chondrified vertebral bodies; they are composed by a dense fibrous tissue, the anulus fibrosus, and by a central zone surrounding the notochord, the nucleus pulposus (Christ and Wilting, 1992).

Vertebrae and intervertebral disks arise from the ventralmost part of the somite, the sclerotome. The somites condense from the unsegmented paraxial mesoderm as epithelial spheres that are not immediately polarized along the dorsoventral (DV) or mediolateral (ML) axis (Aoyama and Asamoto, 1988; Ordahl and Le Douarin, 1992). These polarities are established when the somites reach stage III–IV according to the nomenclature proposed by Ordahl (1993), somite I being the last somite formed at the considered stage. The somite differentiates along the DV axis into the mesenchymal sclerotome ventrally and the epithelial dermomyotome dorsally. Sclerotome and dermomyotome can be further subdivided according to the fate of each area. The dermis of the back originates from the dermatome and the medial moiety of the myotome yields the epaxial muscles, while hypaxial muscles develop from its lateral part (Ordahl and Le Douarin, 1992). Each vertebra is formed by the posterior half and the anterior half of two consecutive somites, thus defining segmental units similar to parasegments in insects. The sclerotome can be subdivided into four parts along the AP and ML axes. The cranial-medial part of the sclerotome, which is located near to the notochord and the ventral part of the neural tube, forms the vertebral body while the intervertebral disc forms from the caudal-medial quarter. The caudal-lateral part gives rise to most of the neural arches, the pedicles of the vertebra, and the somite-derived part of the ribs (Christ and Wilting, 1992). The anterior and posterior halves of the sclerotome have distinct properties that are revealed when neural crest cells migrate and motoneurons extend axons: motor fibers and neural crest cells migrate through the cranial half of the sclerotome and avoid the caudal somitic half; this process is responsible for the metameric distribution of the sensory and sympathetic ganglia (Keynes and Stern, 1984; Kalcheim and Teillet, 1989; Teillet *et al.,* 1987).

The problem of the formation of the spinous process had not been envisaged until our group showed that it develops according to a mechanism largely distinct from that which controls the rest of the vertebra (Takahashi *et al.,* 1992; Monsoro-Burq *et al.,* 1994, 1995, 1996; Watanabe *et al.,* 1998).

II. Gene Expression in the Developing Somitic Cells

In this chapter, we review the data showing that differentiation of cartilage from somitic cells involves distinct molecular pathways in the vertebral body and neural arches on the one hand, and in the spinous process domain on the other hand.

These two parts of the vertebra arise from different groups of cells that we will further designate as "deep" somitic cells for the former and "superficial" somitic cells for those which form the spinous process and develop underneath the ectoderm.

A. Pax *Gene Expression in the Sclerotome*

Two genes of the Pax family, *Pax1* and *Pax9,* are expressed in the sclerotome. In mouse and chick embryos, *Pax1* is detected soon after segmentation. In chick embryos, *Pax1* expression has been first detected in the epithelial somites II (Barnes *et al.,* 1996) or IV (Borycki *et al.,* 1997); at these stages, it labels the ventral part of the somites and the cells contained in the somitocoele. *Pax1* messenger RNA as well as the corresponding protein are detected in virtually all the mesenchymal cells of the sclerotome in young somites and later on become progressively restricted to the caudal and medial part of the sclerotome. When chondrification takes place in the vertebral body, *Pax1* is still expressed in the intervertebral disk (Deutsch *et al.,* 1988; Wallin *et al.,* 1994; Barnes *et al.,* 1996). Mutants of the *Pax1* gene such as the *undulated* mutants, corresponding to various altered alleles of this gene, lack vertebral bodies and intervertebral disks, as well as the proximal part of the ribs. In contrast, the neural arches are almost normally formed, suggesting a key role for *Pax1* in the development of the ventral rather than the lateral and dorsal parts of the vertebra (Chalepakis *et al.,* 1991; Wallin *et al.,* 1994).

Pax9 is first detected at stage 12HH (Hamburger and Hamilton, 1951) in chick embryos (Goulding *et al.,* 1994). The *Pax 9* domain of expression overlaps partly with that of *Pax1,* in the intervertebral disk anlage (Peters *et al.,* 1995).

Because the fine organization of the somite is controlled by environmental cues, many *in vitro* studies have been dedicated to the search for inducers of somitic differentiation, especially for chondrogenesis. These studies point to a prominent role of the ventral axial tissues, the notochord and the ventral part of the neural tube, in sclerotome induction (reviewed by Hall, 1977), whereas the ectoderm was shown to be critical for dermomyotome formation (Gallera, 1966; Swalla and Solursh, 1984; Kenny-Mobbs and Thorogood, 1987).

Experimental manipulations in the chick embryo have allowed further *in vivo* analysis of DV patterning of the somite. (1) Early survival of the sclerotome and the medial part of the dermomyotome depend on the notochord and the neural tube (Teillet and Le Douarin, 1983; Rong *et al.,* 1992). The absence of notochord and floor plate results in the lack of *Pax1*-positive sclerotomal cells and of *Pax3*-expressing dermomyotomal cells, essentially due to cell death (Teillet *et al.,* 1998). The most mature somites, in which myotomal cells are well engaged in their differentiation pathway, form only muscle (Rong *et al.,* 1992; Pourquié *et al.,* 1993; Koseki *et al.,* 1993; Dietrich *et al.,* 1993). The notochord-secreted protein Sonic hedgehog (SHH) is able to rescue the phenotype of notochord- and neural tube-ablated embryos (Teillet *et al.,* 1998). (2) The notochord and the floor plate of the neural tube have a ventralizing influence on the somite. If a supernumerary notochord or floor plate is implanted between the neural tube and the somites, the *Pax1-Pax9* domain is enlarged, at the expense of the dermomyotome. Ectopic lateral vertebral cartilage develops while back muscles and dermis are absent (Pourquié *et al.,* 1993; Brand-Saberi *et al.,* 1993; Goulding *et al.,* 1994; Ebensperger *et al.,* 1995). This effect of extra ventral axial structures on *Pax1* expression can be mimicked by the amino-terminal part of the secreted protein SHH when it is applied on paraxial mesoderm isolated in organotypic *in vitro* cultures (Fan and Tessier-Lavigne, 1994; Fan *et al.,* 1995) or when it is overexpressed *in vivo* by the injection of retroviral constructs (Johnson *et al.,* 1994) or by the graft of SHH-producing cells (Hirsinger *et al.,* 1997; Watanabe *et al.,* 1998). Together, these data point to a role for the ventral axial structures, via SHH signaling, in somite survival and in further development of the sclerotome.

B. *Msx* Gene Expression in the Subectodermal Mesenchyme

We have been interested in the Msx family of genes because of their expression in the somite-derived dorsal mesenchyme. *Msx* genes are homeobox-containing genes isolated in vertebrates (Robert *et al.,* 1989; Hill *et al.,* 1989) as homologs of the *Drosophila* gene *Msh* (Gerhing, 1987; Isshiki *et al.,* 1997; Lord *et al.,* 1995). To date, they have been described in all species studied, from diblastic organisms to humans (Davidson, 1995 for a review). There are two genes recorded in birds and humans, namely, *Msx1* and *Msx2* (Takahashi and Le Douarin, 1990; Robert *et al.,* 1991; Ivens *et al.,* 1990; Jabs *et al.,* 1993), three in mice (Wang *et al.,* 1996; Shimeld *et al.,* 1996), and four in zebrafish (Akimenko *et al.,* 1995). Embryonic expression of *Msx* genes is found in many locations where interactions between an epithelium of ectodermal origin and a mesenchyme are active, such as in the limb bud (Ros *et al.,* 1992 and references therein), the craniofacial rudiments (Takahashi *et al.,* 1991; MacKenzie *et al.,* 1991a; Mina *et al.,* 1995), the tooth bud (MacKenzie *et al.,* 1991b; Jowett *et al.,* 1993), and the skin appendages (Noveen *et al.,* 1995). They are also expressed in other organs such as the heart (Chan-Thomas *et al.,* 1993), in the septum where they are associated with neural crest-derived cells. Expression in adult animals is correlated with epithelium–mesenchyme tissue interactions such as in the mouse uterus (Pavlova *et al.,* 1994). *Msx* genes are also implicated in regenerative phenomena in the amphibian limb (Simon *et al.,* 1995) and in the fins in fishes (Akimenko *et al.,* 1995).

The mesenchyme located on the dorsomedial line above the neural tube is formed by cells migrating from the somites between the roof plate of the neural tube and the superficial ectoderm (Fig. 1A). In chick and quail embryos, at stage 17HH (E2.5), at the cervical level, the somite is differentiated into sclerotome and dermomyotome; however, the dorsal mesenchyme has not yet begun to form

2. Duality of Molecular Signaling in Vertebral Chondrogenesis

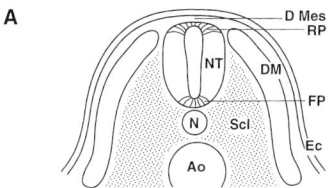

Figure 1 (A) Scheme of a transverse section at the trunk level of an E3 chick embryo. The neural tube (NT) is surrounded by somite-derived structures, the sclerotome (Scl, dotted area) and the dermomyotome (DM). D Mes indicates the space where the dorsal mesoderm will accumulate. Ec, Ectoderm; FP, floor plate; N, notochord; Ao, aorta; RP, roof plate. (B) Timetable of the formation of the dorsal mesenchyme. The dorsal mesenchyme develops according to an anterior to posterior gradient: at stage 17HH, even at the most rostral levels of the spinal cord, no mesenchymal cells are present above the neural tube. From stage 18HH onward, the dorsal mesenchyme forms at the cervical levels first, then above more caudal parts of the spinal cord. AP, Anteroposterior; HH, according to Hamburger and Hamilton chick developmental table (1951).

(Figs. 1B, 2A). At stage 19HH, mesenchymal cells have migrated dorsally to the neural tube at the level of the anterior spinal cord; at the wing level, they just begin to do so. From stage 20HH onward, a few cell layers have accumulated above the roof plate (Fig. 2B,C) from cervical down to trunk levels. The mesenchymal cells express *Msx1* and *Msx2* genes from the time they become dorsally positioned (Fig. 2D–I) but stop to do so as they differentiate into cartilage. The superficial ectoderm and roof plate also express these genes.

C. Origin and Fate of the Dorsal Mesenchyme

We analyzed the fate of the dorsal mesenchyme *in ovo* by using quail–chick chimeras: The first cells covering the neural tube at the brachial level of an E3 quail embryo were removed together with the overlying ectoderm and grafted isotopically in a stage-matched chick host (Fig. 3A). The chimera was reincubated for 5–6 more days, sectioned, and stained with the QCPN monoclonal antibody which binds a perinucleolar antigen in quail cells. The grafted quail cells were subsequently found in the superficial ectoderm overlying the operated area, in the meninges above the roof plate, and they formed the mediodorsal part of the vertebra, that is, the tips of the joining neural arches and the spinous process (Fig. 3B,C and Takahashi *et al.*, 1992). The dermis is of chick origin indicating that it was formed

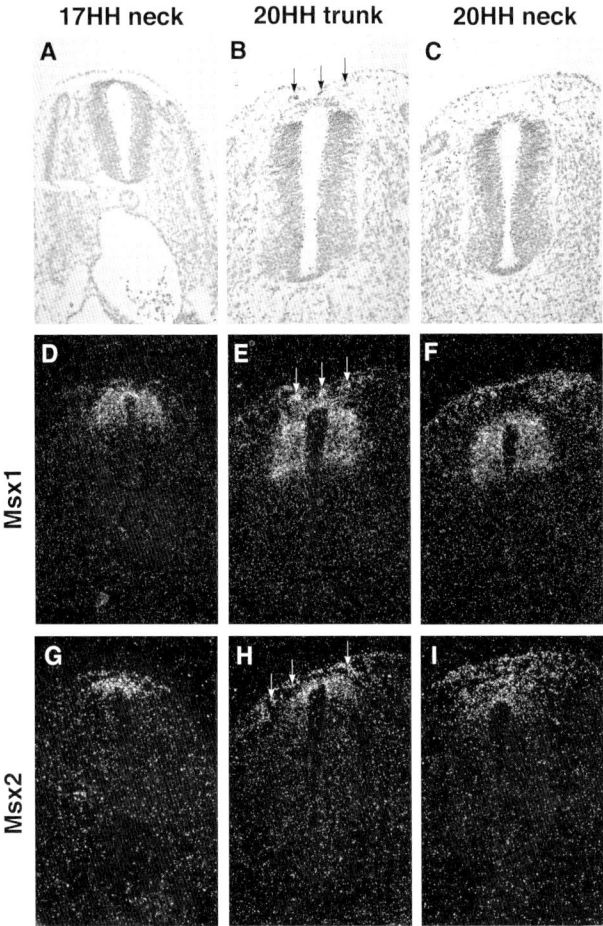

Figure 2 Progressive accumulation of dorsal mesenchymal cells. At stage 17HH, no mesenchymal cell has migrated dorsally even at the more rostral levels of the spinal cord (A, D, G). From stage 18HH onward, cells are progressively localized between the ectoderm and the roof plate, according to a rostral to caudal gradient (B, trunk level; C, neck level). As soon as they are located dorsally, the cells express *Msx1* (E, F, arrows) and *Msx2* (H, I, arrows) genes. The roof plate of the neural tube and the ectoderm also express both genes (see D–I).

by migration from the dermomyotome after the stage of the operation. Thus, the dorsal *Msx*-expressing cells give rise to several tissues that differentiate under the ectoderm, among which are chondrocytes of the dorsal part of the vertebra. This result indicates a heterogeneity of gene expression in the vertebra precursor cells, since ventral and lateral sclerotomal cells that yield the vertebral body and neural arches do not express *Msx* genes but, in contrast to the dorsal chondrogenic cells,

8. Duality of Molecular Signaling in Vertebral Chondrogenesis

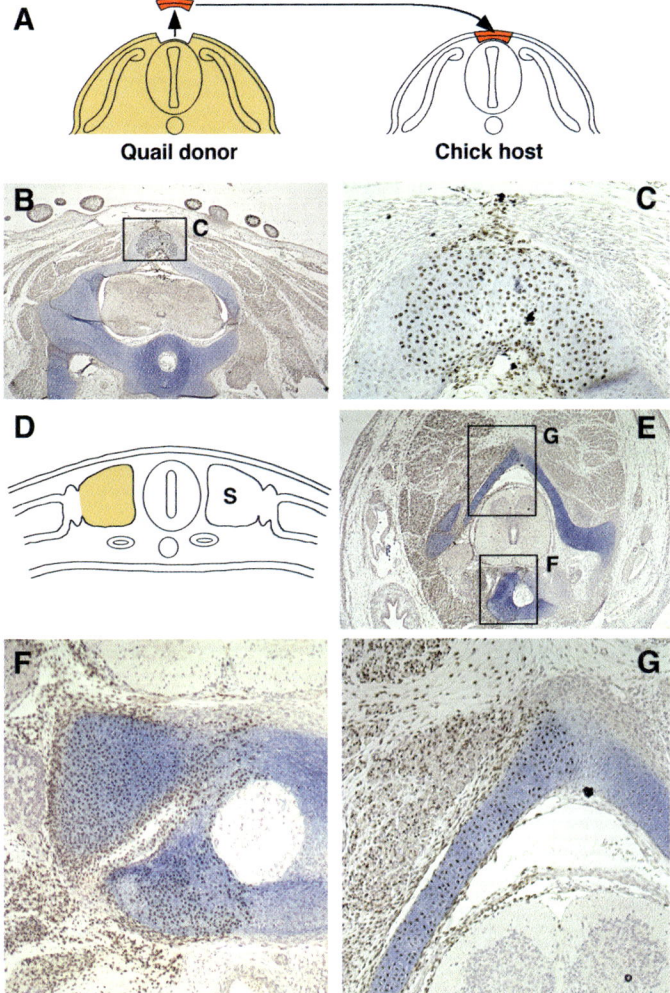

Figure 3 (A–C) The fate of the first cells located dorsally to the roof plate was analyzed by removing this territory together with its overlying ectoderm from a 36 somite stage quail embryo and grafting it isotopically into a stage 17HH chick host (A). Six days after the operation (B, C) the quail cells, evidenced with the QCPN monoclonal antibody (brown) are found in the dorsal meninges, the ectoderm, and in the dorsal part of the vertebra (C shows higher magnification of the boxed area in B). The dermis is not formed by quail but by chick cells indicating that its progenitors migrate from the dermomyotome on top of the neural tube at stages later than 18HH. (D–G) The somitic origin of the dorsal part of the vertebra was confirmed by the unilateral graft of quail paraxial mesoderm into a chick host (D). S, Somite. The vertebra and the other somitic derivatives analyzed at E9 are entirely formed by QCPN positive quail cells on one side of the sagittal plane (E–G). The dorsalmost part of the neural arch and half of the spinous process on this side are of quail origin thus they are formed by somite-derived cells (F, G show higher magnification of the boxed areas in E).

50 Anne-Hélène Monsoro-Burq and Nicole Le Douarin

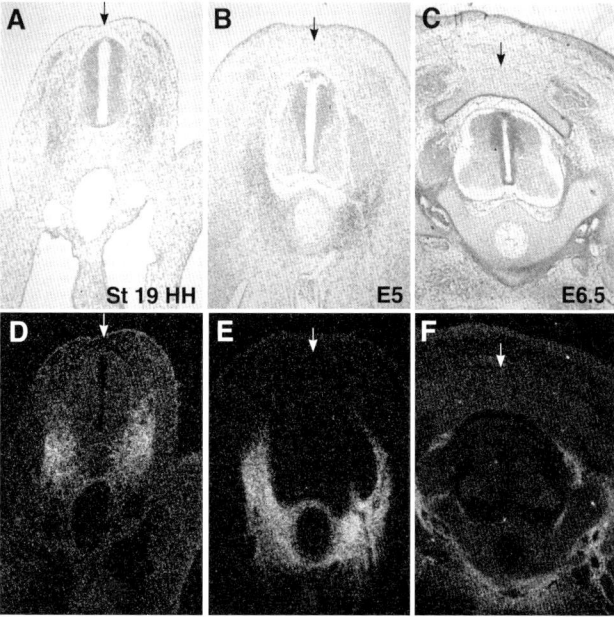

Figure 4 Expression of *Pax1* in the somites. Transverse sections of quail embryos at E3 to E6.5. From early stages of somite development (stage 19HH, A), *Pax1* labels the sclerotome (D). The dorsal mesenchyme does not express *Pax1* at early stages (D, arrow) nor later on (B, C, E, F arrows). When the vertebral cartilage differentiates (C), *Pax1* expression is resticted in the perichondral area of the vertebrate body and neural arches. In addition it is expressed also in the intervertebral disks (not shown). The spinous process begins to differentiate at E6.5 in quail embryos (C, arrow) and at E7.5 in chicks (not shown): *Pax1* is not expressed dorsally at those stages either (F, arrow).

express *Pax1* and *Pax9* genes. A systematic examination in the avian embryos of *Pax1* gene expression reveals that it is detected in almost all sclerotomal cells at early stages, as already reported by Barnes *et al.* (1996) and Borycki *et al.* (1997) (Fig. 4A,D), later on becomes rapidly restricted to the vertebral body and intervertebral disk precursors, and is never detected dorsally, at any stage of development (Fig. 4A–F, arrows). Thus, the dorsal part of the vertebra originates from cells that switch off *Pax1* expression early on and express *Msx1* and *Msx2* genes as soon as they are located dorsally, whereas the ventral and lateral vertebral precursors still express *Pax1*.

Given the dorsal location of the *Msx*-expressing cells, and knowing that, at cephalic levels, neural crest cells have skeletogenic potential and also express the *Msx* markers (Takahashi and Le Douarin, 1990; Takahashi *et al.*, 1991), we wondered if the dorsal part of the vertebra could be derived from a few trunk neural crest cells that would remain in a dorsal position, the alternative being that the entire vertebra actually originates from the somite. The somitic origin of the spinous process was confirmed by the outcome of unilateral grafts of quail somites into chick recipients (Fig. 3D). In such chimeras the whole hemivertebra was formed by quail cells, with the limit between the grafted tissue and the host cartilage being the sagittal plane (Fig. 3E–G).

III. The Dorsal Neural Tube Grafted Ectopically Promotes Dorsal Cartilage Formation in a Subectodermal Position

The roof plate of the neural tube expresses several genes encoding signaling molecules of the TGFβ family such as *Bmp4* (bone morphogenetic protein 4) (Fig. 5), *Bmp5, Bmp7, Activin B* (Liem *et al.*, 1995, 1997), and *Dorsalin-1* (Basler *et al.*, 1993). During chick development, *Bmp4* transcripts are detected as early as the primitive streak stages. At E2 and E3, *Bmp4* is present at the tip of the neural folds, then in the roof plate and at lower levels in the dorsal ectoderm (Fig. 5A,C and Watanabe and Le Douarin, 1996). Later on, the expression of *Bmp4* in the roof plate

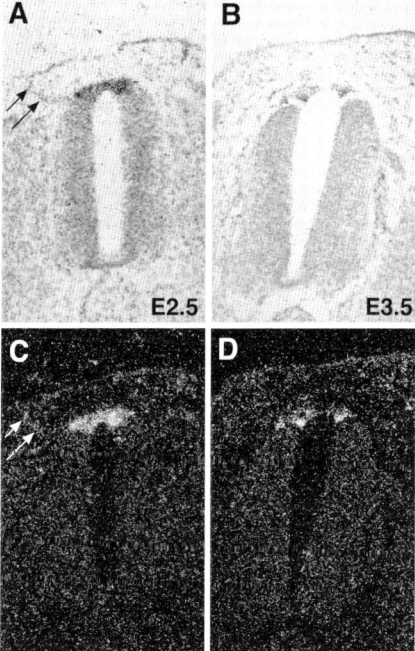

Figure 5 *Bmp4* expression in the dorsal domain. *Bmp4* transcripts are detected in the roof plate and the ectoderm during the formation of the dorsal mesenchyme (stage 19HH, A, C). A weak signal is found in the mesenchyme itself (C, arrows). Later on, the expression diminishes in the roof plate and the overlying tissues (B, D).

diminishes (Fig. 5B,D). The dorsal mesenchyme weakly expresses the *Bmp4* gene at E3 (Fig. 5C).

Bone morphogenetic proteins (BMP) have primarily been isolated from bone extracts as potent cartilage and bone inducers in adult muscle and connective tissues (Urist, 1965; Rosen and Thies, 1992, for a review). Developmental studies have recently focused on BMPs, especially BMP4 and its antagonists, during early DV patterning of the embryos (Hemmati-Brivanlou *et al.*, 1994; Sasai *et al.*, 1995; Zimmerman *et al.*, 1996; Piccolo *et al.*, 1996; Jones *et al.*, 1996; Suzuki *et al.*, 1997), but also on other members of the family, implicated in the formation of many organs and tissues (Hogan, 1996 for a review). In a few developmental systems such as the tooth bud or the rhombencephalic neural crest, it has been shown that BMP4 induces *Msx* gene expression (Vainio *et al.*, 1993; Graham *et al.*, 1994).

We analyzed the expression of *Msx* and *Bmp4* genes during the development of the neural tube and noted that they overlap. *Msx* transcripts are initially detected at primitive streak stages (Suzuki *et al.*, 1991). In stage 10HH (10 somites) chick embryos, near Hensen's node, *Msx1* and *Msx2* are strongly expressed in the sinus rhomboïdalis then in the lateral walls of the neural plate but are not detected in the floor plate–notochord complex (Fig. 6A). When the neural tube closes, *Msx* gene expression becomes progressively restricted to the tip of the neural folds (Fig. 6B). When the neural tube is closed, the roof plate strongly expresses both *Msx* genes as does, although at lower levels, its overlying ectoderm (Fig. 2). Thus, the somitic cells migrating dorsally begin to express *Msx* genes at high levels when they are located between those two Bmp4 and *Msx*-expressing ectoderm-derived layers.

The role of the neural tube on *Msx* gene expression in the paraxial mesodermal cells was analyzed in chick embryos, by grafting a quail neural tube fragment between the host's neural tube and the paraxial mesoderm (Fig. 6C). During normal development, *Msx* gene expression is not detected in the somites and their derivatives located laterally to the neural tube (Fig. 2). Two days after surgery, the ectoderm had healed above the operated area and the grafts were surrounded by host mesenchymal cells (Fig. 6D,F). Above the quail neural tube, chick host cells expressed *Msx2* (Fig. 6E) provided that the neural tube had been grafted in its normal DV orientation. If only the dorsal third of the neural tube, containing the

---

Figure 6 Role of the roof plate in *Msx* gene expression by the dorsal mesenchyme. (A, B) Just rostral to Hensen's node of stage 10HH chick embryos, *Msx* genes are expressed in the lateral walls of the neural plate, with the exception of the notochord–floor plate primordium (A, *Msx1* gene expression); more rostrally, the expression of *Msx* genes in the neural tube becomes progressively restricted to the tips of the neural folds (B, *Msx1* gene expression). (C) The role of the neural tube was tested by ectopic grafts of the dorsal third including the roof plate (a), of the whole (b), or of the ventral third (c) of a quail neural tube, laterally to the chick host neural tube. The roof plate is represented in red (a, b). Two days after the operation, the ectoderm has healed above the graft, which is surrounded by chick host mesenchyme (D, F, arrows). The normal pattern of *Msx2* gene expression is observed above the host neural tube (C, chick host neural tube; Q, quail grafted neural tube fragment). Ectopic expression of *Msx2* has been induced by the roof plate or the entire neural tube grafts (E, arrow), provided that the

2. Duality of Molecular Signaling in Vertebral Chondrogenesis

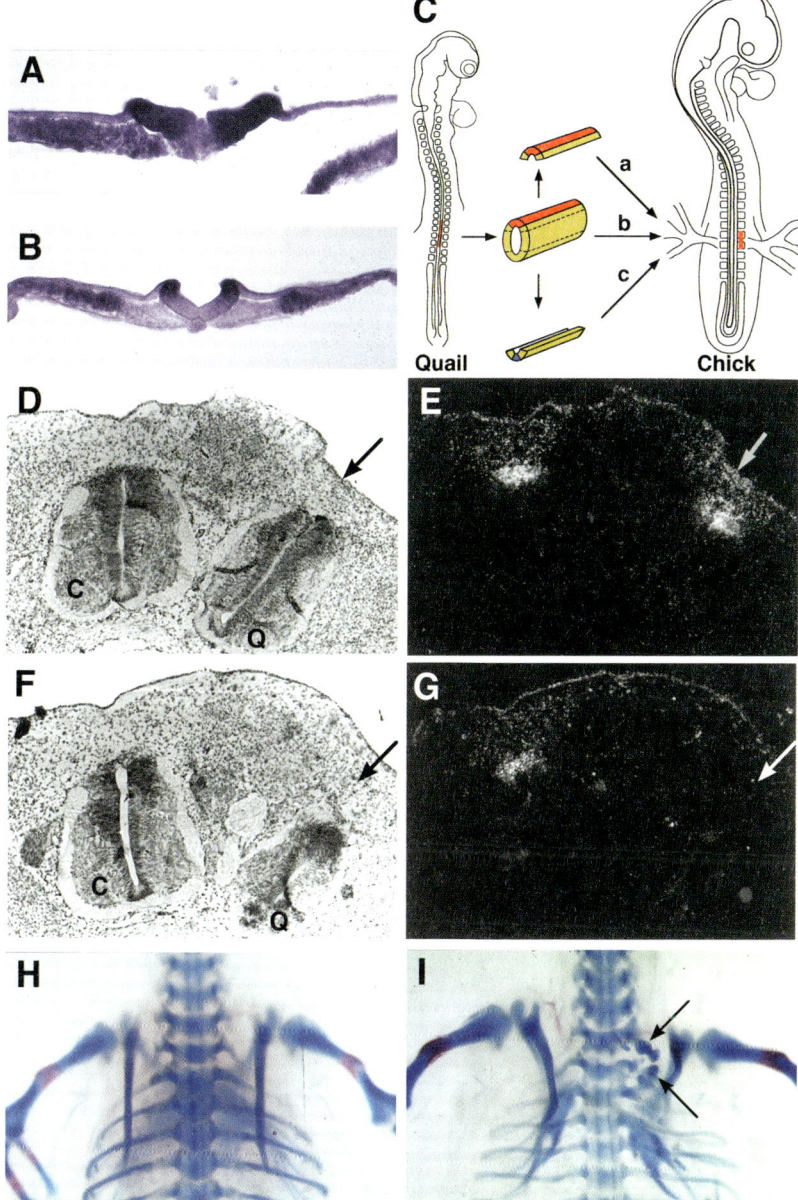

roof plate area is grafted in its normal DV orientation. The graft of the ventral third (F) of the neural tube does not induce *Msx2* expression (G, arrow). (H, I) Skeletal preparations at E9. (H) Normal vertebral column of a control embryo (H, dorsal view). (I) After the graft of a roof plate, ectopic islets of cartilage differentiate above the operated area; they can be considered as equivalent to ectopic spinous process pieces (arrows). (D, E, H, I: from Takahashi *et al.*, 1992.)

roof plate, had been inserted, the chick mesenchyme and the ectoderm covering the graft were also induced to express *Msx2;* however deeply inserted grafts, located far from the ectoderm did not induce this ectopic expression (not shown). When the ventral third of the quail neural tube had been grafted, no *Msx* ectopic expression was observed either (Fig. 6F,G). The differentiation of this area was analyzed in skeleton preparations, 7 days after the operation (at E9), that is, at a stage when the whole vertebra is chondrified (Fig. 6H: normal embryo skeleton). After the graft of a roof plate, one can see that lateral to the vertebral column, above the operated area, ectopic cartilage pieces have differentiated under the ectoderm (Fig. 6I, arrows). In contrast, the graft of the ventral third of the neural tube, which is not followed by any ectopic induction of *Msx2*, does not induce superficial cartilage formation (not shown). Thus, the dorsal part of the neural tube is able to induce ectopically the formation of a dorsal *Msx2*-expressing domain within which subcutaneous cartilage eventually develops (Takahashi *et al.*, 1992). We consider those chondrified elements as equivalent to the spinous process.

IV. Ventral Axial Structures Prevent Subectodermal Cartilage Formation

A. The Notochord

Several studies have demonstrated that the ectopic lateral graft of a notochord or a floor plate results in the induction of ectopic lateral cartilage pieces (Pourquié *et al.*, 1993; Brand-Saberi *et al.*, 1993; Goulding *et al.*, 1994), revealing the sclerotome-inducing activity of the ventral axial structures. We analyzed the effect of the notochord on the development of the dorsal mesenchyme by placing it in a dorsal sagittal position, between the roof plate and the ectoderm, in the nonsegmented region of a stage 12HH chick embryo (Fig. 7A). Control experiments were designed in order to verify that the inserted tissue does not mechanically prevent tissue interactions between the roof plate and the ectoderm. Control grafts consisted of inserting a neutral obstacle such as a piece of thin hair, a silastic membrane, or a paraformaldehyde-fixed notochord dorsally to the just-closed neural tube. This

---

Figure 7 Effect of the notochord and of the floor plate on the dorsal mesenchyme. (A) Dorsal grafts of the notochord (N', yellow) are realized through a slit made in the ectoderm, above the just-closed roof plate (RP, red). (B) The floor plate (FP, blue) is placed in contact to the dorsal ectoderm by rotating the grafted neural tube (yellow) dorsoventrally by 180° thus bringing the roof plate (RP, red) ventrally. (C, F, I) Control experiments consist in grafting a formalin-fixed notochord or a neutral obstacle dorsal to the host roof plate (I, arrowhead indicates the grafted araldite membrane): this operation perturbs neither the expression of *Msx2* (C) nor the development of the vertebra (F: E9 embryo, the arrows indicate the operated area), of the neural tube and of the somite derivatives (I: immunolabeling with an antineurofilament antibody). In contrast, the dorsal grafts of a notochord (D, arrowhead) or the

2. Duality of Molecular Signaling in Vertebral Chondrogenesis

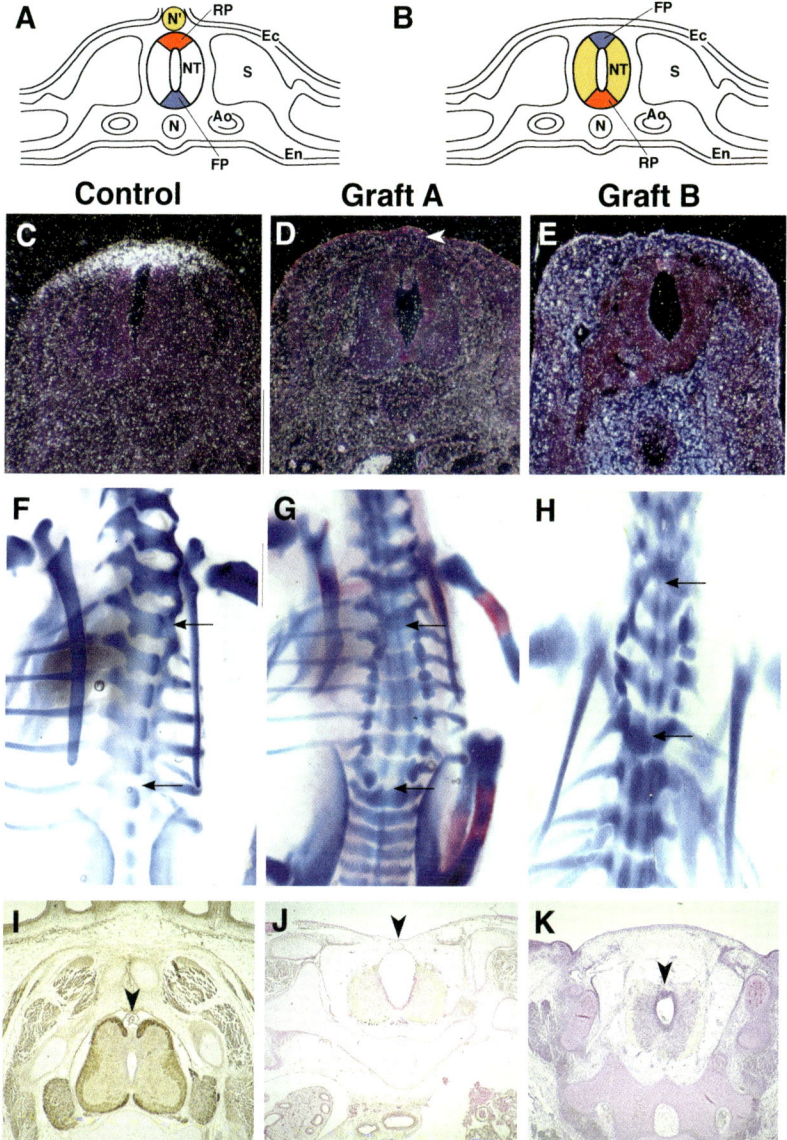

presence of a floor plate (E) result in the absence of *Msx2* expression in the dorsal mesenchyme that has migrated around the graft at E4 (D, E) and prevent the development of the dorsal part of the vertebra over the length of the operated area at E9 (G, H, arrows). At this stage, under the dorsal ectoderm, a loose mesenchyme is formed (J, K), in a place where it should have formed cartilage and skin. This differentiation is inhibited by the notochord and the floor plate (arrowheads), while the ventral and lateral parts of the vertebra are normally formed (J, K). N, Notochord; N', grafted notochord; NT, neural tube; Ec, ectoderm; En, endoderm; S, somite; Ao, aorta. (C, D, F, G, I, J: from Monsoro-Burq *et al.*, 1994, 1995; E, H: from Takahashi *et al.*, 1992).

operation did not perturb early gene expression (not shown) nor vertebral morphogenesis (Fig. 7F,I). In contrast, when a notochord was grafted dorsally in the same position, the expression of *Msx1* and *Msx2* genes was inhibited by the graft, in the ectoderm and in the dorsal mesenchyme (Fig. 7D). In the neural tube, *Msx2* expression, which is normally restricted to the roof plate, was completely abolished, *Msx1* was inhibited in the roof plate and the dorsalmost parts of the alar plates, whereas, more laterally, residual *Msx 1* expression is still detected in the ventralmost part of the alar plates (not shown). Vertebral cartilage development was analyzed: *in toto* skeletal preparations revealed a large deficit in the dorsal part of the vertebrae extending along the whole grafted area (Fig. 7G). Sections through the operated region showed that the spinous process was absent and that the grafted notochord was surrounded by a few dorsal mesenchymal cell layers (Fig. 7J, arrow). Their differentiation into chondrogenic tissue was inhibited by the notochord graft. The ventral and lateral parts of the vertebra develop normally (Monsoro-Burq *et al.*, 1994, 1995).

B. The Floor Plate

The next step of our investigation was to analyze the effect of the floor plate on the differentiation of the dorsal mesenchyme. This was done by rotating the neural tube dorsoventrally by 180° at E2, thus placing the floor plate dorsally in contact to the *Msx* gene-expressing ectoderm (Fig. 7B). This was followed, at E4, by the complete absence of *Msx 2* gene expression dorsally, in the mesenchyme that migrated between the floor plate and the ectoderm, and in the ectoderm itself (Fig. 7E). At a time when the vertebral differentiation is normally completed (E9), the vertebral body and the lateral parts of the neural arches were normally formed in the operated embryos, but the dorsal part of the vertebrae was missing as it was observed after the graft of a notochord dorsally (Fig. 7H,K and Takahashi *et al.*, 1992). In conclusion, when the ventral cartilage inducers, namely the floor plate and the notochord, are placed dorsally, they do not enhance dorsal cartilage formation, but rather inhibit the expression of the early dorsal mesenchymal markers, reduce mesenchymal cell proliferation, and inhibit the differentiation of the superficial cartilage.

We then asked the question as to whether the superficial differentiation of cartilage could occur independently of the ventral axial organs. To test this hypothesis, we ablated the notochord and the floor plate in the region lying rostral to Hensen's node in stage 13HH chick embryos (Fig. 8A). In similar experiments, the expression domain of *Pax3*, a gene normally expressed in the roof and alar plates, is extended ventrally (Goulding *et al.*, 1993) and the neural tube lacks ventral horns containing motoneurons. In our hands, the resulting neural tube exhibited the same rounded morphology (Fig. 8B), with dorsally a thinner area close to the ectoderm

2. Duality of Molecular Signaling in Vertebral Chondrogenesis   57

(Fig. 8B, open arrow); the dorsal root ganglia were fused under the neural tube (arrowhead). There was no expansion of *Msx2* domain ventrally, the gene being expressed only in the part of the neural tube close to the ectoderm (Fig. 8C). The few mesenchymal cells located between the neural tube and the superficial ectoderm, as well as the ectoderm itself, expressed *Msx2* (Fig. 8C, arrows). Similarly, recent experiments have shown that the interneurons of the alar plate of the neural tube remain dorsal after an early notochord removal (Liem *et al.,* 1997), indicating that the dorsal localization of several cells types in the neural tube—roof plate cells and interneurons—does not depend on ventral signals. However, when the embryos were allowed to survive longer (up to E8), in spite of the presence of *Msx2*-positive mesenchymal cells dorsal to the roof plate, no cartilaginous differentiation was observed, neither ventrally nor superficially (Fig. 8D and Monsoro-Burq *et al.,* 1994). Experiments from our laboratory have shown that the notochord and the ventral neural tube, through SHH signaling, are necessary for early survival of the entire somite and in particular for the chondrogenic lineage of the somite (Teillet et Le Douarin, 1983; Rong *et al.,* 1992; Teillet *et al.,* 1998). Thus, although the *Msx2* gene was expressed dorsally independently of the influence of the notochord–floor plate complex, these tissues are required for the differentiation of the ventral as well as of the dorsal parts of the vertebra.

V. Molecular Pathways Leading to Chondrogenesis in the Vertebra

A. Sonic Hedgehog Protein Inhibits Subectodermal Chondrogenesis

The secreted protein Sonic hedgehog (SHH) can mimic the effects of the notochord and the floor plate on early somitic survival *in ovo* (Teillet *et al.,* 1998), and on *Pax1* and sclerotome induction *in vitro* (Fan and Tessier-Lavigne, 1994) and *in vivo* (Johnson *et al.,* 1994). In order to compare the action of SHH with the effects of the floor plate and the notochord on dorsal cartilage differentiation, quail fibroblasts (QT6) stably transfected with chick *Shh* cDNA (SHH-QT6) or control QT6 cells (C-QT6) as described in Duprez *et al.* (1998) were implanted dorsal to the roof plate (Fig. 9A). At E5, at the level of the grafted SHH-QT6 cells, the dorsal mesenchyme was reduced in size when compared to embryos grafted with C-QT6, and *Msx* gene expression was prevented as it was by notochord grafts (not shown). The *Bmp4* gene was also strongly downregulated by both notochord or SHH-QT6 dorsal grafts (not shown, see Watanabe *et al.,* 1998). As a control, we grafted SHH-QT6 cells laterally to the neural tube (Fig. 9B) and found that the *Pax1* expression domain was extended as already described after notochord grafting (Brand-Saberi *et al.,* 1993; Goulding *et al.,* 1994).

Vertebral morphogenesis was not affected after control cell grafts (Fig. 10A),

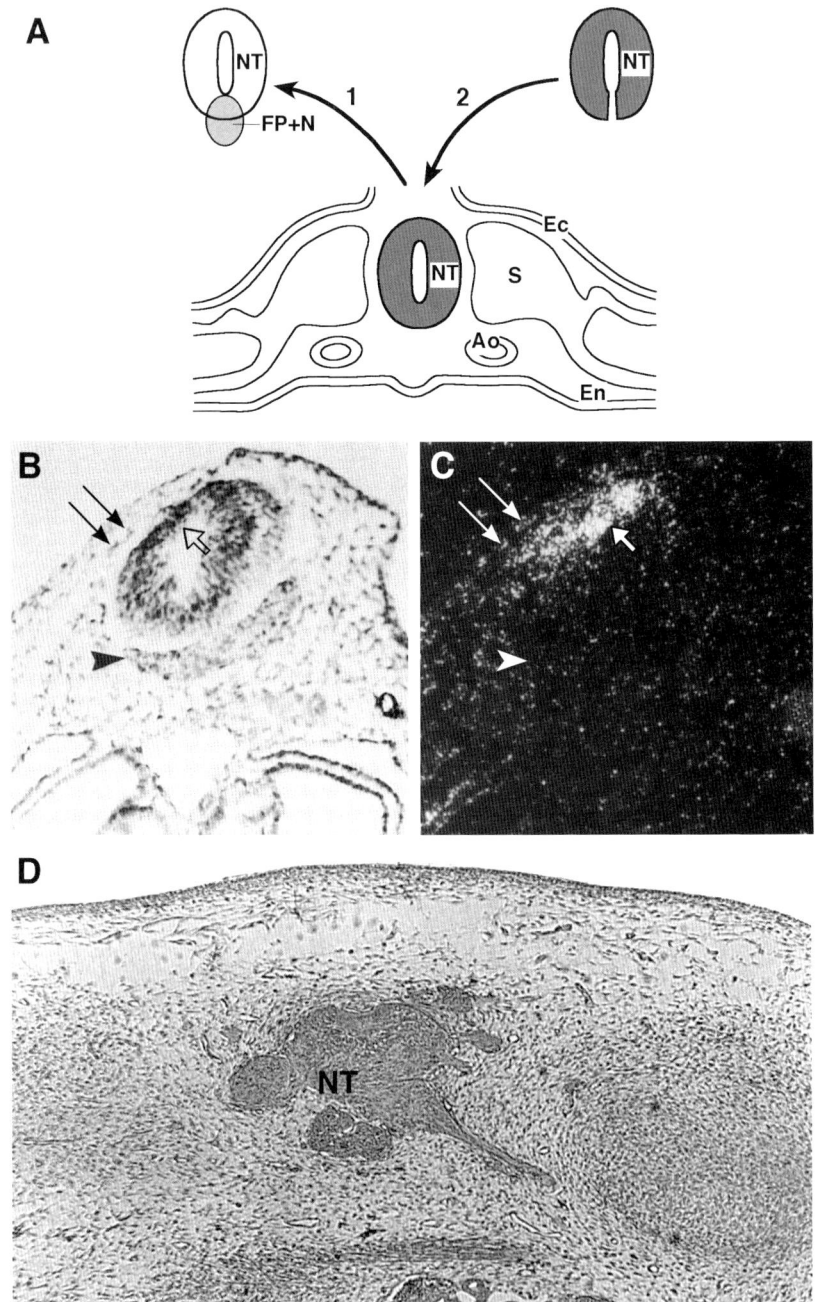

2. Duality of Molecular Signaling in Vertebral Chondrogenesis 59

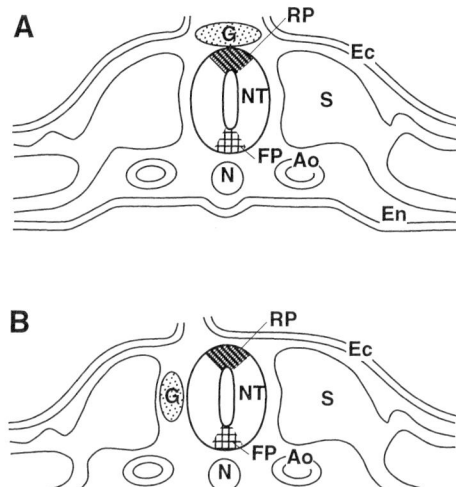

Figure 9 Schematic representation of the grafts of cell pellets. Recombinant quail cells are harvested and pellets (G) are implanted either dorsally (A) or laterally (B) to the neural tube (NT). Ec, Ectoderm; En, endoderm; N, notochord; Ao, aorta; RP, roof plate; FP, floor plate; S, somite.

whereas the SHH-QT6-implanted embryos presented largely open vertebrae (Fig. 10B) as observed after a notochord graft. SHH-QT6 grafts implanted laterally to the neural tube (Fig. 9B) resulted in the expansion of chondrogenesis inducing the fusion of the lateral parts of the neural arches of consecutive vertebrae (Fig. 10C) as already observed with lateral notochord grafts (Pourquié et al., 1993)

Figure 8 Role of the notochord on the development of the dorsal domain of the vertebra. (A) The so-called notochord ablation realized in young (stage 10HH) embryos corresponds to the separation of the notochord–floor plate complex (FP+N) (see Fig. 6A) from the rest of the neural tube (NT) (see Catala et al., 1996). The entire neural tube together with the notochord–floor plate complex was removed from the chick host (white) (arrow 1) and replaced by a quail neural tube (gray), from which the notochord–floor plate complex had been similarly ablated and which was subsequently grafted in its normal DV orientation into the chick (arrow 2). (B) Two days after the operation, the grafted neural tube exhibits a round shape with a thinner roof-plate-like area under the ectoderm (B, open arrow) which strongly expresses *Msx2* gene (C, dark field), and the dorsal root ganglia are fused under the neural primordium (B, arrowhead). A few mesenchymal cells have migrated above the roof-plate-like area (B, arrows); they express *Msx2* gene (C). At E7.5, the neural tube devoid of floor plate (NT) is deeply affected, and no cartilage has differentiated around it: the *Msx2* expression observed earlier on in the dorsal mesenchyme, does not, in these conditions, elicit the development of superficial cartilage. Thus, the notochord–floor plate complex is necessary for the development of the whole vertebra. Ec, Ectoderm; En, endoderm; S, somite; Ao, aorta. (B–D: from Monsoro-Burq et al., 1994.)

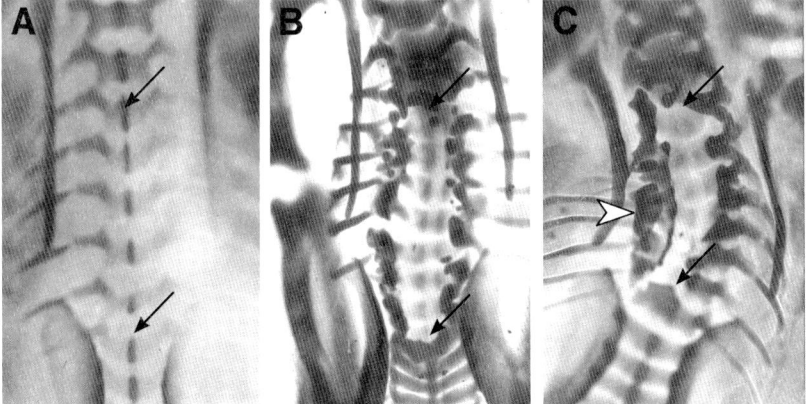

Figure 10 Effect of grafts of Sonic hedgehog (SHH)-expressing cells on development of the vertebra. (A) The graft of control cells does not alter vertebral formation (arrows indicate the operated area). (B) Dorsal grafts of SHH-producing cells inhibit the fusion of the neural arches and the formation of the superficial cartilage forming the spinous processes over the length of the operated domain (arrows). (C) Lateral grafts of SHH-expressing cells also inhibit the differentiation of the subectodermal cartilage (arrows) and induce overdevelopment of the lateral parts of the vertebrae resulting in vertebral fusions (indicated by the white arrowhead). (From Watanabe et al., 1998.)

together with the absence of the dorsal parts of the vertebrae, suggesting a long-distance action of the SHH protein produced by the graft.

In summary, the secreted factor SHH mimics the effects of the notochord and floor plate since it inhibits the expression of the dorsal mesenchyme early marker genes (*Msx1, Msx2, Bmp4*) and the differentiation of subcutaneous cartilage (Watanabe et al., 1998).

B. BMP4 Protein Promotes Dorsal Mesenchyme Development and Superficial Vertebral Cartilage Differentiation

We have tested BMP4 action on vertebral morphogenesis. In chick embryos, *Msx1*, *Msx2*, and *Bmp4* expression overlap in the dorsal part of the neural tube and in the ectoderm (Figs. 2 and 5, and Monsoro-Burq et al., 1996). At later stages (E3), *Bmp4* expression in the dorsal ectoderm extends laterally above the dorsal mesenchymal area (not shown) but at E4, it is not expressed above background in the dorsal mesenchyme (Fig. 5).

In order to analyze the effect of BMPs on the development of the dorsal part of the vertebra *in vivo*, we grafted clumps of quail fibroblasts transfected with various *Bmp* constructs into chick embryos at stage 12–13 HH. Because, in vertebrates, BMP2 and BMP4 are 94% identical in the active part of the molecule and recognize the same receptor complex, we used constructs containing the genes

2. Duality of Molecular Signaling in Vertebral Chondrogenesis 61

encoding either one or the other protein as well as their controls: a quail fibroblast cell line secreting human BMP2 protein (hBMP2-Q2bn) versus cells transfected with the antisense construct (C-Q2bn) (Duprez et al., 1996a) or QT6 quail fibroblasts transiently transfected with mouse *Bmp4* cDNA in the RCAS virus (mBMP4-QT6) versus nontransfected QT6 cells (QT6) (Pourquié et al., 1996; Watanabe and Le Douarin, 1996). The operated embryos were fixed at E3–E5 for gene expression analysis and E9–E10 for the observation of vertebral differentiation.

The effect of ectopic grafts of BMP-producing cells was examined by implanting the clumps of transfected fibroblasts laterally, between the neural tube and the unsegmented paraxial mesoderm (Fig. 9B). The expression of *Msx1* and *2* genes was recorded, and dermomyotomal and sclerotomal structures were identified by the expression of the *Pax3* and *Pax1* markers, respectively. After BMP2/4 grafts, *Msx1* and *2* mRNA were detected lateral to the neural tube in the somitic area as early as E3 (not shown). Sections at E4 show that *Msx* expression was induced around the graft, in the paraxial somitic tissues (Fig. 11E–H). In most cases, this induction occurred in close vicinity to the ectoderm (Fig. 11E,G arrows, F,H), but it could also be observed in deep positions, ventral to the grafted clumps of BMP2-producing cells (Fig. 11E,G arrowhead). In contrast, the expression of the normal somite markers was deeply perturbed: *Pax1* expression was abolished on the grafted side, from E3 onward (Fig. 11C,D) and *Pax3* transcripts could be completely absent at the most affected levels of the operated area (Fig. 11B). At E9–E10, control cell grafts did not modify vertebral chondrogenesis (Fig. 12A). After implantation of BMP2/4-producing cells, two situations could be distinguished according to the location of the graft. (1) Deeply implanted cells inhibited the formation of the lateral parts of the vertebral arches, the dorsal area being not affected (Fig. 12B). This result was probably related to the early inhibition of *Pax 1* expression on the operated side. (2) Larger grafts, encompassing deep lateral as well as superficial lateral positions, resulted in the deficit in lateral vertebral cartilage as described above, together with the fusion of the dorsal parts of several vertebrae and in the differentiation of islets of cartilage in the dermis, under the ectoderm (Fig. 12C,D).

We analyzed the effect of the overexpression of BMP2/4 on the dorsal mesenchyme, by grafting the BMP2/4-releasing cells dorsally to the neural tube in the nonsegmented part of 12- to 16-somite stage chick embryos (Fig. 9A). At E4–E5, the clumps of BMP2-QB2n or BMP4-QT6 cells were found surrounded by an overdeveloped dorsal mesenchyme (Fig. 13A,C). These host mesenchymal cells expressed *Msx* genes at high levels (Fig. 13D) when compared to stage-matched controls. This effect was observed from E4 to E6, that is, just before the onset of chondrogenic differentiation. Based on their morphology and on the expression of several marker genes, the neural tube and the lateral and ventral parts of the somitic-derived structures were not perturbed (Fig. 13B and Monsoro-Burq et al., 1996). Thus, providing extra BMP2/4 in a dorsal position resulted in the increase in number of dorsal mesenchymal cells, as well as in a higher and prolonged

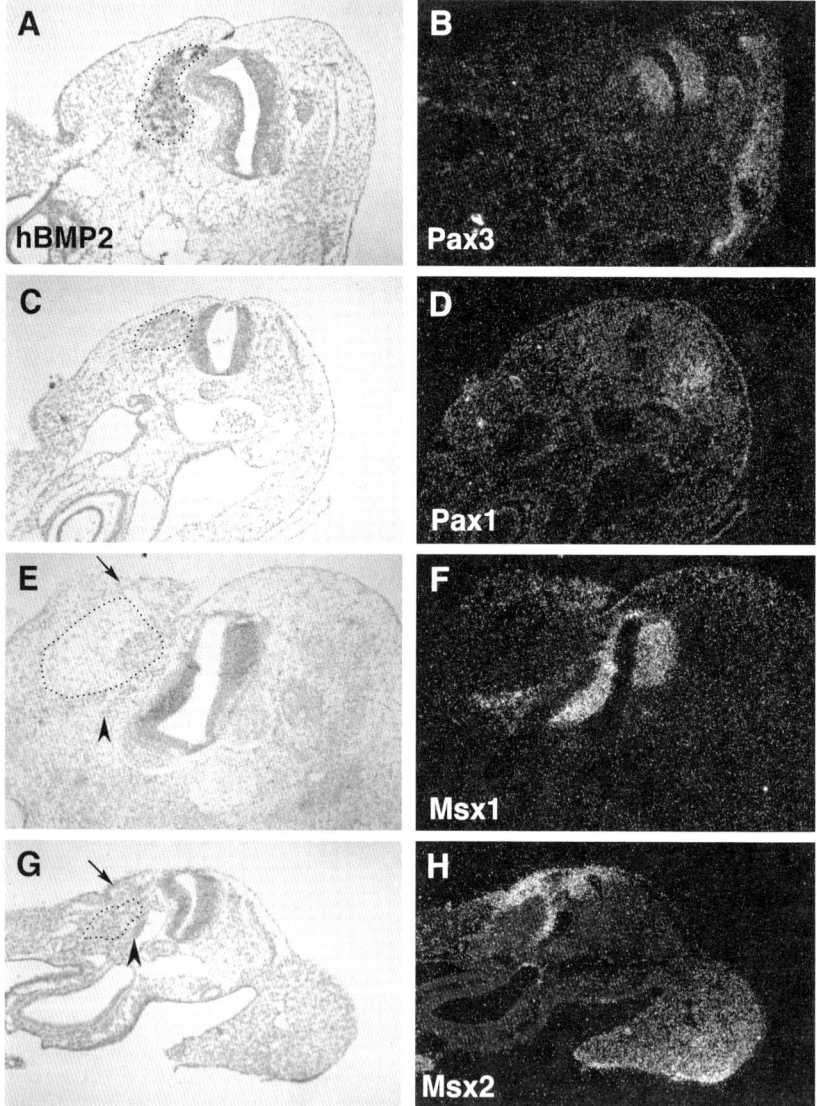

Figure 11 Effect of lateral grafts of BMP-producing cells on somitic gene expression. Radioactive *in situ* hybridization has been used to detect gene expression in transverse sections of E3–E4 grafted embryos (A, C, E, G, bright fields; B, D, F, H, dark fields). The BMP2-expressing cells are grafted lateral to the neural tube at E2 (see Fig. 9B). At E4, the cells remain grouped laterally (A, C, E, G, framed areas) and strongly express *hBmp2* transcripts (A, *hBMP2* transcripts are visualized as the framed dark area). This operation deeply perturbs gene expression in the somite: *Pax3* gene expression (B) as well as the dermomyotomal structure (A) can be completely abolished, as well as *Pax1* gene expression as soon as 24 hr after the graft (C, D, E3 embryo). In contrast, *Msx1* (E, F) and *Msx2* (G, H) expression is induced around the clumps of cells, most often in the vicinity of the ectoderm (arrow) but also in deep positions, ventral to the graft (E–H, arrowheads). (From Monsoro-Burq *et al.*, 1996.)

2. Duality of Molecular Signaling in Vertebral Chondrogenesis 63

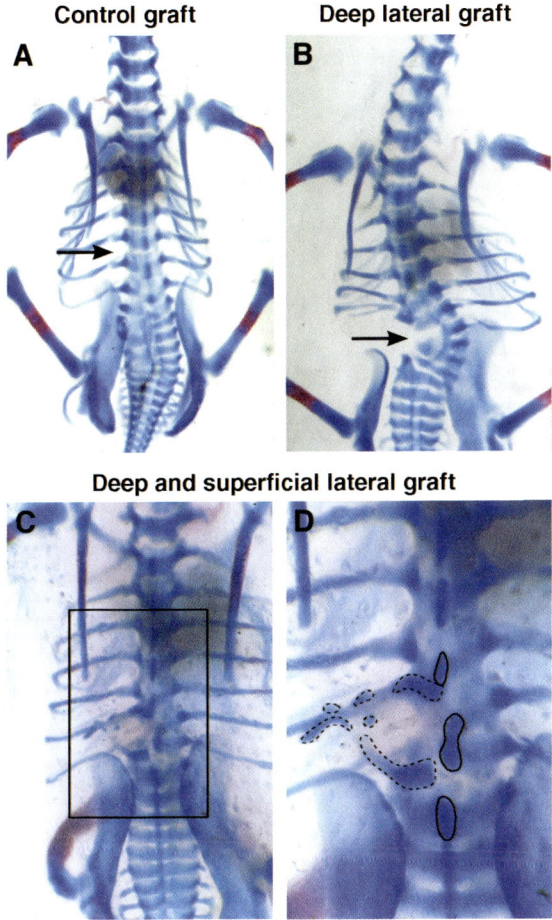

Figure 12 Effect of lateral grafts of BMP-expressing cells on development of the vertebra. The lateral (A) graft of control cells does not perturb the development of the vertebrae (arrow: operated site). In contrast, the lateral graft of BMP2-producing cells in a deep position between the neural tube and the paraxial mesoderm prevents the development of the lateral part of the neural arches (B, arrow) (the spinous processes are still present). If the graft encompasses both deep and superficial positions, the lateral parts of the vertebrae are affected as above. In addition, differentiation of ectopic islets of superficial cartilage takes place in the dermis (C, D, dots), and spinous processes of consecutive vertebrae fuse (D, underlined). (B–D: from Monsoro-Burq et al., 1996.)

expression of *Msx* genes. At E9–E10, while the vertebral column of embryos that received control non-BMP-producing cells was unaffected (not shown), after the dorsal graft of BMP-producing cells at E2, vertebral fusions were observed along the length of the operated area. Dissection of the vertebral column revealed that the fusions were restricted to the dorsal parts of the vertebrae, while their ventral and lateral parts were normal (not shown). Dorsal grafts performed at E3 (stage

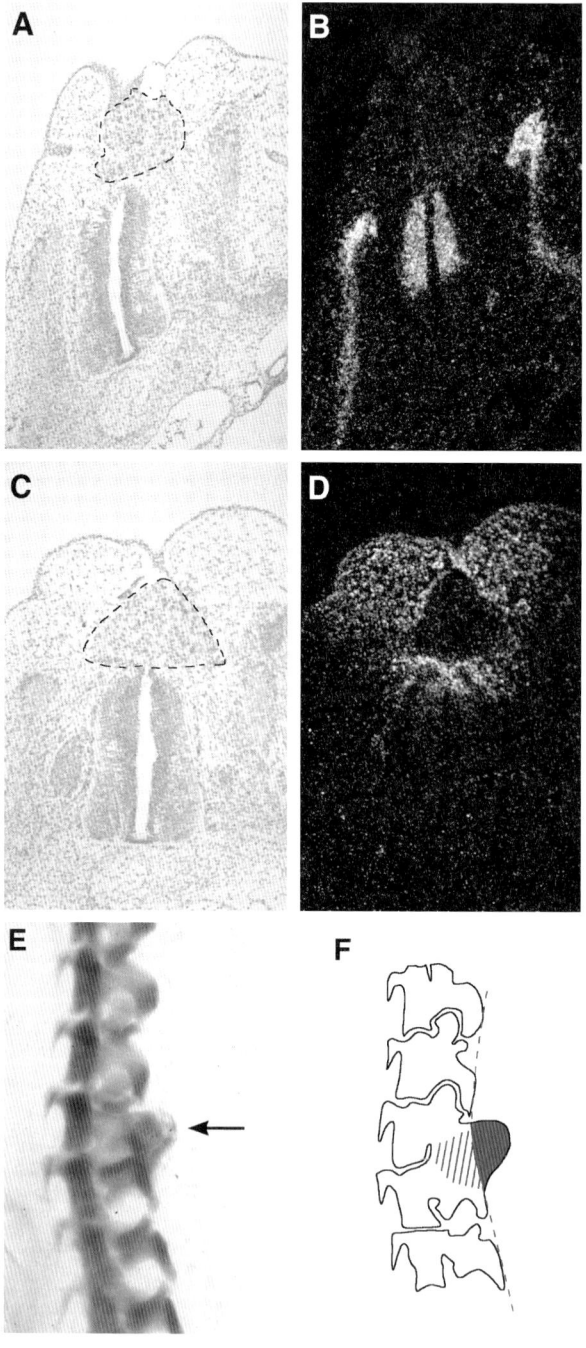

19HH), that is, at a stage when the dorsal mesenchyme begins to migrate dorsally, showed even more pronounced effects of the morphogen with fused areas that were more extended and associated with a large piece of cartilage, whose morphology is reminiscent of a hypertrophied spinous process (Fig. 13E,F).

In conclusion, administration of exogenous BMP2 or 4 undoubtedly mimics the action of the dorsally produced endogenous BMP4, in promoting dorsal *Msx* gene expression and superficial cartilage differentiation. Moreover, we show here that BMPs can inhibit ventrolateral cartilage formation and thus antagonize the action of the ventral sclerotome inducers.

VI. Conclusions: A Novel Model for Vertebra Development

The role of the dorsal part of the neural tube on the morphogenesis of the dorsal domain of the somites has been described in a few classic studies: the hemisection of the dorsal part of the neural tube in chick embryos results in the absence of the dorsal part of the vertebra (Strudel, 1955), and in fish, the ectopic lateral graft of the roof plate is followed by the development of an extra dorsal fin (Watterson *et al.,* 1954). The studies described above reveal a duality in the molecular signals involved in the development of a vertebra. Two domains of vertebral precursors can be distinguished according to their gene expression and response to ventral and dorsal signals. The sclerotomal cells located in a deep position in the embryo respond to the notochord, to the floor plate, and to SHH, by maintaining *Pax1* expression and differentiating into deep cartilaginous structures that are located at a distance from the ectoderm (the vertebral bodies, the ventral and lateral parts of the neural arches, and the pedicles of the ribs) (Pourquié *et al.,* 1993; Goulding *et al.,* 1994; Johnson *et al.,* 1994; Hirsinger *et al.,* 1997; Watanabe *et al.,* 1998); the differentiation of these cells is inhibited by ectopic BMPs implanted ventrally (Monsoro-Burq *et al.,* 1996; Tonegawa *et al.,* 1997). The second domain is formed by the dorsal mesenchyme. Our labeling experiments have distinguished two steps in the migration of mesenchymal cells of somitic origin in the space between the roof plate and the superficial ectoderm. A first step has taken place at E3 (stage of the grafting experiments) and concerns the cells endowed with chondrogenic fate. The second step takes place later on, with the migration of cells from the

---

Figure 13 Effect of dorsal grafts of BMP-producing cells. BMP2-producing cells are grafted dorsally to the roof plate (see Fig. 9A). Two days after grafting, the cells remain localized dorsal to the roof plate (A, C, framed areas). The normal development of the neural tube and of the lateral parts of the somite is not altered as shown, for example, by *Pax3* expression (B). Dorsally, the subectodermal mesenchyme is overdeveloped around the graft, and expresses *Msx* genes at a higher level than in stage-matched controls (D, *Msx2* gene expression). At E9, dorsal grafts, here realized at E3, result in large fusions of the dorsal part of the vertebrae (E, F, see hatched area), with the overdevelopment of the dorsal spinous processes (E, F, red area). (From Monsoro-Burq *et al.,* 1996.)

dermatome. These cells cover the differentiating cartilage of the spinous process. In the mouse embryo, the dermatome precursor cells also express the *Msx1* gene, this expression is downregulated when the cells begin to express the dermis marker *Dermo 1* (D. Houzelstein and B. Robert, personal communication, 1999). The chondrogenic precursors migrating dorsally between the neural tube and the ectoderm express *Msx* genes whereas they do not express *Pax1;* they proliferate and differentiate under the influence of the roof plate and superficial ectoderm as well as experimentally introduced BMP2/4-producing cells (Takahashi *et al.,* 1992; Monsoro-Burq *et al.,* 1996). In contrast, their development is inhibited by ectopically grafted ventral structures, the notochord, the floor plate, or by SHH-producing cells (Monsoro-Burq *et al.,* 1994, 1995; Takahashi *et al.,* 1992; Watanabe *et al.,* 1998).

A model of the antagonistic molecular mechanisms responsible for vertebral development is presented in Fig. 14. At early stages of somite differentiation, the ventral structures, via SHH, exert their action on survival of the whole somite (Teillet *et al.,* 1998). Moreover, they maintain *Pax 1* expression (blue domain) in virtually all sclerotomal cells (dark blue arrows) (Wallin *et al.,* 1994; Barnes *et al.,* 1996; M.-A. Teillet and N. Le Douarin, personal observations, 1998). At these stages, the ectopic implantation of BMP2- or BMP4-producing cells, at the level of the unsegmented paraxial mesoderm or of the last segmented somites, inhibits *Pax1* expression (red bar) and further cartilage differentiation of the deeply located parts of the vertebra (Monsoro-Burq *et al.,* 1996; see also Tonegawa *et al.,* 1997) (Fig. 14A). This inhibitory effect of BMP4 on SHH signaling is similar to that exerted during neural tube DV polarization: *in vitro* and *in vivo* experiments have shown that BMP2/4 can overcome the ventralizing influence of SHH on the differentiation of neural structures such as the motoneurons (Liem *et al.,* 1995; Monsoro-Burq *et al.,* 1996). Similarly, BMP4 protein produced by the ectoderm and the roof plate could restrict SHH influence during somite DV development (red-dotted arrow), as it has been shown during the ML polarization of the somitic mesoderm (Hirsinger *et al.,* 1997).

During later stages of normal somitic development, *Pax 1* expression is rapidly restricted to the most medial part of the sclerotome (Deutsch *et al.,* 1988; Wallin *et al.,* 1994; Barnes *et al.,* 1996). The chondrogenic cells migrating dorsally to form the dorsal mesenchyme (yellow), from stage HH18 onward, do not express the *Pax1* gene, whereas they are induced to express *Msx1* and *Msx2* (open circles) as soon as they are located dorsally, under the inductive influence of the roof plate, and possibly the ectoderm, via BMP4 signals (red) (Fig. 14B). Overexpression of BMP2/4 in this dorsal domain leads to a hypertrophy of the dorsal mesenchyme and enhances *Msx* genes expression and cartilage differentiation of the dorsal parts of the vertebra, that is, the dorsalmost parts of the neural arches and the spinous processes. Because several studies have demonstrated the inhibitory effect of the ectoderm on the development of cartilage, in favor of muscle and dermis differentiation (Gallera, 1966; Swalla and Solursh, 1984; Kenny-Mobbs and Thorogood, 1987), we propose that the dorsal BMP4, *Msx1* and *Msx2* expressing

2. Duality of Molecular Signaling in Vertebral Chondrogenesis   67

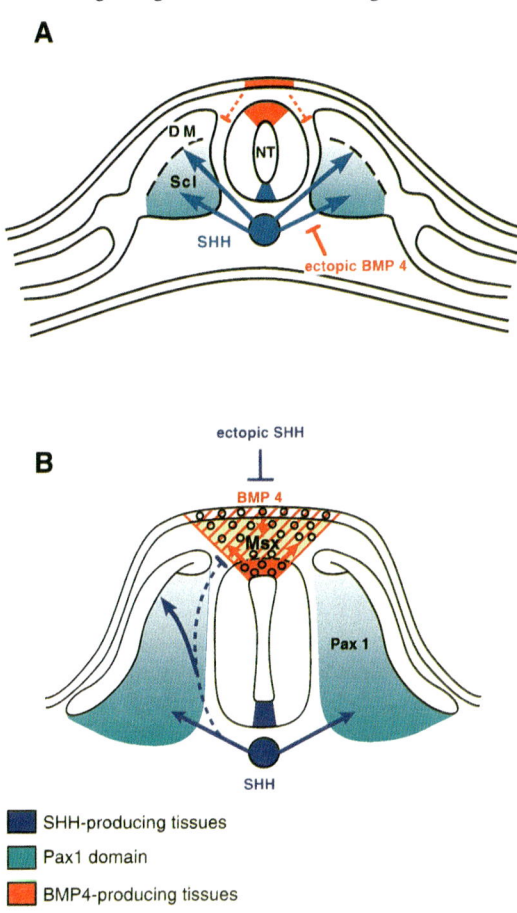

Figure 14 Regulatory pathways controlling the development of the vertebra. Early in somite development, the notochord and the floor plate (dark blue) play a critical role on the survival of the whole somite (DM, dermomyotome; Scl, sclerotome), through the action of the SHH protein (blue arrows) and, later on, on the differentiation of the sclerotome. Lateral grafts of BMP-producing cells at this stage (red bar) deeply perturb gene expression in the somite and subsequently the differentiation of the lateral parts of the vertebra. At E2, the endogenous BMP4 protein, produced dorsally by the roof plate and the ectoderm, may restrict SHH long-distance influence (dotted red bar). At E3, during the development of the dorsal mesenchyme, BMP4 protein produced mainly by the roof plate (red, and red hatches) induce expression of *Msx* genes (open circles) in the dorsal domain. This results in creating an area (yellow) within which superficial cartilage can differentiate, just under the ectoderm, thus allowing the development of the spinous processes leading to the fusion of the neural arches. This developmental pathway can be completely abolished by grafting SHH-producing structures (blue bar). The ventrally produced SHH protein may delimitate the lateral expansion of the dorsal domain (blue dotted bar).

domain defines a local environment where subcutaneous cartilage is able to form by counteracting the antichondrogenic effect of the ectoderm. The already recognized cartilage inducers—notochord, floor plate, or SHH—not only do not fulfill this task, but, when grafted ectopically in a dorsal position at E2, they inhibit the expression of *Bmp4* in the ectoderm and in the roof plate as early as E3, suggesting that the absence of *Msx* gene expression in the neural tube, the ectoderm, and later on in the dorsal mesenchyme is due to the lack of BMP4 protein in this area (Monsoro-Burq *et al.*, 1994, 1995; Watanabe *et al.*, 1998). In these experimental conditions, the development of the dorsal mesenchyme is deeply perturbed: it develops poorly, does not express *Msx* and *Bmp4* genes, and remains as a loose mesenchyme that is not engaged in the chondrogenic nor in any other characteristic differentiation pathway.

The role of the BMP4-positive ectoderm on the induction of the dorsal domain remains unclear: expression of *Msx* genes can be induced in the ectoderm by the roof plate, as seen after ectopic grafting of the neural tube (Takahashi *et al.*, 1992) or by BMP4 (Watanabe and Le Douarin, 1996); however, when the roof plate is located far from the ectoderm, this induction does not occur and the dorsal mesenchyme remains negative for *Msx* genes (Takahashi *et al.*, 1992; Monsoro-Burq *et al.*, 1994). Moreover, after the implantation of BMP-producing cells, induction of *Msx* genes is observed between the ectoderm and the graft in most cases, but also sometimes around the clump of grafted cells far from the ectoderm, suggesting that the ectoderm produces a permissive factor that is dispensable if very high levels of BMPs are provided. However, the differentiation of BMP-induced ectopic cartilage has been observed only in subectodermal positions (Monsoro-Burq *et al.*, 1996; Watanabe and Le Douarin, 1996).

These studies demonstrate a role for BMP4 in vertebral chondrogenesis in addition to that already recorded in limb development (Duprez *et al.*, 1996b; Zou *et al.*, 1997) during embryonic life. This effect on the formation of the subcutaneous cartilage, via *Msx* gene induction, is specifically triggered by BMP2 and BMP4 members of the TGFβ family; other dorsally expressed secreted factors, such as WNTs, do not appear to participate in this developmental pathway (Takahashi *et al.*, 1996). During embryogenesis, the BMPs are essential for gastrulation, DV polarization, and a variety of developmental events; unfortunately, the BMP4 $-/-$ homozygous mutant mice die very early in development and do not allow the analysis of vertebral formation (Winnier *et al.*, 1995). In the light of the present experiments showing that BMP4 plays a role in the development of bony structures in subectodermal locations, in places where muscle or dermis are supposed to form, a disease such as fibrodisplasia ossificans progressiva (FOP) can be interpreted as the postnatal reactivation in ectopic locations of a normal developmental mechanism. It is interesting to note that in the patients presenting ectopic ossification in the major body muscles, the first sites to ossify and fuse are generally the vertebral dorsal spine and the back muscles, the disease progressing from the dorsal to the ventral side (Smith, 1997).

2. Duality of Molecular Signaling in Vertebral Chondrogenesis

In the development of the dorsal mesenchyme, *Msx* genes are downstream targets of BMP4. Interestingly *Msx1* was recently shown to be directly regulated by BMP4 and to mimic some of the ventralizing effects of BMP4 (Maeda *et al.*, 1997). Mutations in *Msx1* or *Msx2* genes either generated in the mouse or recorded in humans are characterized by a number of craniofacial abnormalities such as premature bone fusions in the skull (craniosynostosis) (Satokata and Maas, 1994; Jabs *et al.*, 1993). However, no gross abnormality of the spine has been described. This could be due to the redundant expression of these two genes in the dorsal mesenchymal domain: the generation of double mutant mice should allow this point to be analyzed.

The function of *Msx* genes has been tested in cell culture: transfection of *Msx1*, but not of *Msx2* gene specifically blocks myoblast terminal differentiation as well as *MyoD* gene expression and the cells are maintained in an undifferentiated state (Song *et al.*, 1992). The same inhibitory effect on *MyoD* expression was obtained after human gene (*MSX1*) transfer in fibroblastic mouse cells (Woloshin *et al.*, 1995). Thus, *Msx1* seems to be implicated in the negative regulation of muscle cell development. Thus one can hypothesize that dorsally, above the roof plate, *Msx1* gene expression might prevent muscle formation in the sagittal plane, allowing other cell types to differentiate, such as dermis and cartilage. Moreover, the role of *Msx* genes in skeletogenic cell cultures has been analyzed: *Msx1* gene expression is recorded in all the bone-related cell types analyzed in the study, whereas *Msx2* gene expression is found in the cells expressing the osteocalcin gene (Hoffmann *et al.*, 1994). This is in agreement with the hypothesis of a role for *Msx* genes during skeletal development, with possibly common as well as nonoverlapping roles for *Msx1* and *Msx2*.

Taken together, the results described above and an increasing amount of data from other laboratories suggest highly regulated and dynamic interactions between trunk embryonic tissues during somite DV polarization and differentiation. Complex interactions have been described during DV patterning of the dermomyotomal structures. The neural tube and the ectoderm are known to enhance muscle and dermis formation (Gallera, 1966; Swalla and Solursh, 1984; Christ *et al.*, 1992; Kenny-Mobbs and Thorogood, 1987; Bober *et al.*, 1994; Cossu *et al.*, 1996) through secreted diffusible factors (Fan and Tessier-Lavigne, 1994; Stern *et al.*, 1995; Buffinger and Stockdale, 1995; Stern *et al.*, 1997; Marcelle *et al.*, 1997). However, several of these studies demonstrated a role for the ventral axial structures in promoting myogenesis (Stern *et al.*, 1995; Buffinger and Stockdale, 1995; Pownall *et al.*, 1996), and most of them can be accounted for by the role of SHH on somitic cell survival (Teillet *et al.*, 1998). Moreover, those inductive interactions occur during precise developmental windows; for example, the dorsal neural tube, which has a potent myogenic-inducing activity on early somites, was shown to exert a negative effect on the myogenic differentiation of already-committed somites (Buffinger and Stockdale, 1995). The latter inhibitory effect can be interpreted as a result of the *Bmp4-Msx* gene activation in this area, that would block myogenesis. The general model for somite DV patterning, resulting from these studies, is that (1) high doses of ventralizing factors induce sclerotome and inhibit dermomyotome formation; (2) dorsal signals induce the dermomyotome; and (3) low doses of ventralizing factors act synergistically with dorsalizing factors and induce myotome (Dietrich *et al.*, 1997; Reshef *et al.*, 1998). This model can now be completed by adding the mediodorsal BMP4-*Msx* pathway, providing a way to promote chondrogenesis, in a subectodermal location, when other developmental pathways normally occurring in the somitic structures (via *Pax1* and *Pax3* expression) are inhibited (Monsoro-Burq *et al.*, 1996).

Acknowledgments

The QCPN monoclonal antibody was obtained from the Developmental Studies Hybridoma Bank maintained by the Department of Pharmacology and Molecular Sciences, Johns Hopkins University School of Medicine, Baltimore, MD, and the Department of Biological Sciences, University of Iowa, Iowa City, IA, under contract N01-HD-6-2915 from the NICHD. The authors thank Marie-Aimée Teillet, Catherine Ziller, and Delphine Duprez for critical reading of the manuscript, Martine Bontoux for excellent technical assistance, Sophie Gournet for the drawings, Françoise Viala and Hélène San Clémente for the illustration work, and Chrystèle Guilloteau for the preparation of the manuscript. This work was supported by the Centre National de la Recherche Scientifique (CNRS), the Association pour la Recherche contre le Cancer (ARC), and the Ligue, the Fondation pour la Recherche Médicale (FRM).

References

Akimenko, M. A., Johnson, S. L., Westerfield, M., and Ekker, M. (1995). Differential induction of four *msx* homeobox genes during fin development and regeneration in zebrafish. *Development* **121,** 347–357.
Aoyama, H., and Asamoto, K. (1988). Determination of somite cells: Independence of cell differentiation and morphogenesis. *Development* **104,** 15–28.
Barnes, G. L., Hsu, C. W., Mariani, B. D., and Tuan, R. S. (1996). Chicken *Pax-1* gene: Structure and expression during embryonic somite development. *Differentiation* **61,** 13–23.
Basler, K., Edlund, T, Jessel, T. M., and Yamada, T. (1993). Control of cell pattern in the neural tube—regulation of cell differentiation by *dorsalin-1*, a novel TGF-β family member. *Cell* **73,** 687–702.
Bober, E., Brand-Saberi, B., Ebensperger, C., Wilting, J., Balling, R., Paterson, B. M., Arnold, H. H., and Christ B. (1994). Initial steps of myogenesis in somites are independant of influence from axial structures. *Development* **120,** 3073–3082.
Borycki, A. G., Strunk, K. E., Savary, R., and Emerson, C. P., Jr. (1997). Distinct signal/response mechanisms regulate *pax1* and *QmyoD* activation in sclerotomal and myotomal lineages of quail somites. *Dev. Biol.* **185,** 185–200.
Brand-Saberi, B., Ebensperger, C., Wilting, J., Balling, R., and Christ, B. (1993). The ventralizing effect of the notochord on somite differentiation in chick embryos. *Anat. Embryol.* **188,** 239–245.
Buffinger, N., and Stockdale, F. E. (1995). Myogenic specification of somites is mediated by diffusible factors. *Dev. Biol.* **169,** 96–108.
Burke, A. C., Nelson, C. E., Morgan, B. A., and Tabin, C. (1995). *Hox* genes and the evolution of vertebrate axial morphology. *Development* **121,** 333–346.

Catala, M., Teillet, M.-A., De Robertis, E. D., and Le Douarin, N. M. (1996). A spinal cord fate map in the avian embryo: While regressing, Hensen's node lays down the notochord and floor plate thus joining the spinal cord lateral walls. *Development* **122,** 2599–2610.

Chalepakis, G., Fritsch, R., Fickenscher, H., Deutsch, U., Goulding, M., and Gruss, P. (1991). The molecular basis of the *undulated/Pax-1* mutation. *Cell* **66,** 873–884.

Chan-Thomas, P. S., Thompson, R. P., Robert, B., Yacoub, M. H., and Barton, P. J. R. (1993). Expression of homeobox genes *Msx-1* (*Hox-7*) and *Msx-2* (*Hox-8*) during cardiac development in the chick. *Dev. Dyn.* **197,** 203–216.

Christ, B., and Wilting, J. (1992). From somites to vertebral column. *Ann. Anat.* **174,** 23–32.

Christ, B., Brand-Saberi, B., Grim, M., and Wilting, J. (1992). Local signalling in dermomyotomal cell type specification. *Anat. Embryol.* **186,** 505–510.

Cossu, G., Kelly, R., Tajbakhsh, S., Di Donna, S., Vivarelli, E., and Buckingham, M. (1996). Activation of different myogenic pathways: myf5 is induced by the neural tube and MyoD by the dorsal ectoderm in mouse paraxial mesoderm. *Development* **122,** 429–437.

Davidson, D. (1995). The function and evolution of *Msx* genes: Pointers and paradoxes. *Trends Genet.* **11,** 405–411.

Deutsch, U., Dressler, G. R., and Gruss, P. (1988). *Pax-1*, a member of a paired box homologous murine gene family, is expressed in segmented structures during development. *Cell* **53,** 617–625.

Dietrich, S., Schubert, F. R., and Gruss, P. (1993). Altered *Pax* gene expression in murine notochord mutants: The notochord is required to initiate and maintain ventral identity in the somite. *Mech. Dev.* **44,** 189–207.

Dietrich, S., Schubert, F. R., and Lumsden, A. (1997). Control of dorsoventral pattern in the chick paraxial mesoderm. *Development* **124,** 3895–3908.

Duprez, D., Kostakopoulou, K., Francis-West, P., Tickle, C., and Brickell, P. M. (1996a). Activation of *Fgf-4* and *HoxD* gene expression by BMP-2 expressing cells in the developing chick limb. *Development* **122,** 1821–1828.

Duprez, D., Bell, E. J., Richardson, M. K., Archer, C. W., Wolpert, L., Brickell, P. M., and Francis-West, P. H. (1996b). Overexpression of BMP-2 and BMP-4 alters the size and shape of developing skeletal elements in the chick. *Mech. Dev.* **57,** 145–157.

Duprez, D., Fournier-Thibault, C. and Le Douarin, N. M. (1998). Sonic hedgehog induces proliferation of committed skeletal muscle cells in the chick limb. *Development* **125,** 495–505.

Ebensperger, C., Wilting, J., Brand-Saberi, B., Mizutani, Y., Christ, B., Balling, R., and Koseki, H. (1995). *Pax-1*, a regulator of sclerotome development is induced by notochord and floor plate signals in avian embryos. *Anat. Embryol.* **191,** 297–310.

Fan, C. M., and Tessier-Lavigne, M. (1994). Patterning of mammalian somites by surface ectoderm and notochord: Evidence for sclerotome induction by a hedgehog homolog. *Cell* **79,** 1175–1186.

Fan, C. M., Porter, J. A., Chiang, C., Chang, D. T., Beachy, P. A., and Tessier-Lavigne, M. (1995). Long-range sclerotome induction by *Sonic hedgehog:* Direct role of the amino-terminal cleavage product and modulation by the cyclic AMP signaling pathway. *Cell* **81,** 457–465.

Gallera, J. (1966). Mise en évidence du rôle de l'ectoblaste dans la différenciation des somites chez les Oiseaux. *Rev. Suisse Zool.* **73,** 492–503.

Gehring, W. J. (1987). The homeobox: Structural and evolutionary aspects. *In* "Molecular Approaches to Developmental Biology," pp. 115–129. Alan R. Liss, New York.

Goulding, M. D., Lumsden, A., and Gruss, P. (1993). Signals from the notochord and floor plate regulate the region-specific expression of two *Pax* genes in the developing spinal cord. *Development* **117,** 1001–1016.

Goulding, M., Lumsden, A., and Paquette, A. J. (1994). Regulation of *Pax-3* expression in the dermomyotome and its role in muscle development. *Development* **120,** 957–971.

Graham, A., Francis-West, P., Brickell, P., and Lumsden, A. (1994). The signalling molecule BMP4 mediates apoptosis in the rhombencephalic neural crest. *Nature* **372,** 684–686.

Hall, B. K. (1977). Chondrogenesis of the somitic mesoderm. *Adv. Anat. Embryol. Cell Biol.* **53,** 1–50.

Hamburger, V., and Hamilton, H. L. (1951). A series of normal stages in the development of the chick embryo. *J. Morphol.* **88,** 49–92.
Hemmati-Brivanlou, A., Kelly, O. G. and Melton, D. A. (1994). Follistatin, an antagonist of activin, is expressed in the Spemann organizer and displays direct neuralizing activity. *Cell* **77,** 283–295.
Hill, R. E., Jones, P. F., Rees, A. R., Sime, C. M., Justice, M. J., Copeland, N. G., Jenkins, N. A., Graham, E., and Davidson, D. R. (1989). A new family of mouse homeobox-containing genes: Molecular structure, chromosomal location, and developmental expression of *Hox-7.1*. *Genes. Dev.* **3,** 26–37.
Hirsinger, E., Duprez, D., Jouve, C., Malapert, P., Cooke, J., and Pourquié, O. (1997). Noggin acts downstream of Wnt and Sonic Hedgehog to antagonize BMP4 in avian somite patterning. *Development* **124,** 4605–4614.
Hoffmann, H. M., Catron, K. M., Van Wijnen, A. J., McCabe, L. R., Lian, J. B., Stein, G. S., and Stein, J. L. (1994). Transcriptional control of the tissue-specific, developmentally regulated osteocalcin gene requires a binding motif for the Msx family of homeodomain proteins. *Proc. Natl. Acad. Sci. U.S.A.* **91,** 12887–12891.
Hogan, B. L. M. (1996). Bone morphogenetic proteins in development. *Curr. Opin. Genet. Dev.* **6,** 432–438.
Isshiki, T., Takeichi, M., and Nose, A. (1997). The role of the *msh* homeobox gene during *Drosophila* neurogenesis: Implication for dorsoventral specification of the neuroectoderm. *Development* **124,** 3099–3109.
Ivens, A., Flavin, N., Williamson, R., Dixon, M., Bates, G., Buckingham, M., and Robert, B. (1990). The human homeobox gene HOX7 maps to chromosome 4p16.1 and may be implicated in Wolf-Hirschhorn syndrome. *Hum. Genet.* **84,** 473–476.
Jabs, E. W., Müller, U., Li, X., Ma, L., Luo, W., Haworth, I. S., Klisak, I., Sparkes, R., Warman, M. L., Mulliken, J. B., Snead, M. L. and Maxson R. (1993). A mutation in the homeodomain of the human *MSX2* gene in a family affected with autosomal dominant craniosynostosis. *Cell* **75,** 443–450.
Johnson, R. L., Laufer, E., Riddle, R. D., and Tabin, C. (1994). Ectopic expression of *Sonic hedgehog* alters dorsal–ventral patterning of somites. *Cell* **79,** 1165–1173.
Jones, C. M., Dale, L., Hogan, B. L. M., Wright C. V. E., and Smith J. C. (1996). Bone Morphogenetic protein 4 (BMP4) acts during gastrula stages to cause ventralization of *Xenopus* embryos. *Development* **122,** 1545–1554.
Jowett, A. K., Vainio, S., Ferguson, M. W., Sharpe, P. T., and Thesleff, I. (1993). Epithelial–mesenchymal interactions are required for *msx-1* and *msx-2* gene expression in the developing murine molar tooth. *Development* **117,** 461–470.
Kalcheim, C., and Teillet, M.-A. (1989). Consequences of somite manipulation on the pattern of dorsal root ganglion development. *Development* **106,** 85–93.
Kenny-Mobbs, T., and Thorogood, P. (1987). Autonomy of differentiation in avian branchial somites and the influence of adjacent tissues. *Development* **100,** 449–462.
Kessel, M., and Gruss, P. (1991). Homeotic transformations of murine vertebrae and concomitant alteration of *Hox* codes induced by retinoic acid. *Cell* **67,** 89–104.
Keynes, R. J., and Stern, C. D. (1984). Segmentation in the vertebrate nervous system. *Nature* **310,** 786–789.
Kieny, M., Mauger, A., and Sengel, P. (1972). Early regionalization of somitic mesoderm as studied by the development of axial skeleton of the chick embryo. *Dev. Biol.* **28,** 142–161.
Koseki, H., Wallin, J., Wilting, Y., Kispert, A., Ebensperger, C., Herrmann, B. G., Christ, B., and Balling, R. (1993). A role for *Pax-1* as a mediator of notochordal signals during the dorsoventral specification of vertebrae. *Development* **119,** 649–660.
Krumlauf, R. (1994). *Hox* genes in vertebrate development. *Cell* **78,** 191–201.
Liem, K. F., Tremml, G., Roelink, H., and Jessell, T. M. (1995). Dorsal differentiation of neural plate cells induced by BMP-mediated signals from epidermal ectoderm. *Cell* **82,** 969–979.

2. Duality of Molecular Signaling in Vertebral Chondrogenesis

Liem, K. F., Tremml, G., and Jessell, T. M. (1997). A role for the roof plate and its resident TGFβ-related proteins in neuronal patterning in the dorsal spinal cord. *Cell* **91**, 127–138.

Lord, P. C. W., Lin, M.-H., Hales, K. H., and Storti, R. V. (1995). Normal expression and the effects of ectopic expression of the *Drosophila muscle segment homeobox (msh)* gene suggest a role in differentiation and patterning of embryonic muscles. *Dev. Biol.* **171**, 627–640.

MacKenzie, A., Ferguson, M. W. J., and Sharpe, P. T. (1991a). *Hox-7* expression during murine craniofacial development. *Development* **113**, 601–611.

MacKenzie, A., Leeming, G. L., Jowett, A. K., Ferguson, M. W. J., and Sharpe, P. T. (1991b). The homeobox gene *Hox-7.1* has specific regional and temporal expression patterns during early murine cranio facial embryogenesis, especially tooth development *in vivo* and *in vitro*. *Development* **111**, 269–285.

Maeda, R., Kobayashi, A., Sekine, R., Lin, J.-J., Kung, H.-F., and Maéno, M. (1997). *Xmsx-1* modifies mesodermal tissue pattern along dorsoventral axis in *Xenopus laevis* embryo. *Development* **124**, 2553–2560.

Marcelle, C., Stark, M. R., and Bronner-Fraser, M. (1997). Coordinate actions of BMPs, Wnts, Shh and Noggin mediate patterning of the dorsal somite. *Development* **124**, 3955–3963.

Mina, M., Gluhak, J., Upholt, W.B., Kollar, E. J., and Rogers, B. (1995). Experimental analysis of *Msx-1* and *Msx-2* gene expression during chick mandibular morphogenesis. *Dev. Dyn.* **202**, 195–214.

Monsoro-Burq, A. H., Bontoux, M., Teillet, M. A., and Le Douarin, N. M. (1994). Heterogeneity in the development of the vertebra. *Proc. Natl. Acad. Sci. U.S.A.* **91**, 10435–10439.

Monsoro-Burq, A. H., Bontoux, M., Vincent, C., and Le Douarin, N. M. (1995). The developmental relationships of the neural tube and the notochord: short and long term effects of the notochord on the dorsal spinal cord. *Mech. Dev.* **53**, 157–170.

Monsoro-Burq, A. H., Duprez, D., Watanabe, Y., Bontoux, M., Vincent, C., Brickell, P., and Le Douarin, N. M. (1996). The role of bone morphogenetic proteins in vertebral development. *Development* **122**, 3607–3616.

Noveen, A., Jiang, T. X., Ting-Berreth, S. A., and Chuong, C. M. (1995). Homeobox genes *Msx-1* and *Msx-2* are associated with induction and growth of skin appendages. *J. Invest. Dermatol.* **104**, 711–719.

Ordahl, C. P. (1993). Myogenic lineages within the developing somite. *In* "Molecular Basis of Morphogenesis" (M. Bernfield, Ed), pp. 165–176. Wiley, New York.

Ordahl, C. P., and Le Douarin, N. M. (1992). Two myogenic lineages within the developing somite. *Development* **114**, 339–353.

Pavlova, A., Boutin, E., Cunha, G., and Sassoon, D. (1994). *Msx-1 (Hox-7.1)* in the adult mouse uterus: Cellular interactions underlying regulation of expression. *Development* **120**, 335–346.

Peters, H., Doll, U., and Niessing, J. (1995). Differential expression of the chicken *Pax-1* and *Pax-9* gene: *In situ* hybridization and immunohistochemical analysis. *Dev. Dyn.* **203**, 1–16.

Piccolo, S., Sasai, Y., Lu, B., and De Robertis, E. M. (1996). Dorsoventral patterning in *Xenopus*: Inhibition of ventral signals by direct binding of chordin to BMP-4. *Cell* **86**, 589–598.

Pourquié, O., Coltey, M., Teillet, M. A., Ordahl, C., and Le Douarin, N. M. (1993). Control of dorsoventral patterning of somitic derivatives by notochord and floor plate. *Proc. Natl. Acad. Sci. U.S.A.* **90**, 5242–5246.

Pourquié, O., Fan, C. M., Coltey, M., Hirsinger, E., Watanabe, Y., Bréant, C., Francis-West, P., Brickell, P., Tessier-Lavigne, M., and Le Douarin, N. M. (1996). Lateral and axial signals involved in avian somite patterning: A role for BMP4. *Cell* **84**, 461–471.

Pownall, M. E., Strunk, K. E., and Emerson, C. (1996). Notochord signals control the transcriptional cascade of myogenic bHLH genes in somites of quail embryos. *Development* **122**, 1475–1488.

Reshef, R., Maroto, M., and Lassar, A. B. (1998). Regulation of dorsal somitic cell fates: BMPs and Noggin control the timing and pattern of myogenic regulator expression. *Genes Dev.* **12**, 290–303.

Robert, B., Sassoon, D., Jacq, B., Gerhing, W., and Buckingham, M. (1989). *Hox-7*, a mouse homeobox gene with a novel pattern of expression during embryogenesis. *EMBO J.* **8**, 91–100.

Robert, B., Lyons, G., Simandl, B. K., Kuroiwa, A., and Buckingham, M. (1991). The apical ectodermal ridge regulates *Hox-7* and *Hox-8* gene expression in developing chick limb buds. *Genes Dev.* **5,** 2363–2374.
Rong, P. M., Teillet, M. A., Ziller, C., and Le Douarin, N. M. (1992). The neural tube/notochord complex is necessary for vertebral but not limb and body wall striated muscle differentiation. *Development* **115,** 657–672.
Ros, M. A., Lyons, G., Kosher, R. A., Upholt, W. B., Coelho, C. N., and Fallon, J. F. (1992). Apical ridge dependent and independent mesodermal domains of GHox-7 and GHox-8 expression in chick limb buds. *Development* **116,** 811–818.
Rosen, V., and Thies, R. S. (1992). The BMP proteins in bone formation and repair. *Trends Genet.* **8,** 97–102.
Sasai, Y., Lu, B., Steinbeisser, H., and De Robertis, E. M. (1995). Regulation of neural induction by the Chd and Bmp-4 antagonistic patterning signals in *Xenopus*. *Nature* **376,** 333–336.
Satokata, I., and Maas, R. (1994). *Msx-1* deficient mice exhibit cleft palate and abnormalities of craniofacial and tooth development. *Nat. Genet.* **6,** 348–•••.
Shimeld, S. M., McKay, I. J., and Sharpe, P. T. (1996). The murine homeobox gene *Msx-3* shows highly restricted expression in the developing neural tube. *Mech. Dev.* **55,** 201–210.
Simon, H. G., Nelson, C., Goff, D., Laufer, E., Morgan, B. A., and Tabin, C. (1995). Differential expression of myogenic regulatory genes and *Msx-1* during dedifferentiation and redifferentiation of regenerating Amphibian limbs. *Dev. Dyn.* **202,** 1–12.
Smith, R. (1997). 49th ENMC-sponsored International Workshop Report: Fibrodysplasia (Myositis) Ossificans Progressiva (FOP). *Neuromuscular Disorders* **7,** 407–410.
Song, K., Wang, Y., and Sassoon, D. (1992). Expression of *Hox-7.1* in myoblasts inhibits terminal differentiation and induces cell transformation. *Nature* **360,** 477–481.
Stern, H. M., Brown, A. M. C., and Hauschka, S. D. (1995). Myogenesis in paraxial mesoderm: Preferential induction by dorsal neural tube and by cells expressing *Wnt-1*. *Development* **121,** 3675–3686.
Stern, H. M., Lin-Jones, J., and Hauschka, S. D. (1997). Synergistic interactions between bFGF and a TGF-β family member may mediate myogenic signals from the neural tube. *Development* **124,** 3511–3523.
Strudel, G. (1955). L'action morphogène du tube nerveux et de la chorde sur la différenciation des vertèbres et des muscles vértebraux chez l'embryon de poulet. *Arch. Anat. Microsc. Morphol. Exp.* **44,** 209–235.
Suzuki, H. R., Padanilam, B. J., Vitale, E., Ramierez, F., and Solursh, M. (1991). Repeating developmental expression of G-Hox 7, a novel homeobox-containing gene in the chicken. *Dev. Biol.* **148,** 375–388.
Suzuki, A., Kaneko, E., Ueno, N., and Hemmati-Brivanlou, A. (1997). Regulation of epidermal induction by BMP2 and BMP7 signaling. *Dev. Biol.* **189,** 112–122.
Swalla, B. J., and Solursh, M. (1984). Epithelial enhancement of connective tissue differentiation in explanted somites. *J. Embryol. Exp. Morphol.* **79,** 243–255.
Takahashi, Y., and Le Douarin, N. M. (1990). cDNA cloning of a quail homeobox gene and its expression in neural crest-derived mesenchyme and lateral plate mesoderm. *Proc. Natl. Acad. Sci. U.S.A.* **87,** 7482–7486.
Takahashi, Y., Bontoux, M., and Le Douarin, N. M. (1991). Epithelio–mesenchymal interactions are critical for Quox-7 expression and membrane bone differentiation in the neural crest-derived mandibular mesenchyme. *EMBO J.* **10,** 2387–2393.
Takahashi, Y., Monsoro-Burq, A. H., Bontoux, M., and Le Douarin, N. M. (1992). A role for Quox-8 in the establishment of the dorsoventral pattern during vertebrate development. *Proc. Natl. Acad. Sci. U.S.A.* **89,** 10237–10241.
Takahashi, Y., Tonegawa, A., Matsumoto, K., Ueno, N., Kuroiwa, A., Noda, M., and Nifuji, A. (1996). BMP-4 mediates interacting signals between the neural tube and skin along the dorsal midline. *Genes Cells* **1,** 775–783.

Teillet, M.-A., and Le Douarin, N. M. (1983). Consequences of neural tube and notochord excision on the development of the peripheral nervous system in the chick embryo. *Dev. Biol.* **98,** 192–211.

Teillet, M.-A., Kalcheim, C., and Le Douarin, M. N. (1987). Formation of the dorsal root ganglia in the avian embryo: Segmental origin and migratory behavior of neural crest progenitor cells. *Dev. Biol.* **120,** 329–347.

Teillet, M. A., Watanabé, Y., Jeffs P., Duprez, D., Lapointe, F., and Le Douarin, N. M. (1998). Sonic Hedgehog is required for survival of both myogenic and chondrogenic somitic lineages. *Development* **125,** 2019–2030.

Tonegawa, A., Funayama, N., Ueno, N., and Takahashi, Y. (1997). Mesodermal subdivision along the mediolateral axis in chicken controlled by different concentrations of BMP-4. *Development* **124,** 1975–1984.

Urist, M. R. (1965). Bone: Formation by autoinduction. *Science* **150,** 893–899.

Vainio, S., Karavanova, I., Jowett, A., and Thesleff, I. (1993). Identification of BMP-4 as a signal mediating secondary induction between epithelial and mesenchymal tissues during early tooth development. *Cell* **75,** 45–58.

Wallin, J., Wilting, J., Koseki, H., Fritsch, R., Christ, B., and Balling, R. (1994). The role of *Pax-1* in axial skeleton development. *Development* **120,** 1109–1121.

Wang, W., Chen, X., Xu, H., and Lufkin, T. (1996). *Msx3:* A novel murine homologue of the *Drosophila msh* homeobox gene restricted to the dorsal embryonic central nervous system. *Mech. Dev.* **58,** 203–215.

Watanabe, Y., and Le Douarin, N. M. (1996). A role for BMP-4 in the development of subcutaneous cartilage. *Mech. Dev.* **57,** 69–78.

Watanabe, Y., Duprez, D., Monsoro-Burq, A.-H., Vincent, C., and Le Douarin, N. M. (1998). Two domains in vertebral development antagonistically regulated by SHH and BMP4. *Development* **125,** 2631–2639

Watterson, R. L., Lowler, I., and Fowler, B. J. (1954). The role of the neural tube and notochord in development of the axial skeleton of the chick. *Am. J. Anat.* **95,** 337–399.

Winnier, G., Blessing, M., Labosky, P. A., and Hogen, B. L. M. (1995). Bone morphobenetic protein-4 is required for mesoderm formation and patterning in the mouse. *Genes Dev.* **9,** 2105–2116.

Woloshin, P., Song, K., Degnin, C., McNeill Killary, A., Goldhamer, D. J., Sassoon, D., and Thayer, M. J. (1995). MSX1 inhibits MyoD expression in fibroblast $\times$ 10T$^{1}/_{2}$ cell hybrids. *Cell* **82,** 611–620.

Zimmerman, L. B., De Jesùs-Escobar, J. M., and Harland, R. M. (1996). The Spemann organizer signal noggin binds and inactivates bone morphogenetic protein 4. *Cell* **86,** 599–606.

Zou, H., Weiser, R., Massagué, J., and Niswander, L. (1997). Distinct roles of type I bone morphogenetic protein receptors in the formation and differentiation of cartilage. *Genes Dev.* **11,** 2191–2203.

3
Sclerotome Induction and Differentiation

Jennifer L. Dockter
Department of Anatomy
University of California, San Francisco
San Francisco, California 94143

I. Introduction
II. Induction: Definitions
III. Anatomical and Morphological Description of Sclerotome Development
IV. Genetics
V. Nonmolecular Approaches
   A. *In Vivo* Experiments
   B. *In Vitro* Work
   C. Early Inducer Searches
VI. The Molecular Approach
   A. Molecular Markers of Sclerotome
   B. *In Vivo* Experiments
   C. *In Vitro* Experiments and a Candidate Inducer
VII. *Shh*, cAMP, and Sclerotome
VIII. Issues Concerning *Shh* as the Sclerotome Inducer
   A. The Diffusibility of *Shh*
   B. *Shh* Mutant Mouse Embryos Develop with Mesenchymal Sclerotome, but without Vertebrae
IX. Initiation of the Sclerotome
X. *Shh*, *Noggin*, and Somite Chondrogenesis
XI. Bone Morphogenetic Proteins and Sclerotome Development
XII. Determination of the Sclerotome
XIII. Epithelial–Mesenchymal Transitions and Sclerotome Formation
XIV. Dorsal–Ventral Patterning of the Somite
XV. Interactions between the Somitic Components
XVI. Discussion
XVII. Summary
   References

I. Introduction

The distinguishing feature of the subphylum Vertebrata, in fact, the feature that gives these animals their phylic name, is the presence of a bony or cartilaginous vertebral column. This chapter will discuss the early embryonic events that precede the formation of the cartilaginous model of the vertebral column, which is derived from the group of cells known as the sclerotome. While the sclerotome contributes to the vertebral column, the ribs, and the acromion of the scapula, most experiments have focused on the vertebral column, so the development of the rib and scapula will not specifically be discussed in this chapter, although many events in sclerotome development are presumably common to both derivatives. The first part of this review will discuss inquiries into the influence of tissues surrounding the sclerotome, while the second part will discuss the advances in determining the specific molecular nature of these influences.

The earliest experiments in sclerotome development consisted mainly of *in vivo* embryo manipulations, using chicks and salamanders as model systems, grafting somites with or without other embryonic structures, and assaying for the presence or absence of cartilage. Later experiments used a diverse field of species, continuing with the chick model, but expanding to include fish and mice. In the 1950s and 1960s, *in vitro* assays were developed to complement the *in vivo* studies. The search for inducing factors began as well during this period and was heavily pursued during the 1970s, with a strong focus on components of the perinotochordal extracellular matrix. As molecular markers of the sclerotome such as *Pax-1* were discovered with the advent of advanced molecular techniques in the 1980s, the separation of the concepts of sclerotome induction and patterning and somite chondrogenesis began and a fresh outlook on the roles of inducing organs and molecules was established.

II. Induction: Definitions

Induction is a term classically used to describe the influence of one tissue on the differentiation of another (Holtfreter, 1968; Slack, 1983; Gurdon, 1987). Two types of induction are thought to operate within the embryo: permissive and instructive. Permissive induction creates a condition that allows a tissue to differentiate as would be its normal tendency. In this case, the tissue already is programmed and simply needs the right conditions to express its phenotype. Instructive induction, however, supposes a naive tissue, which is then instructed by an inductive event to follow one particular differentiation pathway out of several possibilities. That inductions do occur in the embryo is not under dispute, but the exact nature of the mechanisms and phenomena underlying inductions is still not very well understood.

Early development of the sclerotome derived cartilages appears to involve two induction steps: (1) induction of the ventral somite cells to form the sclerotomic mesenchyme, which expresses sclerotome (but not cartilage) specific markers, and (2) the induction of these sclerotome cells to then form chondrocytes and express cartilage specific (and not sclerotome specific) markers such as matrix genes. For clarity, the induction of the somite to form sclerotome will be referred to as sclerotome induction and the induction of the somite (sclerotome) to form cartilage will be referred to as somite chondrogenesis. The term sclerotome development will be used as a broad description for the general processes involved in the embryonic genesis of the sclerotome derived cartilages (mainly the vertebral cartilages). The third induction step in sclerotome development, the relatively late event of induction of endochondral bone formation, will not be discussed in this chapter as the signals inducing this process in sclerotome derived cartilages are not well understood.

III. Anatomical and Morphological Description of Sclerotome Development

That the somite contains cells which will form the cartilaginous precursors of the vertebrae, intervertebral disks, the ribs, and the scapular acromion, as well as some of the adult cartilages of these tissues, has been known or at least suspected for centuries (the pre-twentieth century work on this subject is well described in Chap. 9 by Brand-Saberi and Christ, this volume, and readers who are interested are referred to that chapter). The role of the somite and the specific contributions of sclerotome in the development of these axial skeletal structures has been experimentally confirmed in the past 20 years with the advent of the chick–quail chimera system, which allowed researchers to directly follow groups of cells and their descendants as opposed to previous works, which were based on careful morphological and anatomical analyses (Le Douarin, 1973; Chevallier, 1975; Christ and Wilting, 1992). The chief concern of this section is to briefly describe the anatomical development of the sclerotome. Because the most detailed work has been done in the chick, the general descriptions here are based on chick development; detailed morphological descriptions of mammalian vertebral column development can be found in Dawes (1930) and Verbout (1985).

The sclerotome forms from an epithelial–mesenchymal transition undergone by the cells of the ventromedial epithelial somite (Fig. 1). If the most newly formed somite is designated at being at stage I in its development, the next most cranial at stage II, and so on (Ordahl, 1993), the prospective sclerotome cells appear from a somite when it reaches approximately stage IV (Christ and Ordahl, 1995). Mouse somites differ from chick somites in this aspect; the ventromedial cells of the mouse somite do not appear to pass through an epithelial configuration (Ostrovsky et al., 1988). In the chick embryo, the mesenchymous somitocoele cells also contribute to the sclerotome (Huang et al., 1994), ultimately giving rise to parts of the rib and intervertebral discs. Whether chick or mouse, the rest of the somite remains as an epithelium, the dermomyotome, the lateral portion of which

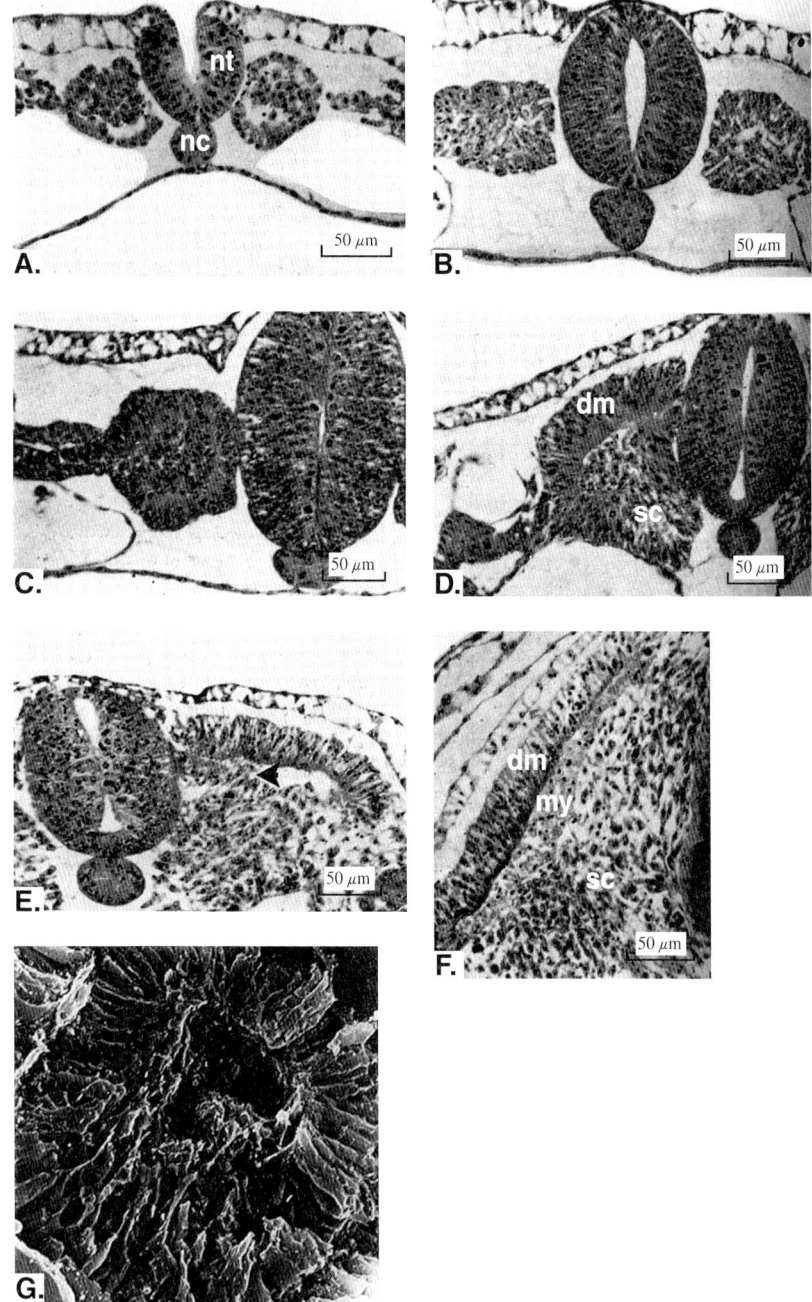

3. Sclerotome Induction and Differentiation

gives rise to hypaxial skeletal muscle (Christ et al., 1974, 1977; Ordahl and Le Douarin, 1992). The medial portion gives rise to a structure called the myotome (Christ et al., 1978; Ordahl and Le Douarin, 1992; Denetclaw et al., 1997), which will form the epaxial skeletal muscle. The myotome is anatomically located between the dermomyotome and the sclerotome. In the zebrafish the sclerotome is not closely apposed to the notochord and comprises only a small number of ventromedial cells of the somite (Morin-Kensicki and Eisen, 1997).

As the sclerotome cells divide and mature, some cells migrate to surround the notochord (Williams, 1910; Jacob et al., 1975). The mechanisms underlying the initiation and coordination of this migration are not well understood, although there is evidence that extracellular matrix molecule production by both the sclerotome cells themselves and the notochord are important for the migration of sclerotome cells in vitro (Newgreen et al., 1986; Lash et al., 1987; Sanders et al., 1988). That this apparent migratory effect is in fact due in part to active cell migration and is not simply an effect of extensive cell division expanding the sclerotome was demonstrated by the use of cytochalasin D, which inhibits sclerotome cell motility, but not cell division (Chernoff and Lash, 1981). Swelling of hyaluronic acid in the extracellular matrix, which would push the cells apart from one another, has also been hypothesized to cause sclerotome cells to end up next to the notochord (Solursh et al., 1979). In addition, sclerotome cells, despite their mesenchymal status, appear to be polarized, which may affect their movements at early and later stages (Trelstad, 1977).

The sclerotome is heterogeneous in its cell distribution. Differences in sclerotome density can be seen in the mediolateral plane, with cell density highest at the lateral border, which will form the arcual process, the costal process, and the pedicle of the vertebra (Verbout, 1985; Christ and Wilting, 1992). In addition to a medial-lateral difference in the distribution of sclerotome cells, a cranial–caudal

---

Figure 1 Formation of the sclerotome. (A–F) Stages of somite differentiation: 5-μm transverse sections stained with iron hematoxylin. Gaps between mesoderm and ectoderm/endoderm are fixation artifacts. All sections are from chick embryos; the dorsal side is toward the top of the figure. The somites are staged according to Ordahl (1993). HH refers to the Hamburger–Hamilton staging system (Hamburger and Hamilton, 1951). Note the changes in the sclerotome. (A) Stage II somite, HH stage 10 embryo. (B) Stage I somite, HH stage 14 embryo. (C) Stage VII somite, HH stage 14 embryo. (D) Stage XII somite, HH stage 14 embryo. (E) Stage XVI somite, HH stage 14 embryo. The arrowhead marks the growing myotome. (F) Stage XX somite, HH stage 18 embryo. dm, Dermomyotome; my, myotome; nc, notochord; nt, neural tube; sc, sclerotome. (G) A scanning electron micrograph of a cross-fracture of a stage IX somite from a chick embryo at HH stage 12 of development. Viewed from the medial side, the cells of the ventral (lower) portion of the somite are elongate, flattened, and surrounded by considerable extracellular space. Cells of the dorsal (upper) half of the somite are columnar and closely applied to one another. (A–F reprinted from O'Hare, 1972a; G reprinted from Solursh et al., 1979. Both reprinted by permission from The Company of Biologists Ltd.)

difference is also seen. Early after sclerotome formation, von Ebner's fissure (von Ebner, 1888), a furrow which divides the lateral portions of each sclerotome into cranial and caudal halves (Williams, 1910; Williams, 1942; Christ and Wilting, 1992), appears. The mechanism for formation of this fissure is unknown. Cells are more densely packed in the caudal half and are more loosely arranged in the cranial half (Wilting *et al.*, 1994), the half through which neural crest cells will migrate (Bronner-Fraser, 1986; Serbedzija *et al.*, 1989; Sanders, 1997). It has been thought that von Ebner's fissure marked the eventual intervertebral boundary, however the story appears to be more complicated (Bagnall *et al.*, 1988; Stern *et al.*, 1988; Bagnall and Sanders, 1989; Christ and Wilting, 1992; Goldstein and Kalcheim, 1992). This chapter will not discuss the concept of *Neugliederung* ("re-" or "new-" segmentation), in which the vertebrae and ribs are thought to ultimately become comprised from two somites' sclerotomes (the reader is referred to Chap. 9 by Brand-Saberi and Christ and Chap. 5 by Stern and Vasiliauskas, this volume, for in-depth treatments of that subject).

The sclerotome is also heterogeneous with regard to cell division and cell death. The caudal halves of sclerotome appear to divide more frequently than the cranial halves, at least at the beginning stages of sclerotome development (Wilting *et al.*, 1994), although mitotic cells can be found throughout the sclerotome (Langman and Nelson, 1968; Shapiro, 1992). These differences in the rate of cell division in the halves of the sclerotome may account for the differences in cell density. Analysis of patterns of cell death in the sclerotome demonstrates that initially, the sclerotome contains no dying cells. Apoptotic cells then appear throughout the ventral sclerotome, ultimately becoming restricted to the cranial half of the ventral sclerotome (Sanders, 1997). Apoptotic cells essentially vanish from the sclerotome by the time the cells begin to differentiate as cartilage. What role this cell death plays in sclerotome and vertebral formation is unknown.

At approximately the time that ventral sclerotome cells begin to differentiate as chondrocytes, they begin to condense in a radial arrangement around the notochord (Williams, 1910; Williams, 1942). Cell condensation is a multistep process necessary for chondrogenesis, and appears to involve extracellular signaling molecules as well as extracellular matrix proteins (Tacchetti *et al.*, 1992; Hall and Miyake, 1995). A sagittal section of an embryo at this point will show thin dense discs of cells, which are the future intervertebral discs, and larger, looser areas, which will ultimately form the vertebral bodies. As sclerotome condensation proceeds, the outlines of the vertebral shapes appear, beginning with the ventral structures and then proceeding in a ventral-to-dorsal sequence. The neural arches are the last structures to become easily distinguishable; interestingly the dorsal most portion (the dorsal spinous process) of the vertebrae appears to take a different developmental pathway than the other vertebral elements (Monsoro-Burq *et al.*, 1994, 1996). (Interested readers are referred to Chap. 8 by Monsoro-Burq and Le Douarin, this volume, for further reading on the subject.)

In the chick, morphologically discernible cartilage differentiation begins at around day 4, the earliest day (HH stage 25; HH refers to the Hamburger–Hamilton staging system; Hamburger and Hamilton, 1951) that collagen type II protein can be detected (Von der Mark *et al.,* 1976); the alternative splice variants of mRNAs for collagen type II can be found before protein staining and appear to define subsets of sclerotome cells (Hayashi *et al.,* 1986; Cheah *et al.,* 1991; Sandell *et al.,* 1991; Ng *et al.,* 1993). Antibody staining to collagen type II at this point is restricted to perinotochordal cells, consistent with the pattern of cell condensation. Alcian blue staining of individual vertebrae is not seen until 2 days later, at embryonic day 6 (ED6) (HH stage 29; Shapiro, 1992). The vertebrae continue from this point to development into mature cartilage elements, with endochondral bone formation beginning at about ED13, again, starting ventrally in the bodies of the vertebrae and then spreading upward to ossify the neural arches last (Shapiro, 1992).

Although not discussed in detail here, the divergent somitic organization and development of different vertebrate species suggests that there may not be a universal solution to sclerotome development. Although a general developmental strategy may be employed across all vertebrates to build up their body plan, variations have probably been adopted by each species for their own needs, and caution needs to be exercised when making mechanistic generalizations from one species to another.

IV. Genetics

There are a number of genetic mouse mutants that affect skeletal development (Grüneberg, 1963; Theiler, 1988). Although the genes affected in these mutations are largely unknown, some of these mutations are probably in cartilage matrix genes or genes controlling ossification, while some appear to specifically affect early developmental processes. There are two main classes of mutations affecting early development: ones that appear to affect the notochord and neural tube primarily (so that the sclerotome and cartilage defects are perhaps secondary effects) and ones that appear to have a direct effect on the somites or sclerotome itself. Within each class, this chapter will only touch on the best known and studied mutants; for a more comprehensive list, the reader is referred to Grüneberg's classic, *The Pathology of Development* (Grüneberg, 1963).

One of the most heavily studied mutants that appears to affect primarily notochord development is *Danforth's short-tail* (*Sd*), a semidominant mutation (Grüneberg, 1958, 1963; Theiler, 1988). Heterozygote mice have tails that are shortened to different lengths and that are sometimes kinked; the lumbar vertebrae ossify from two centers that then meet, leading to a wider vertebral body, which sometimes ossifies prematurely. Homozygote mice have bifid lumbar vertebral bodies. In the heterozygous state, the mice initially form a notochord that then degenerates in a cranial–caudal direction, ultimately disappearing altogether. Homozygous mice never have a notochord in the tail region at all, except for a few small groups of cells. The specific gene defect in this mouse is not known, but the mutant phenotype suggests the importance of the notochord in sclerotome development.

The *T*, or *Brachyury*, mouse is another mutant that affects the notochord (Grüneberg, 1963; Theiler, 1988). Heterozygotes have tails that are shortened or completely lacking; homozygotes die during the embryonic phase without forming the caudal end of the embryo. Again, the most severe defects are in the caudal region, consisting of split, double, or absent vertebral bodies. Bifida of the arches may be present. The notochord of these animals is of normal length, but ultimately becomes incorporated into the ventral neural tube or the gut.

Based on these morphological aspects, it might be expected that the skeletal defects seen in *T*/+ mice are due to the misdevelopment of the notochord. However, a clever set of *in vitro* recombination experiments demonstrated that the skeletal abnormalities seen in *T* mice appear to be in part problems with the somites themselves (Bennett, 1958). Somites from homozygous *T* mice do not give rise to cartilage when recombined with *T/T* notochord. Wild-type mouse somites combined with *T/T* notochord do give rise to cartilage, whereas wild-type notochord and *T/T* somites do not. Homozygous *T* anterior limb buds are capable of giving rise to cartilage in culture, so the problem with *T* mice is not a general cartilage formation defect. The anomalies seen in vertebral development in *T* mice therefore seem to rise from defects within in the somites themselves, despite the fact that it is the notochord which has the most obvious histological and anatomical developmental defects. Of course, some early effects from the notochord on the somites that normally renders them sensitive to cartilage induction by the notochord could be what has been disrupted in these embryos.

Probably the best studied mutant in the class thought to affect the sclerotome itself is the *undulated* mutation. There are three currently known alleles of *undulated* (in order of increasing severity): *undulated* (*un*), *undulated extensive* (*un^{ex}*), and *Undulated short-tail* (*Un^s*) (Grüneberg, 1953, 1963; Theiler, 1988). A fourth allele, *undulated minimal*, is extinct. All display similar phenotypes: reduction in size, but not number, of the caudal and lumbar vertebrae, along with bifid, abnormally formed, or absent vertebral bodies in these areas. Embryologically, it is seen that sclerotome mesenchyme cranial to von Ebner's fissure is reduced in size. The *undulated* locus has been shown to contain a point mutation in the gene for the transcription factor *Pax-1* that alters its affinity for DNA, indicating a role for *Pax-1* in sclerotome development (Balling *et al.*, 1988). The other alleles carry large deletions that most likely destroy more than just the *Pax-1* locus (Dietrich and Gruss, 1995). Sequencing of these mutants may uncover other potential genes or elements relevant to sclerotome development.

Experiments crossing *un* and *Danforth's short-tail* mice demonstrated by genetics a relationship between the notochord and skeletal development, and in particular a role for the notochord in *Pax-1* expression (Koseki *et al.*, 1993), a topic that will be discussed later in Section VI.B. Mice heterozygous for *Sd* and *Un^s* had a much more severe phenotype than mice carrying either mutation alone. These mice had an almost complete loss of vertebral bodies in the thoracolumbosacral regions and the vertebral bodies of the cervicothoracic regions were reduced in size. This compounding of the vertebral phenotypes suggested that the genes affected in each of these mutations were part of a common pathway involved in vertebral development. However, while genetics can provide evidence for interacting pathways, until the components affected by these mutations are identified and their relationships tested by other means, definite relationships cannot be drawn.

V. Nonmolecular Approaches

A. *In Vivo* Experiments

A number of *in vivo* experiments provided the starting point for examination of somite chondrogenesis even though they were undertaken to examine other issues such as the importance of innervation and cell type determination within the somites (reviewed in Hall, 1977). The experiments fall into two basic categories: chorioallantoic grafting of somites with or without other structures and extirpation and grafting of embryonic structures. Independent of the type of experimental manipulation however, these experiments pointed to the importance of the neural tube and notochord in formation of cartilage by the somite.

Grafting of tissues to the chick chorioallantoic membrane (CAM) has long been a favorite method to test the development of tissues outside their normal environment and a way to coculture tissues without taking them out of the embryo. Tissues placed on the CAM are well supplied with blood, providing them with nutrients and circulating factors such as hormones, and will develop quite normally (Saunders, 1996). A large number of researchers have taken this approach with respect to somites (Hoadley, 1925; Murray and Selby, 1933; Williams, 1942; Watterson *et al.,* 1954; Avery *et al.,* 1955, 1956; O'Hare, 1972a; Kenny-Mobbs and Thorogood, 1987). The basic protocol involves taking somites at different ages and placing them on the CAM of an older (usually day 9) embryo. The chorioallantoic membrane is then harvested after a certain period of time and the grafted area analyzed histologically for cartilage formation.

Two major observations can be made from CAM grafting experiments. The first is the observation that the inclusion of axial structures (notochord or neural tube) was necessary for somite chondrogenesis in the assay (Hoadley, 1925; Williams, 1942; Fowler and Watterson, 1953; Watterson *et al.,* 1954; Avery *et al.,* 1955, 1956; Kenny-Mobbs and Thorogood, 1987). Somites that would not produce cartilage when grafted onto the CAM alone, would do so when axial tissues were included in the graft (Fig. 2). The second observation is that the older that a somite was when placed in this assay, the more frequently it showed the ability to differentiate into cartilage autonomously (Fowler and Watterson, 1953; Watterson

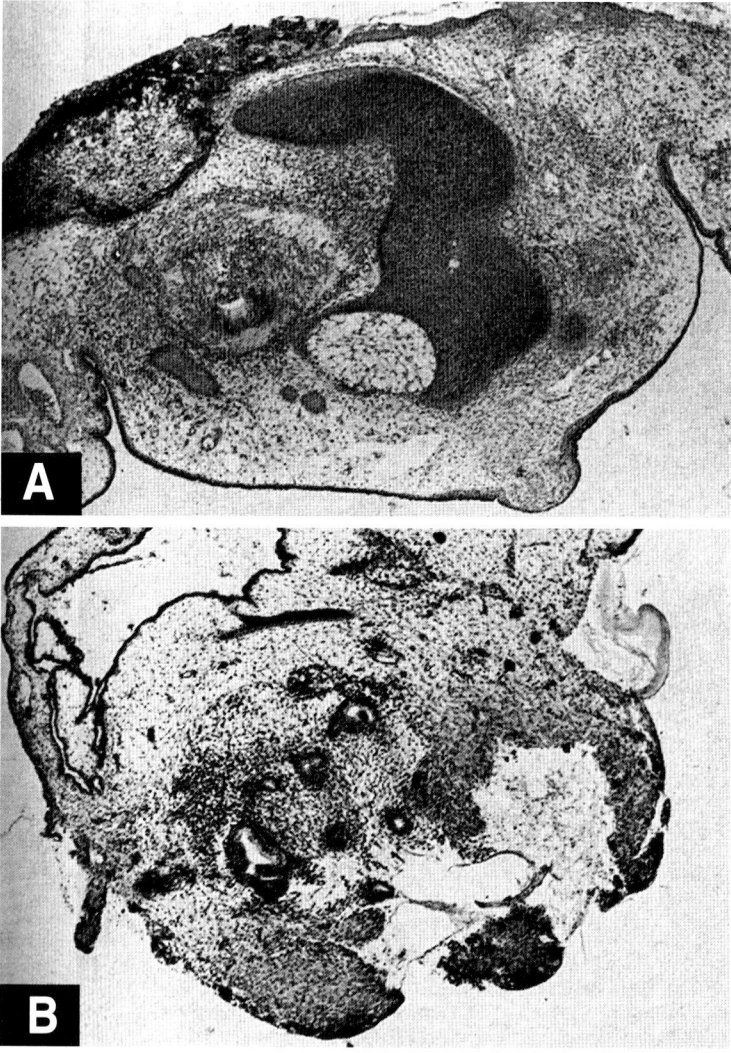

Figure 2 Axial structures are required for somite chondrogenesis in a chorioallantoic grafting assay. Sections of a chick chorioallantoic membrane after grafting of chick donor tissues. (A) Half a vertebra forms following transplantation of one row of somites plus the neural tube and notochord isolated from 17- to 26-somite donors. Magnification: ×43. (B) Vertebral cartilage fails to form when one row of somites (and adjacent nephrotome) is isolated from the neural tube and notochord of 17- to 26-somite donors. Magnification: ×39. (Reprinted from R. L. Watterson, I. Fowler, and B. J. Fowler; The role of the neural tube and notochord in development of the axial skeleton of the chick. *Am. J. Anat.* Copyright © 1954 Wiley-Liss. Reprinted by permission of Wiley-Liss, Inc., a subsidiary of John Wiley & Sons, Inc.)

et al., 1954; Avery *et al.*, 1956; O'Hare, 1972a; Kenny-Mobbs and Thorogood, 1987).

Together, these results were interpreted to mean that axial structures could induce somite chondrogenesis (and thus were probably necessary for that induction during normal development) and that there was a minimum required length of time for somite exposure to axial structures for this induction to occur, after which, exposure was no longer necessary for the induction or development of cartilage. Older somites would autonomously produce cartilage on the CAM because they had had the required period of exposure *in vivo* before being dissected out for experimentation. Younger somites had not been exposed long enough *in vivo* and thus required the further presence of an inducing tissue (neural tube or notochord) in the assay. These basic conclusions would essentially form the basis for all subsequent investigations into sclerotome and vertebral cartilage development and would be supported by *in vitro* experiments as well (see below).

An interesting exception to these results were the experiments of Murray and Selby (1933), who reported that grafts of a single somite, strips of somites, or even unsegmented paraxial mesoderm would give rise to cartilage even without co-grafting of neural tube or notochord, although the presence of those tissues greatly increased the incidence of cartilage formation. Similar results were obtained by Hoadley (1925), but all of his grafts appeared to include notochord as well as somites. All of the grafts in Murray and Selby's experiments also included endoderm and ectoderm; the other CAM grafting experiments either did not include these tissues (Avery *et al.*, 1955, 1956; O'Hare, 1972a; Kenny-Mobbs and Thorogood, 1987) or it is not explicitly stated whether they were removed (Williams, 1942; Fowler and Watterson, 1953; Watterson *et al.*, 1954). Whether or not the presence of these tissues accounts for the difference in results between these and other studies is not clear; studies specifically testing the effect of ectoderm and endoderm on somite chondrogenesis have produced conflicting results (Seno and Büyüközer, 1958; O'Hare, 1972b; Kenny-Mobbs and Thorogood, 1987). The ectoderm has however been assigned a role in the development of the dorsal structures of the somite (Fan and Tessier-Lavigne, 1994; Dietrich *et al.*, 1997; Sosic *et al.*, 1997; Reshef *et al.*, 1998). It is also possible that Murray and Selby's observations were the result of fortuitously choosing somites which would autodifferentiate or were the result of including some nephric tissue, which has the ability to form cartilage (Lash, 1963a,b).

Murray and Selby also noticed the following result; grafted tissues were recovered with greater frequency in chimeric embryos containing grafts of somites made with axial organs than in those that were made without. They interpreted this result as evidence for the axial organs having a role in promoting the survival of somite cells (Murray and Selby, 1933). This result and conclusion are important to keep in mind when considering hypotheses for the specific mechanisms underlying the notochord and neural tube influences on sclerotome induction and somite chondrogenesis, particularly when we discuss the *Sonic hedgehog* molecule later in Section VI.C. *Sonic hedgehog* has been proposed as an inducer of sclerotome, but in fact its role is more likely that of maintenance, survival, and/or proliferative factor (see below and Johnson *et al.*, 1994; Fan *et al.*, 1995; Teillet *et al.*, 1998).

The CAM assays tested whether axial organs were sufficient to induce somite chondrogenesis in an *in vivo* environment, but did not test their sufficiency in a normal somitic developmental setting, nor could they test the necessity of the axial organs in normal sclerotome and cartilage development. It took *in vivo* ectopic grafting and extirpation experiments to address these questions and to explore the individual roles of neural tube and notochord (Fig. 3) (reviewed in Strudel, 1967 and Hall, 1977).

That the neural tube and notochord are each individually sufficient to induce/support somite chondrogenesis was demonstrated by a variety of experiments. Embryos containing a neural tube but no notochord (and vice versa) at the level of already formed somites form some vertebral cartilage (Watterson *et al.*, 1954; Strudel, 1955, 1967; Teillet and Le Douarin, 1983; Rong *et al.*, 1992). Grafts of ectopic neural tube/neural tube floor plate or notochord, whether placed dorsal (Pourquié *et al.*, 1993), or lateral to the somites (Watterson *et al.*, 1954) induce ectopic cartilage formation. Similar ectopic grafting experiments also demonstrated that the notochord was sufficient to induce sclerotome mesenchyme formation at the expense of dorsal somitic derivatives (Brand-Saberi *et al.*, 1993; Goulding *et al.*, 1994); if this is true sclerotome induction is not clear.

That the axial organs are necessary for vertebral chondrogenesis *in vivo* was demonstrated by the observation that extirpation of the neural tube and notochord together (at either segmental plate or formed somite levels) led to a complete lack of vertebral cartilage (Strudel, 1952; Watterson *et al.*, 1954; Strudel, 1955; Teillet and Le Douarin, 1983; Rong *et al.*, 1992). In fact, simply slitting the paraxial mesoderm away from the axial organs appears to adversely affect sclerotome development, demonstrating that these organs are probably important for sclerotome induction (Rong *et al.*, 1992; Hirano *et al.*, 1995). The apparent increase in somitic cell death seen in embryos having undergone either type of operation (Teillet and Le Douarin, 1983; Christ *et al.*, 1992; Rong *et al.*, 1992; Hirano *et al.*, 1995) supports the concept that part of the role of the axial organ in vertebral development is to keep the cells alive.

Extirpation experiments uncovered what appear to be complementary roles for the neural tube and notochord in vertebral development. Complete removal of the neural tube only at the level of already formed somites leads to either a lack of cartilage (Fowler and Watterson, 1953) or the formation only of centrumlike masses of vertebral cartilage that encircle the notochord with the neural arches absent (Watterson *et al.*, 1954; Strudel, 1967; Teillet and Le Douarin, 1983; Rong *et al.*, 1992), indicating that the neural tube is necessary for normal vertebral formation. Removal of the neural tube does not seem to impair sclerotome formation, however (Teillet and Le Douarin, 1983; Christ *et al.*, 1992), indicating that the role of the neural tube is in somite morphogenesis and not sclerotome induction.

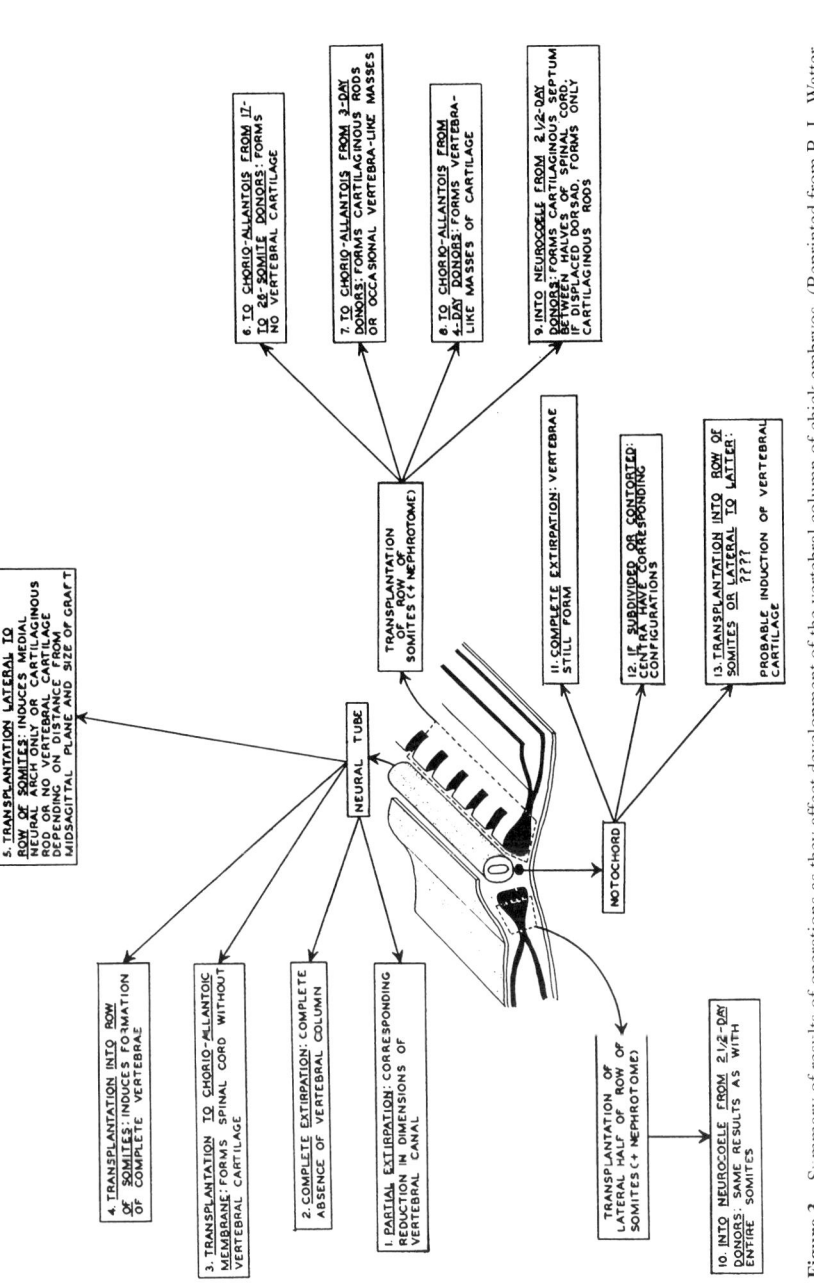

Figure 3 Summary of results of operations as they affect development of the vertebral column of chick embryos. (Reprinted from R. L. Watterson, I. Fowler, and B. J. Fowler; The role of the neural tube and notochord in development of the axial skeleton of the chick. *Am. J. Anat.* Copyright © 1954 Wiley-Liss. Reprinted by permission of Wiley-Liss, Inc., a subsidiary of John Wiley & Sons, Inc.)

Removal of the notochord at the level of already formed somites does not appear to greatly affect vertebral cartilage formation (Fowler and Watterson, 1953; Watterson et al., 1954; Strudel, 1955; Teillet and Le Douarin, 1983; Rong et al., 1992). However, sclerotome formation and/or sclerotome cell survival appear to be adversely affected in embryos notochordectomized at unsegmented paraxial mesoderm levels (van Straaten and Hekking, 1991; Pourquié et al., 1993), indicating that the role of the notochord may be in sclerotome induction, initial sclerotome cell proliferation, and/or sclerotome cell survival *in vivo,* and is perhaps not as important once the sclerotome has been formed or for somite chondrogenesis.

Experiments on the influence of axial structures on sclerotome-derived cartilage formation performed using *Ambystoma* (salamander) embryos confirmed the importance of the neural tube and notochord, but gave them different roles than the chick experiments did. Extirpation of the notochord in *Ambystoma* demonstrated that cartilage would form in its absence, but that the cartilage formed was unsegmented, implicating the notochord in the segmentation, but not the induction of the vertebral cartilage (Kitchin, 1949; Holtzer, 1952a). This is in contrast to notochordectomized chick embryos in which the notochord does appear to be essential for cartilage formation, at least at the segmental plate level (see above). Like chick notochords, salamander notochords have the ability to induce somite chondrogenesis (Avery et al., 1955).

A number of experiments in *Ambystoma* ultimately led to assigning to the neural tube the primary role in inducing and patterning the vertebral cartilage in this species. Embryos in which the neural tube had been extirpated frequently failed to form vertebral cartilage, indicating that the neural tube was essential for cartilage formation, unlike the notochord (Holtzer, 1952b; Holtzer and Detwiler, 1953). Experiments involving extirpation of the neural tube in fish gave similar results (Watterson, 1952).

The ability of the neural tube to induce and pattern somite chondrogenesis in *Ambystoma* was confirmed by a series of grafting experiments. Ectopically implanted somites required the presence of neural tube in order to form cartilage, confirming the ability of this tissue to induce somite chondrogenesis (Holtzer and Detwiler, 1954). Ectopic implantation of the neural tube or displacement of the neural tube dorsad led to the formation of relatively normal vertebrae surrounding the exogenous tube (Fig. 4) (Holtzer, 1951, 1952a; Detwiler and Holtzer, 1956).

Somites implanted into the lumen of the neural tube would form individual normal neural arches around each half of the bisected tube (Holtzer and Detwiler, 1954; Detwiler and Holtzer, 1956), as they would when a piece of neural tube was implanted into the somites (Holtzer and Detwiler, 1953). These experiments lead to the conclusion that not only could the neural tube induce chondrogenesis, it could pattern it as well. Similar results with these manipulations were also obtained with chick embryos (Fig. 4) (Watterson et al., 1954), but were not given the patterning interpretation.

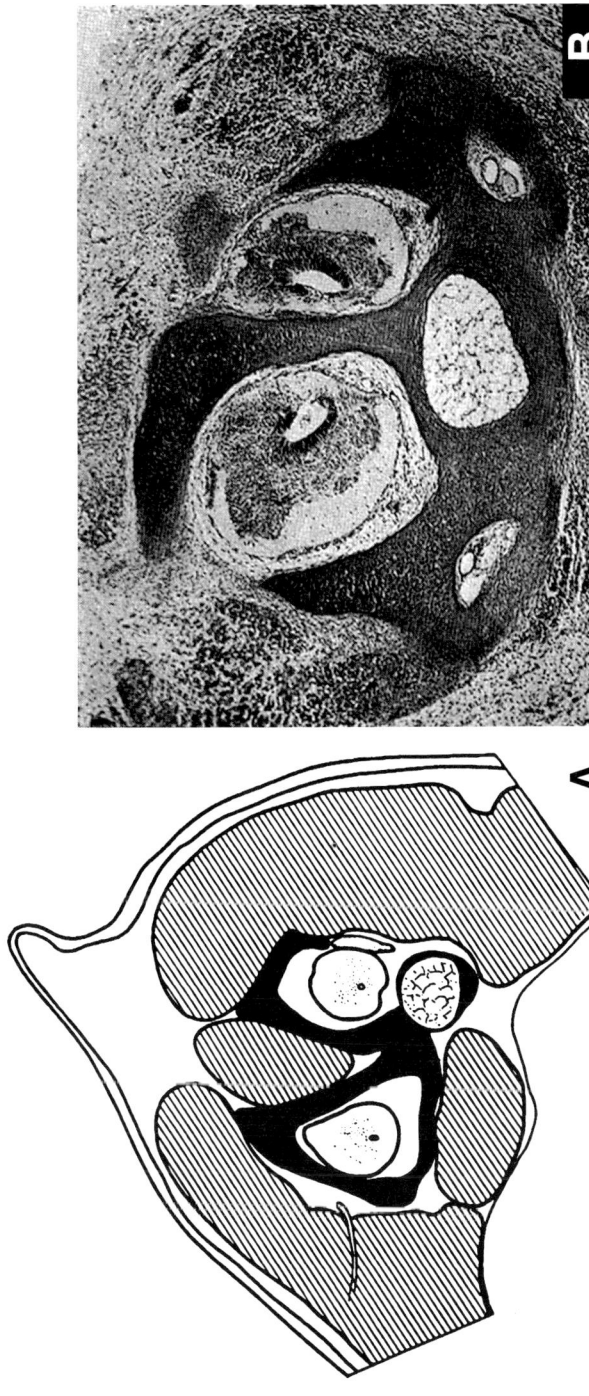

Figure 4. The neural tube has cartilage inducing abilities *in vivo*. (A) Drawing showing normal neural arches (right) and a supernumerary induced arch system (left) in an *Ambystoma* embryo surrounding a cord that was grafted adjacent to the normal. Note absence of notochord in the supernumerary system. Black, cartilage; lined area, muscle. Magnification: ×33. (B) Transplantation of a row of chick somites (and adjacent nephrotome) into the neurocoele of a chick embryo. Each lateral half of the host spinal cord has reformed a closed tube. Between these two closed halves of the host spinal cord the donor somites have formed a complete cartilaginous septum. Magnification: ×43. (A reprinted from S. R. Detwiler and H. Holtzer; The developmental dependence of the vertebral column upon the spinal cord in the urodeles. *J. Exp. Zool.* Copyright © 1956 Wiley-Liss. B reprinted from R. L. Watterson, I. Fowler, and B. J. Fowler; The role of the neural tube and notochord in development of the axial skeleton of the chick. *Am. J. Anat.* Copyright © 1956 Wiley-Liss. Both reprinted by permission of Wiley-Liss, Inc., a subsidiary of John Wiley & Sons, Inc.)

Holtzer was the first to observe that induction of cartilage by the neural tube could occur at a distance, but that induction by notochord could not, a conclusion that presaged later conclusions from *in vitro* work (Holtzer, 1952b; Avery *et al.*, 1955). Another interesting observation from the *Ambystoma* work was that progressive reductions in the size of the neural tube led to reductions in the width of the spinal canal, indicating a sort of dose-dependent relationship for cartilage formation between the neural tube and the somite (Holtzer, 1952b). A similar observation was made for chick embryos (Fowler and Watterson, 1953; Watterson *et al.*, 1954); the significance of this result in either system is not well understood.

In summary, differences can be seen between the roles of the the axial organs in cartilage induction and development in chick and *Ambystoma*. As described above, the notochord appears to be important for segmenting the cartilage in salamanders, but not for inducing it, in contrast to the chick, where the role of the notochord appears to be for induction and/or survival of the sclerotome, with its influence probably occurring in the segmental plate. In both systems, the notochord can ectopically induce cartilage. The neural tube appears to be essential for somite chondrogenesis in both systems and has the ability to induce ectopic cartilage formation. In *Ambystoma*, the neural tube also appears to be patterning the cartilage formed; whether it has this role in the chick as well is not known.

B. *In Vitro* Work

In addition to the *in vivo* experiments, a number of *in vitro* studies have also ascribed a role for neural tube and notochord in somite chondrogenesis (Lash, 1968a; Hall, 1977). But these experiments went beyond simple induction and began to investigate issues such as localization of the inducer, the specific characteristics of the inducing tissues, and the chondrogenic capabilities of somites at different stages of their development. *In vitro* experiments also allowed experimental investigation into mouse sclerotome development that could not be performed *in vivo* because embryonic mice are not surgically accessible like chicks.

In vitro experiments allowed even more specific definition and characterization of somite chondrogenesis inducers (Table I). Many researchers confirmed the ability of the neural tube and notochord to induce somite chondrogenesis in young somites, even across species (Grobstein and Parker, 1954; Grobstein, 1955; Lash *et al.*, 1957; Holtzer, 1960; Cooper, 1965). Tests of a wide variety of live tissues including muscle, pancreas, and lung demonstrated that really only the neural tube, notochord, and embryonic cartilage were capable of inducing cartilage in somites (Grobstein and Holtzer, 1955; Avery *et al.*, 1956; Lash *et al.*, 1957; Cooper, 1965); an interesting exception was the mesonephros (Lash *et al.*, 1957) which itself was shown to be capable of forming cartilage (Lash, 1963a, b). The effect of ectoderm *in vitro*, as discussed previously, is a matter of some controversy (O'Hare, 1972b; Kenny-Mobbs and Thorogood, 1987). Furthermore, fixed tissues or inert sub-

3. Sclerotome Induction and Differentiation

Table 1 The Frequency of Cartilage Formation in Chick Somites Cultured with Various Tissues and Objects[a]

	Negative	Cartilage
Living tissues		
Notochord (3 days)	13	245
Spinal cord–ventral half (3 days)	12	240
Mesonephros (3 days)	35	7
Metanephros (12 days to posthatching)	40	12
Adrenal (1 day posthatching)	20	0
Bone (14 days, femur)	20	0
Bone marrow (guinea pig)	30	0
Cartilage (10, 12 days)	20	0
Liver (16, 17 days)	36	1
Lung (14, 17 days)	20	1
Muscle (14 days)	20	0
Nerve (15, 17 days, sciatic)	20	0
Pancreas (7, 9 days)	18	0
Small intestine (17 days)	21	0
Spinal cord–dorsal half (3 days)	20	0
Spinal ganglia (9 days)	20	0
Fixed tissues		
Bone (14 days, femur, alcohol fixed)	20	0
Bone (14 days, femur, frozen–thawed)	20	0
Bone marrow (guinea pig, alcohol fixed)	30	0
Cartilage (12 days, formalin fixed)	20	0
Cartilage (14 days, frozen–thawed)	20	0
Cartilage (14 days, alcohol fixed)	20	0
Liver (guinea pig, alcohol fixed)	20	0
Metanephros (guinea pig, alcohol fixed)	20	0
Muscle (12 days, formalin fixed)	22	0
Nerve (15, 17 days, sciatic, formalin fixed)	18	0
Notochord (3 days, formalin fixed)	21	0
Notochord (3 days, glycerinated)	22	0
Spinal cord (3 days, formalin fixed)	20	0
Others		
Glass beads	20	0
Wood	20	0
Cauterization	20	0
Somite controls (stages 17, 18, 19)	289	2

[a] Reprinted by permission of Academic Press from Lash *et al.* (1957).

stances such as wood would not induce chondrogenesis in cultured somites (Grobstein and Holtzer, 1955; Avery *et al.*, 1956; Lash *et al.*, 1957).

Careful studies were performed to further characterize the inducing tissues. The notochord could function as an inducer independent of its stage of differentiation (Cooper, 1965), but the notochord cells needed to undergo vacuolation while in culture to induce. Preculture of the notochord alone would cause it to lose its inducing ability (Lash et al., 1957).

The ability of cartilage to induce did depend on its state of differentiation and whether it came from chick or mouse embryos (Cooper, 1965). It should be noted that other attempts at using embryonic cartilage as an inducing source were not successful (Grobstein and Holtzer, 1955; Avery et al., 1956; Lash et al., 1957), but that difference may be due to the preparation of the tissue and/or the specific culture conditions used. Cooper (1965) used chondrocytes dissected from growth plates and found that it was only the younger, nonhypertrophied chondrocytes that were capable of induction. Similar to the notochord however, the inducing chondrocytes did need to undergo hypertrophy during their time in culture to be able to induce somite chondrogenesis, indicating that extracellular matrix molecules might be important.

The tissue best characterized during this time period was the spinal cord. Inducing activity in the spinal cord was localized specifically to the ventral half (Grobstein and Holtzer, 1955; Lash et al., 1957); dorsal–ventral differences in the inducing capability of the spinal cord had been seen in its influence on the metanephros (Grobstein, 1955). In contrast to the notochord, spinal cord tissue could be cultured alone before being placed with somites and retain its inducing abilities (Lash et al., 1957). The spinal cord also retains its inducing capability *in vitro* at embryonic ages from the time of sclerotome induction and somite chondrogenesis up to times that are long after vertebral cartilage formation has taken place: up to embryonic day 18 in the chick (Tremaine and Hall, 1979) and up to day 13 in the mouse (Grobstein and Holtzer, 1955). Such observations may indicate that the activity of the spinal cord is required throughout sclerotome development and not simply for somite chondrogenesis. Interestingly, while fixed neural tube is not longer capable of inducing *in vitro* somite chondrogenesis, lethally X-radiated neural tube is (O'Hare, 1972c).

As well as characterizing the inducing tissues, the *in vitro* experiments also allowed examination of the behaviors of the somites themselves. Experiments dissociating somites and then exposing them to notochord or neural tube demonstrated that destruction of the organization of the somite did not relieve the cells of the axial organ requirement for somite chondrogenesis (Stockdale et al., 1961). Conversely, this experiment also demonstrates that the cells need not be in an organized somite to be able to respond to axial organ influences. That the cartilage in any whole somite induction experiment probably does come from the sclerotome was shown by dissecting the somites into their component structures (dermomyotome and sclerotome) and testing their individual responses to the notochord (Cheney and Lash, 1981). Only the sclerotome responded to the notochord by forming nodules, increasing accumulation of glycosaminoglycans, increasing synthesis of proteoglycans, and increasing DNA synthesis, all hallmarks of differentiating cartilage (Marzullo and Lash, 1967; Lash, 1968b; Gordon and Lash, 1974).

As had been seen with the CAM assays (see above), the *in vitro* assays demonstrated the ability of somites from older embryos to form cartilage without the presence of a inducing structure (Grobstein and Holtzer, 1955; Avery et al., 1956; Kenny-Mobbs and Thorogood, 1987). An apparently contradictory result was obtained with somites from chick at stages 15–18 (Hamburger and Hamilton, 1951), where the more posterior (younger) somites showed greater chondrogenic potential than the more anterior (older) ones (Lash, 1967, 1968b). It is difficult to compare all of these studies without knowing whether exactly the same stage somites from the same stage embryos were taken. These results had been interpreted as demonstrating that the inducer was needed for a certain length of time (a minimum length of time for induction had been defined *in vitro;* Lash et al., 1957) after which it was not needed any more. Merely manipulating the cell culture conditions, however (such as changing serum source, using more somitic tissue/culture, covering the culture with paraffin oil, and changing potassium ion concentrations), could enhance somite chondrogenesis and eliminate totally the need for the presence of spinal cord or notochord to elicit somite chondrogenesis from young somites (Ellison et al., 1969; Ellison and Lash, 1971; Lash et al., 1973), suggesting that the somites were determined early for cartilage development and that the role of the neural tube and notochord was to permit the somite to form the cartilage which it was already programmed to become.

Both the notochord and neural tube were capable of inducing somite chondrogenesis transfilter (Lash et al., 1957; Cooper, 1965; Flower and Grobstein, 1967). Spinal cord induction of cartilage, however, appeared to be fundamentally different from that of notochord. Whether the inducing tissue and the somites were in direct contact or transfilter, the nodules formed by spinal cord induction generally had intervening tissue while the nodules formed by notochord were generally directly apposed to the notochord (Fig. 5) (Avery et al., 1955, 1956; Lash et al., 1957; Cooper, 1965; Flower and Grobstein, 1967). As previously stated, the *Ambystoma* experiments (Holtzer, 1952b; Avery et al., 1955) had provided early evidence of this phenomenon. It has been suggested that this effect is due to a difference in the amount of tissue (and thus inductive factor) that the somites are exposed to (Flower and Grobstein, 1967), consistent with the "dose-dependent" effect of the neural tube seen *in vivo* (Holtzer, 1952b; Fowler and Watterson, 1953; Watterson et al., 1954). The significance during normal development and the underlying mechanism(s) of this difference between neural tube and notochord induction are still not understood.

One important paradigm to come out of the *in vitro* work was the idea of a secreted inductive factor, which was implied by the fact that axial organ induction of somite chondrogenesis could occur transfilter (i.e., with no physical contact between the inducing and responding tissues). The presence of a secreted, soluble inducing molecule would make a search for the inducing molecules easier. In fact, the establishment of the *in vitro* somite chondrogenesis assay made the search for the inducer possible since conditions, concentrations, and components of the system could then be controlled and varied.

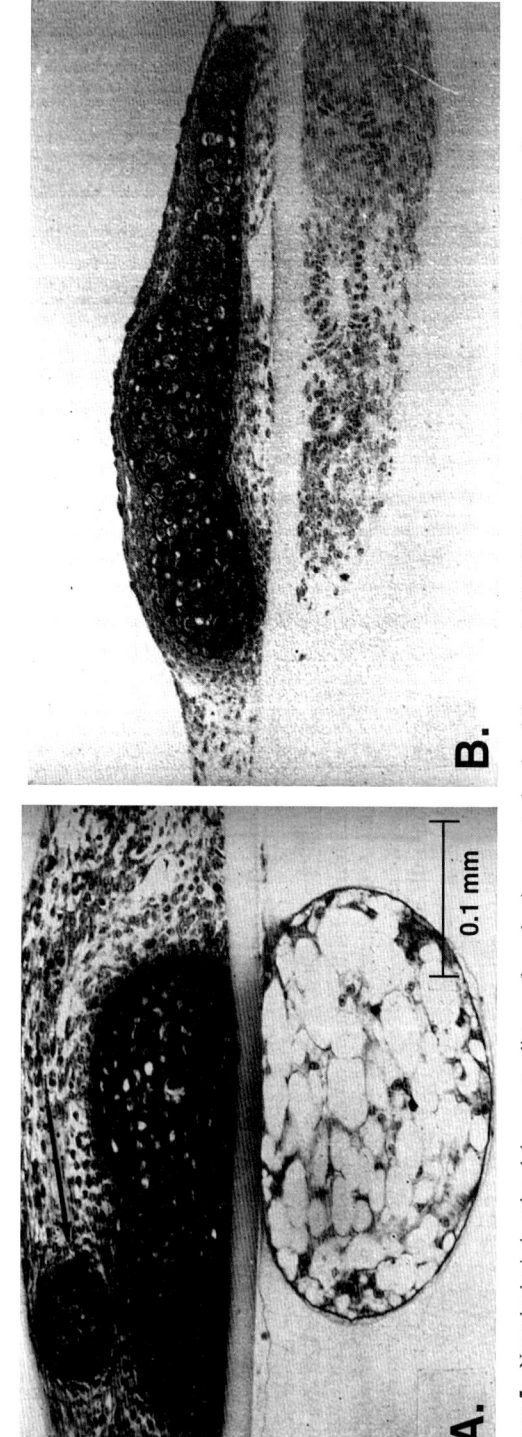

Figure 5 Neural tube induced nodules are at a distance from the tissue; notochord nodules are not. (A) A 120-hr culture of stage 32 chick notochord transfilter to mouse somites. Hypertrophied notochord cells remain intact *in vitro* and are bounded by a metachromatic sheath. Induced cartilage differentiated both in contact with and displaced from the filter. The nodule in the latter position (arrow) is surrounded by a perichondrium. Magnification: ×200. (B) A 120-hr culture of 9-day embryonic mouse spinal cord (ventral half) transfilter to mouse somites. Induced cartilage is characteristically displaced from the filter. Magnification: ×160. (Reprinted by permission of Academic Press from Cooper, 1965.)

C. Early Inducer Searches

Lash *et al.* (1957) made one of the earliest attempts to identify potential inducing substances by staining the filter in their culture assays for the presence of DNA, RNA, and polysaccharides; positive staining was not revealed for any of these substances. More successful subsequent attempts to purify the inducing agent involved making extracts from the neural tube and notochord. Acid extracted fractions from chick 2- to 3-day embryonic spinal cord and notochord gave a low molecular weight "nucleotide" fraction capable of inducing cartilage nodule formation by somites in culture and a "sugar–phosphate" fraction that was not (Lash *et al.*, 1962). The low molecular weight fraction (eluted off of a charcoal–celite column) could be further purified by ion-exchange column and was shown subsequently to promote uptake of ^{35}S-labeled sodium sulfate (presumed to be in the sulfated proteoglycans of differentiating cartilage matrix) (Fig. 6) (Hommes *et al.*, 1962; Zilliken, 1967). Strudel, in the same year, made extracts of neural tube and notochord by homogenization and centrifugation. The resulting supernatant could

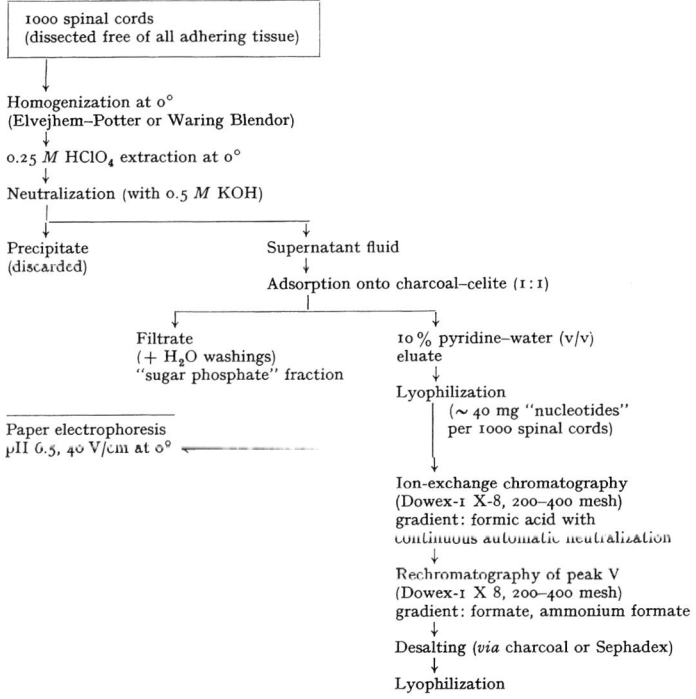

Figure 6 An early chondrogenic inducer purification scheme. (Reprinted from *Biochim. Biophys. Acta* **56**; F. A. Hommes, G. Van Leeuwen, and F. Zilliken; Induction of cell differentiation. II. The isolation of a chondrogenic factor from embryonic chick spinal cords and notochords; 320–325; copyright © 1962, with permission from Elsevier Science.)

induce chondrogenesis in somite culture. Not only would this extract induce cartilage in somites that would not self-differentiate, but it would also cause somites that would self-differentiate to form more nodules than they would have without extract (Strudel, 1962, 1963). These experiments were important for demonstrating that inductive activity of spinal cord and notochord could be separated from the tissues themselves; however, it was difficult to fully purify the responsible substance.

Investigators then turned to examining components within the cartilage matrix or perinotochordal matrix for their effects on somite chondrogenesis (reviewed in Hall, 1986 and Vasan, 1987). Previous experiments had demonstrated that notochord and spinal cord would secrete matrix proteins such as collagen when in contact with somites (Kvist and Finnegan, 1970a, b; Cohen and Hay, 1971; Minor, 1973; Trelstad *et al.*, 1973) and that these proteins appear to enter the sclerotome matrix, making them good candidates for possibly inducing the sclerotome to make cartilage. Chondromucoprotein (a substance containing a number of proteoglycans including proteochondroitin sulfate) extracted from embryonic cartilage and purified proteoglycans could stimulate *in vitro* somite chondrogenesis as measured by visual appearance and by glycosaminoglycan accumulation (Kosher *et al.*, 1973; Lash and Vasan, 1978; Belsky *et al.*, 1980). Collagen and procollagen could also induce *in vitro* somite chondrogenesis, so the effect was not restricted to proteoglycans (Kosher and Church, 1975; Lash and Vasan, 1978).

More direct tests of the spinal cord and notochord extracellular matrices also implicated them in somite chondrogenesis. Several investigators used enzyme treatments to demonstrate the importance of an intact perinotochordal extracellular matrix in *in vitro* somite chondrogenesis induction assays. Treatment of the notochord with trypsin, hyaluronidase, chondroitinase ABC, or chondroitinase AC resulted in the loss of the ability of the notochord to stimulate sulfated glycosaminoglycan accumulation in somites. Importantly, this effect was reversible; if the notochord was enzyme treated and then allowed to recover, the stimulating ability also recovered (O'Hare, 1972c; Kosher and Lash, 1975; Lash and Kosher, 1975; Strudel, 1975a,b).

Inquiries into extracellular matrix molecules as the inducers of somite chondrogenesis slowed down after the late 1970s as the advent of molecular cloning techniques and the identification of sclerotome specific markers led to alternative strategies in the search for the molecular identity of sclerotome inducers.

VI. The Molecular Approach

The molecular advancements in developmental biology have allowed the detailed examination of early embryonic events through the examination of the expression of developmentally regulated genes. Although molecular marker evaluation has become increasingly popular, the activation of a whole developmental program is

3. Sclerotome Induction and Differentiation

more complex than just the turning on of a few molecular markers. Analysis of the expression of a set of genetically and functionally important genes for a specific developmental process can allow us to monitor inductive steps that operate to orchestrate differentiation pathways.

A. Molecular Markers of Sclerotome

The cloning of the mammalian *Pax* genes has thoroughly changed our approach to uncovering the mechanisms that control somitic pattern (Table II) (Dahl et al., 1997). The *Pax* genes are a family of transcription factors containing the conserved paired-homeobox DNA binding domain. Different, specific *Pax* gene family members are expressed in different domains of the somite, making them extremely useful for investigating influences on somitic pattern. The two *Pax* genes that serve as specific sclerotome markers are *Pax-1* and *Pax-9*. *Pax-1* mRNA is not

Table II. Vertebrate *Pax* Genes[a]

Pax genes	Basic Structure			Localization		Mouse Mutant		Human syndrome
	Paired box	OP	Homeobox	Mouse	Human	Natural	Targeted	
Group I								
Pax1	N— ■■■■	—O—	—C	2	20p11	Undulated: un/un^{ex}/Un^S	Not published	Spina bifida (?)
Pax9				12	14q12-q13	Not known	Not published	Not known
Group II								
Pax2	N—▨▨▨—	—O—□—	—C	19	10q25	Pax2^{1Neu}	Yes	Renal coloboma syndrome
Pax8				2	2q12-q14	Not known	Not known	Not known
Pax5				4	9p13	Not known	Yes	Not known
Group III								
Pax3	N—▨▨▨—	—O—□—	—C	1	2q35	Splotch	Not known	Waardenburg Syndrome I, III
Pax7				4	1p36.2	Not known	Yes	Not known
Group IV								
Pax4	N—▥▥▥—	—□—	—C	6	7q32	Not known	Yes	Not known
Pax6				2	11p13	Small eye	Not known	Aniridia Peter's anomaly

[a] Vertebrate *Pax* genes have been classified into four different groups (I–IV). Members within a group are characterized by a specific assembly of three structural motifs: the paired box, the homeobox, and the octapeptide (OP). Group I is formed by *Pax-1* and *Pax9* because neither gene has a paired type homeobox. Group II is formed by *Pax2*, *Pax5*, and *Pax8*, which contain a partial homeobox with only the first α-helix present. Group III contains *Pax3* and *Pax7*, which possess the complete set of motifs. Group IV genes (*Pax4* and *Pax6*) do not contain the octapeptide. Reprinted from E. Dahl, H. Koseki, and R. Balling; Pax genes and organogenesis. *BioEssays*. Copyright © 1997 Wiley-Liss, Reprinted by permission of Wiley-Liss, Inc., a subsidairy of John Wiley & Sons, Inc.)

expressed in the presomitic mesoderm, but is expressed in the ventral part of the somite that prefigures the sclerotomal mesenchyme (Deutsch et al., 1988; Ebensperger et al., 1995; Borycki et al., 1997). As development continues, Pax-1 mRNA becomes restricted to the caudal cells of the somite, with expression eventually becoming even more restricted to the cells of the intervertebral discs; expression is never seen in the dorsal sclerotome (Fig. 7A,B) (Deutsch et al., 1988; Wallin et al., 1994; Peters et al., 1995; Müller et al., 1996). The chick differs from the mouse in that Pax-1 mRNA expression is also detected in the chondrocytes of immature vertebral bodies (Peters et al., 1995). As described in Section IV on genetics, there is a natural mouse mutant at the Pax-1 locus called undulated, the phenotype of which indicates that Pax-1 has a function in sclerotome development (Dietrich et al., 1993; Wallin et al., 1994; Dietrich and Gruss, 1995). Pax-1, while specifically expressed in the sclerotomal compartment of the somite, is not somite restricted in its expression; Pax-1 mRNA is also expressed in the developing pectoral and pelvic girdles and in the developing limbs (Timmons et al., 1994; Peters et al., 1995) as well as in some nonchondrogenic tissues (Peters et al., 1995; Müller et al., 1996).

In contrast to Pax-1, Pax-9 mRNA is not expressed in the chick somite until the sclerotomal mesenchyme has actually formed, around somite stage IV–V (Müller et al., 1996). Its expression is highest in the dorsolateral sclerotome, in contrast to Pax-1 (Fig. 7C,D; Müller et al., 1996). Pax-9 mRNA expression eventually becomes restricted to the caudal half of the sclerotome and later, the intervertebral discs (Peters et al., 1995; Müller et al., 1996). Like Pax-1, Pax-9 mRNA is

Figure 7 Embryonic expression of various sclerotome markers. (A, C) Cross sections at the thoracic level of stage 22 embryos showing Pax-1 expression (A) in almost all sclerotome cells (sc), whereas Pax-9 expression (C) is found mainly in dorsolaterally located sclerotome cells. dm, Dermomyotome: n, notochord; nt, neural tube. Bar=100 μm. (B, D) Whole mount in situ hybridization of Pax-1 mRNA (B) in quail and Pax-9 (D) in chick embryos. (B) Stage 22 embryo showing Pax-1 expression in five pharyngeal pouches (I–V) and the sclerotomes. Expression is decreased in the five most cranial sclerotomes (arrow). (D) Stage 22 embryo. Pax-9 is strongly expressed in four pharyngeal pouches (I–IV) and the somites of the tail bud region. More cranial sclerotomes show Pax-9 transcripts at lower levels. Note the narrow segmental strips in the caudal sclerotome halves. Arrows indicate Pax-9 expression in the hind limb bud mesenchyme and the visceral arch mesenchyme. (E, F) M-twist mRNA expression. (E) Transverse section of a 9.5 days postcoitum (p.c.) mouse embryo showing M-twist expression in the sclerotome. (F) 12 p.c. mouse embryo. Labeling is seen in the superficial mesenchyme of the developing face, the lingua, the skin, and the ventral part of the vertebrae. (G, H) Scleraxis mRNA expression. (G) Expression in the sclerotome of a day 12.5 p.c. mouse embryo, transverse section. Transcripts can be seen throughout the sclerotome and limb buds, as well as in the body wall and trachea. fl, Forelimb bud; h, heart; hl, hind limb bud; n, neural tube; t, trachea. (H) Parasaggital section of day 12.5 p.c. mouse embryo showing expression in the intervertebral discs (id), chest wall (cw), diaphragm (d), hind paw (hp), heart valves (hv), mandible (mn), pelvis, (p), and tongue (t). (A–D reprinted from Müller et al., 1996; E, F reprinted from Wolf et al., 1991; both reprinted by permission of Academic Press. G,H reprinted by permission of The Company of Biologists Ltd. from Cserjesi et al., 1995.)

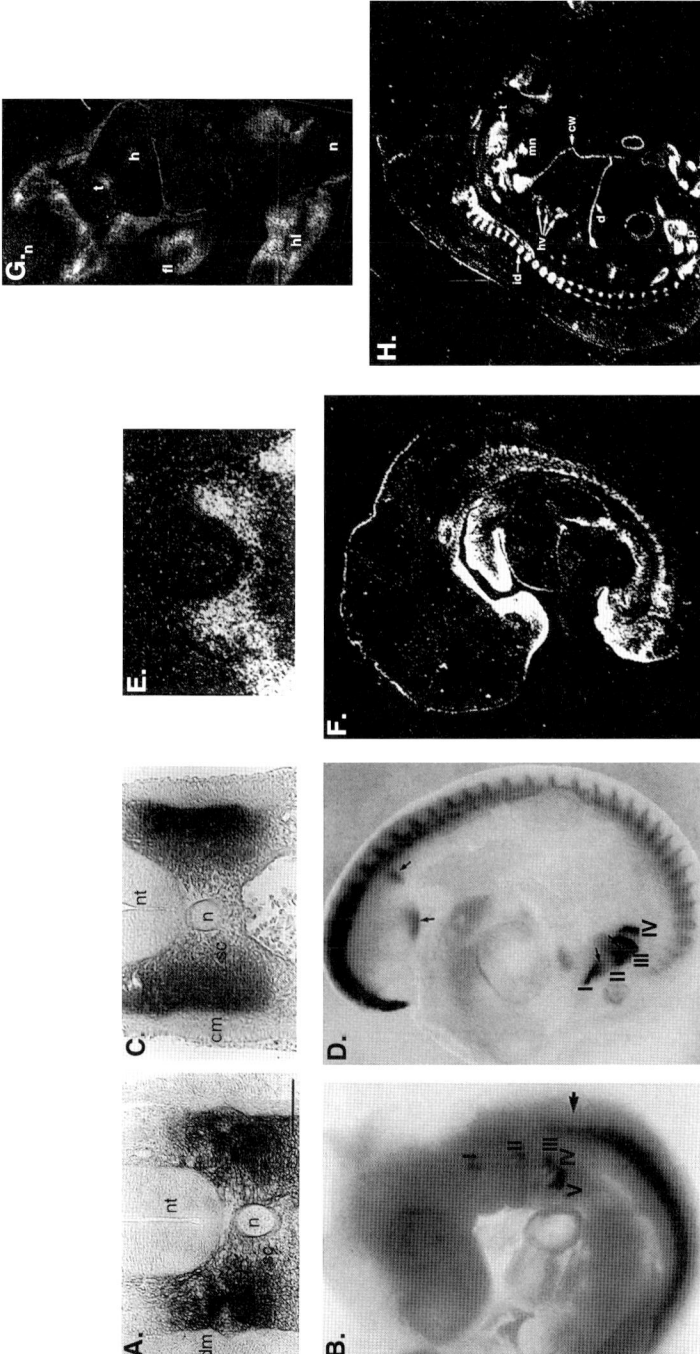

expressed only in the sclerotome of the somite, but is not restricted to solely somitic expression. Expression of *Pax-9* mRNA is also seen in the appendicular skeleton and the pharyngeal endoderm (Peters *et al.,* 1995; Müller *et al.,* 1996). Whether *Pax-9* has a functional role in sclerotome development is not yet known.

Another *Pax* family member frequently used in analyzing somite development is *Pax-3*. While it is a dermomyotome marker rather than a sclerotome marker, it is important to mention it here because it is frequently used in studies on dorsoventral somite patterning. In the mouse and the chick, *Pax-3* mRNA is expressed in the presomitic mesoderm and the somites (Goulding *et al.,* 1991; Williams and Ordahl, 1994). *Pax-3* mRNA expression diminishes in the ventral portion of the somite as the sclerotome forms, but expression in the dorsal dermomyotome is retained. *Pax-3* mRNA also marks the migrating limb muscle precursor cells from the lateral half of the somite (Williams and Ordahl, 1994).

Aside from the *Pax* genes, there are two other genes commonly employed as sclerotome markers. *M-twist,* the mouse homolog of the *Drosophila twist* gene, is a basic helix–loop–helix transcription factor expressed at both the protein and mRNA levels in the sclerotome and dermatome as well as other, nonsomitic mesodermal tissues (Fig. 7E,F) (Wolf *et al.,* 1991; Füchtbauer, 1995; Gitelman, 1997). mRNA for *M-twist* can be found in presomitic mesoderm and epithelial somites (Füchtbauer, 1995), but antibody studies do not detect *M-twist* protein until somite stage V–VI (Gitelman, 1997). *M-twist* knockout mice have increased cell death in the sclerotome indicating that the function of *M-twist* in sclerotome may be to promote growth and/or survival of those cells (Chen and Behringer, 1995).

Another basic helix–loop–helix protein used as a sclerotome marker is *scleraxis*. *Scleraxis* mRNA is expressed in the sclerotome of somites that have already begun to compartmentalize (Cserjesi *et al.,* 1995). Expression is initially in the lateral sclerotome, but becomes restricted to the ventromedial region (Fig. 7G,H). *Scleraxis* mRNA continues to be expressed in prevertebrae and the intervertebral discs as development proceeds. *Scleraxis* mRNA is also expressed in the developing maxillofacial cartilages, limbs, and ribs (Cserjesi *et al.,* 1995). *Scleraxis'* role(s) and function(s) in sclerotome development are unknown at this point.

A traditionally used marker for the sclerotome is the binding of peanut agglutinin, a lectin that preferentially binds $Gal_{\beta1-3}$ GalNAc residues (Lotan *et al.,* 1975). Peanut agglutinin binds specifically to the caudal portion of the sclerotome (Bagnall and Sanders, 1989), the half of the sclerotome through which neural crest cells do not migrate. In contrast to the previously described sclerotome markers, the peanut agglutinin binding molecule(s) is thought to function in segmentation of the neural crest and not directly in sclerotome development. A variety of other genes are expressed in the sclerotome (Orr-Urtreger and Lonai, 1992; Monsoro-Burq *et al.,* 1994; Wright *et al.,* 1995; Healy *et al.,* 1996; Winnier *et al.,* 1997) and antibodies (George-Weinstein *et al.,* 1988; Mark *et al.,* 1989) have been developed against sclerotome, but these are not widely used as sclerotome markers.

A common feature among these sclerotome markers is that they are frequently

3. Sclerotome Induction and Differentiation   103

expressed in chondrogenic regions derived from embryonic origins other than the sclerotome, and in nonchondrogenic tissues as well. No marker that is strictly sclerotome restricted has yet been found.

B. *In Vivo* Experiments

Using the *Pax* genes as markers, recent workers have been able to examine one of the earliest steps in the formation of embryonic vertebral cartilage, the induction of sclerotome from the somite (as measured by marker gene expression). Again, the approaches are a combination of *in vivo* grafting, extirpation, and *in vitro* coculture, but now the assays are not carried to the end point of chondrogenesis. Mutant mice that lack a notochord develop without activating the *Pax-1* gene in the ventral somite; the *Pax-3* gene remains expressed throughout the entire somite (Dietrich *et al.*, 1993; Koseki *et al.*, 1993). These results provided strong genetic evidence that activation of this functionally important gene requires input from the notochord.

Approaching this issue from an experimental embryology perspective, chick embryologists have shown that ablation of the notochord at the presomitic level leads the somitic mesoderm to develop with dermomyotomal characteristics (Goulding *et al.*, 1994; Ebensperger *et al.*, 1995). In these embryos, the somitic mesoderm expresses the dermomyotome-specific gene *Pax-3* and does not express the sclerotomal marker *Pax-1,* indicating that the first step of sclerotomal development is dependent on the notochord. Furthermore, they showed that ectopic notochord can induce ectopic *Pax-1* expression (Brand-Saberi *et al.*, 1993; Goulding *et al.*, 1994; Ebensperger *et al.*, 1995).

C. *In Vitro* Experiments and a Candidate Inducer

With the advent of molecular markers, an *in vitro* system much like the one established by Grobstein and Parker (1954) could then be used to study the mechanism(s) of this early inductive event. By employing molecular markers as indicators for the commencement of a specific somitic developmental program, an *in vitro* approach allows correlation with developmental program *in vivo*. Such an *in vitro* assay also allows characterization of the biochemical properties of the signal(s) and the molecular identification of the signal(s). Using three dimensional collagen gels (Tessier-Lavigne *et al.*, 1987), Fan and Tessier-Lavigne (1994) performed a short-term culture assay to investigate early somitic patterning events by assaying for expression of molecular markers known to have functions in the somite. Using this experimental paradigm, it was shown that mouse presomitic mesoderm was naive as it did not spontaneously express assayable levels of *Pax-1* or *Pax-3* transcripts in culture. In contrast, presomitic mesoderm cocultured with different embryonic tissues would express dermomyotome or sclerotome specific markers. The sclerotomal markers, *Pax-1, M-twist* (Fan and Tessier-Lavigne, 1994), and *Pax-9* (C.-M. Fan, 1998, personal communication), were upregulated in the presomitic mesodermal explant in the presence of the notochord or the floor plate of the neural tube while the dermomyotome markers *Pax-3* and *Pax-7* were upregulated in the presence of ectoderm.

Using mRNA *in situ* hybridization, it was observed that the sclerotome inducing signal from the notochord and ventral neural tube appears to act over a long distance to induce the sclerotomal marker expression in the absence of any other embryonic tissues (Fan and Tessier-Lavigne, 1994). Induction of sclerotome markers in this assay system could occur transfilter, supporting the notion that the signal is a diffusible one, consistent with the *in vitro* culture results on somite chondrogenesis discussed previously. Coculture of mature sclerotome and presomitic mesoderm further demonstrated that sclerotome cells do not possess a homeogenetic inductive activity, that is, the ability to induce the naive presomitic cells to adopt the sclerotomal fate; this is different than the ability of cartilage to induce somite chondrogenesis (Cooper, 1965). Furthermore, it was determined that within 6 hr. of coincubation with the notochord, expression of *Pax-1* is induced and that this induction can occur in the presence of cyclohexamide, a protein synthesis inhibitor (C. M. Fan *et al.*, 1995 and 1998, unpublished). Similar results on notochord induction of *Pax-1 in vitro* were also obtained independently by Ebensperger *et al.* (1995) and Müller *et al.* (1996). Thus, the activation of the sclerotomal program appears to require direct input of a long-range diffusible factor secreted by the notochord and/or the floor plate of the neural tube. Interestingly, Müller *et al.* (1996) found that the notochord is not equal along its cranial–caudal axis in its inducing capabilities. In contrast to thoracic level notochord, preotic notochord, which normally resides next to mesoderm that never expresses *Pax-1* mRNA, is incapable of inducing *Pax-1* in thoracic level unsegmented paraxial mesoderm. The *Pax-1* inducing capabilities of the notochord are also restricted to paraxial mesoderm, head mesoderm does not respond (Müller *et al.*, 1996).

To identify the inducing molecule(s), Fan and Tessier-Lavigne (1994) analyzed known secreted growth factors that were documented to be expressed in the notochord and floor plate during the time of sclerotome specification for their ability to mediate the induction of sclerotomal marker expression. The signaling molecule *Sonic hedgehog* (*Shh*), a homolog of the *Drosophila hedgehog* gene, was known to be expressed in the floor plate and notochord during the time of sclerotome formation (Echelard *et al.*, 1993; Krauss *et al.*, 1993; Riddle *et al.*, 1993; Chang *et al.*, 1994; Marti *et al.*, 1995). Cells expressing Shh can mimic the effect of the notochord and the floor plate to induce sclerotome-specific gene expression in cultured presomitic mesoderm (Fig. 8); (Fan and Tessier-Lavigne, 1994; Fan *et al.*, 1995). In parallel, Johnson *et al.* (1994), showed that in the chick, a retrovirus containing a *Shh* cDNA can direct sclerotome marker expression dorsally when injected into the dorsal portion of the somite. Assaying for both the dorsal

3. Sclerotome Induction and Differentiation 105

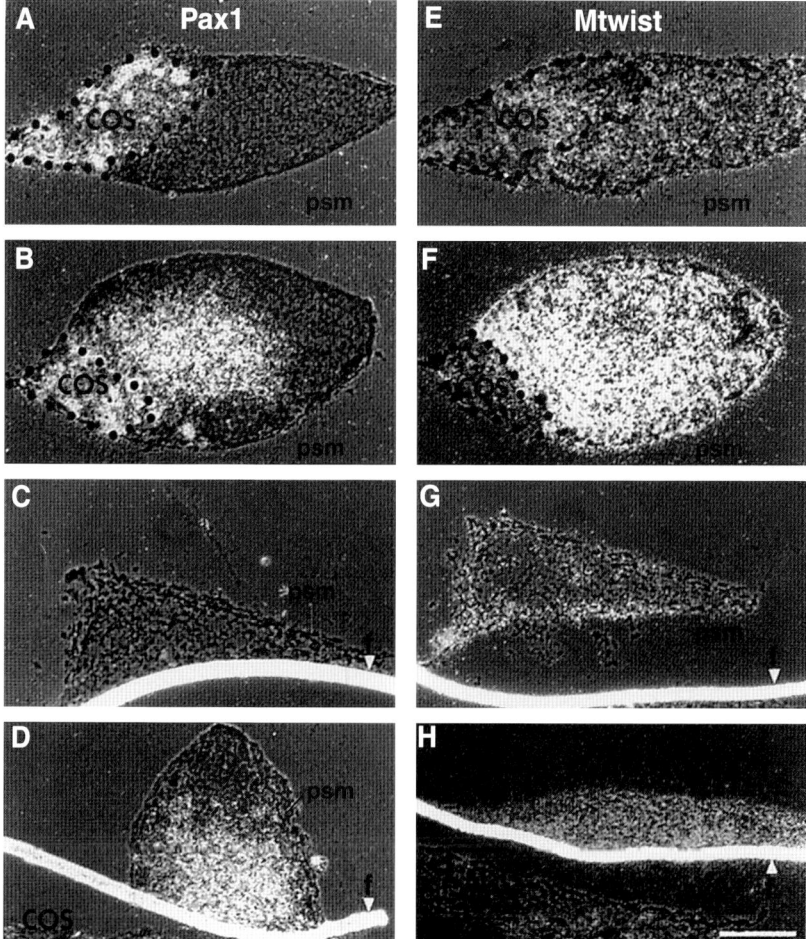

Figure 8 *Shh* can induce sclerotome marker expression. Mouse presomitic mesoderm explants (psm) were placed in collagen gels with aggregates of COS cells that had been transfected with control or *Shh* expression constructs and were cultured for 24 hr. Expression of *Pax-1* (A–D) and *M-twist* (E–H) was assessed in presomitic mesoderm explants cultured wth COS cells (outlined by dots) transfected with a control construct (A, C, E, and G) or a *Shh* expression construct (B, D, F, and H), either in direct contact (A, B, E, and F) or separated (C, D, G, and H) from the COS cells by a nucleopore filter (f) (which sometimes became separated from tissues during sectioning). In A, B, E, and F, two explants of presomitic mesoderm were cocultured with the COS cells. In C, D, G, and H, one (C, G, and H) or two (D) explants of presomitic mesoderm were placed across the membrane from the COS cells. Note the spurious hybridization of both probes to the COS cells. (Reprinted by permission of Cell Press from Fan and Tessier-Lavigne, 1994.)

dermomyotome marker *Pax-3* and ventral sclerotome marker *Pax-1* in these experimental systems, the two groups reached the same conclusion that *Shh* acts in a manner similar to ectopic transplanted notochord and/or floor plate to induce sclerotome at the expense of dermomyotome (Fan and Tessier-Lavigne, 1994; Johnson *et al.*, 1994; Fan *et al.*, 1995).

VII. *Shh*, cAMP, and Sclerotome

The *Sonic hedgehog* signaling pathway appears to have an influence on cAMP levels. In *Drosophila*, genetic evidence indicates that *hedgehog* signaling antagonizes the activity of protein kinase A (Jiang and Struhl, 1995; Li *et al.*, 1995; Pan and Rubin, 1995; Strutt *et al.*, 1995). Since protein kinase A (PKA) activity is positively dependent on the levels of cAMP, it can be inferred that *hedgehog* signaling in *Drosophila* decreases the level of cAMP, although there is no direct evidence for that. In the mouse, increasing the cAMP or PKA levels in mouse presomitic mesoderm results in an inhibition of the ability of Shh to induce *Pax-1* expression, indicating a possible mechanism for the action of *Sonic hedgehog* on somitic mesoderm (Fan *et al.*, 1995), while in the zebrafish, inhibition of PKA leads to an expansion of the sclerotome (Hammerschmidt and McMahon, 1998).

As early as 1976, a link between cAMP levels and notochord-induced somite chondrogenesis was being established. Raising cAMP levels with cAMP analogs inhibited sulfated glycosaminoglycan (GAG) accumulation both in "spontaneous" (self-differentiating) somites and in somites exposed to notochord or collagen (Kosher, 1976). Later experiments measuring cAMP levels in somites revealed that exposure to notochord decreased intracellular cAMP levels in somites (Vasan *et al.*, 1985; Vasan, 1986). It can thus be speculated that the effects of *Shh* on sclerotome induction are mediated by causing a decrease in cAMP levels and PKA activity and that these are important for subsequent somite chondrogenesis.

VIII. Issues Concerning *Shh* as the Sclerotome Inducer

There are two major issues to be concerned about whether *Shh* is the principal sclerotome inducer.

A. The Diffusibility of *Shh*

Shh encodes a 45-kDa secreted protein that is autoproteolytically processed into an N-terminal 19-kDa protein (Shh-N) and a C-terminal 26-kDa protein (Shh-C); the processing activity lies within the C-terminal 26-kDa portion of the precursor (Lee *et al.*, 1994; Porter *et al.*, 1995). Crystal structures of both processed mature

3. Sclerotome Induction and Differentiation 107

polypeptides have been obtained and implicated to contain evolutionarily conserved structural modules with other proteolytic enzymes (Hall *et al.*, 1995, 1997). During processing, the extreme C-terminal cystein residue of Shh-N acquires a cholesterol moiety through its sulfide group and anchors Shh-N on the plasma membrane (Porter *et al.*, 1996). A paradox thus arises: how does a membrane-anchored protein function as a long-range diffusible factor? Since the bacterially produced and purified Shh-N protein, which has undergone neither proteolytic processing nor cholesterol-modification, can induce sclerotome-specific gene expression in cultured presomitic mesoderm at a reasonably low concentration (25 ng/ml) in the presence of cyclohexamide, and since cultured cells expressing properly processed Shh-N can induce sclerotome marker expression in the presomitic mesoderm over a long distance across a filter barrier without direct contact (Fan *et al.*, 1995), one must speculate that an active "release" mechanism must be employed to produce diffusible form(s) of Shh-N. Interestingly, Shh-N fused with a FGF-receptor-R1 transmembrane domain can only act locally and does not function across a filter barrier, whereas processed and cholesterol-linked Shh-N can act over a long distance when expressed by cultured cells (C.-M. Fan, personal communication, 1998). These results further strengthen the hypothesis that Shh-N can act on the sclerotome over a long distance and that the cholesterol-linked Shh-N must somehow become "transported" over some distance.

B. *Shh* Mutant Mouse Embryos Develop with Mesenchymal Sclerotome, but without Vertebrae

Shh knockout mice lack the entire vertebral column except for four or five rib cartilages (Chiang *et al.*, 1996). Although superficially the vertebraless phenotype of *Shh* mutant embryos supports *Shh* being the candidate sclerotome inducer, close examination of early stages of the sclerotome development indicates that sclerotome mesenchyme initially forms normally (Chiang *et al.*, 1996). The fact that remnants of ribs develop is consistent with the initiation of the sclerotomal program, however, the sclerotome is smaller in size and the level of *Pax-1* mRNA expression is drastically reduced. Importantly, the minimal *Pax-1* expression is activated in the somite with correct timing compared to that of the wild-type animal. The failure to maintain *Pax-1* expression and the eventual loss of the sclerotome derivatives in *Shh* knockout embryos makes it likely that *Shh* does not initiate the sclerotome program *in vivo*. Instead, *Shh* appears to be necessary for the survival of the sclerotomal component, as well as other somitic tissues (Teillet *et al.*, 1998). As discussed previously, sclerotome and the subsequent vertebral cartilages do not develop in embryos in which the neural tube and notochord have been removed (Teillet and Le Douarin, 1983; Rong *et al.*, 1992). Heterologous cells expressing *Shh* can rescue these embryos, preventing the somitic cell death and loss of *Pax-1* mRNA expression normally seen in these embryos, and even restoring normal axial skeletal development (Teillet et al., 1998). Thus, the apparent inductive ability of *Shh* may simply be an effect of allowing cells to survive long enough to express *Pax-1; Shh* may also be important for the proliferation of sclerotome cells (Johnson et al., 1994; Fan et al., 1995; Teillet et al., 1998).

IX. Initiation of the Sclerotome

What then is the initiating factor for sclerotome development? One potential sclerotome initiation factor in the mouse is the Noggin protein (McMahon et al., 1998). Noggin is characterized as a secreted dorsalizing factor that can promote the differentiation of both neural tissue and mesoderm in *Xenopus* embryos (Smith and Harland, 1992). The mouse *Noggin* gene is expressed in the notochord at the time of sclerotome formation (McMahon et al., 1998). Homozygous *Noggin* mutant embryos display delayed *Pax-1* gene expression (delayed until approximately stage X somite in contrast to its expression in stage II–III somites in wild-type embryos) and loss of caudal vertebrae (McMahon et al., 1998). Importantly, purified Noggin protein can induce minimal *Pax-1* gene expression in cultured presomitic mesoderm. The concentration of Noggin required to activate *Pax-1* in the presomitic mesoderm *in vitro* is higher than expected from conventional growth factors (100 ng/ml versus 25 ng/ml of Shh-N). Noggin appears to induce *Pax-1* expression independent of Shh-N since in the presence of treatments that abolish the activity of Shh-N, (a function blocking antibody or increased cAMP levels), Noggin can still induce *Pax-1* gene expression in presomitic mesoderm (McMahon et al., 1998). Importantly, a threshold (above 100 ng/ml) but not a linear response curve was observed with increasing concentrations of Noggin applied in culture, consistent with a mode of action that antagonizes an inhibitor of sclerotome differentiation. In contrast, sclerotomal marker expression in the presomitic mesoderm is activated by Shh-N in a linear fashion when the Shh-N concentration is above 25 ng/ml. Although these results would suggest independent parallel pathways in sclerotome induction mediated by *Noggin* and *Shh*, a subthreshold concentration of Noggin (25–50 ng/ml) stimulates the activity of Shh-N in activating *Pax-1* gene expression by approximately 4-fold, suggesting a synergism between the two molecules (McMahon et al., 1998, and see below).

Several facts lead to the speculation that, *in vivo*, *Noggin* acts to initiate low levels of *Pax-1* expression and that Shh-N acts to augment and maintain the level of *Pax 1* expression initiated by *Noggin:* (1) both *Noggin* and *Shh* are expressed in the notochord at the time of sclerotome formation; (2) Noggin alone can induce low levels of *Pax-1* expression; (3) *Noggin* mutant embryos show delayed *Pax-1* expression in the somites; (4) *Shh* mutant embryos display minimal initial *Pax-1* expression but fail to maintain *Pax-1* expression; and (5) Noggin and Shh-N can synergize to stimulate high levels of *Pax-1* expression. In the absence of *Noggin* (*Noggin* mutant embryos), *Shh* can eventually activate the sclerotomal program when its concentration has accumulated to high enough levels *in vivo*, hence the compensatory delayed sclerotome development; keep in mind, however, that *Shh* initiation of the program is not sufficient for normal chondrogenesis because the *Noggin* mutant lacks some vertebrae. On the other hand, in the absence of *Shh* (*Shh* mutant embryos), *Noggin* can initiate minimal *Pax-1* expression and sclerotome mesenchyme formation, however, it alone cannot sustain the sclerotomal program and thus the eventual loss of vertebral development in these embryos. It would be interesting to breed the two mutant mice and see if the effects on vertebral development exhibited by each mutation alone are more severe in the double mutant.

How could these two seemingly independent pathways synergize with each other? The most likely intermediates would be the bone morphogenetic protein (BMP) family of proteins. Noggin is known to antagonize BMP4 via direct binding and preventing BMP4 to interact with its receptors (Zimmerman *et al.*, 1996). *In vitro*, BMP2 and BMP4 can inhibit the sclerotome induction mediated by Noggin, consistent with their antagonistic partnership (McMahon *et al.*, 1998). Interestingly, BMP2 and BMP4 can also counteract the sclerotome inducing activity of Shh-N, as well as the combination of Noggin and Shh-N. Since there are BMP family members expressed in the vicinity of the developing somites (Pourquié *et al.*, 1996; Marcelle *et al.*, 1997), it is reasonable to speculate that BMPs are inhibitors of sclerotome differentiation. The role of *Noggin* would then be to counter the inhibitory effect of BMPs within the somite, thus allowing somites to receive/respond to further patterning influences.

X. *Shh, Noggin,* and Somite Chondrogenesis

It is possible that *Shh* and *Noggin* may play roles in somite chondrogenesis. However, extensive trials with various concentrations and application times of Shh-N alone *in vitro* showed that it is not sufficient to promote sclerotomal cells to further differentiate into cartilage (C.-M. Fan, personal communication, 1998). Even in the presence of high levels of both Noggin and Shh-N, no alcian blue positive cartilage cells are observed *in vitro*. It is intriguing that when complemented with chick embryo extracts, Noggin and Shh-N can induce chondrogenesis *in vitro* (C.-M. Fan, personal communication, 1998). This result strongly argues for the presence of additional trophic factors or inducing molecules within the extract (and thus probably *in vivo* as well) that can promote somite chondrogenesis.

The molecular identities of these inducers of somite chondrogenesis are largely unknown. As discussed previously, a somite chondrogenic inducing activity can be partially purified from neural tube extracts (but was never fully purified and identified) and basement membrane or extracellular matrix proteins can elicit somite chondrogenesis *in vitro*. The notochord and neural tube appear to have slightly different roles in sclerotome development as shown by *in vivo* experiments (although both are capable of supporting chondrogenesis in somites that have already formed the mesenchymous sclerotome) and have different inducing actions on somite chondrogenesis *in vitro* in terms of the types and positions of nodules formed. While the hypothesis has been put forth that these differences simply reflect differences in the amount of substances provided by the two tissues (Flower and Grobstein, 1967) it also seems likely that both the neural tube and notochord have other trophic factors, additional inducing factors, or extracellular matrix components to promote cartilage formation, and that some of these additional factors must be different between the two tissues.

XI. Bone Morphogenetic Proteins and Sclerotome Development

The origin of the name BMPs (bone morphogenetic proteins) implies a role in cartilage/bone formation; the initial activity identified with these proteins was to induce endochondral bone formation (Wozney *et al.,* 1988). Yet, thus far in *in vitro* assay (McMahon *et al.,* 1998) as well as in *in ovo* assay (Monsoro-Burq *et al.,* 1996), BMPs appear to counteract ventral sclerotome differentiation. However, recent results implicate BMPs (particularly BMP5) in positively regulating the morphogenesis of specific cartilage structures (King *et al.,* 1994, 1996; Kingsley, 1994). Essentially nothing is known about how the vertebrae assume their final shape. It is thought that the axial identity of a vertebra (which ultimately determines its final shape) is established by the time that the presegmental plate is formed (Kieny *et al.,* 1972), and that axial level specification is mediated by the *Hox* gene family (reviewed in Maconochie *et al.,* 1996). How this gene expression is translated into the final shape of the vertebrae at each level is unknown. It may be that BMPs are important molecules in the final morphogenetic steps.

Bone morphogenetic proteins may be playing different roles in sclerotome development at different times during embryogenesis. Early, BMPs appear to be inhibitors of ventral sclerotome development, but later they may be involved in the final morphogenesis of the vertebrae. It is of course possible that each BMP family member will display distinct activities, some inhibiting and some promoting sclerotome formation and somite chondrogenesis. There are also likely to be non-BMP family members at work as well. Investigation into the possible role(s) of BMPs in sclerotome development is really just beginning and until more late expressed markers and better assay systems are developed, the role of BMPs in sclerotome development remains uncertain.

XII. Determination of the Sclerotome

Determination of a tissue is usually tested by removing the tissue from its normal embryonic environment and assaying its subsequent differentiation (Slack, 1983).

If the test tissue differentiates true to its normal embryonic fate, the tissue is considered to have been determined at the time of its removal. Many of the early experiments discussed in this chapter involving whole somite removal could be interpreted in terms of sclerotome determination to the cartilage fate.

In many cases, somites of older age would form cartilage when placed by themselves into culture, indicating that these somites were determined for cartilage formation at the time that they were removed from the embryo (assuming that the cartilage formed in these assays was from the sclerotomal cells). Experiments rotating whole somites about their dorsoventral axis demonstrated that somites at stage I or II were capable of reorienting and had formed tissues with normal dorsoventral relationships by 24 hr postoperation, but somites rotated at stage III produced a mesenchyme dorsolateral to the normally oriented dermomyotome and myotome (Aoyama and Asamoto, 1988). This mesenchyme was interpreted as an ectopic sclerotome, although no marker expression studies were done to confirm this. This result was interpreted to mean that early on, the ventral cells of the somite are undetermined, but that by the time the somite has reached stage III, these cells have become determined. However, these experiments, as well as the early chorioallantoic membrane grafting and *in vitro* experiments, do not directly test the determination state of the sclerotome itself because they were performed in a whole somite context, nor did they examine the expression of sclerotome specific markers.

There is a series of more recent experiments examining the determination of the sclerotome alone (Wachtler *et al.*, 1982; Christ *et al.*, 1992; Aoyama, 1993; Dockter and Ordahl, 1998). Aoyama (1993) and Christ *et al.* (1992) confirmed, in part, the whole somite results of Aoyama and Asamoto (1988) by showing that the ventral half or a portion of the ventral half of a stage I somite grafted into a dorsal position is capable of forming the dorsal structures of myotome and dermomyotome, indicating that these cells, which will eventually form sclerotome, are undetermined at the time of somite formation.

Experiments by Wachtler *et al.* (1982) tested actual sclerotome mesenchyme (not just the epithelial ventral cells) from various stage somites by placing the grafts into the developing wing bud of the chick. They found that the sclerotome would form muscle tissue (indicating that the sclerotome was undetermined) until as late as approximately somite stage XX. A series of grafting experiments placing quail sclerotome tissue into a dorsal axial position in a chick host (Fig. 9) have shown that in fact, the chick sclerotome remains undetermined up until the time that the cells begin expressing collagen II and begin differentiating as cartilage (Dockter and Ordahl, 1998). No gene expression was shown to correlate with determination. The grafts should all have been expressing *Pax-1, Pax-9, scleraxis,* and *twist* at the time of grafting, yet fate changes occurred, demonstrating that none of these markers is indicative of, or is sufficient for, determination of sclerotome to cartilage. In all of these experiments, the actual determination state of individual cells within the sclerotome cannot be ascertained; as stated before, the

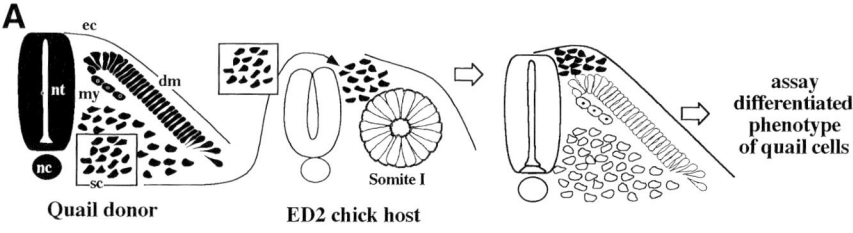

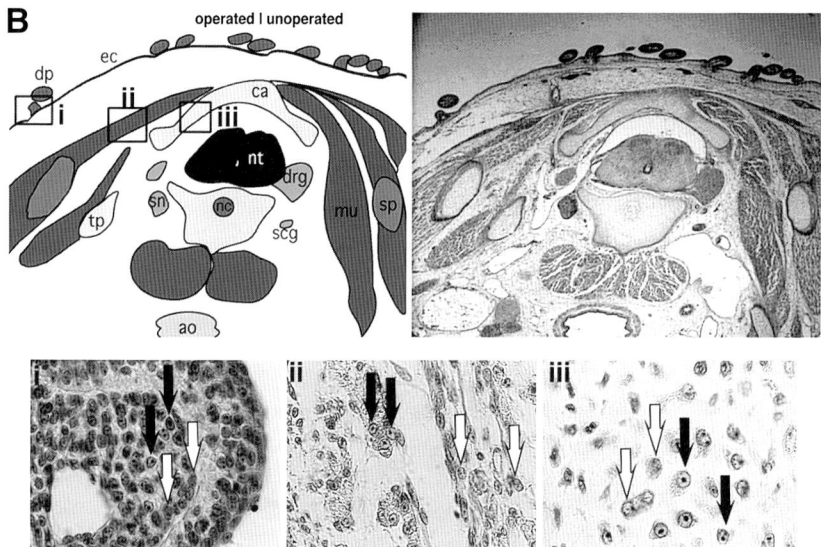

Figure 9 An *in vivo* assay for sclerotome determination. (A) Schematic representation of the experimental design of the dorsal challenge assay for sclerotome determination. All embryos are represented as cross sections. A sclerotome fragment (boxed area) is removed from a quail donor embryo. The donor sclerotome fragment is then placed into an ED2 chick host, underneath the ectoderm and between the neural tube and the stage I somite. The ectoderm heals over the graft, which is now in a dorsal signaling environment. The chimeric embryos are allowed to develop for various periods of time and the differentiated phenotype of the quail cells is analyzed. dm, Dermomyotome; ec, ectoderm; my, myotome; nc, notochord; nt, neural tube; sc, sclerotome. (B) Camera lucida drawing and low power view (5× objective view) of a cross section of a Feulgen stained chimeric chick embryo harvested at ED12, 10 days postgrafting of a quail somite stage XX sclerotome fragment underneath the ectoderm, between the neural tube and stage I somite of an ED2 chick host. The operated side is the left side; note that the operated and unoperated sides are morphologically indistinguishable. The boxes delineate areas containing quail nucleoli shown at greater magnification in panels i, ii, and iii. ao, Aorta; ca, cartilaginous vertebra; dp, dermal papilla; drg, dorsal root ganglion; ec, ectoderm; mu, muscle; nc, notochord; nt, neural tube; scg, sympathetic chain ganglion; sn, spinal nerve; sp, scapula; tp, transverse process. (i–iii) High power (100× objective) views of a dermal papilla (i), muscle fibers (ii), and vertebral cartilage (iii). The darkly staining quail nucleoli can be seen fully integrated into each of these tissues and intermingled with host chick nuclei. Black arrows mark quail nucleoli; white arrows mark chick nuclei. (Reprinted from Dockter and Ordahl, 1998, with permission.)

early embryo may contain some determined chondrogenic precursors before somite formation has occurred (George-Weinstein *et al.*, 1994, 1996).

XIII. Epithelial–Mesenchymal Transitions and Sclerotome Formation

The observation that *Pax-1* expression is initiated in the ventral somite preceding sclerotome mesenchyme formation (Brand-Saberi *et al.*, 1993) suggests that the sclerotomal program can be initiated without going through the epithelial state. Since epithelialization is not observed in paraxial mesoderm cultures *in vitro*, it is presumed that epithelialization of the somite is not absolutely required for sclerotome induction; a conclusion supported by the *paraxis* knockout mouse in which somite epithelialization does not occur, but *Pax-1* mRNA expression does (Burgess *et al.*, 1996; Barnes *et al.*, 1997). This idea is further supported by the analysis of a mouse mutant in which *Dll1*, the mouse homolog of the *Drosophila delta* gene, is inactivated. No somitic epithelium is observed, yet *Pax-9* mRNA is expressed (Hrabe de Angelis *et al.*, 1997). However, these *Dll1* −/− somites appear to develop without the cranial–caudal differences in cell density normally seen in the sclerotome. Together these mutants indicate that although the passage through an epithelial state is not required for sclerotome induction it is likely that a close association between segmentation, epithelialization, and differentiation programs is coordinated for proper allocation of segment specific cells into each somite and the proper spatial organization of the cell types within each somite to assure a properly segmented, organized body plan along the anterior–posterior axis.

XIV. Dorsal–Ventral Patterning of the Somite

How does the somite become patterned along the dorsoventral axis? From previous discussions in this chapter, the ventral neural tube and notochord appear to be the major influences on the ventral cells of the somite while the ectoderm and dorsal neural tube (Spence *et al.*, 1996) play roles in formation of the dorsal derivatives. As discussed above, *Shh* and *Noggin* appear to be notochord produced molecules that induce initial sclerotome formation as determined by marker expression. Consistent with that result, *Shh* has been shown to suppress dermomyotome marker expression in presegmental mesoderm explants (Fan and Tessier-Lavigne, 1994) and *in vivo* (Johnson *et al.*, 1994; Marcelle *et al.*, 1997). *Shh* has also been shown to be capable of inducing muscle marker expression in the somite (Münsterberg *et al.*, 1995), implicating it in myotome formation. What other molecules are patterning the somite?

The other molecules implicated in dorsoventral patterning of somites are the BMPs and the *Wnts* (Fan *et al.*, 1997; Marcelle *et al.*, 1997; Capdevila *et al.*, 1998;

Reshef et al., 1998). Direct application of BMP recombinant proteins in presomitic mesoderm explants does not lead to the activation of dermomyotomal markers (e.g., *Pax-3, Pax-7,* and *Sim-1* mRNAs; Fan *et al.,* 1997), indicating that BMPs alone cannot be responsible for patterning the dorsal somite; there is also evidence that BMPs inhibit the development of the myotome from the dermomyotome in the medial somite (Pourquié *et al.,* 1996; Marcelle *et al.,* 1997). *Noggin,* which is expressed in the dermomyotome (Marcelle *et al.,* 1997), would then be presumed to have the role of inhibiting the direct action of BMP on the dermomyotome (Reshef *et al.,* 1998).

It appears that the role of BMP in the dorsal somite is to induce the *Wnt* family of genes in the dorsal neural tube (Marcelle *et al.,* 1997), which are then the direct dorsalizers of the somite. Heterologous cells expressing Wnt proteins can induce dermomyotome-specific marker expression in presomitic mesoderm explants (Fan *et al.,* 1997). Ectopic expression of *Wnt-1 in vivo* expands dorsal somite marker expression, while suppressing sclerotome marker expression (Marcelle *et al.,* 1997; Capdevila *et al.,* 1998). Interestingly, *Wnt-1* also appears to be capable of inducing *Noggin* in the somite (Reshef *et al.,* 1998).

The current model then (Fig. 10) for dorsoventral patterning of the somite is that BMPs produced in the dorsal neural tube induce *Wnts* in the dorsal neural tube which then induce/maintain the dermomyotome, likely by inducing other Wnt family members in the dorsal lip of the dermomyotome (Marcelle *et al.,* 1997), and suppress the sclerotome fate. *Shh* and *Noggin* produced by the neural tube and notochord counteract BMP functions ventrally to promote sclerotome development and repress dermomyotome development. Wnt proteins and Shh promote myotome formation from the dermomyotome (Stern *et al.,* 1995). Myotome formation is suppressed by BMPs (Pourquié *et al.,* 1996), and so *Shh* and *Wnts* (and possibly *Noggin*) act as counteragents to relieve the repression effect of BMPs and allow dermomyotomal cells to proceed to myotomal fate. How these pathways might interact intracellularly is not understood.

XV. Interactions between the Somitic Components

Most studies to date have focused on the importance and influence of the axial organs on sclerotome development. However, the skeletal elements must eventually coordinate with the muscular derivatives of the somite to form the functional unit of each vertebral segment. While myotome produced members of the fibroblast growth factor family have been postulated to be necessary for the development of rib chondrocytes (Grass *et al.,* 1996), essentially nothing is known about how the other compartments of the somite influence sclerotome development. It would be very surprising if there was not important cross talk between the dermomyotome, myotome, and sclerotome.

3. Sclerotome Induction and Differentiation

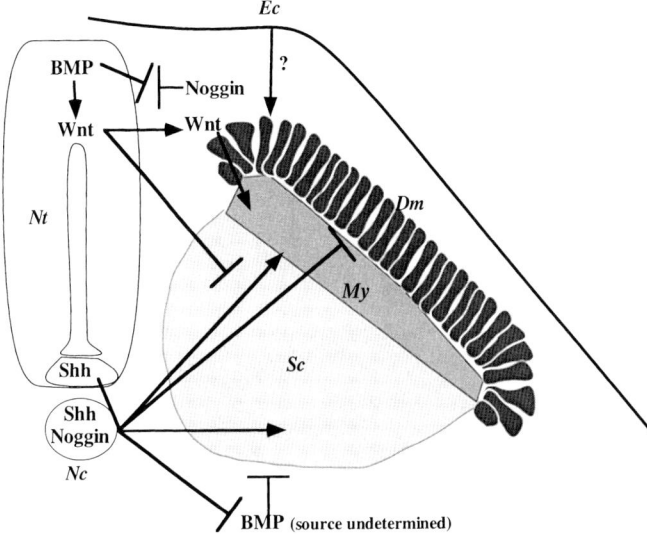

Figure 10 A model of interactions in dorsal–ventral patterning of the somite. A schematic cross section of a representative chick embryo somite. Arrows indicate positive inductions, while lines with bars represent inhibitory actions. Dorsal is at the top of the figure. The ectoderm is known to promote dermomyotome formation, but the nature of the signal(s) is unknown (?). BMPs produced in the dorsal neural tube induce Wnts in the dorsal neural tube that act on the dorsal lip of the dermomyotome to induce Wnt-11, which then acts to promote myotome formation (arrow). BMPs are known to inhibit myogenesis, but Noggin produced by the dorsal dermomyotome lip is thought to block that inhibition. Wnts (probably from the neural tube) also appear to inhibit sclerotome formation. The floor plate of the neural tube and the notochord produce Sonic hedgehog (Shh), which promotes myotome formation and inhibits dermomyotome. Together with Noggin produced by the notochord, Shh promotes sclerotome and may counteract Wnts and BMPs that would inhibit sclerotome formation. The embryonic source of sclerotome inhibiting BMPs has not been determined. (This figure was compiled from data in Münsterberg *et al.*, 1995; Pourquié *et al.*, 1996; Fan *et al.*, 1997; Marcelle *et al.*, 1997; Capedevila *et al.*, 1998; Reshef *et al.*, 1998.) Dm, Dermomyotome; Ec, ectoderm; My, myotome, Nc, notochord; Nt, neural tube; Sc, sclerotome.

XVI. Discussion

This chapter has focused on the tissues and molecules which appear to induce events in sclerotome development. It is important to consider what induction really means with regard to the sclerotome (Lash, 1963b, 1968a; Holtzer, 1968) and the fact that we do not understand the mechanisms underlying these events. As discussed above, the neural tube and notochord appear to induce sclerotome formation and somite chondrogenesis. Two molecules, *Shh* and *Noggin,* have been identified as the possible mediators of sclerotome induction by these tissues (Fan and Tessier-Lavigne, 1994; McMahon *et al.*, 1998). The simple conclusion would be that the notochord and neural tube produce Shh and Noggin proteins, which are received by the ventral somite cells, the result of which is that these cells turn on sclerotome markers (instructive induction). It has been argued however, that because manipulating *in vitro* culture conditions can eliminate the need for the presence of spinal cord or notochord to elicit somite chondrogenesis from young somites (Ellison *et al.*, 1969; Ellison and Lash, 1971), somites are determined early for cartilage development and that the role of the neural tube and notochord is to permit the somite to form the cartilage that it was already programmed to do (permissive induction). Consistent with this hypothesis, the chick notochord and *Shh* appear to be important for the survival of somite cells (Gordon and Lash, 1974; Fan *et al.*, 1995; Teillet *et al.*, 1998). It could be that what we interpret as induction by the notochord (the same could be true for the neural tube) is simply the notochord keeping already determined sclerotome cells alive long enough, or causing enough proliferation for them to form cartilage (George-Weinstein *et al.*, 1988, 1994, 1996), although it has not been ruled out that *Shh* has a more direct role in *Pax-1* expression.

It has also been suggested that the role of the axial tissues and/or endo- and ectoderm is actually a mechanical one, stabilizing the cells of the somite to allow extracellular matrix production, which then allows for differentiation (Seno and Büyüközer, 1958; Ellison *et al.*, 1969; Kenny-Mobbs and Thorogood, 1987), which would be another type of permissive induction. These differing proposed mechanisms (direct effects on gene expression, promotion of survival, and mechanical stabilization) for induction are not mutually exclusive; the induction of sclerotome or somite chondrogenesis could be the result of the combination of all of these mechanisms.

One concern about any inductive event is whether the defined active concentrations of an inducing factor in an experimental situation are relevant to the normal *in vivo* concentrations of these factors. In the presence of complex extracellular matrix influences and various antagonistic and synergistic signals, the operating concentration of a specific growth factor in the embryonic environment is likely quite different from that provided in an experiment. This is of particular concern when the diffusibility of a molecule is called into question. For example, Shh-N is conjugated to cholesterol and becomes membrane anchored (Porter *et al.*, 1996), and the Wnt family of proteins is notorious for associating with extracellular matrix and/or membrane glycoproteins (Smolich *et al.*, 1993). Techniques for the *in vivo* measurement of bioactive molecules are often unavailable, making it almost impossible to make meaningful comparisons. Therefore, the reputed inducing abilities and functions of molecules should always be considered with a dose of skepticism.

The functional relevance of the activation or expression of induced sclerotomal molecular markers also remains to be investigated. For example, *Pax-1* and *M-twist* are two genes known to be required for correct sclerotome development, yet the exact point(s) or step(s) at which they function within the sclerotome/cartilage developmental program are not yet defined. Furthermore, the relevance of levels of marker expression obtained in induction experiments to events *in vivo* is not clear. Thus, the activation of *Pax-1*, *M-twist*, or *Pax-9* gene expression, although strongly suggestive that at least part of the sclerotome program is initiated, by no means proves that the entire sclerotome program is set in motion, an idea borne out by the phenotypes of mouse mutants (Chiang *et al.*, 1996; McMahon *et al.*, 1998) and studies on the determination of sclerotome (Wachtler *et al.*, 1982; Aoyama and Asamoto, 1988; Christ *et al.*, 1992; Aoyama, 1993; Dockter and Ordahl, 1998).

This discussion is meant to illustrate that the fundamental nature of inductive events in sclerotome development eludes our understanding. Very little is understood about the specific processes, both intracellular and extracellular, that underlie the basic mechanics of sclerotome development. While some molecular players in sclerotome induction have been identified, clearly there are more factors and more inductive steps required for normal sclerotome development than we currently know of.

XVII. Summary

Inductive events in the development of the sclerotome and their possible underlying mechanisms were reviewed from the primary literature. A brief review of morphological and anatomical aspects of sclerotome development was given. The importance of the notochord and neural tube in sclerotome induction and somite chondrogenesis *in vivo* and *in vitro* was established. The functions and patterns of expression of different sclerotome markers were discussed. *Shh* and *Noggin* were discussed as two molecules produced by the neural tube and notochord that appear to maintain and initiate the sclerotome, respectively. While the abilities of the axial organs and *Shh* and *Noggin* to induce sclerotome marker expression in the somite was not disputed, the exact nature of these inductions was discussed with regard to possible effects on gene expression, effects on cell survival, and physical effects on the cells and it was argued that the fundamental nature of inductive events in the sclerotome is still unknown.

Acknowledgments

The author would like to acknowledge the guidance of Dr. Charles Ordahl, in whose laboratory this chapter was written. She would also like to acknowledge Dr. James Lash, Dr. Bodo Christ, Dr. Frank Stockdale, and Dr. Sarah Larkin for critical readings of the manuscript. She would also like to thank Dr. Chen-Ming Fan for allowing citation of unpublished results and for written contributions. The author would also like to acknowledge The Company of Biologists Ltd., Academic Press, Elsevier Science, Wiley-Liss Inc., and Cell Press for figure reprint permissions.

References

Aoyama, H. (1993). Developmental plasticity of the prospective dermatome and the prospective sclerotome region of an avian somite. *Dev. Growth Differ.* **35**, 507–519.

Aoyama, H., and Asamoto, K. (1988). Determination of somite cells: Independence of cell differentiation and morphogenesis. *Development* **104**, 15–28.

Avery, G., Chow, M., and Holtzer, H. (1955). Comparison of the salamander and chick notochords in the differentiation of somitic tissues. *Anat. Rec.* **122**, 444.

Avery, G., Chow, M., and Holtzer, H. (1956). An experimental analysis of the development of the spinal column. V. Reactivity of chick somites. *J. Exp. Zool.* **132**, 409–426.

Bagnall, K. M., and Sanders, E. J. (1989). The binding pattern of peanut lectin associated with sclerotome migration and the formation of the vertebral axis in the chick embryo. *Anat. Embryol.* **180**, 505–513.

Bagnall, K. M., Higgins, S. J., and Sanders, E. J. (1988). The contribution made by a single somite to the vertebral column: Experimental evidence in support of resegmentation using the chick–quail chimera model. *Development* **103**, 69–85.

Balling, R., Deutsch, U., and Gruss, P. (1988). *undulated,* a mutation affecting the development of the mouse skeleton, has a point mutation in the paired box of *Pax 1*. *Cell* **55**, 531–535.

Barnes, G. L., Alexander, P. G., Hsu, C. W., Mariani, B. D., and Tuan, R. S. (1997). Cloning and characterization of chicken *Paraxis:* A regulator of paraxial mesoderm development and somite formation. *Dev. Biol.* **189**, 95–111.

Belsky, E., Vasan, N. S., and Lash, J. W. (1980). Extracellular matrix components and somite chondrogenesis: A microscopic analysis. *Dev. Biol.* **79**, 159–180.

Bennett, D. (1958). *In vitro* study of cartilage induction in *T/T* mice. *Nature* **181**, 1286.

Borycki, A. G., Strunk, K. E., Savary, R., and Emerson, C. P. (1997). Distinct signal/response mechanisms regulate *pax1* and *QmyoD* activation in sclerotomal and myotomal lineages of quail somites. *Dev. Biol.* **185**, 185–200.

Brand-Saberi, B., Ebensperger, C., Wilting, J., Balling, R., and Christ, B. (1993). The ventralizing effect of the notochord on somite differentiation in chick embryos. *Anat. Embryol.* **188**, 239–245.

Bronner-Fraser, M. (1986). Analysis of the early stages of trunk neural crest migration in avian embryos using monoclonal antibody HNK-1. *Dev. Biol.* **115**, 44–55.

Burgess, R., Rawls, A., Brown, D., Bradley, A., and Olson, E. N. (1996). Requirement of the *paraxis* gene for somite formation and musculoskeletal patterning. *Nature* **384**, 570–573.

Capdevila, J., Tabin, C., and Johnson, R. L. (1998). Control of dorsoventral somite patterning by Wnt-1 and β-catenin. *Dev. Biol.* **193**, 182–194.

Chang, D. T., Lopez, A., von Kessler, D. P., Chiang, C., Simandel, B. K., Zhao, R., Seldon, M. F., Fallon, J. F., and Beachy, P. A. (1994). Products, genetic linkage and limb patterning activity of a mouse *hedgehog* gene. *Development* **120**, 3339–3353.

Cheah, K. S. E., Lau, E. T., Au, P. K. C., and Tam, P. P. L. (1991). Expression of the mouse α1(II) collagen gene is not restricted to cartilage during development. *Development.* **111**, 945–953.

Chen, Z.-F., and Behringer, R. R. (1995). *twist* is required in head mesenchyme for cranial neural tube morphogenesis. *Genes Dev.* **9**, 686–699.

Cheney, C. M., and Lash, J. W. (1981). Diversification within embryonic chick somites: Differential response to notochord.*Dev. Biol.* **81**, 288–298.

Chernoff, E. A. G., and Lash, J. W. (1981). Cell movement in somite formation and development in the chick: Inhibition of segmentation. *Dev. Biol.* **87**, 212–219.

Chevallier, A. (1975). Role du mesoderme somitique dans le developpement de la cage thoracique de l'embryon d'oiseau. I. Origine du segment sternal et mecanismes de la differenciation des cotes. *J. Embryol. Exp. Morphol* **33**, 291–311.

Chiang, C., Litingtung, Y., Lee, E., Young, K. E., Croden, J. L., Westphal, H., and Beachy, P. A.

3. Sclerotome Induction and Differentiation

(1996). Cyclopia and defective axial patterning in mice lacking *Sonic hedgehog* gene function. *Nature* **383**, 407–413.
Christ, B., and Ordahl, C. P. (1995). Early stages of chick somite development. *Anat. Embryol.* **191**, 381–396.
Christ, B., and Wilting, J. (1992). From somites to vertebral column. *Ann. Anat.* **174**, 23–32.
Christ, B., Jacob, H. J., and Jacob, M. (1974). Uber den ursprung der flugelmuskulatur. Experimentelle untersuchungen mit wachtel- und huhnerembryonen. *Experientia* **30**, 1446–1449.
Christ, B., Jacob, H. J., and Jacob, M. (1977). Experimental analysis of the origin of the wing musculature in avian embryos. *Anat. Embryol.* **150**, 171–186.
Christ, B., Jacob, H. J., and Jacob, M. (1978). On the formation of the myotomes in avian embryos. An experimental and scanning electron microscope study. *Experientia* **34**, 514–516.
Christ, B., Brand-Saberi, B., Grim, M., and Wilting, J. (1992). Local signalling in dermomyotomal cell type specification. *Anat. Embryol.* **186**, 505–510.
Cohen, A. M., and Hay, E. D. (1971). Secretion of collagen by embryonic neuroepithelium at the time of spinal cord–somite interaction. *Dev. Biol.* **26**, 578–605.
Cooper, G. W. (1965). Induction of somite chondrogenesis by cartilage and notochord: A correlation between inductive activity and specific stages of cytodifferentiation. *Dev. Biol.* **12**, 185–212.
Cserjesi, P., Brown, D., Ligon, K. L., Lyons, G. E., Copeland, N., Gilbert, D. J., Jenkins, N. A., and Olson, E. N. (1995). Scleraxis: A basic helix-loop-helix protein that prefigures skeletal formation during mouse embryogenesis. *Development* **121**, 1099–1110.
Dahl, E., Koseki, H., and Balling, R. (1997). *Pax* genes and organogenesis. *BioEssays* **19**, 755–765.
Dawes, B. (1930). The development of the vertebral column in mammals, as illustrated by its development in *Mus musculus*. *Philos. Trans. R. Soc. London (Biol.)* **218**, 115–170.
Denetclaw, W. F., Christ, B., and Ordahl, C. P. (1997). Location and growth of epaxial myotome precursor cells. *Development* **124**, 1601–1610.
Detwiler, S. R., and Holtzer, H. (1956). The developmental dependence of the vertebral column upon the spinal cord in the urodeles. *J. Exp. Zool.* **132**, 299–310.
Deutsch, U., Dressler, G. R., and Gruss, P. (1988). *Pax 1*, a member of a paired box homologous murine gene family, is expressed in segmented structures during development. *Cell* **53**, 617–625.
Dietrich, S., and Gruss, P. (1995). *undulated* phenotypes suggest a role of *Pax-1* for the development of vertebral and extravertebral structures. *Dev. Biol.* **167**, 529–548.
Dietrich, S., Schubert, F. R., and Gruss, P. (1993). Altered *Pax* gene expression in mouse notochord mutants: The notochord is required to initiate and maintain ventral identity in the somite. *Mech. Dev.* **44**, 189–207.
Dietrich, S., Schubert, F. R., and Lumsden, A. (1997). Control of dorsoventral pattern in the chick paraxial mesoderm. *Development* **124**, 3895–3908.
Dockter, J. L., and Ordahl, C. P. (1998). Determination of sclerotome to the cartilage fate. *Development* **125**, 2113–2124.
Ebensperger, C., Wilting, J., Brand-Saberi, B., Mizutani, Y., Christ, B., Balling, R., and Koseki, H. (1995). *Pax 1*, a regulator of sclerotome development is induced by notochord and floor plate signals in avian embryos. *Anat. Embryol.* **191**, 297–310.
Echelard, Y., Epstein, D. J., St.-Jacques, B., Shen, L., Mohler, J., McMahon, J. A., and McMahon, A. (1993). Sonic hedgehog, a member of a family of putative signaling molecules, is implicated in the regulation of CNS polarity. *Cell* **75**, 1417–1430.
Ellison, M. L., and Lash, J. W. (1971). Environmental enhancement of *in vitro* chondrogenesis. *Dev. Biol.* **26**, 486–496.
Ellison, M. L., Ambrose, E. J., and Easty, G. C. (1969). Chondrogenesis in chick embryo somites *in vitro*. *J. Embryol. Exp. Morphol.* **21**, 331–340.
Fan, C.-M., and Tessier-Lavigne, M. (1994). Patterning of mammalian somites by surface ectoderm and notochord: Evidence for sclerotome induction by a hedgehog homolog. *Cell* **79**, 1175–1186.
Fan, C.-M., Porter, J. A., Chiang, C., Chang, D. T., Beachy, P. A., and Tessier-Lavigne, M. (1995).

Long-range sclerotome induction by sonic hedgehog: Direct role of the amino-terminal cleavage product and modulation by the cyclic AMP signaling pathway. *Cell* **81**, 457–465.
Fan, C.-M., Lee, C. S., and Tessier-Lavigne, M. (1997). A role for WNT proteins in induction of dermomyotome. *Dev. Biol.* **191**, 160–165.
Flower, M., and Grobstein, C. (1967). Interconvertibility of induced morphogenetic responses of mouse embryonic somites to notochord and ventral spinal cord. *Dev. Biol.* **15**, 193–205.
Fowler, I., and Watterson, R. L. (1953). The role of the neural tube in development of the axial skeleton of the chick. *Anat. Rec.* **117**, 555–556.
Füchtbauer, E. M. (1995). Expression of *M-Twist* during postimplantation development of the mouse. *Dev. Dyn.* **204**, 316–322.
George-Weinstein, M., Decker, C., and Horwitz, A. (1988). Combinations of monoclonal antibodies distinguish mesenchymal, myogenic, and chondrogenic precursors of the developing chick embryo. *Dev. Biol.* **125**, 34–50.
George-Weinstein, M., Gerhart, J. V., Foti, G. J., and Lash, J. W. (1994). Maturation of myogenic and chondrogenic cells in the presomitic mesoderm of the chick embryo. *Exp. Cell Res.* **211**, 263–274.
George-Weinstein, M., Gerhart, J., Reed, R., Flynn, J., Callihan, B., Mattiacci, M., Miehle, C., Foti, G., Lash, J. W., and Weintraub, H. (1996). Skeletal myogenesis: The preferred pathway of chick embryo epiblast cells *in vitro*. *Dev. Biol.* **173**, 279–291.
Gitelman, I. (1997). Twist protein in mouse embryogenesis. *Dev. Biol.* **189**, 205–214.
Goldstein, R. S., and Kalcheim, C. (1992). Determination of epithelial half-somites in skeletal morphogenesis. *Development* **116**, 441–445.
Gordon, J. S., and Lash, J. W. (1974). *In vitro* chondrogenesis and cell viability. *Dev. Biol.* **36**, 88–104.
Goulding, M. D., Chalepakis, G., Deutsch, U., Erselius, J. R., and Gruss, P. (1991). Pax-3, a novel murine DNA binding protein expressed during early neurogenesis. *EMBO J.* **10**, 1135–1147.
Goulding, M. D., Lumsden, A., and Paquette, A. J. (1994). Regulation of *Pax-3* expression in the dermomyotome and its role in muscle development. *Development* **120**, 957–971.
Grass, S., Arnold, H.-H., and Braun, T. (1996). Alterations in somite patterning of *Myf-5*-deficient mice: A possible role for FGF-4 and FGF-6. *Development* **122**, 141–150.
Grobstein, C. (1955). Inductive interaction in the development of the mouse metanephros. *J. Exp. Zool.* **130**, 319–339.
Grobstein, C., and Holtzer, H. (1955). *In vitro* studies of cartilage induction in mouse somite mesoderm. *J. Exp. Zool.* **128**, 333–357.
Grobstein, C., and Parker, G. (1954). *In vitro* induction of cartilage in mouse somite mesoderm by embryonic spinal cord. *Proc. Soc. Exp. Biol. Med.* **85**, 477–481.
Grüneberg, H. (1953). Genetical studies on the skeleton of the mouse. XII. The development of undulated. *J. Genet.* **52**, 441–462.
Grüneberg, H. (1958). Genetical studies on the skeleton of the mouse. XXII. The development of Danforth's short-tail. *J. Embryol. Exp. Morphol.* **6**, 124–128.
Grüneberg, H. (1963). "The Pathology of Development. A Study of Inherited Skeletal Disorders in Animals." Blackwell, Oxford.
Gurdon, J. B. (1987). Embryonic induction—molecular prospects. *Development* **99**, 285–306.
Hall, B. K. (1977). Chondrogenesis of the somitic mesoderm. *Adv. Anat. Embryol. Cell Biol.* **53**, 3–47.
Hall, B. K. (1986). Initiation of chondrogenesis from somitic, limb and craniofacial mesenchyme: Search for a common mechanism. *In* "Somites in Developing Embryos" (R. Bellairs, D. A. Ede, and J. W. Lash, Eds.), pp. 247–259. Plenum, New York.
Hall, B. K., and Miyake, T. (1995). Divide, accumulate, differentiate: Cell condensation in skeletal development revisited. *Int. J. Dev. Biol.* **39**, 881–893.
Hall, T. M. T., Porter, J. A., Beachy, P. A., and Leahy, D. J. (1995). A potential catalytic site revealed by the 1.7A crystal structure of the amino-terminal signaling domain of Sonic hedgehog. *Nature* **378**, 212–216.

Hall, T. M., Porter, J. A., Young, D. E., Koonin, E. V., Beachy, P. A., and Leahy, D. J. (1997). Crystal structure of a Hedgehog autoprocessing domain: Homology between Hedgehog and self-splicing proteins. *Cell* **91**, 85–97.
Hamburger, V., and Hamilton, H. L. (1951). A series of normal stages in the development of the chick embryo. *J. Morphrol.* **88**, 49–92.
Hammerschmidt, M., and McMahon, A. P. (1998). The effect of pertussis toxin on zebrafish development: A possible role for inhibitory G-proteins in hedgehog signaling. *Dev. Biol.* **194**, 166–171.
Hayashi, M., Ninomiya, Y., Parsons, J., Hayashi, K., Olsen, B. R., and Trelstad, R. L. (1986). Differential localization of mRNAs of collagen types I and II in chick fibroblasts, chondrocytes, and corneal cells by *in situ* hybridization using cDNA probes. *J. Cell Biol.* **102**, 2302–2309.
Healy, C., Uwanohgo, D., and Sharpe, P. T. (1996). Expression of the chicken Sox9 gene marks the onset of cartilage differentiation. *Ann. N.Y. Acad. Sci.* **785**, 261–262.
Hirano, S., Hirako, R., Kajita, N., and Norita, M. (1995). Morphological analysis of the role of the neural tube and notochord in the development of somites. *Anat. Embryol.* **192**, 445–457.
Hoadley, L. (1925). The differentiation of isolated chick primordia in chorio-allantoic grafts. II. The effect of the presence of the spinal cord, i.e., innervation, on the differentiation of the somitic region. *J. Exp. Zool.* **42**, 143–162.
Holtfreter, J. (1968). Mesenchyme and epithelia in inductive and morphogenetic processes. *In* "Epithelial-Mesenchymal Interactions" (R. Fleischmajer and R. E. Billingham, Eds.), pp. 1–30. Williams & Wilkins, Baltimore.
Holtzer, H. (1951). Morphogenetic influence of the spinal cord on the axial skeleton and musculature. *Anat. Rec.* **109**, 373–374.
Holtzer, H. (1952a). An experimental analysis of the development of the spinal column: The dispensability of the notochord. *J. Exp. Zool.* **121**, 573–591.
Holtzer, H. (1952b). An experimental analysis of the development of the spinal column. I. Response of pre-cartilage cells to size variations of the spinal cord. *J. Exp. Zool.* **121**, 121–148.
Holtzer, H. (1960). Aspects of chondrogenesis and myogenesis. *In* "Synthesis of Molecular and Cellular Structure" (D. Rudnick, Ed.), pp. 35–87. Ronald Press, New York.
Holtzer, H. (1968). Induction of chondrogenesis: A concept in quest of mechanisms. *In* "Epithelial–Mesenchymal Interactions" (R. Fleischmajer and R. E. Billingham, Eds.), pp. 152–164. Williams & Wilkins, Baltimore.
Holtzer, H., and Detwiler, S. R. (1953). An experimental analysis of the development of the spinal column. III. Induction of skeletogenous cells. *J. Exp. Zool.* **123**, 335–366.
Holtzer, H., and Detwiler, S. R. (1954). The dependence of somitic differentiation on the neural axis. *Anat. Rec.* **118**, 390.
Hommes, F. A., Van Leeuwen, G., and Zilliken, F. (1962). Induction of cell differentiation. II. The isolation of a chondrogenic factor from embryonic chick spinal cords and notochords. *Biochim. Biophys. Acta* **56**, 320–325.
Hrabe de Angelis, M., McIntyre II, J., and Gossler, A. (1997). Maintenance of somite borders in mice requires the *Delta* homologue *Dll1*. *Nature* **386**, 717–721.
Huang, R., Zhi, Q., Wilting, J., and Christ, B. (1994). The fate of somitocoele cells in avian embryos. *Anat. Embryol.* **190**, 243–250.
Jacob, M., Jacob, H. J., and Christ, B. (1975). The early differentiation of the perinotochordal connective tissue. A scanning and transmission electron microscopic study on chick embryos. *Experientia* **31**, 1083–1086.
Jiang, J., and Struhl, G. (1995). Protein kinase A and hedgehog signaling in *Drosophila* limb development. *Cell* **80**, 563–572.
Johnson, R. L., Laufer, E., Riddle, R. D., and Tabin, C. (1994). Ectopic expression of sonic hedgehog alters dorsal–ventral patterning of somites. *Cell* **79**, 1165–1174.
Kenny-Mobbs, T., and Thorogood, P. (1987). Autonomy of differentiation in avian brachial somites and the influence of adjacent tissues. *Development* **100**, 449–462.

Kieny, M., Mauger, A., and Sengel, P. (1972). Early regionalization of the somitic mesoderm as studied by the development of the axial skeleton of the chick embryo. *Dev. Biol.* **28,** 142–161.

King, J. A., Marker, P. C., Seung, K. J., and Kingsley, D. M. (1994). BMP5 and the molecular, skeletal, and soft-tissue alterations in short ear mice. *Dev. Biol.* **166,** 112–122.

King, J. A., Storm, E. E., Marker, P. C., Dileone, R. J., and Kingsley, D. M. (1996). The role of BMPs and GDFs in development of region-specific skeletal structures. *Ann. N.Y. Acad. Sci.* **785,** 70–79.

Kingsley, D. M. (1994). What do BMPs do in mammals? Clues from the mouse short-ear mutation. *Trends Genet.* **10,** 16–21.

Kitchin, I. C. (1949). The effects of notochordectomy in *Amblystoma mexicanum. J. Exp. Zool.* **112,** 393–415.

Koseki, H., Wallin, J., Wilting, J., Mizutani, Y., Kispert, A., Ebensperger, C., Herrmann, B. G., Christ, B., and Balling, R. (1993). A role for *Pax-1* as a mediator of notochordal signals during the dorsoventral specification of vertebrae. *Development* **119,** 649–660.

Kosher, R. A. (1976). Inhibition of "spontaneous," notochord-induced, and collagen-induced *in vitro* somite chondrogenesis by cyclic AMP derivatives and theophylline. *Dev. Biol.* **53,** 265–276.

Kosher, R. A., and Church, R. L. (1975). Stimulation of *in vitro* somite chondrogenesis by procollagen and collagen. *Nature* **258,** 327–330.

Kosher, R. A., and Lash, J. W. (1975). Notochordal stimulation of *in vitro* somite chondrogenesis before and after enzymatic removal of perinotochordal materials. *Dev. Biol.* **42,** 362–378.

Kosher, R. A., Lash, J. W., and Minor, R. R. (1973). Environmental enhancement of *in vitro* chondrogenesis. IV. Stimulation of somite chondrogenesis by exogenous chondromucoprotein. *Dev. Biol.* **35,** 210–220.

Krauss, S., Concordet, J.-P., and Ingham, P. W. (1993). A functionally conserved homolog of the *Drosophila* segment polarity gene hedgehog is expressed in tissues with polarizing activity in zebrafish embryos. *Cell* 1431–1444.

Kvist, T. N., and Finnegan, C. V. (1970a). The distribution of glycosaminoglycans in the axial region of the developing chick embryo. I. Histochemical analysis. *J. Exp. Zool.* **175,** 221–240.

Kvist, T. N., and Finnegan, C. V. (1970b). The distribution of glycosaminoglycans in the axial region of the developing chick embryo. II. Biochemical analysis. *J. Exp. Zool.* **175,** 241–257.

Langman, J., and Nelson, G. R. (1968). A radioautographic study of the development of the somite in the chick embryo. *J. Embryol. Exp. Morphol.* **19,** 217–226.

Lash, J. (1963a). Studies on the ability of embryonic mesonephros explants to form cartilage. *Dev. Biol.* **6,** 219–232.

Lash, J. W. (1963b). Tissue interaction and specific metabolic responses: Chondrogenic induction and differentiation. *In* "Cytodifferentiation and Macromolecular Synthesis" (M. Locke, Ed.), pp. 235–259. Academic Press, New York.

Lash, J. W. (1967). Differential behavior of anterior and posterior embryonic chick somites *in vitro. J. Exp. Zool.* **165,** 47–56.

Lash, J. W. (1968a). Phenotypic expression and differentiation: *In vitro* chondrogenesis. *In* "The Stability of the Differentiated State" (H. Ursprung, Ed.), pp. 17–24. Springer-Verlag, New York.

Lash, J. W. (1968b). Somitic mesenchyme and its response to cartilage induction. *In* "Epithelial–Mesenchymal Interactions" (R. Fleishmajer and R. E. Billingham, Eds.), pp. 165–172. Williams & Wilkins, Baltimore.

Lash, J., and Kosher, R. A. (1975). Perinotochordal proteoglycans and somite chondrogenesis. *In* "Extracellular Matrix Influences on Gene Expression" (H. C. Slavkin and R. C. Greulich, Eds.), pp. 671–676. Academic Press, New York.

Lash, J. W., and Vasan, N. S. (1978). Somite chondrogenesis *in vitro:* Stimulation by exogenous extracellular matrix components. *Dev. Biol.* **66,** 151–171.

Lash, J., Holtzer, S., and Holtzer, H. (1957). An experimental analysis of the development of the spinal column. *Exp. Cell Res.* **13,** 292–303.

3. Sclerotome Induction and Differentiation

Lash, J. W., Hommes, F. A., and Zilliken, F. (1962). Induction of cell differentiation. I. The *in vitro* induction of vertebral cartilage with a low-molecular-weight tissue component. *Biochim. Biophys. Acta* **56**, 313–319.

Lash, J. W., Rosene, K., Minor, R. R., Daniel, J. C., and Kosher, R. A. (1973). Environmental enhancement of *in vitro* chondrogenesis. III. The influence of external potassium ions and chondrogenic differentiation. *Dev. Biol.* **35**, 370–375.

Lash, J. W., Linask, K. K., and Yamada, K. M. (1987). Synthetic peptides that mimic the adhesive recognition signal of fibronectin: Differential effects on cell–cell and cell–substratum adhesion in embryonic chick cells. *Dev. Biol.* **123**, 411–420.

Le Douarin, N. M. (1973). A biological cell labelling technique and its use in experimental embryology. *Dev. Biol.* **30**, 217–222.

Lee, J. J., Ekker, S. C., von Kessler, D. P., Porter, J. A., Sun, B. I., and Beachy, P. A. (1994). Autoproteolysis in hedgehog protein biogenesis. *Science* **266**, 1528–1537.

Li, W., Ohlmeyer, J. T., Lane, M. E., and Kalderon, D. (1995). Function of protein kinase A in hedgehog signal transduction and *Drosophila* imaginal disc development. *Cell* **80**, 553–562.

Lotan, R., Skultelsky, E., Danon, D., and Sharon, N. (1975). The purification, composition, and specificity of the anti-T lectin from peanut (*Arachis hypogaea*). *J. Biol. Chem.* **250**, 8518–8523.

McMahon, J. A., Takada, S., Zimmerman, L. B., Fan, C.-M., Harland, R. M., and McMahon, A. P. (1998). Noggin-mediated antagonism of BMP signaling is required for growth and patterning of the neural tube and somite. *Genes Dev.* **12**, 1438–1452.

Maconochie, M., Nonchev, S., Morrison, A., and Krumlauf, R. (1996). Paralogous *Hox* genes: Function and regulation. *Annu. Rev. Genet.* **30**, 529–556.

Marcelle, C., Stark, M. R., and Bronner-Fraser, M. (1997). Coordinate actions of BMPs, Wnts, Shh and noggin mediate patterning of the dorsal somite. *Development* **124**, 3955–3963.

Mark, M. P., Butler, W. T., and Ruch, J.-V. (1989). Transient expression of a chondroitin sulfate-related epitope during cartilage histomorphogenesis in the axial skeleton of fetal rats. *Dev. Biol.* **133**, 475–488.

Marti, E., Takada, R., Bumcrot, D. A., Sasaki, H., and McMahon, A. P. (1995). Distribution of Sonic hedgehog peptides in the developing chick and mouse embryo. *Development* **121**, 2537–2547.

Marzullo, G., and Lash, J. W. (1967). Acquisition of the chondrocytic phenotype. *In* "Experimental Biology and Medicine. Morphological and Biochemical Aspects of Cytodifferentiation" (E. Hagen, W. Wechsler, and P. Zilliken, Eds.), pp. 213–218. Karger, Basel.

Minor, R. R. (1973). Somite chondrogenesis: A structural analysis. *J. Cell Biol.* **56**, 27–50.

Monsoro-Burq, A.-H., Bontoux, M., Teillet, M.-A., and Le Douarin, N. M. (1994). Heterogeneity in the development of the vertebra. *Proc. Natl. Acad. Sci. U.S.A.* **91**, 10435–10439.

Monsoro-Burq, A., Duprez, D., Watanabe, Y., Bontoux, M., Vincent, C., Brickell, P., and Le Douarin, N. (1996). The role of bone morphogenetic proteins in vertebral development. *Development* **122**, 3607–3616.

Morin-Kensicki, E. M., and Eisen, J. S. (1997). Sclerotome development and peripheral nervous system segmentation in embryonic zebrafish. *Development* **124**, 159–167.

Müller, T. S., Ebensperger, C., Neubuser, A., Koseki, H., Balling, R., Christ, B., and Wilting, J. (1996). Expression of avian *Pax1* and *Pax9* is intrinsically regulated in the pharyngeal endoderm, but depends on environmental influences in the paraxial mesoderm. *Dev. Biol.* **178**, 403–417.

Münsterberg, A. E., Kitajewski, J., Bumcrot, D. A., McMahon, A. P., and Lassar, A. B. (1995). Combinatorial signaling by Sonic hedgehog and Wnt family members induces myogenic bHLH gene expression in the somite. *Genes Dev.* **9**, 2911–2922.

Murray, P. D. F., and Selby, D. S. (1933). Chorio-allantoic grafts of single somites and of the unsegmented paraxial region of the two-day chick embryo. *J. Anat.* **67**, 563–572.

Newgreen, D. F., Scheel, M., and Kastner, V. (1986). Morphogenesis of sclerotome and neural crest in avian embryos. *In vivo* and *in vitro* studies on the role of notochordal extracellular material. *Cell Tissue Res.* **244**, 299–313.

Ng, L. J., Tam, P. P. L., and Cheah, K. S. E. (1993). Preferential expression of alternatively spliced mRNAs encoding type II procollagen with a cysteine-rich amino-propeptide in differentiating cartilage and nonchondrogenic tissues during early mouse development. *Dev. Biol.* **159,** 403–417.

O'Hare, M. J. (1972a). Differentiation of chick embryo somites in chorioallantoic culture. *J. Embryol. Exp. Morphol.* **27,** 215–228.

O'Hare, M. J. (1972b). Chondrogenesis in chick embryo somites grafted with adjacent and heterologous tissues. *J. Embryol. Exp. Morphol.* **27,** 229–234.

O'Hare, M. J. (1972c). Aspects of spinal cord induction of chondrogenesis in chick embryo somites. *J. Embryol. Exp. Morphol.* **27,** 235–243.

Ordahl, C. P. (1993). Myogenic lineages within the developing somite. *In* "Molecular Basis of Morphogenesis" (M. Berfield, Ed.), pp. 165–176. Wiley, New York.

Ordahl, C. P., and Le Douarin, N. M. (1992). Two myogenic lineages within the developing somite. *Development* **114,** 339–353.

Orr-Urtreger, P., and Lonai, P. (1992). Platelet-derived growth factor-A and its receptor are expressed in separate, but adjacent cell layers of the mouse embryo. *Development* **115,** 1045–1058.

Ostrovsky, D., Sanger, J. W., and Lash, J. W. (1988). Somitogenesis in the mouse embryo. *Cell Differ.* **23,** 17–26.

Pan, D., and Rubin, G. M. (1995). cAMP-dependent protein kinase and hedgehog act antagonistically in regulating decapentaplegic transcription in *Drosophila* imaginal discs. *Cell* **80,** 543–552.

Peters, H., Doll, U., and Niessing, J. (1995). Differential expression of the chicken *Pax-1* and *Pax-9* gene: *In situ* hybridization and immunohistochemical analysis. *Dev. Dyn.* **203,** 1–16.

Porter, J. A., von Kessler, D. P., Ekker, S. C., Yong, K. E., Lee, J. J., Moses, K., and Beachy, P. A. (1995). The product of hedgehog autoproteolytic cleavage active in local and long-range signaling. *Nature* **374,** 363–366.

Porter, J. A., Young, K. E., and Beachy, P. A. (1996). Cholesterol modification of hedgehog signaling proteins in animal development. *Science* **274,** 255–259.

Pourquié, O., Coltey, M., Teillet, M., Ordahl, C., and Le Douarin, N. M. (1993). Control of dorsoventral patterning of somitic derivatives by notochord and floor plate. *Proc. Natl. Acad. Sci. U.S.A.* **90,** 5242–5246.

Pourquié, O., Fan, C.-M., Coltey, M., Hirsinger, E., Watanabe, Y., Breant, C., Francis-West, P., Brickell, P., Tessier-Lavigne, M., and Le Douarin, N. M. (1996). Lateral and axial signals involved in avian somite patterning: A role for BMP4. *Cell* **84,** 461–471.

Reshef, R., Maroto, M., and Lassar, A. B. (1998). Regulation of dorsal somitic cell fates: BMPs and Noggin control the timing and pattern of myogenic regulator expression. *Genes Dev.* **12,** 290–303.

Riddle, R. D., Johnson, R. L., Laufer, E., and Tabin, C. (1993). Sonic hedgehog mediates the polarizing activity of the ZPA. *Cell* **75,** 1401–1416.

Rong, P. M., Teillet, M., Ziller, C., and Le Douarin, N. M. (1992). The neural tube/notochord complex is necessary for vertebral but not limb and body wall striated muscle differentiation. *Development* **115,** 657–672.

Sandell, L. J., Morris, N., Robbins, J. R., and Goldring, M. B. (1991). Alternatively spliced type II procollagen mRNAs define distinct populations of cells during vertebral development: Differential expression of the amino-propeptide. *J. Cell Biol.* **114,** 1307–1319.

Sanders, E. J. (1997). Cell death in the avian sclerotome. *Dev. Biol.* **192,** 551–563.

Sanders, E. J., Prasad, S., and Cheung, E. (1988). Extracellular matrix synthesis is required for the movement of sclerotome and neural crest cells on collagen. *Differentiation* **39,** 34–41.

Saunders, J. W. (1996). Operations on limb buds of avian embryos. *In* "Methods in Cell Biology" (M. Bronner-Fraser, Ed.), pp. 125–145. Academic Press, San Diego.

Seno, T., and Büyüközer, I. (1958). Cartilage formation in somite grafts of chick blastoderm. *Proc. Natl. Acad. Sci. U.S.A.* **44,** 1274–1284.

Serbedzija, G. N., Bronner-Fraser, M., and Fraser, S. E. (1989). A vital dye analysis of the timing and pathways of avian trunk neural crest cell migration. *Development* **106,** 809–816.

3. Sclerotome Induction and Differentiation

Shapiro, F. (1992). Vertebral development of the chick embryo during days 3–19 of incubation. *J. Morphol.* **213**, 317–333.

Slack, J. M. W. (1983). "From Egg to Embryo; Determinative Events in Early Development." Cambridge Univ. Press, Cambridge.

Smith, W. C., and Harland, R. M. (1992). Expression cloning of noggin, a new dorsalizing factor localized to the Spemann organizer in *Xenopus* embryos. *Cell* **70**, 829–840.

Smolich, B. D., McMahon, J. A., McMahon, A. P., and Papkoff, J. (1993). Wnt family proteins are secreted and associated with the cell surface. *Mol. Biol. Cell* **4**, 1267–1275.

Solursh, M., Fisher, M., Meier, S., and Singley, C. T. (1979). The role of extracellular matrix in the formation of the sclerotome. *J. Embryo. Exp. Morphol.* **54**, 75–98.

Sosic, D., Brand-Saberi, B., Schmidt, C., Christ, B., and Olson, E. N. (1997). Regulation of *paraxis* expression and somite formation by ectoderm- and neural tube-derived signals. *Dev. Biol.* **185**, 229–243.

Spence, M. S., Yip, J., and Erickson, C. A. (1996). The dorsal neural tube organizes the dermamyotome and induces axial myocytes in the avian embryo. *Development* **122**, 231–241.

Stern, C. D., Fraser, S. E., Keynes, R. J., and Primmett, D. R. N. (1988). A cell lineage analysis of segmentation in the chick embryo. *Development* **104** (Suppl.), 231–244.

Stern, H., Brown, A. M., and Hauschka, S. (1995). Myogenesis in paraxial mesoderm: Preferential induction by dorsal neural tube and cells expressing Wnt-1. *Development* **121**, 3675–3686.

Stockdale, F., Holtzer, H., and Lash, J. (1961). An experimental analysis of the development of the spinal column. VII. Response of dissociated somite cells. *Acta Embryol. Morphol. Exp.* **4**, 40–46.

Strudel, G. (1952). Influene morphogène du tube nerveux sur la différenciation de la colonne vertèbrale. *C.R. Soc. Biol., Paris.* **147**, 132–133.

Strudel, G. (1955). L'action morphogène du tube nerveux et de la corde sur la différenciation des vertèbres et des muscles vertébraux chez l'embryon de poulet. *Arch. Anat. Micr. Morph. Exp.* **44**, 209–235.

Strudel, G. (1962). Induction de cartilage *in vitro* par l'extrait de tube nerveux et de chorde de l'embryon de Poulet. *Dev. Biol.* **4**, 67–86.

Strudel, G. (1963). Autodifférenciation et induction de cartilage à partir de mésenchyme somitique de Poulet cultivé *in vitro. J. Embryol. Exp. Morphol.* **11**, 399–412.

Strudel, G. (1967). Some aspects of organogenesis of the chick spinal column. *In* "Experimental Biology and Medicine. Morphological and Biochemical Aspects of Cytodifferentiation" (E. Hagen, W. Wechsler, and P. Zilliken, Eds.), pp. 183–198. Karger, Basel.

Strudel, G. (1975a). Control of the phenotypic vertebral cartilage differentiation by the periaxial extracellular material. *In* "Extracellular Matrix Influences on Gene Expression" (H. C. Slavkin and R. C. Greulich, Eds.), pp. 655–670. Academic Press, New York.

Strudel, G. (1975b). Periaxial extracellular material and vertebral chondrogenesis. *In* "Protides of the Biological Fluids," pp. 51–58. Pergamon, Oxford.

Strutt, D. I., Wiersdorff, V., and Mlodzik, M. (1995). Regulation of furrow progression in the *Drosophila* eye by cAMP-dependent protein kinase A. *Nature* **373**, 705–708.

Tacchetti, C., Tavella, S., Dozin, B., Quarto, R., Robino, G., and Cancedda, R. (1992). Cell condensation in chondrogenic differentiation. *Exp. Cell Res.* **200**, 26–33.

Teillet, M.-A., and Le Douarin, N. M. (1983). Consequences of neural tube and notochord excision on the development of the peripheral nervous system in the chick embryo. *Dev. Biol.* **98**, 192–211.

Teillet, M.-A., Watanabe, Y., Jeffs, P., Duprez, D., Lapointe, F., and Le Douarin, N. M. (1998). Sonic hedgehog is required for survival of both myogenic and chondrogenic somitic lineages. *Development* **125**, 2019–2030.

Tessier-Lavigne, M., Placzek, M., Lumden, A. G. S., Dodd, J., and Jessell, T. M. (1987). Chemotropic guidance of developing axons in the mammalian central nervous system. *Nature* **336**, 775–778.

Theiler, K. (1988). Vertebral malformations. *Adv. Anat. Embryol. Cell Biol.* **112**, 1–97.

Timmons, P. M., Wallin, J., Rigby, P. W. J., and Balling, R. (1994). Expression and function of *Pax-1* during development of the pectoral girdle. *Development* **120**, 2773–2785.

Trelstad, R. L. (1977). Mesenchymal cell polarity and morphogenesis of chick cartilage. *Dev. Biol.* **59**, 153–163.

Trelstad, R. L., Kang, A. H., Cohen, A. M., and Hay, E. D. (1973). Collagen synthesis *in vitro* by embryonic spinal cord epithelium. *Science* **179**, 295–297.

Tremaine, R., and Hall, B. K. (1979). Retention during embryonic life of the ability of avian spinal cord to induce somitic chondrogenesis *in vitro*. *Acta Anat.* **105**, 78–85.

van Straaten, H. W. M., and Hekking, J. W. M. (1991). Development of floor plate, neurons and axonal outgrowth pattern in the early spinal cord of the notochord-deficient chick embryo. *Anat. Embryol.* **184**, 55–63.

Vasan, N. S. (1986). Somite chondrogenesis: Extracellular matrix production and intracellular changes. *In* "Somites in Developing Embryos" (R. Bellairs, D. A. Ede, and J. W. Lash, Eds.), pp. 237–246. Plenum, New York.

Vasan, N. S. (1987). Somite chondrogenesis: The role of the microenvironment. *Cell Differ.* **21**, 147–159.

Vasan, N., Lamb, K. M., and Heick, A. E. (1985). Somite chondrogenesis: Alterations in cyclic AMP levels and proteoglycan synthesis. *Cell Differ.* **16**, 229–234.

Verbout, A. J. (1985). The development of the vertebral column. *Adv. Anat. Embryol. Cell Biol.* **90**, 1–122.

Von der Mark, H., Von der Mark, K., and Gay, S. (1976). Study of differential collagen synthesis during development of the chick embryo by immunofluorescence. *Dev. Biol.* **48**, 237–249.

von Ebner, E. (1888). Urwirbel und Neugliederung der Wirbelsaule. *Sitzungsber. Akad. Wiss. Wien. Math. Naturwiss. Kl. Abt. 3* **97**, 194–206.

Wachtler, F., Christ, B., and Jacob, H. J. (1982). Grafting experiments on determination and migratory behaviour of presomitic, somitic and somatopleural cells in avian embryos. *Anat. Embryol.* **164**, 369–378.

Wallin, J., Wilting, J., Koseki, H., Fritsch, R., Christ, B., and Balling, R. (1994). The role of *Pax-1* in axial skeleton development. *Development* **120**, 1109–1121.

Watterson, R. L. (1952). Neural tube extirpation in *Fundulus heteroclitus* and resultant neural arch defects. *Biol. Bull.* **102**, 310.

Watterson, R. L., Fowler, I., and Fowler, B. J. (1954). The role of the neural tube and notochord in development of the axial skeleton of the chick. *Am. J. Anat.* **95**, 337–399.

Williams, B. A., and Ordahl, C. P. (1994). Pax-3 expression in segmental mesoderm marks early stages in myogenic cell specification. *Development* **120**, 785–796.

Williams, J. L. (1942). The development of cervical vertebrae in the chick under normal and experimental conditions. *Am. J. Anat.* **71**, 153–175.

Williams, L. W. (1910). The somites of the chick. *Am. J. Anat.* **11**, 55–100.

Wilting, J., Kurz, H., Brand-Saberi, B., Steding, G., Yang, Y. X., Hasselhorn, H.-M., Epperlein, H.-H., and Christ, B. (1994). Kinetics and differentiation of somite cells forming the vertebral column: Studies on human and chick embryos. *Anat. Embryol.* **190**, 573–581.

Winnier, G. E., Hargett, L., and Hogan, B. L. M. (1997). The winged helix transcription factor *MFH1* is required for proliferation and patterning of paraxial mesoderm in the mouse embryo. *Genes Dev.* **11**, 926–940.

Wolf, C., Thisse, C., Stoetzel, C., Thisse, B., Gerlinger, P., and Perrin-Schmitt, F. (1991). The *M-twist* gene of *Mus* is expressed in subsets of mesodermal cells and is closely related to the *Xenopus X-twi* and the *Drosophila twist* genes. *Dev. Biol.* **143**, 363–373.

Wozney, J. M., Rosen, V., Celeste, A. J., Mitsock, L. M., Whittiers, M. J., Kriz, R. W., Hewick, R. M., and Wang, E. A. (1988). Novel regulators of bone formation: Molecular clones and activities. *Science* **242**, 1528–1534.

Wright, E., Hargrave, M. R., Christiansen, J., Cooper, L., Kun, J., Evans, T., Gangadharan, U., Greenfield, A., and Koopman, P. (1995). The *Sry*-related gene *Sox9* is expressed during chondrogenesis in mouse embryos. *Nat. Genet.* **9,** 15–20.

Zilliken, F. (1967). Notochord induced cartilage formation in chick somites. Intact tissue versus extracts. *In* "Experimental Biology and Medicine. Morphological and Biochemical Aspects of Cytodifferentiation" (E. Hagen, W. Wechsler, and P. Zilliken, Eds.), pp. 199–212. Karger, Basel.

Zimmerman, L. B., De Jesus-Escobar, J. M., and Harland, R. M. (1996). The Spemann organizer signal noggin binds and inactivates bone morphogenetic protein 4. *Cell* **86,** 599–606.

4
Genetics of Muscle Determination and Development

Hans-Henning Arnold and Thomas Braun
Department of Cell and Molecular Biology
Institute of Biochemistry and Biotechnology
Technical University of Braunschweig
38106 Braunschweig, Germany

I. Introduction
II. The Origin of Skeletal Muscles in Vertebrate Organisms
III. The MyoD Family of Muscle-Specific Transcription Factors
   A. Temporal Expression during Mouse Embryogenesis Is Distinct for Each Gene of the Myogenic Basic Helix–Loop–Helix Family
   B. Targeted Inactivations of Myogenic Factor Genes Reveal Their Roles in Mouse Myogenesis
   C. The Role of MDFs in Adult Muscle
   D. Collaboration of MDFs with Other Transcription Factors in Myogenesis
   E. Transcriptional Control of MDF Gene Expression
IV. Development of the Limb Musculature
V. Structures Surrounding the Somites and Signals Emanating from Them Affect Myogenesis
VI. The Myogenic Potential of Mesoderm: Induction or Derepression?
   References

Skeletal muscles in vertebrates develop from somites as the result of patterning and cell type specification events. Here, we review the current knowledge of genes and signals implicated in these processes. We discuss in particular the role of the myogenic determination genes as deduced from targeted gene disruptions in mice and how their expression may be controlled. We also refer to other transcription factors which collaborate with the myogenic regulators in positive or negative ways to control myogenesis. Moreover, we review experiments that demonstrate the influence of tissues surrounding the somites on the process of muscle formation and provide model views on the underlying mechanisms. Finally, we present recent evidence on genes that play a role in regeneration of muscle in adult organisms.

I. Introduction

For decades, skeletal muscle cells have been a preferred model system to study processes that regulate many aspects of cell differentiation. Using skeletal muscle cells in culture it has been recognized that muscle formation involves at least two discernable steps referred to as commitment of mesodermal progenitors to myoblasts and subsequent differentiation of committed myoblasts to contractile myotubes. It was also established that proliferation and terminal differentiation of skeletal muscle cells are mutually exclusive events as cell cycle arrest appears to be a prerequisite for myoblasts in order to undergo differentiation.

The identification of four skeletal muscle specific basic helix–hoop–helix (bHLH) transcription factors that are involved in the control of both steps, the commitment and differentiation was seminal to the current understanding of skeletal myogenesis. With the detection of these pivotal myogenic regulators, it has become possible to approach the molecular mechanisms underlying muscle formation not only in cultured cells *in vitro* but also during embryogenesis *in vivo*. Not unexpectedly, it was found more recently that additional transcription factors, in particular those of the MEF2 family of MADS-box proteins, collaborate with bHLH proteins to activate muscle-specific gene transcription, whereas other proteins inhibit myogenic determination factors (MDFs) to prevent myogenesis at inappropriate sites. One also begins to understand how the expression of the crucial myogenic bHLH proteins may be induced and maintained during muscle development and how external signals may influence these processes.

In this article we will discuss the genetic information on skeletal muscle formation with particular emphasis on the individual role of myogenic bHLH factors and their interaction with other transcriptional regulators. We also review the current knowledge of how the expression of these genes in muscle forming regions may be controlled by influences of adjacent structures and putative signals which embark on progenitors of skeletal muscle cells in various regions of the body.

II. The Origin of Skeletal Muscles in Vertebrate Organisms

Somites are the main source of muscle forming cells in the vertebrate embryo. They arise in a craniocaudal direction on both sides of the neural tube by segmentation of the paraxial mesoderm that is embedded between axial structures, neural tube and notochord, and lateral structures, intermediate and lateral mesoderm (reviewed by Christ and Ordahl, 1995). Upon their formation somites differentiate into dorsal epithelial dermomyotome, intermediate myotome, and ventral mesenchymal sclerotome. In addition to the dorsoventral subdivision there is a mediolateral patterning that takes place. Isotypic grafting experiments between quail and chick embryos that allowed to follow the fate of early somitic cells have convincingly shown that early somites already consist of different regions destined to give rise to distinct derivative structures (Brand-Saberi *et al.*, 1996; Christ and Ordahl, 1995; Christ and Wilting, 1992; Ordahl and Christ, 1997). Although this regionalization already exists in the epithelial somites long before overt cell differentiation, cell lineages from these diverse regions are not yet irreversibly determined but depend on environmental conditions (Gamel *et al.*, 1995).

The dorsomedial quadrant of somites develops into the epaxial back muscles

(Denetclaw et al., 1997; Ordahl and Le Douarin, 1992) and the dermis of the back (Chevallier et al., 1977; Christ et al., 1983), while the dorsolateral quadrant contributes to hypaxial muscles of the ventral body wall, the limbs, and occipitally the tongue (Christ et al., 1977; Christ and Ordahl, 1995; Jacob et al., 1978, 1979; Ordahl and Le Douarin, 1992; Noden, 1983). Thus, two different territories of the somites are the source of most skeletal muscles of the body. Only a few precursor cells of extraocular muscles are derived from the prechordal mesoderm (Wachtler et al., 1984). Both ventral quadrants and the somitocoel generate the chondrogenic and fibroblastic cell lineages of the sclerotome that medially will form the vertebral column and laterally the ribs (Huang et al., 1996).

While dorsoventral and mediolateral patterning of paraxial mesoderm and newly formed somites is not discernable by morphology, it may be visualized by specific patterns of gene expression. Indeed, regionalized expression of some genes not only appears to correlate with somite regions but may contribute to the determination of specific cell lineages within these compartments together with signals provided by the surrounding tissues. Thus, the concept of somite compartmentalization has been supported on the molecular level by the identification of several genes that exhibit regionalized expression patterns. Among those patterning genes are transcription factors of the Hox- Pax-, and bHLH gene families. Whereas Hox genes are involved in the specification of regional identity along the anterior–posterior (A–P) axis (Gruss and Kessel, 1991), Pax- and bHLH genes appear to play a role in early determination and differentiation steps of cells within each individual somite. In particular, the development of skeletal muscle has become an intensely studied model system to investigate how these genes may be involved in the generation of somitic derivatives.

III. The MyoD Family of Muscle-Specific Transcription Factors

Early cell fusion experiments demonstrated that nuclei of various nonmuscle cell types can be programmed to activate muscle-specific genes in heterocaryons with differentiated muscle cells (Blau et al., 1983; Wright, 1984). These observations suggested that freely diffusible and dominant-acting regulators exist that are capable of changing the activity pattern of liver cell nuclei and those of other cell types. Indeed, four transcription factors were subsequently identified that are able to generate the skeletal muscle phenotype when overexpressed in 10T1/2 fibroblasts or in various other cell lines. This remarkable capacity was first demonstrated for MyoD (Davis et al., 1987) and later for myogenin (Braun et al., 1989a; Edmondson and Olson, 1989; Wright et al., 1989), Myf-5 (Braun et al., 1989b), and MRF4 (Rhodes and Konieczny, 1989). The latter was independently isolated as herculin (Miner and Wold, 1990) and Myf-6 in humans (Braun et al., 1990a).

The four proteins of this family are structurally related to each other by a highly homologous bHLH domain that mediates sequence-specific DNA-binding and dimerization with widely expressed bHLH proteins of the E2A gene family. Little sequence conservation among the four proteins exists outside of the bHLH domain suggesting that their common myogenic properties may depend primarily on the conserved motif. However, individual and specific features may involve other regions of the proteins. Many investigators have shown that the members of this family of MDFs act as gene-specific transcriptional activators of muscle genes by binding to the frequently occurring cis-acting DNA control element, referred to as E-box (for review: Olson, 1990, 1993; Weintraub *et al.*, 1991a; Buckingham, 1992; Olson and Klein, 1994; Arnold and Braun, 1996). It was also demonstrated that activation of the transcriptional machinery requires nonconserved transactivation domains located in the NH2- and/or COOH-termini of the various myogenic factors in addition to the bHLH domain (Braun *et al.*, 1990b; Mak *et al.*, 1992; Weintraub *et al.*, 1991a, b; Winter *et al.*, 1992).

Expression of the MyoD-related genes is strictly limited to skeletal muscle cells excluding cardiac and smooth muscle. Myf-5, however, has also been shown to be expressed in the developing neural tube and defined domains of the brain (Tajbakhsh and Buckingham, 1995; Tajbakhsh *et al.*, 1994). The role of neuronal Myf-5 expression is unknown. In established muscle cell lines and primary muscle cells in culture MyoD and Myf-5 are expressed in proliferating myoblasts prior to overt differentiation but continue to be present at reduced levels in differentiated myocytes. In contrast, myogenin is not present in myoblasts but begins to accumulate at the onset of differentiation. MRF4 is only synthesized after prolonged culture of most established cell lines. These observations led to the concept that MyoD and Myf-5 may play a role in early events of myogenesis, possibly the determination of myoblasts, whereas myogenin and probably MRF4 appear to be involved in the terminal differentiation program of muscle cells. By enlarge these presumed roles have been confirmed *in vivo* by targeted gene disruption in mice that lack one or several members of the MyoD gene family (see below). It must be kept in mind, however, that bHLH genes which are active during myoblast determination may also contribute to differentiation when present at sufficiently high concentrations. Thus, the processes of determination and differentiation may not be seen as two entirely separate steps but overlapping events on the way to muscle.

A. Temporal Expression during Mouse Embryogenesis Is Distinct for Each Gene of the Myogenic Basic Helix–Loop–Helix Family

The spatiotemporal expression patterns of the four myogenic regulators during mouse development revealed an apparent order in which these genes are activated in the various muscle forming regions of the embryo. *In situ* hybridizations demonstrated that each of the four genes is activated for a defined developmental period in muscle cells and their progenitors in somites, visceral arches, and limb buds (Bober *et al.*, 1991; Hinterberger *et al.*, 1991; Ott *et al.*, 1991; Sassoon *et al.*,

4. Genetics of Muscle Determination and Development 133

1989). Transcripts accumulate first in cranial somites and successively in all somites along the rostrocaudal axis following the natural developmental gradient of somitogenesis. The same anteroposterior sequence applies for their expression in muscle precursor cells of the visceral arches and limb buds.

Myf-5 transcripts appear prior to those of the other three myogenic factor genes in newly formed somites of mouse embryos at day 8 postcoitum (p.c.). The expression begins in the dorsomedial quadrant of the epithelial somites and later spreads to the entire myotomes and all body muscles (Ott *et al.*, 1991; Tajbakhsh *et al.*, 1996a). Maximal levels of Myf-5 mRNA and protein are reached between day 9.5 and 10.5 p.c. and then decline rapidly in a craniocaudal direction. This transient expression pattern is not seen for MyoD and myogenin during prenatal life. Myf-5 gene activation in limb buds and branchial arches also precedes that of the other family members, albeit by a smaller time margin. Myogenin mRNA is first detectable in rostral somites of day 8.5 p.c. embryos, approximately 12 hr after Myf-5 when myotomal muscles begin to form and muscle-specific marker genes start to be expressed. Its expression continues at a relatively high level until birth and declines thereafter. MRF4 (Myf-6) expression follows myogenin activation in somites with a delay of about 12 hr. Similar to Myf-5, this gene is also transcribed transiently in somites for only 2 days of development (Bober *et al.*, 1991). In contrast to Myf-5 and myogenin, MRF4 transcripts are undetectable in branchial arches and limb buds as well as in axial skeletal muscles between day 9 and 15 of embryogenesis. In fetal mice from day 16 onward, MRF4 transcripts reappear in all skeletal muscles and constitute the major component among the MDFs until and after birth. Interestingly, the reappearance of MRF4 transcripts in this biphasic expression pattern coincides approximately with the formation of secondary muscle fibers and the onset of motor innervation. MyoD expression begins in myotomes of mature somites, and in myoblasts of the limb buds and facial muscles at day 10.5 p.c. Like myogenin, its expression continues throughout prenatal life. Although the late onset of MyoD expression *in vivo* does not appear to support the assumption of early functions in myogenesis as predicted from cultured myoblasts, it should be pointed out that the epistasis of the four genes in a genetic sense cannot be deduced from the temporal order of their expression. Even though each gene of the MyoD family follows a distinct temporal activation pattern, there is considerable overlap in the expression of these four genes.

B. Targeted Inactivations of Myogenic Factor Genes Reveal Their Roles in Mouse Myogenesis

Multiple *in vitro* studies suggest that all members of the MyoD family have similar biochemical properties and myogenic activities. Most impressively, forced expression of any of the four proteins in 10T1/2 cells results in muscle cells that are essentially indistinguishable, regardless of the factor used to generate them. Moreover, DNA-binding to E-boxes, dimerization with E2A-type bHLH proteins, and transactivation of target genes appear largely identical among all family members (Braun and Arnold, 1991). A possible explanation for these observations may come from the fact that considerable auto- and cross-activation occurs in cells when one myogenic bHLH protein is overexpressed (Braun *et al.,* 1989a; Thayer *et al.,* 1989). Thus, individual contributions of single factors are difficult to assess in this kind of experimental setup. The major biochemical similarity and overlap of functions of the four muscle-specific bHLH proteins have raised the question of why do four myogenic regulatory genes exist for apparently similar purposes and why have they been conserved during evolution. The answer may at least in part lie at the different parameters of the *in vivo* situation which are not necessarily recapitulated in tissue culture experiments. First, the developmental time of activation is different for each of the four MyoD-related genes and also the duration of their expression; second, the level of MDF transcripts and proteins may be different for each transcription factor and together with distinct threshold requirements may result in specific patterns of gene activation; third, the myogenic bHLH factor genes may be activated in different cells with the effect that the distinct molecular context determines the activity patterns of MDFs in different cell lineages.

The state of the art allows one to mutate single genes in mice of otherwise identical genetic background. Targeted gene inactivations have been performed for each of the four myogenic bHLH trancription factor genes with interesting and partly unexpected results. In general the obtained mouse mutants confirmed the fundamental and cell-autonomous importance of these genes for the development of skeletal muscles and defined roles for individual members in commitment of progenitors, myoblast proliferation, and terminal differentiation as well as in the maturation and regeneration of skeletal muscle under pathophysiological conditions. Moreover, combinations of inactivated MDF genes revealed that some of their functions overlap, while others are more specific. Entirely surprising, the mouse gene knockouts also resulted in non cell-autonomous effects indicating cell interactions within the early somites that have not been recognized before. The combined observations in the various knockout mice as discussed below allow us to propose an epistasis model that represents the genetic relationship of the relevant MDF genes and other potentially relevant genes for myogenesis (Fig. 1).

Figure 1 Schematic representation of an epistasis model of MDF genes and other genes potentially involved in myogenesis. According to this model myogenic cells can be generated independently by Myf-5 and MyoD. Either gene activity leads to the activation of myogenin and subsequently to activation of contractile protein (CP) expression. Expression of MRF4 may also be dependent on a myogenic bHLH transcription factor and act downstream of MyoD and Myf-5, possibly in a similar way as myogenin. Putative signals activating MyoD and Myf-5 gene expression include wnts and Sonic hedgehog but are most likely different for both genes. Expression of MyoD appears to require either Myf-5 or the paired box transcription factor Pax-3 directly or indirectly. The Myf-5 derived myotomal cells are necessary for rib development, probably by generating an essential growth factor. The early (Myf-5) myotomal cells can also give rise to the entire set of body muscles. The arrows indicate genetic relationships but not direct biochemical activities.

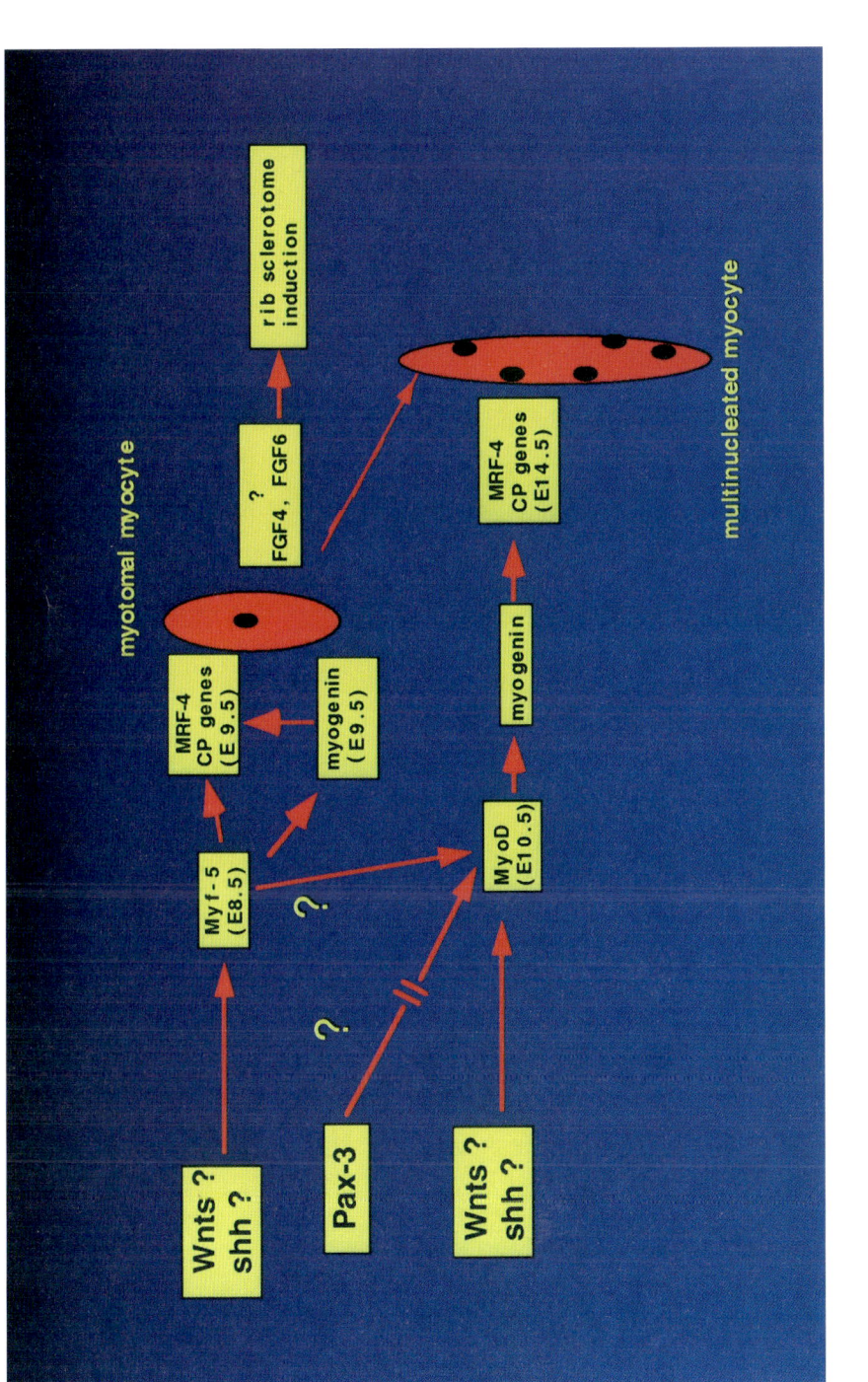

1. Homozygous Myf-5 Mutant Mice Show Abnormal Somite Development and a Severe Rib Defect

Two independent Myf-5 mutant alleles have been generated by homologous recombination in mice. Both carry insertions in the first exon that disrupt the open reading frame of the bHLH domain and therefore yield no functional Myf-5 protein. Tajbakhsh et al. (1996a) inserted the *nls-LacZ* reporter gene, while Braun *et al.* (1992) introduced only the *PGK-neo* gene for selection. Both mutations are recessive lethal in the perinatal period. The most obvious defect in homozygous mutants is a severe truncation of all ribs with the major distal parts lacking. The absence of a functional rib cage results in the inability of mutant mice to breathe causing death shortly after birth. Surprisingly, skeletal muscles of newborn mutants appear morphologically normal and express the normal set of muscle-specific marker genes including the myogenic factors MyoD, myogenin, and MRF4.

According to these results Myf-5 seems dispensable for the development of skeletal muscle but plays a crucial role for rib formation (Arnold and Braun, 1993). In view of the fact that Myf-5 expression has never been observed in somites outside of the myotome, the failure to develop complete ribs that are derived from the lateral sclerotome was entirely unexpected. How then may this phenotype arise? A partial answer to this question comes from developmental studies which show that early somitogenesis is impaired in homozygous mutant embryos (Braun *et al.*, 1994). While wild-type and heterozygous embryos develop myotomes between day 8.5 and 10.5 p.c., the homozygous littermates form no myotomal cells until day 10.5 p.c when MyoD expression ensues. Interestingly, the expression of the myogenin gene normally beginning at day 8.5 p.c. in cranial somites is also not seen in mutants suggesting that it may directly or indirectly require Myf-5 for its activation. The ribs develop from cells located in the lateral part of the sclerotome which, unlike the medial part giving rise to the vertebral column, stays in close contact with the early myotome. Because Myf-5 is exclusively expressed in the myotome and not in the sclerotome the rib defect must be explained by a non cell autonomous mechanism. The anatomical situation in the somites at the time when the rib blastema begins to form suggests that signals produced in the myotome may be required for permissive or inductive interactions in order to allow complete rib development from sclerotomal precursors. The putative signals and the nature of the suspected interactions have not been defined to date. Members of the fibroblast growth factors (FGF) family of growth factors seemed good candidates, because they are known to be involved in morphogenetic events and some of them are expressed in the early myotome (Grass *et al.*, 1996). However, targeted inactivation of several FGF genes, such as *FGF4, FGF5, FGF6,* and *FGF7* have not shown the rib phenotype (Floss *et al.*, 1997 and references therein). Functional redundancy among the FGFs expressed in somites may be the reason for this result or alternatively, entirely different signaling molecules may mediate the myotome–sclerotome interaction. It should be noted, however, that later myotomal cells formed under the influence of MyoD are unable to substitute for the early myotome and cannot rescue the rib phenotype. Interestingly, the platelet derived growth factor alpha (PDGFα) receptor knockout also results in a mild rib phenotype (Soriano, 1997). This receptor is expressed in the sclerotome, while its ligand PDGFα is expressed in the myotome. Thus, disruption of a myotome-sclerotome signaling pathway can disturb rib formation.

Targeting the β-galactosidase gene into the Myf-5 locus has allowed one to follow the fate of Myf-5 minus cells in the embryo (Tajbakhsh *et al.*, 1996b). These studies revealed that myotomal cells normally expressing Myf-5 fail to adopt the muscle phenotype and their appropriate location in the absence of Myf-5 protein. They migrate aberrantly, away from the central myotome into the ventral sclerotome and the dorsal dermatome and express scleraxis and Dermo-1, marker genes for chondrogenic and epidermal cells, respectively. This indicates that Myf-5 is required to convert multipotential stem cells in somites to the myogenic fate. In the absence of Myf-5 these cells differentiate into other somitic derivatives according to their local environment. Therefore, *Myf-5* clearly qualifies as a determination gene for myotomal muscle cells. Despite the lack of Myf-5, mutant mice have apparently normal muscles at birth. When MyoD begins to be expressed in wild type, MyoD and myogenin positive cells can also be detected in somites of Myf-5 null mutants (Braun *et al.*, 1994). Obviously, MyoD expression can occur independently of Myf-5 and the expressing cells go on to give rise to normal skeletal muscles. Consequently, one has to assume that two temporally distinct entry points for myogenesis exist, one initiated by Myf-5 which is necessary to establish the early myotome and another one marked by the expression of MyoD causing the formation of myotomal cells at a later stage. Whether both genes act in the same mesodermal stem cell or determine independent myotomal muscle cell lineages has been a matter of debate.

In embryonal stem cells (ES) triggered to differentiate into skeletal muscles *in vitro*, the MyoD gene and the *LacZ* gene inserted into the Myf-5 locus appear to be activated in different cells in a mutually exclusive manner (Braun and Arnold, 1996). Even in Myf-5 null ES cells the *MyoD* gene never seems activated in the same cell which also transcribes the Myf-5 locus. Thus, a simple model of genetic hierarchy or a default mode in which *MyoD* substitutes for lacking *Myf-5* within one cell seems unlikely, rather both genes appear to establish two independent myogenic cell lineages that can compensate for each other. This view is consistent with the observation that Myf-5 and MyoD expressing cells initially segregate in different regions of the myotome, presumably under the influence of different local cues (Cossu *et al.*, 1995, 1996).

Nonoverlapping expression domains were also found with antibodies to MyoD and Myf-5 (Braun and Arnold, 1996; Smith *et al.*, 1994). On the other hand, at later developmental stages myotomal cells are found that apparently express both, MyoD and Myf-5 simultaneously (Tajbakhsh *et al.*, 1996a). The inconsistency of these observations has not been resolved but may be due to the *in vitro* culture conditions of the ES cell, whereas *in vivo* the environmental conditions in the embryo change and may eventually relax the strict lineage separation of originally Myf-5 and MyoD-derived myogenic populations. Whatever the truth may be, it is clear from the Myf-5 knockout mouse model that the delayed myotome formation initiated by MyoD expression gives rise to the full complement of skeletal muscles including slow and fast fiber types. However, the possible patterning role of the early myotome as indicated by the rib phenotype in Myf-5 mutant mice may not be provided by MyoD derived muscle cells.

2. MyoD Mouse Mutants Develop Normal Skeletal Muscles through Compensation by Myf-5

Targeting the MyoD gene by a mutant allele in mice generates no obvious muscle phenotype in heterozygous or homozygous animals during early development (Rudnicki *et al.*, 1992). Motility and evoked movements of mutants are indistinguishable from wild-type. Histology of muscles from various body regions, expression of numerous muscle-specific marker genes, and the ultrastructure of sarcomeres appear unaltered. The lack of muscle defects indicates that MyoD is not essential to promote normal skeletal muscle development consistent with the idea that other members of the MyoD family may substitute for MyoD under these conditions. Indeed, MRF4 and myogenin are expressed at normal levels, while Myf-5 transcript levels are somewhat elevated in heterozygous and drastically increased in homozygous mutants. This is in marked contrast to the normal downregulation of Myf-5 expression in wild-type animals at the time of MyoD activation, suggesting that Myf-5 gene expression is normally suppressed either directly or indirectly by MyoD in a dosage-dependent manner. The elevated and sustained expression of Myf-5 in the absence of MyoD may well explain the lack of any obvious muscle phenotype in mutant animals. However, it should be mentioned that homozygous MyoD mutants in mixed litters survive the weaning period at a lower frequency suggesting that under competitive conditions a fraction of MyoD mutants dies postnatally. Thus, a mild phenotype may have escaped detection. More recently, it was shown that the adult MyoD mutants have a clear deficit in muscle regeneration which indicates that MyoD has a function in satellite cells or muscle precursor cells that are recruited following muscle injuries (Megeney *et al.*, 1996). Additionally, MyoD mutant mice display a slight delay of limb muscle development suggesting that Myf-5 expression in the limb is initially not sufficient for normal progression of myogenesis (Kablar *et al.*, 1997).

The issue of why MyoD does not appear essential for myogenesis has been answered unequivocally in double mutant mice that lack both MyoD and Myf-5 (Rudnicki *et al.*, 1993). While each gene disruption individually results in apparently normal skeletal muscles, both genes inactivated together lead to the complete absence of any skeletal muscles in mutant animals. Interestingly, these animals not only lack myofibers but also any mononucleated muscle precursor cells

4. Genetics of Muscle Determination and Development   139

as judged by histology and desmin-staining which also detects myoblasts. The muscle-forming regions are partly devoid of cells or populated by loose mesenchyme which is entirely free of muscle-specific transcripts including the four myogenic bHLH mRNAs. The lack of myocytes is not restricted to muscles originating in somites but also involves those derived from prechordal plate mesoderm, such as the extraocular muscles. Therefore, Myf-5 or MyoD is absolutely essential to establish or maintain cells of the myogenic lineage in all parts of the body. These results also demonstrate that Myf-5 and MyoD have overlapping functions despite their distinct temporal expression in somites and elsewhere. That the functional redundancy is not complete, however, is illustrated by two observations. First, the rib deficiency caused by the absence of Myf-5 is not rescued by the expression of MyoD; second, muscle formation is sensitive to the concentration of Myf-5 and this concentration dependence is not seen for MyoD. MyoD null mutants carrying only one active Myf-5 allele producing half of the wild-type level of Myf-5 have severely reduced skeletal muscle masses. In contrast, Myf-5 null mutants with one active MyoD allele develop normal skeletal muscle. Therefore, Myf-5 and MyoD have either intrinsicly different activities or their threshold levels for full biological activity are different.

While the MyoD/Myf-5 double knockout phenotype provides clear evidence that either Myf-5 or MyoD is required to generate myoblasts, it is not known how these transcription factors mediate determination of mesodermal progenitors to myoblasts, maintain their propagation, or both. With the possible exception of the myogenin gene (Buchberger *et al.,* 1994; Hollenberg *et al.,* 1993) no target genes for Myf-5 or MyoD are known in myoblasts that would provide clues for the early functions of these myogenic factors. Since neither myogenin nor MRF4 is expressed in MyoD/Myf-5 double mutants, it cannot be excluded formally that both genes would also be capable to generate the myogenic lineage. However, under normal circumstances they are clearly not necessary to make muscle precursor cells as we will discuss below.

3. The Myogenin Gene Is Necessary for Myocyte Differentiation during Mouse Development

In muscle cell lines in culture activation of the myogenin gene strictly correlates with the onset of differentiation suggesting that it may have an important role in this process. This proposition has been tested *in vivo* through targeted gene inactivation of the myogenin gene in mice (Hasty *et al.,* 1993; Nabeshima *et al.,* 1993). Homozygous mutants are born immobile and decease shortly after birth. They have mild rib malformations and their muscles are reduced in mass. More importantly, these mutant mice contain a drastically reduced density of myofibers with mononucleated cells replacing most of the mature muscle cells. The muscle defect includes the diaphragm which is nonfunctional most likely causing the perinatal lethality. The total number of nuclei in the muscle forming regions of homozygous myogenin mutant animals at birth appears to be approximately the same as in wild type. However, most muscle-specific marker genes are not expressed in mutants, suggesting that these cells represent undifferentiated myoblasts as they do express virtually normal levels of MyoD. Interestingly, MRF4 is markedly reduced indicating that MyoD expression is entirely independent of myogenin, whereas MRF4 gene activation may be influenced by myogenin. Alternatively, the reduced level of MRF4 in myogenin mutants may only reflect the absence of differentiated skeletal muscle cells and may not be directly linked to the lack of myogenin. Taken together these observations indicate that cells can enter the myogenic lineage in the absence of myogenin but apparently are unable to differentiate and activate most of the typical, sarcomeric muscle genes.

Significantly, neither MyoD nor Myf-5 that are presumably present at normal levels can rescue the differentiation defect which seems to reflect distinct intrinsic properties of myogenin, or specificity brought about by its temporal pattern or level of expression. Indeed myogenin transcript levels appear high, particularly when compared to Myf-5. The strict requirement of myogenin for the generation of functional muscle fibers does not apply to the early developmental stages (Venuti et al., 1995). The early events of somite differentiation leading to primary myogenesis occur normally in myogenin mutant embryos. However, when secondary myofibers form in wild type very little muscle formation takes place in the mutants, suggesting that myogenin affects secondary myogenesis more severely than primary myogenesis. The reason why myotomes initially form and myoblasts appear in all muscle-forming regions may be that MRF4 is expressed in a temporally overlapping pattern with myogenin in the myotome at early stages but it is downregulated during subsequent development. Thus, a transient functional substitution of lacking myogenin by MRF4 may rescue primary muscle cells but not secondary ones. A myogenin-MRF4 double null mutant should clarify this issue.

The unique role of myogenin in skeletal muscle differentiation *in vivo*, despite its overlapping expression pattern with MyoD and Myf-5 in muscle precursors and their descendants, has been corroborated in double homozygous null mutants lacking MyoD and myogenin, or Myf-5 and myogenin (Rawls et al., 1995). These double mutant mice show embryonic and perinatal phenotypes characteristic of the combined defects observed in mice mutants of each gene alone. Thus, the functions of myogenin do not overlap with those of Myf-5 and MyoD consistent with the view that myogenin acts in a genetic pathway downstream of MyoD and Myf-5 (see Fig. 1).

Elegant studies with mice carrying myogenin targeted into the Myf-5 locus attempted to address the question of intrinsic functional redundancy among the myogenic bHLH transcription factors (Wang et al., 1996). Myogenin knock-in mice expressing myogenin under the transcriptional control of the *Myf-5* gene are viable and do not show the rib cage truncation of Myf-5 null mutants. Moreover, the delay of myotome formation observed in homozygous Myf-5 mutants is also not present. This indicates that myogenin can rescue the defect in skeletal morpho-

4. Genetics of Muscle Determination and Development

genesis as well as early steps of myogenesis. Therefore, both proteins share overlapping intrinsic properties. However, when myogenin knock-in mice were crossed with MyoD-null or myogenin-null mutants in order to analyze later contributions of the myogenin knock-in allele on myogenesis, it became apparent that it cannot completely rescue skeletal muscle formation (Wang and Jaenisch, 1997). Homozygous myogenin knock-in mice lacking MyoD die shortly after birth owing to reduced skeletal muscle including the diaphragm. Therefore, myogenin expressed from the Myf-5 locus is not able to completely replace the function of Myf-5 in muscle development, although it can establish the early myogenic lineage to some extent. Similarly, early expressed myogenin from the Myf-5 locus cannot rescue the myogenin $(-/-)$ phenotype in double homozygous mutants, as these mice largely lack differentiated muscle cells similar to the myogenin null mutation itself.

Taken together, the various results from knockout and knock-in experiments argue for a genetic pathway in which the four myogenic bHLH proteins perform distinct functions, albeit with some overlap, in particular MyoD and Myf-5. The observations that certain aspects of myogenesis may be uniquely associated with individual myogenic factors, while some of their features appear interchangeable may be discussed in two ways. The unique function of each myogenic bHLH transcription factor may be due to its distinct spatiotemporal expression pattern including different levels of expression or alternatively to unique properties in their transcriptional activation function. Both possibilities are, however, not mutually exclusive.

Let us consider the various scenarios provided by the genetically manipulated mouse models. Myogenin targeted in the Myf-5 locus is apparently capable of rescuing the rib truncation and replacing Myf-5 function in early myotome formation suggesting functional redundancy between both proteins. However, it is important to realize that under these conditions the level of myogenin almost certainly exceeds that of wild type as four active alleles are present. That expression levels of these transcription factors may be critical can be inferred from the mouse carrying myogenin knocked into the Myf-5 locus on a myogenin null background. These mice exert the myogenin null muscle phenotype with a deficiency in myoblast differentiation. Obviously, myogenin expressed under Myf-5 control is not sufficient to rescue muscle differentiation. The most simple explanation seems to be that the level of myogenin is not high enough, because the Myf-5 promotor is downregulated during late embryonic and fetal development. A more complicated explanation may invoke some sort of essential collaboration of Myf-5 and myogenin that would implicate distinct intrinsic properties of either protein. This is however unlikely, because myoblasts do differentiate efficiently in the absence of Myf-5. Another possibility to explain the drastically reduced muscle masses in Myf-5/myogenin knock-in mice which also lack MyoD or myogenin is that myogenin may not be as efficient as Myf-5 in recruiting muscle precursor cells into the myogenic lineage and/or maintaining the lineage. However, reduced numbers of myoblasts in the early myotome under these conditions have actually not been demonstrated. If this would be the case, one could assume that myogenin is indeed intrinsically less potent than Myf-5 or MyoD in determining the myogenic cell lineage. Since the transcriptional activation domains of the myogenic bHLH factors are very divergent, they may in fact exert different efficiencies in promoting the various steps of myogenesis (Asakura et al., 1993). In line with this view, it has been shown that myogenin is drastically less efficient than Myf-5 or MyoD in activating gene transcription in repressive chromatin (Gerber et al., 1997). To clarify the individual roles of the nonconserved transactivation domains among the myogenic bHLH factors it would be necessary to swap the respective domains between the different factors which has not yet been performed in vivo.

Wang and Jaenisch (1997) proposed that the total level of MDFs rather than individual intrinsic specificities may be critical in determining the different states of muscle differentiation. In this scenario the role of Myf-5 and MyoD would simply be to activate myogenin solely to increase the total level of myogenic bHLH factors thereby promoting terminal muscle differentiation. In this context it is interesting that myoblasts of myogenin null mutants which express MyoD are able to differentiate in vitro, although they essentially fail to do so in vivo (Nabeshima et al., 1993). Thus, MyoD seems capable of triggering differentiation like myogenin under certain conditions. Why then do myogenin knockout mice lack most of the differentiated muscle fibers? A possible explanation invokes a repressor function in vivo that silences MyoD in the animal but is released under culture conditions. This repressor activity may be a growth factor present in the muscle forming regions of the embryo. For instance, it is known that several FGFs, transforming growth factor (TGF-β) and nerve growth factor (NGF) are expressed in muscle and all of these proteins suppress myoblasts differentiation in vitro. Alternatively, the dissociation of cells alone may also release a repressor activity. Whatever the putative repressor may be, it is clear that MyoD should be more sensitive to it than myogenin. So far no molecule has been described that exhibits such selective inhibitory activity on MyoD versus myogenin. It will be interesting to test whether the myogenin null mutant phenotype can be rescued in mutants lacking either growth factors or their receptors on myoblasts.

4. Gene Disruptions Do Not Reveal a Decisive Role for MRF4 in Myogenesis

The expression pattern of MRF4 during mouse embryogenesis differs from that of the other family members in several respects. First, in the myotome MRF4 is only expressed in a short transient wave between day 9.0 and 11.5 p.c. and disappears thereafter, unlike the other bHLH factors that continue to be expressed throughout myotomal muscle formation. Second, MRF4 is not expressed during the early formation of head muscles derived from branchial arches or limb muscles derived from migratory dermamyotomal cells and therefore is not at all involved in the generation of these muscle groups. Third, MRF4 expression is massively upregulated in late fetal muscle and eventually becomes the predominant myogenic bHLH

4. Genetics of Muscle Determination and Development

transcription factor expressed in adult musculature. On the basis of its expression pattern, it has been proposed that MRF4 may regulate skeletal muscle maturation and aspects of adult myogenesis.

In order to determine the function of MRF4, three independent mouse mutant alleles have been generated in three different laboratories (Braun and Arnold, 1995; Patapoutian *et al.,* 1995; Zhang *et al.,* 1995). While all three mutants exert a very mild muscle phenotype with only subtly reduced expression of a subset of muscle-specific genes, the overall phenotypes are surprisingly different. The MRF4 germ line mutation obtained in the Olson group (Zhang *et al.,* 1995) results in viable mice with multiple rib anomalies including bifurcations, fusions, and supernumerary processes. The muscles in adult mice are fairly normal but express elevated levels of myogenin. Therefore, the authors suggest that MRF4 is required for the downregulation of myogenin that normally occurs after birth. They also suggested that the drastic increase of myogenin may compensate for the lack of MRF4.

The second MRF4 allele generated in the Wold laboratory (Patapoutian *et al.,* 1995) results in a recessive perinatal lethal phenotype with incomplete penetrance. The homozygous mutant mice contain grossly normal muscle at birth but show severe rib defects. The truncated ribs fail to attach to the sternum causing strong respiratory distress and death at birth. This rib defect, however, is not as severe as in the Myf-5 knockout and appears more similar to the myogenin null rib phenotype. Most interestingly, Wold's mutant mice exert a transient deficit in myotome development that corresponds in time with the first wave of MRF4 expression in somites beginning around E9 and ending around E11. During this period expression of Myf-5, myogenin, and MyoD as well as downstream muscle-specific genes are significantly reduced in the homozygous mutant relative to wild type. Prior to this period and following the time of somitic MRF4 expression formation of myotomal muscles appear essentially unaltered. These observations led Patapoutian *et al.* to suggests a model of triphasic expansion of the myotome with each phase being dependent on a different MDF. According to this hypothesis the first myotome arises at the onset of Myf-5 expression, the second phase is triggered by MRF4, and the third phase is mediated by the expression of MyoD. While independent initiation of myogenesis by Myf-5 and MyoD is well supported by experimental data, the assessment of a MRF4 mediated phase of myotomal expansion is complicated by the fact that Myf-5 expression fully overlaps with this period. Moreover, targeted disruption of the *MRF4* gene apparently affects transcription of the *Myf-5* gene that is located approximately 8.5 kb downstream on the same mouse chromosome (Braun and Arnold, 1995; Olson *et al.,* 1996). In compound heterozygous animals carrying the mutated *MRF4* gene on one chromosome and the mutated *Myf-5* gene on the other chromosome it was formally shown that small deletions within the *MRF4* gene or the insertion of the pgk-neo selection cassette downregulate the expression of Myf-5 by a cis-acting mechanism (Floss *et al.,* 1996; Yoon *et al.,* 1997). In the most severe case this results in a complete phenocopy of Myf-5 knockout mice with extensive rib truncations (Fig. 2). While this is an interesting finding with respect to the transcriptional

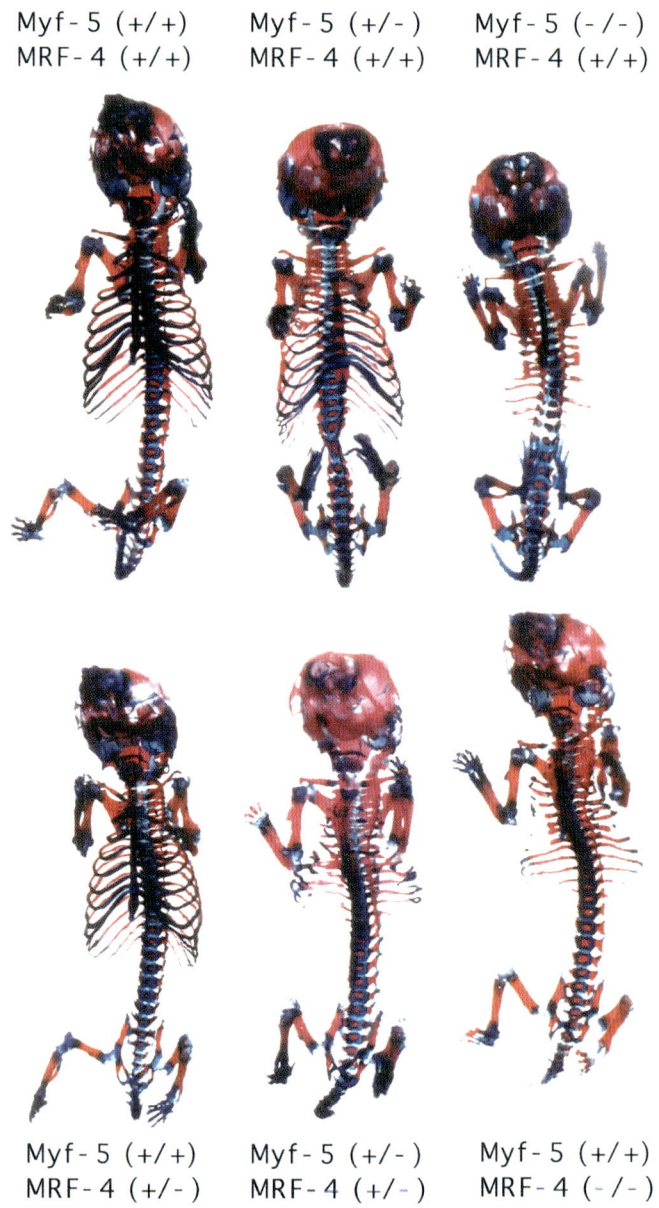

Figure 2 Ventral views of alizarin stained skeletons of newborn wild-type, MRF4(−/−), Myf-5(−/−), MRF4(−/+), Myf-5(−/+) and double heterozygous Myf-5(−/+)/ MRF4(−/+) mutant mice. Note the severe rib truncations in double heterozygous mutants similar to the Myf-5 null phenotype. Heterozygous mutants for either gene develop normal ribs. No other bone or cartilage defects are observed.

regulation of the *Myf-5* gene, it makes it difficult to judge the individual contribution of MRF4 in myotome formation.

The downregulation of Myf-5 transcription was most pronounced in the third MRF4 mutant allele generated by Braun and Arnold (1995) leading essentially to the phenotype of Myf-5 null mice. However, despite the lack of MRF4 and most of Myf-5 these mice develop fairly normal muscle in most parts of the body with only a moderate reduction of deep back muscles. Since this MRF4 mutant does not express any MDFs in somites at least until day 10.5 p.c. when MyoD is activated, this result underscores the enormous plasticity of myogenesis in temporal terms. Unfortunately, this result also fails to answer the question of what the specific role of MRF4 in the myotome is.

C. The Role of MDFs in Adult Muscle

Due to early lethality or no striking muscle phenotype the MDF knockout mutants provide little information on MDF functions in adult muscle. In particular for MyoD and Myf-5 the postulated importance for transcriptional activation of sarcomeric protein genes remains unproven, because these null mutations do not reveal major effects on the expression of muscle-specific genes (Rudnicki *et al.,* 1992; Braun *et al.,* 1992). Likewise, in MRF-4 mutants only the embryonic myosin heavy chain gene appears downregulated, while most other structural proteins are expressed normally (Braun and Arnold, 1995; Patapoutian *et al.,* 1995). Even an essential role of myogenin for muscle gene activation cannot be assessed unequivocally in the mutant, as myocytes isolated from the mutant express apparently most muscle genes. For MyoD deficient mice a pathological response to muscle damage was recently demonstrated (Megeney *et al.,* 1996). Apparently, these mice exert augmented muscle stem cell renewal and a reduction of satellite cell differentiation leading to the severe deficiency of skeletal muscle regeneration on freeze crush injury of limb muscles or in dystrophic mdx mutant mice. A role of MyoD in muscle regeneration is further supported by its transcriptional activation when quiescent satellite cells begin to proliferate in muscle injuries. In MyoD mutants these cells fail to exit the cell cycle and enter the differentiation program.

A number of growth factors have been implicated in the activation of satellite cell proliferation and MyoD expression (reviewed by Grounds and Yablonka-Reuveni, 1993). Neutralizing antibodies to bFGF, IGF-1, LIF, and hepatic growth factor or scatter factor (HGF/SF) inhibit muscle regeneration (Lefaucheur and Sebille, 1995a) and supplementation of these growth factors increases muscle fiber density and size after injury (Lefaucheur and Sebille, 1995b). The importance of FGF-6 for activation of satellite cells and MyoD expression was recently demonstrated by targeted gene disruption in the mouse (Floss *et al.,* 1997). FGF-6 exhibits expression predominantly in the myogenic lineage. Wild-type mice upregulate FGF-6 expression after skeletal muscle injuries and completely restore the damage within a week or two. In contrast, in FGF-6($-/-$) mutant mice the number of MyoD and myogenin expressing activated satellite cells is significantly reduced resulting in a severe regeneration defect, and fibrosis and myotube degeneration. Similarly, FGF-6($-/-$) mutants crossed with mdx mice lead to a marked dystrophic phenotype in skeletal muscles of the double homozygous mice with myotube degeneration, accumulation of mononuclear cells, and large collagen deposits. The regeneration defect is apparently not due to a reduced pool of quiescent satellite cells but presumably to the inability to activate them suggesting that FGF-6 is a critical component to stimulate, attract, or activate satellite cells for differentiation (Floss *et al.,* 1997).

D. Collaboration of MDFs with Other Transcription Factors in Myogenesis

Numerous *in vitro* studies demonstrated that MDFs interact with additional proteins to acquire high affinity DNA binding to the canonical E-box and possibly muscle-specific transcriptional activity (Murre *et al.,* 1989; Braun and Arnold, 1991). Proteins that modulate MDF functions include the E2A gene products and their relatives which act positively, and members of the Id, twist, and I-mf families of HLH proteins which affect MDFs negatively (for reviews see: Arnold and Braun, 1996; Yun and Wold, 1996; Molkentin and Olson, 1996). Whether E2A proteins constitute the actual dimerization partners of MDFs *in vivo* is still an open question, as E2A knockout mice have no muscle phenotype (Zhuang *et al.,* 1994). However, a number of E2A-like proteins, such as ITF-1, ITF-2, and HEB have been identified that may substitute for E2A (Henthorn *et al.,* 1990; Hu *et al.,* 1992).

1. Potentially Negative Regulators of MDF Function

Helix–loop–helix (HLH) proteins which may negatively control MDF actions are mostly not expressed in the myotome but in other somite compartments and elsewhere in the embryo. Evidence that they may inhibit myogenesis is mainly based on their expression patterns *in vivo* and functional studies in cell culture or in the test tube. The Id proteins heterodimerize with E2A proteins or MDFs and prevent them from DNA binding, because they lack the basic DNA binding domain (Jen *et al.,* 1992). Id proteins accumulate predominantly in proliferating mesenchymal and epithelial tissues and decline when differentiation ensues. In contrast to the developing nervous system and the heart, Id is not expressed appreciably in early epithelial somites or later in the myotome (Evans and O'Brien, 1993; Wang *et al.,* 1992). The actual role of Id proteins *in vivo* is still unclear, as no gene inactivations have yet been performed.

The bHLH transcription factor twist is expressed throughout the presomitic mesoderm and in the neural tube. In differentiating somites twist transcripts are present in the sclerotome and the lateral plate mesoderm but are excluded from the

4. Genetics of Muscle Determination and Development 147

dermomyotome. Two mechanisms have been proposed by which twist inhibits myogenesis in tissue culture cells. First, it competes for E-proteins by dimerization and thereby inhibits high-affinity DNA binding of MDFs. Second, it prevents transactivation by direct interference with MEF2/MDF complexation (Hebrok *et al.*, 1994; Spicer *et al.*, 1996). The expression pattern of twist and its ability to inhibit muscle-specific transcription *in vitro* may argue for a repressor function in myogenesis of vertebrates. In *Drosophila*, however, twist apparently plays a different role (Baylies and Bate, 1996). Here, high levels of twist seem to be required for somatic myogenesis, while it blocks the formation of other mesodermal derivatives. Expression of twist in the fly ectoderm drives these cells into myogenesis. On this basis Baylies and Bate (1996) proposed that in *Drosophila* twist regulates mesodermal differentiation and propels a subset of mesodermal cells into somatic muscle cell lineages. A similar role of twist for the subdivision of mesoderm in vertebrates appears unlikely, since skeletal muscle development in twist mutants has not been reported to be affected. However, this analysis may be complicated by the early lethality of twist mutant mice at E11.5 (Chen and Behringer, 1995). I-mf protein may be another factor that prevents inappropriate myogenesis. This molecule is highly expressed in the sclerotome. *In vitro* it inhibits transactivation by MyoD family members and represses myogenesis. I-mf associates with MyoD and retains it in the cytoplasm, presumably by masking the nuclear localization signal. I-mf can also interfere with the DNA binding activity of the myogenic bHLH regulators (Chen *et al.* 1996).

2. MEF2 Transcription Factors Positively Cooperate with MDFs in Myogenesis

Many promoters or enhancers of muscle-specific genes contain AT-rich binding sites for MADS-box transcription factors. The four members of the MEF2 family, MEF2A–D, have been shown to contribute to the activation of muscle genes in transfection experiments in cell culture (for review: Ludolph and Konieczny, 1995). Moreover, coexpression of MEF2A–D together with MDFs substantially increases the efficiency of myogenic conversion of nonmuscle cells (Kaushal *et al.*, 1994; Molkentin *et al.*, 1995). A claim that MEF2 proteins alone are sufficient to induce myogenesis in nonmuscle cells (Kaushal *et al.*, 1994) was not confirmed by others (Molkentin *et al.*, 1995; Yun and Wold, 1996). Two mechanisms may explain the muscle-promoting activity of MEF2 transcription factors. First, MEF2 may directly bind to muscle-specific promotors/enhancers and activate transcription via its transactivation domain. This mode of action has been demonstrated for the myogenin gene which contains essential MEF2 binding sites that are necessary for corrrect spatiotemporal myogenin expression in the mouse embryo and may help to enhance myogenin gene transcription by an autoregulatory loop (Edmondson *et al.*, 1992; Yee and Rigby, 1993; Buchberger *et al.*, 1994). Second, MEF2 proteins can interact with MDF/E12 heterodimers, possibly as obligatory cofactors of myogenesis. This idea has been suggested since both the interaction of MEF2 proteins with MDFs and the muscle-specific transactivation function of myogenic bHLH factors depend on the same conserved alanine and threonine residues within the bHLH domain of MDFs (Molkentin et al., 1995).

Even if this is only circumstantial evidence, MEF2 proteins are good candidates for the putative cofactors critical for the induction of the myogenic program. This sort of cooperation may also explain how muscle-specificity can be established for genes that lack any MDF binding sites, as heteromeric MEF2–MDF-E protein complexes can bind to promotors either through the bHLH domain or the MEF2 MADS box thereby eliminating the need for E-boxes (Molkentin et al., 1995; Kaushal et al., 1994). MEF2 transcription factors are expressed in many tissues including CNS, skeletal muscle, smooth, and cardiac muscle, and MEF2 related binding activities are even more widespread (Lyons et al., 1995). Therefore, cell type specific gene activation by MEFs may always depend on interactions with tissue-restricted transcription factors. The precise role of MEF2 proteins for skeletal myogenesis *in vivo* is yet unclear. Disruption of the *MEF2C* gene in mice results in a cardiac phenotype without apparent defects in skeletal muscle development (Lin et al., 1997). Other MEF2 gene mutations have not been reported so far. The single *D-Mef2* gene in *Drosophila* is necessary for the differentiation of all muscle tissues including heart and somatic muscles (Lilly et al., 1995). Whether this reflects a fundamental difference between flies and vertebrates or is due to the multiple *MEF2* genes in higher eukaryotes needs further investigations.

E. Transcriptional Control of MDF Gene Expression

A comprehensive view of skeletal muscle development will not only depend on understanding the functions of MDF proteins but also on the knowledge of how MDF gene expression is regulated during embryogenesis. A genetic hierarchy controlling myogenesis can be deduced at least in part from the various mouse knockout phenotypes. According to these it appears clear that expression of the myogenin gene requires the activity of Myf-5 or MyoD as one of the essential upstream transcription factors. Promoter analysis of the myogenin gene in transgenic mice confirmed the importance of an E-box binding site within the promoter for correct spatiotemporal myogenin expression (Yee and Rigby, 1993; Cheng et al., 1993). These transgene studies also revealed that binding of MEF2 transcription factors is crucial for myogenin gene activation both *in vitro* and *in vivo* (Buchberger et al., 1994; Cheng et al., 1993). No other essential enhancer sites have been identified. Thus, transcription of the myogenin gene seems to be controlled by relatively few transcription factors, most critically myogenic bHLH and MEF2 MADS-box proteins. Conversely, myogenin does not play a role for the expression of Myf-5 and MyoD, since both genes are active in myogenin null mutants. These observations place myogenin genetically downstream of Myf-5 and MyoD.

4. Genetics of Muscle Determination and Development

Despite the clear temporal order of Myf-5 and MyoD gene activation in muscle forming cells of the mouse embryo, the epistatic relationship between both genes appears not so simple. The original observation that MyoD is activated in muscle precursor cells of Myf-5 null mutants at approximately the correct developmental time suggested that MyoD expression occurs independent of Myf-5 (Braun *et al.*, 1994). This notion is supported by the normal expression of a *LacZ* transgene driven by a 6-kb MyoD promoter fragment in Myf-5 null mutants (Kablar *et al.*, 1997). In MRF4 mutants MyoD activation also seems to be correct in time and space indicating that neither Myf-5 nor MRF4 act upstream of MyoD (Braun and Arnold, 1995; Patapoutian *et al.*, 1995). Enhancers of the human and murine MyoD genes have been identified that drive high level expression in transgenic animals but no transcription factors binding to the cis-regulatory elements have yet been identified (Faerman *et al.*, 1995; Goldhamer *et al.*, 1995; Tapscott *et al.*, 1992).

An interesting new perspective on *MyoD* gene regulation was proposed recently by the analysis of Myf-5/Pax3 double homozygous mutant mice (Tajbakhsh *et al.*, 1997). These mutants lack body muscles and fail to activate the *MyoD* gene which therefore seems to be dependent on either Myf-5 or Pax3. In line with this mutant phenotype it was also shown that ectopic expression of Pax3 or Pax7 results in transcriptional activation of the *MyoD* gene even in neural ectoderm (Maroto *et al.*, 1997). Moreover, myogenesis promoting signals from surface ectoderm or wnt and sonic hedgehog can induce somitic expression of Pax3 and Pax7 concomitant with Myf-5 and prior to MyoD. Whether the paired-box transcription factors Pax3 or Pax7 activate the *MyoD* gene directly or indirectly is yet unclear, as no binding sites for these transcription factors have been recognized in the enhancers controlling MyoD gene expression.

Transcriptional regulation of the *MRF4* and *Myf-5* genes during mouse development is a complex issue, as both genes are closely linked on chromosome 10 approximately 8.5 kb apart in the same transcriptional orientation. Despite the close physical linkage both genes exert disparate spatiotemporal patterns during development but common cell type-specific expression. Thus, it should be interesting and informative to unravel the organization and utilization of individual gene regulatory elements within this gene locus. Reporter constructs driven by 5.5-kb and 6.5-kb fragments of the Myf-5 and MRF4 promoters, respectively, support expression in a subset of muscles in transgenic mice but fail to recapitulate the entire embryonic pattern. In particular the early expression in somites and limb buds cannot be achieved with these promoter fragments (Patapoutian *et al.*, 1993).

Successful attempts to reveal the correct developmental regulation of the *Myf-5* gene have been reported recently using yeast artificial chromosomes (YACs) carrying the *LacZ* gene in the Myf-5 transcription unit (Zweigerdt *et al.*, 1997). This investigation shows that Myf-5 expression in somites and limb buds is controlled by separate regions which are unusually distant to the gene itself (Fig. 3). Interestingly, these putative enhancers are located upstream of the MRF4 gene with no

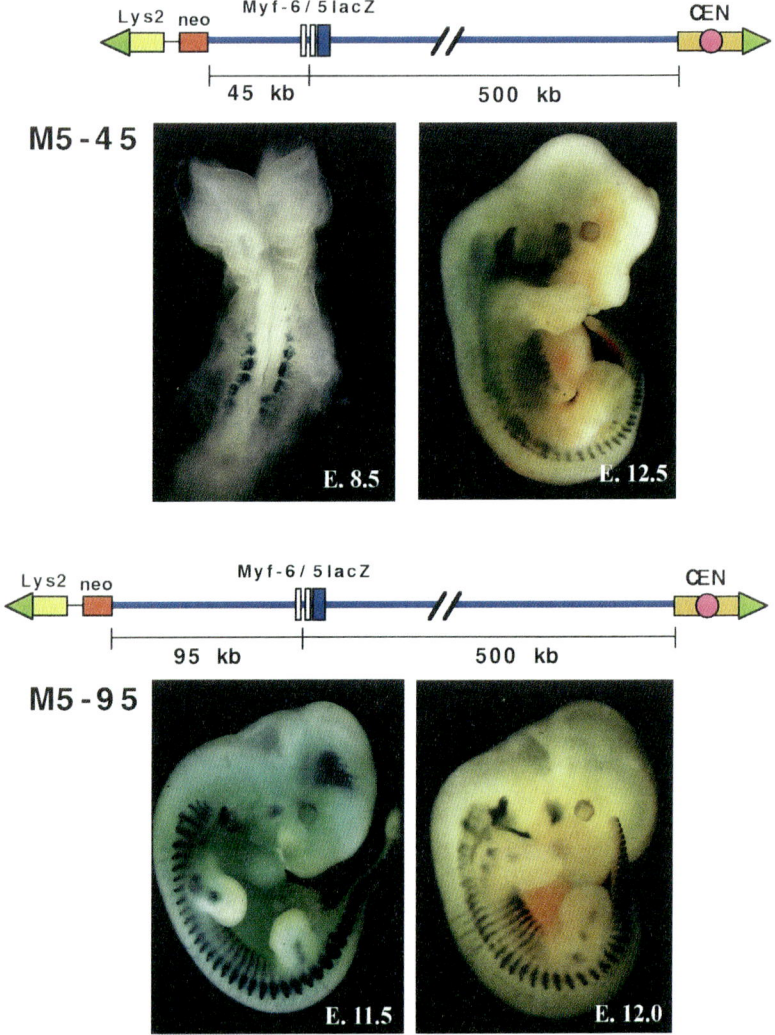

Figure 3 Chimeric mouse embryos expressing Myf-5 driven LacZ from YAC transgenes. The upper panels show the correct Myf-5 expression in somites and branchial arches obtained with the indicated YAC M5-45 containing 45 kb of 5' upsteam sequence and 500 kb downstream of the *Myf-5* gene. The downstream sequence is not required for this expression pattern (H.-H. Arnold, R. Zweigerdt, and T. Braun, 1998, unpublished results). Note the lack of Myf-5 expression in limb buds. The lower panels indicate the correct embryonic expression pattern of Myf-5 including the limb buds obtained with YAC M5-95 which contains 95 kb of 5' upstream sequence. The developmental stages of the analyzed embryos are indicated.

apparent effect on its expression. How this ensulation of enhancer action occurs will require further dissection of the Myf-5 regulatory elements. Apart from the long distance control seen in the Myf-5 locus, regions within the *MRF4* gene also

4. Genetics of Muscle Determination and Development 151

seem to influence Myf-5 transcription. The various MRF4 knockout alleles clearly indicate that alterations of the *MRF4* gene affect Myf-5 expression in cis (Floss *et al.,* 1996; Olson *et al.,* 1996; Yoon *et al.,* 1997). How these effects are mediated is not known but detailed analysis of the involved sequences performed in P. Rigby's laboratory confirm the general notion of their importance (1998, personal communication). The biphasic expression of the *MRF4* gene with a short transient period in early somites and the late phase in all muscles of the fetus and adult animals has been obtained with a 8.5-kb MRF4 promoter fragment in transgenic mice (Pin *et al.,* 1997). Significantly, however, in these transgenics the somitic MRF4 expression was limited to the thoracic and lumbar level suggesting that additional elements may be required to respond to signals that mediate MRF4 transcription in more rostral and caudal somites.

In summary, our understanding of the molecular mechanisms that control MDF gene expression is still limited. Current investigations in many laboratories will soon allow explanations of how various external cues and intrinsic cellular properties collaborate to activate the MDF genes as one of the first, if not the first event in the formation of skeletal muscles.

IV. Development of the Limb Musculature

Precursors of limb musculature are provided by the lateral portion of somites at the level of limb buds (Chevallier *et al.,* 1977; Christ *et al.,* 1977; Williams and Ordahl, 1994). These progenitor cells migrate from the dermomyotome toward the limb but do not express MDFs on route (Sassoon *et al.,* 1989; Braun *et al.,* 1994; Tajbakhsh and Buckingham, 1994). In fact, it may be important that the migrating cells remain undifferentiated which is possibly due to BMP-4 that is produced in lateral plate mesoderm (Pourquié *et al.,* 1996). Indeed, transplantation of lateral dermomyotome into limb buds which removes it from the influence of BMP-4 results in muscle differentiation (Daston *et al.,* 1996). The subdivision of myogenic progenitors in epaxial and hypaxial muscle precursors is part of somite patterning through the adjacent structures (Aoyama and Asamoto, 1988; Ordahl and Le Douarin, 1992). In contrast to the medial myotome, muscle formation in limb buds does not depend on signals from the neural tube and/or the notochord (Rong *et al.,* 1992; Bober *et al.,* 1994a) but involves genes, such as *Pax-3, hepatic growth or scatter factor* (*HGF/SF*), and its receptor c-met that are seemingly less important for epaxial muscle formation. Development of limb musculature proceeds in several steps including de-epithelialization of the lateral dermomyotomal lip, migration of progenitor cells to the limb bud, homing into the prospective muscle regions, myogenic specification of precursors, and differentiation to functional muscle.

Pax-3 initially expressed throughout the paraxial mesoderm becomes later confined to the dermomyotome and appears essential to generate migrating muscle progenitors (Williams and Ordahl, 1994). Splotch mutant mice lacking Pax-3 fail to form limb musculature (Bober *et al.*, 1994b; Goulding *et al.*, 1994; Daston *et al.*, 1996). How Pax-3 affects migratory muscle progenitors is unknown but it is most likely not involved in myogenic determination, as Pax-3 minus cells can form muscle (Daston *et al.*, 1996). The delamination of dermomyotomal cells at limb level appears induced by signals generated in the limb buds (Hayashi and Ozawa, 1995). It involves the HGF/c-met signaling system, as inactivations of both genes in mice result in a similar deficiency of migrating muscle precursor cells (Bladt *et al.*, 1995). In fact, HGF is expressed in limb bud mesenchyme and can induce epitheliomesenchymal transformation of dermomyotomal cells that express the c-met receptor (Brand-Saberi *et al.*, 1996c; Heymann *et al.*, 1996). Whether HGF/SF also directs cell migration of limb precursors is still debated (Heymann *et al.*, 1996; Bladt *et al.*, 1995). It is reasonable to assume that muscle cell migration involves interactions between muscle progenitor cells and cells of the limb bud mesoderm mediated by cell surface and adhesion molecules as well as secreted signals acting as chemoattractants. The prepatterned vascular network between somites and lateral plate mesoderm has been implicated as the migration pathway for limb muscle precursors (Solursh *et al.*, 1987). Antibodies against fibronectin (Brand *et al.*, 1985) or integrin (Jaffredo *et al.*, 1988) inhibit migration of muscle precursor cells. Thus, extracellular matrix including components such as fibronectin, glycosaminoglycans, and hyaloronic acid may be essential (Krenn *et al.*, 1991). Significantly, however, genetic disruption of the integrin complex does not affect limb musculature in chimeric mice casting doubts on the role of fibronectin–integrin interactions for the migration process (Fässler and Meyer, 1995). The recently identified homeobox gene *lbx1* is expressed in migrating muscle precursor cells and may be a candidate transcription factor which activates cell surface protein genes that are involved in the pathfinding process (Jagla *et al.*, 1995). Lbx1 is coexpressed with Pax-3 and c-met in the lateral dermomyotome and limbs but probably acts downstream of Pax-3 (Mennerich *et al.*, 1998). The investigation of the precise role of lbx1 must await mutation of the gene.

Local arrangement of myogenic precursors in the limb muscle blastema and their alignment for fusion into myotubes most likely involves surface molecules which mediate heterophilic and homophilic cell–cell recognition, such as N-CAM, N-cadherin, and M-cadherin. All three proteins are expressed in limb muscle cells and have been shown to be important for muscle cell interactions and cell fusion. N-cadherin was also suggested to play a critical role in the migration of myoblasts, since *in vivo* injections of antibodies or Fab-fragments against the homophilic binding site of N-cadherin lead to aggregation of myoblasts, apparently due to immobilization (Brand-Saberi *et al.*, 1996b). In *Dosophila* a N-cadherin mutation does not alter muscle cell migration (M. Takeichi, Kyoto University, 1997, personal communication). The knockout mutation in mice is not informative in this regard, since homozygous mutant embryos die before migration starts (Radice *et al.*, 1997). Recent experiments with muscle cells from homozygous N-cadherin mutants *in vitro* and *in vivo* do not support an essential role of N-cadherin for myoblast fusion (Charlton *et al.*, 1997), however its function in myoblast migration is still not resolved.

V. Structures Surrounding the Somites and Signals Emanating from Them Affect Myogenesis

Virtually all tissues adjacent to somites have been discussed as important for somite patterning and irreversible commitment of initially multipotent somitic cells to defined fates including prospective epaxial and hypaxial muscles (Aoyama and Asamoto, 1988; Christ and Ordahl, 1995). The various experimental approaches, however, are complex and sometimes yield controversial results.

The important influence of the neural tube on myotome formation or survival of epaxial muscle cells has been demonstrated primarily by ablations and transplantations in chick embryos (Teillet and Le Douarin, 1983; Rong *et al.*, 1992; Pownall *et al.*, 1996; Spence *et al.*, 1996; Xue and Xue, 1996; Bober *et al.*, 1994a). However, myogenic cells arise even in the absence of neural tube, possibly under the influence of dorsal surface ectoderm (Spence *et al.*, 1996; Bober *et al.*, 1994a). While Xue and Xue (1996) implicated the entire neural tube, Spence *et al.* (1996) emphasize the role of the dorsal neural tube alone, but *in vitro* the ventral neural tube appears sufficient to induce muscle (Buffinger and Stockdale, 1995). The mouse open brain mutation which exerts a defect in the development of the dorsal neural tube and lacks epaxial muscles supports a critical role of the dorsal neural tube for medial myotome formation but not for hypaxial muscles (Spörle *et al.*, 1996). The role of the notochord on myogenesis is also controversial, because no effect has been reported in one set of experiments (Teillet and Le Douarin, 1983; Rong *et al.*, 1992), whereas others describe myotome induction (Bober *et al.*, 1994a; Pownall *et al.*, 1996; Xue and Xue, 1996) or repression (Brand-Saberi *et al.*, 1993; Pourquié *et al.*, 1993; Goulding *et al.*, 1994; Xue and Xue, 1996). The explanation for the controversy may be that cells of the dorsomedial quadrant of somites react differently to notochord depending on their developmental stage and degree of maturation (Williams and Ordahl, 1997). Somites of all stages are competent to form chondrocytes instead of muscle and dermis in response to notochord but this competence decreases in more mature somites which therefore develop muscle more efficiently. Thus, notochord can clearly promote chondrogenesis in newly formed somites but older somites acquire resistance to fate changes from muscle to chondrogenic cells. The lateral plate mesoderm represses myotome formation probably due to BMP-4 which is expressed in the lateral plate and can confer inhibition of myogenesis on the medial half of somites (Pourquié *et al.*, 1995, 1996). BMP-4 induces sim-1 expression, a marker for the lateral half of somites, and thus is a candidate signaling molecule for lateral somite specification. BMP-4, however, is also expressed in the dorsal neural tube where its lateralizing effect may be counteracted by factors secreted from the neural tube that restrict sim-1 expression (Pourquié et al., 1996).

The roles of noggin, shh, and wnts as potential antagonists of lateral somite patterning have been examined (Hirsinger et al. 1997). Implantation of noggin expressing cells between the lateral plate and somites results in upregulation of MyoD in the lateral somite. Wnt1 also antagonizes BMP-4 mediated lateralization, possibly by activating noggin expression. The importance of noggin *in vivo* however is still unclear. Follistatin is another molecule which can antagonize BMP and TGFβ activities and seems to play a role in somite compartmentalization in avians (Amthor et al., 1996). Marcelle et al., (1997) showed recently that BMPs expressed in the dorsal neural tube upregulate Wnt-11 in somites, probably indirectly via Wnt1 and Wnt-3a and induce the medial lip of the myotome. GDF-8 or myostatin, is another member of the TGF-β family which is expressed in developing mouse somites and regulates the amount of skeletal muscle cells (McPherron et al., 1997). Myostatin knockout mice develop muscles 2–3 times the size of wild-type mice which is essentially the opposite to what is observed in follistatin knockout mice (Matzuk et al., 1995). Thus, the size of muscles may be controlled by the number of cell divisions in the myotome most likely regulated by a delicate balance of positive and negative factors acting on myoblasts.

Surface ectoderm has also been reported to induce muscle formation (Kenny-Mobbs and Thorogood, 1987; Cossu et al., 1996; Maroto et al., 1997) but other investigators found it entirely dispensable for myogenesis (Buffinger and Stockdale, 1994, 1995; Stern and Hauschka, 1995) or playing only a minor role (Spence et al., 1996). In an investigation of dorsoventral somite patterning Dietrich et al. (1997) showed that signals from the dorsal neural tube and surface ectoderm stimulate development of the dermomyotome, whereas signals from notochord and floor plate ventralize the somite. According to these authors high levels of ventralizing signals override any dorsal information and induce sclerotome which is in contrast to Williams and Ordahl (1997) who demonstrated that myocytes can appear close to notochord in mature somites.

The influence of individual tissues neighboring the somites or isolated signals on myogenesis without the complication of additional structures in the vicinity as it is the case in the embryo has been explored in explants of paraxial mesoderm *in vitro*. The effect of conjugated tissues or applied signaling molecules can then be assessed by RT–PCR (reverse transcriptase–polymerase chain reaction) for the accumulation of muscle-specific marker transcripts. In this type of experiment both the neural tube and the notochord appear to be essential for the activation of myotome-specific genes in explanted segmental plate (Münsterberg and Lassar, 1995; Stern and Hauschka, 1995).

More recently it was demonstrated that bFGF and TGF-β family members derived from the dorsal neural tube synergistically stimulate muscle differentiation in explant cultures (Stern et al., 1997). A two step mechanism for the bFGF/TGFβ1 effect was suggested, because TGFβ1 alone was sufficient during the first

4. Genetics of Muscle Determination and Development 155

12 hr of explant culture, whereas bFGF and the TGFβ like factor were required thereafter in order to induce myogenesis. Which member of the TGFβ family may mediate this effect is unknown but an antibody against TGFβ1–3 was unable to neutralize this TGFβ related activity.

A similar two step mechanism of muscle induction was also proposed by Münsterberg and Lassar (1995) who demonstrated that somites respond differently to axial structures depending on the stage of maturation. Immature somites require the presence of the floor plate/notochord complex to induce muscle-specific bHLH transcription factors, while in later somites the presence of notochord and floor plate is no longer required but the dorsal neural tube alone is sufficient. Surgical ablations in the developing quail embryo support this result (Pownall *et al.,* 1996). Early removal of the notochord at the time when bHLH gene transcription is being initiated results in complete loss of bHLH gene expression. Ablation of the neural tube and notochord in stages when somites have already switched on bHLH gene transcription results in continuing transcription in the most mature somites. Thus, early exposure to ventral midline signals may sensitize somites to respond to the neural tube later (Münsterberg *et al.,* 1995). After a certain time, however, the myotome becomes autonomous with respect to transcription of bHLH genes (Bober *et al.,* 1994a; Pownall *et al.,* 1996). In light of the various observations a unifying hypothesis of the true key regulators controlling muscle induction under physiological conditions seems difficult. Genetic approaches may be needed to identify the relevant signals *in vivo.*

VI. The Myogenic Potential of Mesoderm: Induction or Derepression?

The realization of the myogenic potential in mesoderm may be considered conceptually in two ways. Mesoderm may have an intrinsic tendency to form muscle but is prevented from doing so by suppressive signals acting on it during and after gastrulation. Alternatively, mesoderm may only develop into muscle, if it is induced by positive signals emanating from adjacent structures. There is experimental evidence for both views which are of course not mutually exclusive. The potential role of inductive events for myogenesis has been discussed in the previous section.

Numerous observations in lower vertebrates and invertebrates argue for an intrinsic myogenic potential of mesoderm. For instance, cells from pregastrulation chick or *Xenopus* embryos express MyoD and differentiate into muscle when explanted and cultured at clonal density in serum-free medium (Holtzer *et al.,* 1990; Godsave and Slack, 1991). Similarly, dissociated chick epiblast plated at low density forms muscle but not when cultured as intact epithelium (George-Weinstein *et al.,* 1996). These results suggest that the myogenic program may be inhibited by cell to cell contacts but ensues in the absence of cell interactions. In zebrafish, the homeobox gene *floating head* (*flh*) appears to confer suppression of myogenesis on the midline (Talbot *et al.*, 1995). In the flh mutant, cells which are normally destined to form notochord now produce muscle. During early developmental stages the midline cells of mutant embryos coexpress both, notochord and muscle markers suggesting that flh does not provide inductive cues but prevents myogenesis. In its absence the myogenic program prevails and represses notochord formation. In *Xenopus* MyoD transcripts are found in the oocyte and zygotic expression starts ubiquitously at midblastula transition (Harvey, 1990; Rupp and Weintraub, 1991). Only later MyoD becomes restricted to muscle forming regions in the embryo suggesting that initially myogenesis is inhibited in all nonmuscle forming cells and released from inhibition in prospective muscle cells. Also in *Caenorhabditis elegans* MyoD is initially expressed in many more cells than those that actually will form muscle (Krause *et al.*, 1994). In this organism musculature is determined in descendants of defined blastomers, referred to as EMS, C, D, and AB by inductive interactions of daughter cells (Schnabel, 1994). Ablation of all blastomers except EMS results in the normal number of muscles indicating that under normal circumstances EMS descendants are prevented to form extra muscle and triggered into neuronal and nonmuscle mesodermal cells instead. In the invertebrate ascidian the primary muscle cell lineage also arises cell autonomously dependent on cytoplasmic determinants (Nishida, 1990). The primary muscle precursors express the ascidian homologue of MyoD but some of these cells are fated to produce notochord, obviously by repression of their myogenic capacity (Araki *et al.*, 1994). Segregation of both cell lineages and the downregulation of MyoD coincide with the expression of the brachyury homologue in future notochord cells (Yasuo and Satoh, 1993). Thus, it is a repressive principle that prevents myogenesis and supports notochord development. That myogenic fate may be a ground state in somite development seems supported by the chordate amphioxus, probably the closest relative of vertebrates and a useful model organism for the proximate invertebrate ancestor of vertebrates. The dermal segments of amphioxus, analogues of somites in vertebrates lack the dermomyotome and the sclerotome, as no axial skeleton and migratory somitic cells exist in this organism. The primary cell fate of mesoderm in amphioxus is muscle. Even in midline cells, such as the notochord one finds sarcomeric proteins emphasizing the dominance of this differentiation pathway (Holland *et al.*, 1995, 1997).

Taken together these various observations suggest that muscle formation originally may be a default pathway of mesoderm differentiation in many animal species. Negative regulation of the myogenic determination genes therefore seems important to ascertain that muscles do not form ectopically. It may also be essential in nonmyotomal cells of somites, because of their intrinsic bias toward muscle. A number of genes which can suppress myogenic differentiation are known and have been described above. In summary, the correct formation of muscles in space and time may be the result of an intricate interplay of negative and positive signals. In this view the intrinsic propensity of mesoderm to develop into muscle is controlled by a lasting repression that is only regionally released by activating signals. This then culminates in the stable expression of the myogenic regulators and the differentiated skeletal muscle phenotype.

Acknowledgments

Work in the authors' laboratory was supported by the Deutsche Forschungsgemeinschaft and the Fond der Chemischen Industrie.

References

Amthor, H., Connolloy, D., Patel, K., Brand-Saberi, B., Wilkinson, D. G., Cooke, J., and Christ, B. (1996). The expression and regulation of follistatin and a follistatin-like gene during avian somite compartmentalization and myogenesis. *Dev. Biol.* **178,** 343–362.
Aoyama, H., and Asamoto, K. (1988). Determination of somite cells: Independence of cell differentiation and morphogenesis. *Development* **104,** 15–28.
Araki, I., Saiga, H., Makabe, K. W., and Satoh, N. (1994). Expression of AMD1, a gene for a MyoD-related factor in the ascidian halocynthia-roretzi. *Roux's Arch. Dev. Biol.* **203,** 320–327.
Arnold, H. H., and Braun, T. (1993). The role of Myp-5 in somitogenesis and the development of skeletal muscles in vertebrates. *J. Cell Science* **104,** 957–960.
Arnold, H. H., and Braun, T. (1996). Targeted inactivation of myogenic factor genes reveals their role during myogenesis: A review. *Int. J. Dev. Biol.* **40,** 345–363.
Asakura, A., Fujisawa-Sehara, A., Komiya, T., Nabeshima, Y., and Nabeshima, Y. (1993). MyoD and myogenin act on the chicken myosin light-chain 1 gene as distinct transcriptional factors. *Mol. Cell. Biol.* **13,** 7153–7162.
Baylies, M. K., and Bate, M. (1996). Twist: A myogenic switch in *Drosophila*. *Science* **272,** 1481–1484.
Bladt, F., Riethmacher, D., Isenmann, S., Aguzzi, A., and Birchmeier C. (1995). Essential role for the c-met receptor in the migration of myogenic precursor cells into the limb bud. *Nature* **376,** 768–771.
Blau, H. M., Chiu, C. P., and Webster, C. (1983). Cytoplasmic activation of human nuclear genes in stable heterocaryons. *Cell* **32,** 1171–1180.
Bober, E., Lyons, G. E., Braun, T., Cossu, G., Buckingham, M., and Arnold, H. H. (1991). The muscle regulatory gene, *Myf-6,* has a biphasic pattern of expression during early mouse development. *J. Cell Biol.* **113,** 1255–1265.
Bober, E., Brand-Saberi, B., Ebensperger, C., Wilting, J., Balling, R., Paterson, B. M., Arnold, H. H., and Christ, B. (1994a). Initial steps of myogenesis in somites are independent of influence from axial structures. *Development* **120,** 3073–3082.
Bober, E., Franz T., Arnold, H. H., Gruss P., and Tremblay, P. (1994b). Pax-3 is required for the development of limb muscles: A possible role for the migration of dermomyotomal muscle progenitor cells. *Development* **120,** 603–612.
Brand, B., Christ, B., and Jacob, H. J. (1985). An experimental analysis of the developmental capacities of distal parts of avian leg buds. *Am. J. Anat.* **173,** 321–340.
Brand-Saberi, B., Ebensperger, C., Wilting, J., Balling, R., and Christ, B. (1993). The ventralizing effect of the notochord on somite differentiation in chick embryos. *Anat. Embryol.* **188,** 239–245.
Brand-Saberi, B., Wilting, J., Ebensperger, C., and Christ, B. (1996a). The formation of somite compartments in the avian embryo. *Int. J. Dev. Biol.* **40,** 411–420.
Brand-Saberi, B., Gamel, A. J., Krenn, V., Muller, T. S., Wilting, J., and Christ, B. (1996b). N-cadherin is involved in myoblast migration and muscle differentiation in the avian limb bud. *Dev. Biol.* **178,** 160–173.
Brand-Saberi, B., Muller, T. S., Wilting, J., Christ, B., and Birchmeier, C. (1996c). Scatter factor/

hepatocyte growth factor (SF/HGF) induces emigration of myogenic cells at interlimb level in vivo. *Dev. Biol.* **179**, 303–308.
Braun, T., and Arnold, H. H. (1991). The four human muscle regulatory helix-loop-helix proteins Myf3–Myf6 exhibit similar hetero-dimerization and DNA binding properties. *Nucleic Acids Res.* **19**, 5645–5651.
Braun, T., and Arnold, H. H. (1995). Inactivation of Myf-6 and Myf-5 genes in mice leads to alterations in skeletal muscle development. *EMBO J.* **14**, 1176–1186.
Braun, T., and Arnold, H. H. (1996). *Myf-5* and *myoD* genes are activated in distinct mesenchymal stem cells and determine different skeletal muscle cell lineages. *EMBO J.* **15**, 310–318.
Braun, T., Bober, E., Buschhausen-Denker, G., Kohtz, S., Grzeschik, K. H., Arnold, H. H., and Kotz, S. (1989a). Differential expression of myogenic determination genes in muscle cells: Possible autoactivation by the Myf gene products [published erratum appears in *EMBO J.* (1989) **Dec.;8**(13): 4358]. *EMBO J.* **8**, 3617–3625.
Braun, T., Buschhausen-Denker, G., Bober, E., Tannich, E., and Arnold, H. H. (1989b). A novel human muscle factor related to but distinct from MyoD1 induces myogenic conversion in 10T1/2 fibroblasts. *EMBO J.* **8**, 701–709.
Braun, T., Bober, E., Winter, B., Rosenthal, N., and Arnold, H. H. (1990a). Myf-6, a new member of the human gene family of myogenic determination factors: Evidence for a gene cluster on chromosome 12. *EMBO J.* **9**, 821–831.
Braun, T., Winter, B., Bober, E., and Arnold, H. H. (1990b). Transcriptional activation domain of the muscle-specific gene-regulatory protein myf5. *Nature* **346**, 663–665.
Braun, T., Rudnicki, M. A., Arnold, H. H., and Jaenisch, R. (1992). Targeted inactivation of the muscle regulatory gene *Myf-5* results in abnormal rib development and perinatal death. *Cell* **71**, 369–382.
Braun, T., Bober, E., Rudnicki, M. A., Jaenisch, R., and Arnold, H. H. (1994). MyoD expression marks the onset of skeletal myogenesis in Myf-5 mutant mice. *Development* **120**, 3083–3092.
Buchberger, A., Ragge, K., and Arnold, H. H. (1994). The myogenin gene is activated during myocyte differentiation by preexisting, not newly synthesized transcription factor MEF-2. *J. Biol. Chem.* **269**, 17289–17296.
Buckingham (1992).
Buffinger N., and Stockdale, F. E. (1994). Myogenic specification in somites: Induction by axial structures. *Development* **120**, 1443–1452.
Buffinger, N., and Stockdale, F. E. (1995). Myogenic specification of somites is mediated by diffusible factors. *Dev. Biol.* **169**, 96–108.
Charlton, C. A., Mohler, W. A., Radice, G. L., Hynes, R. O., and Blau, H. M. (1997). Fusion competence of myoblasts rendered genetically null for N-cadherin in culture. *J. Cell Biol.* **138**, 331–336.
Chen, Z. F., and Behringer, R. R. (1995). Twist is required in head mesenchyme for cranial neural tube morphogenesis. *Genes Dev.* **15**, 686–699.
Chen, C.-M., Kraut, N., Groudine, M., and Weintraub, H. (1996). I-mf, a novel myogenic repressor interacts with members of the MyoD family. *Cell* **86**, 731–741.
Cheng, T. C., Wallace, M. C., Merlie, J. P., and Olson, E. N. (1993). Separable regulatory elements governing myogenin transcription in mouse embryogenesis. *Science* **261**, 215–218.
Chevallier, A., Kieny, M., and Mauger, A. (1977). Limb–somite relationship: Origin of the limb musculature. *J. Embryol. Exp. Morphol.* **41**, 245–258.
Christ, B., and Ordahl, C. P. (1995). Early stages of chick somite development. *Anat. Embryol.* **191**, 381–396.
Christ, B., and Wilting, J. (1992). From somites to vertebral column. *Anat. Anz.* **174**, 23–32.
Christ, B., Jacob, H. J., and Jacob, M. (1977). Experimental analysis of the origin of the wing musculature in avian embryos. *Anat. Embryol.* **150**, 171–186.
Christ, B., Jacob, M., and Jacob, H. J. (1983). On the origin and development of the ventrolateral abdominal muscles in the avian embryo. An experimental and ultrastructural study. *Anat. Embryol.* **166**, 87–101.

4. Genetics of Muscle Determination and Development

Cossu, G., Kelly, R., Di Donna, S., Vivarelli, E., and Buckingham, M. (1995). Myoblast differentiation during mammalian somitogenesis is dependent upon a community effect. *Proc. Natl. Acad. Sci. U.S.A.* **92,** 2254–2258.

Cossu, G., Kelly, R., Tajbakhsh, S., Di Donna, S., Vivarelli, E., and Buckingham, M. (1996). Activation of different myogenic pathways: myf-5 is induced by the neural tube and MyoD by the dorsal ectoderm in mouse paraxial mesoderm. *Development* **122,** 429–437.

Daston, G., Lamar, E., Olivier, M., and Goulding, M. (1996). Pax-3 is necessary for migration but not differentiation of limb muscle precursors in the mouse. *Development* **122,** 1017–1027.

Davis, R. L., Weintraub, H., and Lassar, A. B. (1987). Expression of a single transfected cDNA converts fibroblasts to myoblasts. *Cell* **51,** 987-1000.

Denetclaw, W. F., Jr., Christ, B., and Ordahl, C. P. (1997). Location and growth of epaxial myotome precursor cells. *Development* **124,** 1601–1610.

Dietrich, S., Schubert, F. R., and Lumsden, A. (1997). Control of dorsoventral pattern in chick paraxial mesoderm. *Development* **124,** 3895–3908.

Edmondson, D. G., and Olson, E. N. (1989). A gene with homology to the myc similarity region of MyoD1 is expressed during myogenesis and is sufficient to activate the muscle differentiation program [published erratum appears in *Genes Dev.* (1990) **Aug;4**(8):1450]. *Genes Dev.* **3,** 628–640.

Edmondson, D. G., Cheng, T. C., Cserjesi, P., Chakraborty, T., and Olson, E. N. (1992). Analysis of the myogenin promoter reveals an indirect pathway for positive autoregulation mediated by the muscle-specific enhancer factor MEF-2. *Mol. Cell. Biol.* **12,** 3665–3677.

Evans, S. M., and O'Brien, T. X. (1993). Expression of the helix-loop-helix factor Id during mouse embryonic development. *Dev. Biol.* **159,** 485–499.

Faerman, A., Goldhamer, D. J., Puzis, R., Emerson, C. P., Jr., and Shani, M. (1995). The distal human myoD enhancer sequences direct unique muscle-specific patterns of lacZ expression during mouse development. *Dev. Biol.* **171,** 27–38.

Fässler, R., and Meyer, M. (1995). Consequences of lack of $\beta 1$ integrin gene expression in mice. *Genes Dev.* **9,** 1896–1908.

Floss, T., Arnold, H. H., and Braun, T. (1996). Myf-5(m1)/Myf-6(m1) compound heterozygous mouse mutants down-regulate Myf-5 expression and exert rib defects: Evidence for long-range cis effects on Myf-5 transcription. *Dev. Biol.* **174,** 140–147.

Floss, T., Arnold, H. H., and Braun, T. (1997). A role for FGF-6 in skeletal muscle regeneration. *Genes Dev.* **11,** 2040–2051.

Gamel, A. J., Brand-Saberi, B., and Christ, B. (1995). Halves of epithelial somites and segmental plate show distinct muscle differentiation behavior *in vitro* compared to entire somites and segmental plate. *Dev. Biol.* **172,** 625–639.

George-Weinstein, M., Gerhart, J., Reed, R., Flynn, J., Callihan, B., Mattiacci, M., Miehle, C., Foti, G., Lash, J. W., and Weintraub, H. (1996). Skeletal myogenesis: The preferred pathway of chick embryo epiblast cells *in vitro*. *Dev. Biol.* **173,** 729–741.

Gerber, A. N., Klesert, T. R., Bergstrom, D. A., and Tapscott, S. J. (1997). Two domains of MyoD mediate transcriptional activation of genes in repressive chromatin: A mechanism for lineage determination in myogenesis. *Genes Dev.* **11,** 436–450.

Godsave, S. F., and Slack, J. M. W. (1991). Clonal analysis of mesoderm formation in the *Xenopus* embryo. *Development* **111,** 523–530.

Goldhamer, D. J., Brunk, B. P., Faerman, A., King, A., Shani, M., and Emerson, C. P., Jr. (1995). Embryonic activation of the *myoD* gene is regulated by a highly conserved distal control element. *Development* **121,** 637–649.

Goulding, M., Lumsden, A., and Paquette, A. J. (1994). Regulation of Pax3 expression in the dermomyotome and its role in muscle development. *Development* **120,** 957–971.

Grass, S., Arnold, H. H., and Braun, T. (1996). Alterations in somite patterning of Myf-5-deficient mice: A possible role for FGF-4 and FGF-6. *Development Suppl.* **122,** 141–150.

Grounds, M. D., and Yablonka-Reuveni, Z. (1993). Molecular and cellular biology of muscle

regeneration. *In* "Molecular and Cellular Biology of Muscular Dystrophy" (T. Partridge, Ed.), pp. 210–256. Chapman & Hall, London.
Gruss, P., and Kessel, M. (1991). Axial specification in higher vertebrates. *Curr. Opin. Genet. Dev.* **1**, 204–210.
Harvey, R. P. (1990). The *Xenopus* MyoD gene: An unlocalised maternal mRNA predates lineage-restricted expression in the early embryo. *Development* **108**, 669–680.
Hasty, P., Bradley, A., Morris, J. H., Edmondson, D. G., Venuti, J. M., Olson, E. N., and Klein, W. H. (1993). Muscle deficiency and neonatal death in mice with a targeted mutation in the myogenin gene [see comments]. *Nature* **364**, 501–506.
Hayashi, K., and Ozawa, E. (1995). Myogenic cell migration from somites is induced by tissue contact with medial region of the presumptive limb mesoderm in chick embryos. *Development* **121**, 661–669.
Hebrok, M., Wertz, K., and Fuchtbauer, E. M. (1994). M-twist is an inhibitor of muscle differentiation. *Dev. Biol.* **165**, 537–544.
Henthorn, P., Kiledjian, M., and Kadesch, T. (1990). Two distinct transcription factors that bind the immunoglobulin enhancer microE5/kappa 2 motif. *Science* **247**, 467–470.
Heymann, S., Koudrava, Arnold, H. H., Köster, M., and Braun, T. (1996). Regulation and function of SF/HGF during migration of limb muscle precursor cells in chicken. *Dev. Biol.* **180**, 566–578.
Hinterberger, T. J., Sassoon, D. A., Rhodes, S. J., and Konieczny, S. F. (1991). Expression of the muscle regulatory factor MRF4 during somite and skeletal myofiber development. *Dev. Biol.* **147**, 144–156.
Hirsinger, E., Duprez, D., Jouve, C., Malapert, P., Cooke, J., and Pourquié, O. (1997). Noggin acts downstream of Wnt and Sonic Hedgehog to antagonize BMP4 in avian somite patterning. *Development* **124**, 4605–4614.
Holland, L. Z., Pace, D. G., Blink, M. L., Kene, M., and Holland, N. D. (1995). Sequence and expression of amphioxus alkali myosin light chain (Amphi MLC-alk) throughout development: Implications for vertebrate myogenesis. *Dev. Biol.* **171**, 665–676.
Holland, L. Z., Kene, M., Williams, N. A., and Holland, N. D. (1997). Sequence and embryonic expression of the amphioxus engrailed gene (AmphiEn): The metameric pattern of transcription resembles that of its segment–polarity homolog in *Drosophila*. *Development* **124**, 1723–1732.
Hollenberg, S. M., Cheng, P. F., and Weintraub, H. (1993). Use of a conditional MyoD transcription factor in studies of MyoD trans- activation and muscle determination. *Proc. Natl. Acad. Sci. U.S.A.* **90**, 8028–8032.
Holtzer, H., Schultheiss, T., DiLullo, C., Choi, J., Costa, M., Lu, M., and Holtzer, S. (1990). Autonomous expression of the differentiation programs of cells in the cardiac and skeletal myogenic lineages. *Ann. N.Y. Acad. Sci.* **599**, 158–169.
Hu, J. S., Olson, E. N., and Kingston, R. E. (1992). HEB, a helix-loop-helix protein related to E2A and ITF2 that can modulate the DNA-binding ability of myogenic regulatory factors. *Mol. Cell. Biol.* **12**, 1031–1042.
Huang, R., Zhi, Q., Neubuser, A., Muller, T. S., Brand-Saberi, B., Christ, B., and Wilting, J. (1996). Function of somite and somitocoele cells in the formation of the vertebral motion segment in avian embryos. *Acta Anat.* **155**, 231–241.
Jacob, M., Christ, B., and Jacob, H. J. (1978). On the migration of myogenic stem cells into the prospective wing region of chick embryos. A scanning and transmission electron microscope study. *Anat. Embryol.* **153**, 179–193.
Jacob, M., Christ, B., and Jacob, H. J. (1979). The migration of myogenic cells from the somites into the leg region of avian embryos. An ultrastructural study. *Anat. Embryol.* **157**, 291–309.
Jaffredo, T. J., Horwitz, A. F., Buck, C. A., Rong, P. M., and Dieterlen-Lievre, F. (1988). Myoblast migration specifically inhibited in the chick embryo by grafted CSAT hybridoma cells secreting an anti-integrin antibody. *Development* **103**, 431–446.
Jagla, K., Dolle, P., Mattei, M. G., Jagla, T., Schuhbaur, B., Dretzen, G., Bellard, F., and Bellard, M.

4. Genetics of Muscle Determination and Development 161

(1995). Mouse Lbx1 and human LBX1 define a novel mammalian homeobox gene family related to the *Drosophila lady bird* genes. *Mech. Dev.* **53,** 345–356.

Jen, Y., Weintraub, H., and Benezera, R. (1992). Over expression of Id protein inhibits the muscle differentiation program: *In vivo* association of Id with E2A proteins. *Genes Dev.* **6,** 1466–1479.

Kablar, B., Krastel, K., Ying, C., Asakura, A., Tapscott, S., and Rudnicki, M. (1997). MyoD and Myf-5 differentially regulate the development of limb versus trunk skeletal muscle. *Development Suppl.* **124,** 4729–4738.

Kaushal, S., Schneider, J. W., Nadal-Ginard, B., and Mahdavi, V. (1994). Activation of the myogenic lineage by Mef2a, a factor that induces and cooperates with MyoD. *Science* **266,** 1236–1240.

Kenny-Mobbs, T., and Thorogood, P. (1987). Autonomy of differentiation in avian brachial somites and the influence of adjacent tissue. *Development* **100,** 449–462.

Krause, M., Harrison, S. W., Xu, S. Q., Chen, L., and Fire, A. (1994). Elements regulating cell- and stage-specific expression of the *C. elegans* MyoD family homolog hlh-1. *Dev. Biol.* **166,** 133–148.

Krenn, V., Brand-Saberi, B., and Wachtler, F. (1991). Hyaluronic acid influences the migration of myoblasts within the embryonic wing bud. *Am. J. Anat.* **192,** 400–406.

Lefaucheur, J. P., and Sebille, A. (1995a). Basic fibroblastic growth factor promotes *in vivo* muscle regeneration in murine muscular dystrophy. *Neurosci. Lett.* **202,** 121–124.

Lefaucheur, J. P., and Sebille, A. (1995b). Muscle regeneration following injury can be modified *in vivo* by immune neutralization of basic fibroblastic growth factor, transforming growth factor beta-1 or insulin-like growth factor I. *J. Neuroimmunol.* **5,** 85–91.

Lilly, B., Zhao, B., Ranganayakulu, G., Paterson, B. M., Schulz, R. A., and Olson, E. N. (1995). Requirement of MADS domain transcription factor D-MEF2 for muscle formation in *Drosophila*. *Science* **267,** 688–693.

Lin, Q., Schwarz, J., Bucana, C., and Olson, E. N. (1997). Control of mouse cardiac morphogenesis and myogenesis by transcription factor MEF2C. *Science* **276,** 1404–1407.

Ludolph, D. C., and Konieczny, S. F. (1995). Transcription factor families: Muscling in on the myogenic program. *FASEB J.* **9,** 1595–1604.

Lyons, G. E., Micales, B. K., Schwarz, J., Martin, J. F., and Olson, E. N. (1995). Expression of MEF2 genes in the mouse central nervous system suggests a role in neuronal maturation. *J. Neurosci.* **15,** 5727–5738.

McPherron, A. C., Lawler, A. M., and Lee, S.J. (1997). Regulation of skeletal muscle mass in mice by a new TGF-β superfamily member. *Nature* **387,** 83–90.

Mak, K. L., To, R. Q., Kong, Y., and Konieczny, S. F. (1992). The MRF4 activation domain is required to induce muscle-specific gene expression. *Mol. Cell. Biol.* **12,** 4334–4346.

Marcelle, C., Stark, M. R., and Bronner-Fraser, M. (1997). Coordinate actions of BMPs, Wnts, Shh and noggin mediate patterning of the dorsal somite. *Development* **124,** 3955–3963.

Maroto, M., Reshef, R., Munsterberg, A. E., Koester, S., Goulding, M., and Lassar, A. B. (1997). Ectopic Pax-3 activates MyoD and Myf-5 expression in embryonic mesoderm and neural tissue. *Cell* **89,** 139–148.

Matzuk, M. M, Naifang, L., Vogel, H., Sellheyer, K., Roop, D. R., and Bradley, A. (1995). Multiple defects and perinatal death in mice deficient in Follistatin. *Nature* **374,** 360–363.

Megeney, L. A., Kablar, B., Garrett, K., Anderson, J. E., and Rudnicki, M. A. (1996). MyoD is required for myogenic stem cell function in adult skeletal muscle. *Genes Dev.* **10,** 1173–1183.

Mennerich, D., Schäfer, K., and Braun, T. (1998). Pax-3 is necessary but not sufficient for lbx1 expression in myogenic precursor cells of the limb. *Mech. Dev.* **73,** 147–158.

Miner, J. H., and Wold, B. (1990). Herculin, a fourth member of the MyoD family of myogenic regulatory genes. *Proc. Natl. Acad. Sci. U.S.A.* **87,** 1089–1093.

Molkentin, J. D., and Olson, E. N. (1996). Defining the regulatory networks for muscle development. *Curr. Opin. Gen. Dev.* **6,** 445–453.

Molkentin, J. D., Black, B. L., Martin, J. F., and Olson, E. N. (1995). Cooperative activation of muscle gene expression by MEF2 and myogenic bHLH proteins. *Cell* **83,** 1125–1136.

Münsterberg, A. E., and Lassar, A. B. (1995). Combinatorial signals from the neural tube, floor plate and notochord induce myogenic bHLH gene expression in the somite. *Development* **121,** 651–660.

Münsterberg, A. E., Kitajewski, J., Bumcrot, D. A., McMahon, A. P., and Lassar, A. B. (1995). Combinatorial signaling by Sonic hedgehog and Wnt family members induces myogenic bHLH gene expression in the somite. *Genes Dev.* **9,** 2911–2922.

Murre, C., McCaw, P.S., and Baltimore, D. (1989). A new DNA binding and dimerization motif in immunoglobulin enhancer binding, daughterless, MyoD, and myc proteins. *Cell* **10,** 777–783.

Nabeshima, Y., Hanaoka, K., Hayasaka, M., Esumi, E., Li, S., Nonaka, I., and Nabeshima, Y. (1993). Myogenin gene disruption results in perinatal lethality because of severe muscle defect [see comments]. *Nature* **364,** 532–535.

Nishida, H. (1990). Determinative mechanisms in secondary muscle lineages of ascidian embryos: Development of muscle-specific features in isolated muscle progenitor cells. *Development* **108,** 559–568.

Noden, D. M. (1983). The embryonic origins of avian cephalic and cervical muscles and associated connective tissues. *Am. J. Anat.* **167,** 257–276.

Olson, E. N. (1990). MyoD family: A paradigm for development? *Genes Dev.* **4,** 1454–1461.

Olson, E. N. (1993). Regulation of muscle transcription by the MyoD family. The heart of the matter. *Circ. Res.* **72,** 1–6.

Olson, E. N., and Klein, W. H. (1994). bHLH factors in muscle development: Dead lines and commitments; what to leave in and what to leave out. *Genes Dev.* **8,** 1–8.

Olson, E. N., Arnold, H. H., Rigby, P. W., and Wold, B. J. (1996). Know your neighbors: Three phenotypes in null mutants of the myogenic bHLH gene MRF4. *Cell* **85,** 1–4.

Ordahl, C. P., and Christ, B. (1997). Avian somite transplantation: A review of basic methods. *Methods Cell Biol* **52,** 3–27.

Ordahl, C. P., and Le Douarin, N. M. (1992). Two myogenic lineages within the developing somite. *Development* **114,** 339–353.

Ott, M. O., Bober, E., Lyons, G., Arnold, H., and Buckingham, M. (1991). Early expression of the myogenic regulatory gene, *myf-5,* in precursor cells of skeletal muscle in the mouse embryo. *Development* **111,** 1097–1107.

Patapoutian, A., Miner, J. H., Lyons, G. E., and Wold, B. (1993). Isolated sequences from the linked Myf-5 and MRF4 genes drive distinct patterns of muscle-specific expression in transgenic mice. *Development* **118,** 61–69.

Patapoutian, A., Yoon, J. K., Miner, J. H., Wang, S., Stark, K., and Wold, B. (1995). Disruption of the mouse MRF4 gene identifies multiple waves of myogenesis in the myotome. *Development* **121,** 3347–3358.

Pin, C. L., Ludolph, D. C., Cooper, S. T., Klocke, B. J., Merlie, J. P., and Konieczny, S. F. (1997). Distal regulatory elements control MRF4 gene expression in early and late myogenic cell populations. *Dev. Dyn.* **208,** 299–312.

Pourquié, O., Coltey, M., Teillet, M. A., Ordahl, C., and Le Douarin, N. M. (1993). Control of dorsoventral patterning of somitic derivatives by notochord and floor plate. *Proc. Natl. Acad. Sci. U.S.A.* **90,** 5242–5246.

Pourquié, O., Coltey, M., Breant, C., and Le Douarin, N. M. (1995). Control of somite patterning by signals from the lateral plate. *Proc. Natl. Acad. Sci. U.S.A.* **92,** 3219–3223.

Pourquié, O., Fan, C. M., Coltey, M., Hirsinger, E., Wantanabe, Y., Breant, C., Francis West, P., Brikell, P., Tessier-Lavigne, M., and Le Douarin, N. M. (1996). Lateral and axial signals involved in avian somite patterning: A role for BMP4. *Cell* **84,** 461–471.

Pownall, M. E., Strunk, K. E., and Emerson, C. P., Jr. (1996). Notochord signals control the transcriptional cascade of myogenic bHLH genes in somites of quail embryos. *Development* **122,** 1475–1488.

Radice, G. L., Rayburn, H., Matsunami, H., Knudsen, K. A., Takeichi, M., and Hynes, R. O. (1997). Developmental defects in mouse embryos lacking N-cadherin *Dev. Biol.* **181,** 64–78.

4. Genetics of Muscle Determination and Development 163

Rawls, A., Morris, J. H., Rudnicki, M., Braun, T., Arnold, H. H., Klein, W. H., and Olson, E. N. (1995). Myogenin's functions do not overlap with those of MyoD or Myf-5 during mouse embryogenesis [published erratum appears in *Dev. Biol.* (1996) **Mar.15;174**(2):453]. *Dev. Biol.* **172,** 37–50.

Rhodes, S. J., and Konieczny, S. F. (1989). Identification of MRF4: A new member of the muscle regulatory factor gene family. *Genes Dev.* **3,** 2050–2061.

Rong, P. M., Teillet, M. A., Ziller, C., and Le Douarin, N. M. (1992). The neural tube/notochord complex is necessary for vertebral but not limb and body wall striated muscle differentiation. *Development* **115,** 657–672.

Rudnicki, M. A., Braun, T., Hinuma, S., and Jaenisch, R. (1992). Inactivation of MyoD in mice leads to up-regulation of the myogenic HLH gene *Myf-5* and results in apparently normal muscle development. *Cell* **71,** 383–390.

Rudnicki, M. A., Schnegelsberg, P. N., Stead, R. H., Braun, T., Arnold, H. H., and Jaenisch, R. (1993). MyoD or Myf-5 is required for the formation of skeletal muscle. *Cell* **75,** 1351–1359.

Rupp, R. A., and Weintraub, H. (1991). Ubiquitous MyoD transcription at the midblastula transition precedes induction-dependent MyoD expression in presumptive mesoderm of *X. laevis. Cell* **65,** 927–937.

Sassoon, D., Lyons, G., Wright, W. E., Lin, V., Lassar, A., Weintraub, H., and Buckingham, M. (1989). Expression of two myogenic regulatory factors myogenin and MyoD1 during mouse embryogenesis. *Nature* **341,** 303–307.

Schnabel, R. (1994). Autonomy and nonautonomy in cell fate specification of muscle in the *Caenorhabditis elegans* embryo. A reciprocal induction. *Science* **263,** 1449–1452.

Smith, T. H., Kachinsky, A. M., and Miller, J. B. (1994). Somite subdomains, muscle cell origins, and the four muscle regulatory factor proteins. *J. Cell Biol.* **127,** 95–105.

Solursh, M., Drake, C., and Meier, S. (1987). The migration of myogenic cells from somites at the wing level in avian embryos. *Dev. Biol.* **121,** 389–396.

Soriano, P. (1997). The PDGF alpha receptor is required for neural crest cell development and for normal patterning of the somites. *Development* **124,** 2691–2700.

Spence, M. S., Yip, J., and Erickson, C. A. (1996). The dorsal neural tube organizes the dermomyotome and induces axial myocytes in the avian embryo. *Development* **l22,** 231–241.

Spicer, D. B., Rhee, J., Cheung, W. L., and Lassar, A. B. (1996). Inhibition of myogenic bHLH and MEF2 transcription factors by the bHLH protein twist. *Science* **272,** 1476–1480.

Spörle, R., Gunther, T., Struwe, M., and Schughart, K. (1996). Severe defects in the formation of epaxial musculature in open brain (opb) mutant mouse embryos. *Development* **122,** 79–86.

Stern, H. M., and Hauschka, S. D. (1995). Neural tube and notochord promote *in vitro* myogenesis in single somite explants. *Dev. Biol.* **167,** 87–103.

Stern, H. M., Lin-Jones, J., and Hauschka, S. D. (1997). Synergistic interactions between bFGF and a TGF-beta family member may mediate myogenic signals from the neural tube. *Development* **124,** 3511–3523.

Tajbakhsh, S., and Buckingham, M. E. (1994). Mouse limb muscle is determined in the absence of the earliest myogenic factor myf-5. *Proc. Natl. Acad. Sci. U.S.A.* **91,** 747–751.

Tajbakhsh, S., and Buckingham, M. (1995). Lineage restriction of the myogenic conversion factor myf-5 in the brain. *Development Suppl.* **121,** 4077–4083.

Tajbakhsh, S., Bober, E., Babinet, C., Pournin, S., Arnold, H., and Buckingham, M. (1996a). Gene targeting the myf-5 locus with nlacZ reveals expression of this myogenic factor in mature skeletal muscle fibres as well as early embryonic muscle. *Dev. Dyn.* **206,** 291–300.

Tajbakhsh, S., Rocancourt, D., and Buckingham, M. (1996b). Muscle progenitor cells failing to respond to positional cues adopt non-myogenic fates in myf-5 null mice. *Nature* **384,** 266–270.

Tajbakhsh, S., Rocancourt, D., Cossu, G., and Buckingham, M. (1997). Redefining the genetic hierarchies controlling skeletal myogenesis: Pax-3 and Myf-5 act upstream of MyoD. *Cell* **89,** 127–138.

Tajbakhsh, S., Vivarelli, E., Cusella-De Angelis, G., Rocancourt, D., Buckingham, M., and Cossu, G. (1994). A population of myogenic cells derived from the mouse neural tube. *Neuron* **13,** 813–821.

Talbot, W. S., Trevarrow, B., Halpern, M. E., Melby, A. E., Farr, G., Postlethwait, J. H., Jowett, T., Kimmel, C. B., and Kimelman, D. (1995). A homeobox gene essential for zebrafish notochord development. *Nature* **378,** 150–157.
Tapscott, S. J., Lassar, A. B., and Weintraub, H. (1992). A novel myoblast enhancer element mediates MyoD transcription. *Mol. Cell. Biol.* **12,** 4994–5003.
Teillet, M. A., and Le Douarin, N. M. (1983). Consequences of neural tube and notochord excision on the development of the peripheral nervous system in the chick embryo. *Dev. Biol.* **98,** 192–211.
Thayer, M. J., Tapscott, S. J., Davis, R. L., Wright, W. E., Lassar, A. B., and Weintraub, H. (1989). Positive autoregulation of the myogenic determination gene *MyoD1*. *Cell* **58,** 241–248.
Venuti, J. M., Morris, J. H., Vivian, J. L., Olson, E. N., and Klein, W. H. (1995). Myogenin is required for late but not early aspects of myogenesis during mouse development. *J. Cell Biol.* **128,** 563–576.
Wachtler, F., Jacob, H. J., Jacob, M., and Christ, B. (1984). The extrinsic ocular muscles in birds are derived from the prechordal plate. *Naturwissenschaften* **71,** 379–380.
Wang, Y., and Jaenisch, R. (1997). Myogenin can substitute for Myf5 in promoting myogenesis but less efficiently. *Development* **124,** 2507–2513.
Wang, Y., Benezra, R., and Sassoon, D. A. (1992). Id expression during mouse development: A role in morphogenesis. *Dev. Dyn.* **194,** 222–230.
Wang, Y., Schnegelsberg, P. N., Dausman, J., and Jaenisch, R. (1996). Functional redundancy of the muscle-specific transcription factors Myf5 and myogenin. *Nature* **379,** 823–825.
Weintraub, H., Davis, R., Tapscott, S., Thayer, M., Krause, M., Benezra, R., Blackwell, T. K., Turner, D., Rupp, R., Hollenberg, S., Zhuang, Y., and Lassar, A. (1991a). The myoD gene family: Nodal point during specification of the muscle cell lineage. *Science* **251,** 761–766.
Weintraub, H., Dwarki, V. J., Verma, I., Davis, R., Hollenberg, S., Snider, L., Lassar, A., and Tapscott, S. J. (1991b). Muscle-specific transcriptional activation by MyoD. *Genes Dev.* **5,** 1377–1386.
Williams, B. A., and Ordahl, C. P. (1994). Pax-3 expression in segmental mesoderm marks early stages in myogenic cell specification. *Development* **120,** 785–796.
Williams, B., and Ordahl, C. (1997). Emergence of determined myotome precursor cells in the somite. *Development* **124,** 4983–4997.
Winter, B., Braun, T., and Arnold, H. H. (1992). Cooperativity of functional domains in the muscle-specific transcription factor Myf-5. *EMBO J.* **11,** 1843–1855.
Wright, W. E. (1984). Expression of differentiated functions in heterokaryons between skeletal myocytes, adrenal cells, fibroblasts and glial cells. *Exp. Cell Res.* **151,** 55–69.
Wright, W. E., Sassoon, D. A., and Lin, V. K. (1989). Myogenin, a factor regulating myogenesis, has a domain homologous to MyoD. *Cell* **56,** 607–617.
Xue, X. J., and Xue, Z.G. (1996). Spatial and temporal effects of axial structures on myogenesis of developing somites. *Mech. Dev.* **60,** 73–82.
Yasuo, H., and Satoh, N. (1993). Function of vertebrate *T* gene. *Nature* **364,** 582–583.
Yee, S. P., and Rigby, P. W. (1993). The regulation of myogenin gene expression during the embryonic development of the mouse. *Genes Dev.* **7,** 1277–1289.
Yoon, J. K., Olson, E. N., Arnold, H. H., and Wold, B. J. (1997). Different MRF4 knockout alleles differentially disrupt Myf-5 expression: Cis-regulatory interactions at the MRF4/Myf-5 locus. *Dev. Biol.* **188,** 349–362.
Yun, K., and Wold, B. (1996). Skeletal muscle determination and differentiation: Story of a core regulatory network and its context. *Curr. Opin. Cell Biol.* **8,** 877–889.
Zhang, W., Behringer, R. R., and Olson, E. N. (1995). Inactivation of the myogenic bHLH gene MRF4 results in up-regulation of myogenin and rib anomalies. *Genes Dev.* **9,** 1388–1399.
Zhuang, Y., Soriano, P., and Weintraub, H. (1994). The helix-loop-helix gene E2A is required for B cell formation. *Cell* **79,** 875–884.
Zweigerdt, R., Braun, T., and Arnold, H. H. (1997). Faithful expression of the *myf-5* gene during mouse myogenesis requires distant control regions: A transgene approach using yeast artificial chromosomes [In Process Citation]. *Dev. Biol.* **192,** 172–180.

5
Multiple Tissue Interactions and Signal Transduction Pathways Control Somite Myogenesis

Anne-Gaëlle Borycki and Charles P. Emerson, Jr.
Department of Cell and Developmental Biology
University of Pennsylvania School of Medicine
Philadelphia, Pennsylvania 19104-6058

I. Introduction
II. Mesodermal Origins of Skeletal Muscle in Vertebrate Embryos
   A. Origins of Craniofacial Muscles
   B. Origins of Body and Limb Muscles
III. Regulatory Genes That Control Somite Myogenesis
   A. *Myogenic Regulatory Factor* Genes Are Essential for Somite Myogenesis
   B. Additional Somite-Expressed Regulatory Genes
   C. *Pax3* Is an Essential Upstream Gene for Somite Myogenesis
IV. Tissue Interactions That Control Somite Myogenesis
   A. Myogenic Determination Is Acquired during Somite Formation
   B. Axial Tissues Control Epaxial Somite Myogenesis
   C. Notochord Provides Signals for the Initiation of Epaxial Somite Myogenesis
   D. Neural Tube Provides Signals for Maintenance of Epaxial Somite Myogenesis
   E. Surface Ectoderm Signals Control Somite Formation, Competence to Respond to Axial Signals, and Hypaxial Myogenesis
   F. Lateral Mesoderm Provides Inhibitory Signals That Spatially Restrict Epaxial Somite Myogenesis
V. Signaling Molecules and Transduction Pathways Controlling Myogenesis in Somites
   A. Sonic Hedgehog Is an Essential and Sufficient Notochord Signal for Epaxial Myogenesis
   B. Surface Ectoderm Wnts Are Signals for Activation of Hypaxial Myogenesis, and Neural Tube Wnts Are Signals for Maintenance of Epaxial Myogenesis
   C. BMP4 and the Bone Morphogenetic Protein Antagonist, Noggin, Regulate Mediolateral Boundaries of the Epaxial and Hypaxial Somite Domains
   D. Transforming Growth Factors β Have Positive and Negative Functions in Myogenesis
   E. Fibroblast Growth Factors Promote Mesoderm Formation and Have Later Maintenance Functions in Myogenesis
   F. The Notch/Delta Signaling Pathway May Control the Maintenance of the Epithelial Organization and Myogenesis in Somites
VI. Overview and Future Perspectives
   References

I. Introduction

The intricate anatomy of skeletal muscles in vertebrates reflects the complex contractile functions and motor activities of this behaviorally versatile group of organisms. This complexity becomes evident early during the development of vertebrate embryos when myogenic progenitor cells arise from paraxial mesoderm and locate at sites in the body, limbs, head, and neck, where they differentiate to form functional, anatomically defined muscles. During the 1990s, the developmental origins of skeletal muscle progenitor cells have been investigated intensively in vertebrate embryos. This work has been greatly stimulated by the discovery of the *Myogenic Regulatory Factor* (*MRF*) and *Pax* regulatory genes, which are expressed in early myogenic progenitor cells and are genetically essential for the formation and differentiation of myogenic progenitors. These muscle regulatory genes have provided powerful molecular and genetic tools for experimental investigations of the tissue interactions and signal transduction pathways that initiate the earliest steps of somite myogenesis.

In this chapter, we review embryological and molecular studies that have established a central role for developmental signaling mechanisms in the establishment of myogenic cells in the somites of vertebrate embryos, and we evaluate current models of embryonic signaling in the control of skeletal myogenic determination. Most current knowledge of developmental signaling in myogenesis has been contributed by studies of the avian embryo. Recent genetic and embryological research in the mouse embryo has begun to focus on more complex issues of signaling mechanisms that more globally coordinate the processes of myogenesis throughout the embryo, including genes and signaling mechanisms involved in epaxial and hypaxial muscle formation. In the context of the avian and mammalian studies, we also will discuss recent genetic studies of myogenesis in the zebrafish and embryological studies in *Xenopus*. It is important for the reader to keep in mind that substantial information on tissue interactions that control somite myogenesis is now available, although most studies have focused on the formation of epaxial muscles, which are the first myotomal muscles to form in the embryo. The tissue interaction and signaling mechanisms that control the formation of hypaxial, limb, and head muscles are not well understood, although it has become apparent that their formation is subject to different regulation. Research on developmental signal transduction mechanisms is more limited. As will be presented in this review, current evidence provides a general view that somite myogenesis is controlled by complex tissue interactions that expose somitic cells to multiple, independently functioning, and partially redundant signals that promote and inhibit myogenesis. These complex tissue interactions and signal transduction processes apparently have a common genetic outcome, which is the activation of *MRF* genes that commit somitic cells to a myogenic fate. This complexity of interactions and signaling processes, therefore, can be viewed as an adaptive mechanism to exploit

5. Tissue Interactions and Signaling Mechanisms in Somite Myogenesis

the diversity of signaling mechanisms available throughout the embryo to activate a common *MRF* genetic regulatory pathway for muscle differentiation at widely dispersed anatomical sites in the vertebrate body.

II. Mesodermal Origins of Skeletal Muscle in Vertebrate Embryos

A brief discussion of the origins of myogenic cells in the vertebrate embryo will serve as an introduction to a more detailed review of current knowledge of the tissue interactions and developmental signal transduction mechanisms involved in vertebrate skeletal myogenesis. The origins of skeletal muscle in vertebrate embryos have been investigated using a variety of lineage tracing methods. Early work in this field utilized carbon particles as markers and identified somites as the embryonic source of cells for body skeletal muscles in chick embryos (Spratt, 1955). Lineage tracing studies in avian embryos were greatly advanced through the development of the chick–quail chimera assay using quail nucleolar marker (Le Douarin, 1973). This technology has been applied to study the embryonic origins of many tissues including skeletal muscle (Ordahl and Le Douarin, 1992). Recently, molecular tools using *in situ* hybridization with probes of lineage-specific genes, particularly the *Myogenic Regulatory Factor* (MRF), *MyoD* and *Myf5* (Charles de la Brousse and Emerson, 1990; Ott *et al.*, 1991; Pownall and Emerson, 1992), and transgenesis and gene replacement with lacZ reporters under the control of the *MyoD* and *Myf5* transcriptional regulatory elements (Goldhamer *et al.*, 1992, 1995; Pinney *et al.*, 1995; Tajbakhsh *et al.*, 1996a) have made it possible to identify and trace the fates of muscle progenitor cells in the early embryo, when these progenitor cells originate from the mesoderm.

Studies using lineage tracer technologies have provided corroborative evidence that the progenitor cells that give rise to body, neck, and limb muscles arise from somites, which form by the epithelialization of paraxial mesoderm adjacent to the axial neural tube/notochord tissues. Somites form in an anterior to posterior progression during axis formation. Avian embryos have 52–53 somite pairs (Sanders *et al.*, 1986), which form about every 90 min during the first 3 days of development, whereas mammalian embryos form somites during days 8–12 of development (Christ and Ordahl, 1995).

A. Origins of Craniofacial Muscles

Posterior head and neck muscles arise from epaxial cells of the anteriormost cranial somites, and anterior head muscles originate from unsegmented paraxial mesoderm and, for some extraocular muscles, from condensation of prechordal mesoderm (Noden, 1983, 1986; Wachtler and Christ, 1992). The tissue interactions and signal transduction processes that control the formation of head muscles have been less studied, although work in this area is progressing (Noden, 1988, Hacker and Guthrie, 1998).

B. Origins of Body and Limb Muscles

Soon after a somite in the body axis forms as an epithelial ball, cells in the medial and lateral and the dorsal and ventral spatial domains undergo different morphogenetic changes (Fig. 1A). Cells in the ventral aspect undergo an epithelial to mesenchymal transition to give rise to sclerotomal cells that will differentiate to form the cartilage and associated connective tissues of the vertebrae and ribs (Fig. 1A) (Jacob et al., 1975; Christ and Wilting, 1992). Cells in the dorsal aspect remain epithelial and form the dermomyotome. Cells in the medial and lateral aspects of the dermomyotome give rise to the epaxial and hypaxial muscle lineages, respectively (Fig. 1A) (Ordahl and Le Douarin, 1992; Cossu et al., 1996a). The epaxial muscles of the medial dermomyotome form the myotome, a layer of longitudinally oriented, differentiated muscle cells that migrate between the dermomyotome and the sclerotome (Fig. 1A). The myotomal cells form the deep back muscles. The hypaxial muscles derive from the lateral dermomyotome. Cells at the lateral tip of the dermomyotome delaminate and migrate to diverse sites of the embryo to populate and form the abdominal, body wall, and limb muscles (Ordahl, 1993). Interestingly, although medial and lateral somitic cells originate initially from the same presumptive area of the chick epiblast (Hatada and Stern, 1994), by the primitive streak stage (Hamburger–Hamilton (HH) stage 4), lineage tracing experiments show that cells of the medial and lateral somite arise from distinct presumptive areas (Selleck and Stern, 1991). However, it is unknown whether medial and lateral cells are distinctly specified in the paraxial mesoderm, although chick–quail chimera lineage experiments suggest that medial and lateral somite lineages are exchangeable in their fates to form epaxial muscles and hypaxial muscles (Ordahl and Le Douarin, 1992). As will be discussed in detail below, *MRF* genes, which are the earliest lineage-specific genes expressed during myogenesis, are first activated in medial somite cells that give rise to epaxial muscles in both avian and mammalian embryos, prior to dermomyotome formation (Charles de la Brousse and Emerson, 1990; Ott et al., 1991). Hypaxial muscle progenitor cells on the lateral side of the dermomyotome activate *MRF* gene expression later (Tajbakhsh et al., 1996a). Cells from the lateral dermomyotome migrate to the limb bud and continue to express somite genes such as *Pax3*, but do not activate *MRF* gene expression until these cells become located and patterned in the limb and abdominal regions of myogenesis (Buckingham, 1992).

5. Tissue Interactions and Signaling Mechanisms in Somite Myogenesis 169

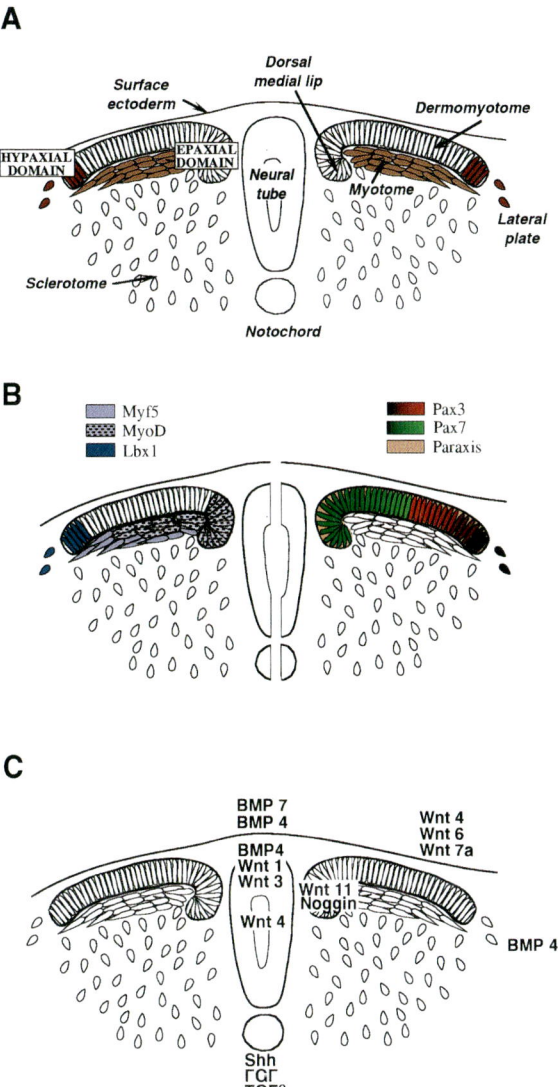

Figure 1 Schematic representation of a cross section of an avian somite showing the different somite domains and lineages (A) involved in somite patterning. (B) Expression domains of the dermomyotome markers *Paraxis* (salmon), *Pax3* (red gradient), *Pax7* (green gradient), *Lbx1* (blue), and the myotome markers *Myf5* (lilac) and *MyoD* (doted lilac). (C) Expression sites of various signaling molecules that have been shown *in vitro* or *in vivo* to control myogenesis.

III. Regulatory Genes That Control Somite Myogenesis

A growing collection of lineage-restricted regulatory and differentiation-specific transcription factor genes have been identified. These genes are expressed in somites, myogenic progenitor cells, and differentiated muscles, are activated in response to signaling, and have provided important molecular markers for genetic and cellular study of myogenesis in the vertebrate embryo. Regulatory genes of particular importance include basic helix–loop–helix (bHLH) transcription factor genes, specifically the myogenic regulatory factors, *MyoD*, *Myf5*, *Myogenin*, and *MRF4* (Sassoon *et al.*, 1989; Hinterberger *et al.*, 1991; Ott *et al.*, 1991), and somite-expressed genes, *Paraxis* (Burgess *et al.*, 1995), *Sim1* (Fan *et al.*, 1996), *Lbx1* (Dietrich *et al.*, 1998), and the *Pax* (paired-homeodomain) family of early embryonic regulators expressed in the neural tube, presomitic mesoderm, and somites (Strachan and Read, 1994).

A. Myogenic Regulatory Factor Genes Are Essential for Somite Myogenesis

The *MRF* genes, *MyoD*, *Myf5*, *Myogenin*, and *MRF4*, encode bHLH transcription factors that are expressed in myogenic progenitor cells and are essential for myogenic determination and differentiation. These genes were originally identified based on their dominant activities to convert nonmuscle cells to skeletal muscle cells in cell culture (Weintraub *et al.*, 1991). Gene expression studies in vertebrate embryos established that two of these genes, *Myf5* and *MyoD*, are expressed in somitic epaxial and hypaxial muscle progenitor cells that arise from somites (Fig. 1B) (Sassoon *et al.*, 1989; Charles de la Brousse and Emerson, 1990; Ott *et al.*, 1991; Pownall and Emerson, 1992; Tajbakhsh *et al.*, 1996a).

In the mouse embryo, *MyoD* and *Myf5* have different temporal expression, suggesting their differential functions in the establishment of muscle lineages. *Myf5* transcripts are first detected at embryonic day (E)8.0 in cells of the dorsal medial quadrant of the somite in the prospective epaxial domain (Ott *et al.*, 1991; Tajbakhsh *et al.*, 1996a). *Myf5* expression is detected at E8.0 in the most anterior somites, and progresses in an anterior to posterior gradient as embryos develop. *Myf5* is subsequently activated at E9.5 in the lateral dermomyotome in the prospective hypaxial domain of interlimb somites. *MyoD* is activated later at E9.75 in this same hypaxial domain of interlimb somites (Sassoon *et al.*, 1989). In the limb, branchial arches, and head muscles, *MyoD* and *Myf5* are activated nearly simultaneously, in what appears to be the same population of myogenic progenitors (Buckingham, 1992; Hacker and Guthrie, 1998). The two additional *MRF* genes, *Myogenin* and *MRF4*, are expressed in myogenic lineages at the time of differentiation, and therefore are downstream of *Myf5* and *MyoD* (Sassoon *et al.*, 1989; Hinterberger *et al.*, 1991).

5. Tissue Interactions and Signaling Mechanisms in Somite Myogenesis 171

Gene targeting studies in mouse have provided definitive evidence for the essential functions of *Myf5* and *MyoD* in the establishment of epaxial and hypaxial progenitor cell lineages in the mammalian embryo, as compound mutants for *MyoD* and *Myf5* genes completely lack muscle progenitors (Rudnicki *et al.*, 1993). However, homozygous *Myf5* or *MyoD* mutant embryos do not exhibit gross abnormalities in muscle formation (Braun *et al.*, 1992; Rudnicki *et al.*, 1992), indicating that they can functionally compensate for each other.

Significantly, epaxial muscle formation is delayed in *Myf5* homozygous mutant mice (Tajbakhsh *et al.*, 1997), consistent with its essential function in determination of the epaxial myogenic lineage. Lineage tracing experiments in *Myf5* mutant embryos, using a *lacZ* reporter gene insertion to knockout the *Myf5* locus, indicate that *Myf5* mutant somite cells accumulate at the medial, epaxial edge of the dermomyotome (Tajbakhsh *et al.*, 1996b). Some mutant cells migrate from this site and locate in sclerotome and dermatome derivatives (Tajbakhsh *et al.*, 1996b), providing evidence that *Myf5* expression is essential for the determination of epaxial somitic cells to the muscle lineage. However, the majority of *lacZ*-expressing somitic cells accumulating at the medial edge of the dermomyotome activate *MyoD* after a 1.5-day delay and proceed with epaxial myogenesis (Braun *et al.*, 1994; Tajbakhsh *et al.*, 1997). In *Myf5*-mutant mice, activation of *MyoD* is delayed in both the epaxial and hypaxial domains of the dermomyotome, indicating that initial activation of *MyoD* in both these somite domains requires *Myf5*, but a later activation of *MyoD* occurs in a *Myf5*-independent fashion. In contrast, in the limbs and head, *MyoD* activation is *Myf5*-independent (Tajbakhsh *et al.*, 1997). These observations already suggest that multiple regulatory mechanisms control *Myf5* and *MyoD* expression in the embryo.

In *MyoD* mutant mice, *Myf5* expression is upregulated (Rudnicki *et al.*, 1992), suggesting that *Myf5* overexpression compensates for the loss of *MyoD* function. The reader should keep in mind that there likely are other, yet-to-be-identified upstream regulatory genes that control *Myf5* and *MyoD* activation and myogenesis. However, these *MRF* genes are activated when somite cells become committed to myogenesis and therefore provide important markers for studies of tissue interactions and signal transduction mechanisms that target the activation of these genes as well as initiate the processes of somite myogenesis.

There are apparent species-specific differences in the developmental timing of expression of *MyoD* and *Myf5* between mammals and other vertebrates. Mouse embryos activate *Myf5* several days prior to *MyoD* in the medial epaxial somite (Buckingham, 1992), whereas in bird and frog embryos *MyoD* and *Myf5* are activated nearly simultaneously in the dorsal medial cells of the epithelial somite (Hopwood *et al.*, 1989, 1991; Pownall and Emerson, 1992). In the frog, *MyoD* and *Myf5* are activated nearly simultaneously in dorsal mesoderm, prior to somite formation (Hopwood *et al.*, 1989, 1991). In the zebrafish, *Myf5* has not yet been cloned, but *MyoD* is activated in early lineages including the pioneer cells (Weinberg *et al.*, 1996). In the avian embryo, *MyoD* is the first *MRF* gene transcribed following somite formation, and *Myf5* expression follows shortly thereafter (Charles de la Brousse and Emerson, 1990; Pownall and Emerson, 1992). In contrast, in the head muscle mesoderm of avian embryos, *Myf5* activation appears to precede *MyoD* (Hacker and Guthrie, 1998).

MyoD and *Myf5* activation in the epaxial domain occurs in three phases during early somite formation (Borycki *et al.*, 1997). These genes are not expressed in somites of early embryos until about the time 10 somite pairs have formed. In contrast, *Pax1*, a sclerotome-specific gene, is actively expressed in the ventral somite from the earliest stages of somite formation, and its expression throughout development always initiates in somite III[1] back from the newest born somite. By HH stage 11, when embryos have 10–12 somite pairs, *MyoD* and *Myf5* are activated, first in the most anterior somites, followed by a wave of activation that rapidly progresses to the most posterior somites (Borycki *et al.*, 1997). By HH stage 14, *MyoD* and *Myf5* become activated in coordination with the formation of each new somite, preceding the expression of differentiation markers (Borycki *et al.*, 1997). In contrast, *Myogenin* and *MRF4* are activated later in more mature somites at the onset of differentiation (Borycki *et al.*, 1997). These findings establish that the expression of *MRF* genes in the epaxial somite domain is precisely regulated during avian embryo development.

B. Additional Somite-Expressed Regulatory Genes

Sim1 encodes a bHLH transcription factor that is the homologue of the *Drosophila single-minded* gene and is specifically expressed in the lateral compartment of the dermomyotome in avian embryos (Pourquié *et al.*, 1996), whereas it is first detected in the lateral dermomyotome, but then becomes restricted to a central region of the dermomyotome in mouse embryos (Fan *et al.*, 1996). A somite phenotype for *Sim1* mutant mice has not yet been reported (Michaud *et al.*, 1998), but this gene provides a useful marker of somite morphogenesis in embryological and genetic studies.

Paraxis is first expressed throughout the somite and then its expression becomes restricted to the dorsal dermomyotome (Fig. 1B) (Burgess *et al.*, 1995). Mouse embryos that are homozygous mutant for *Paraxis* lack epithelial somites and a defined dermomyotome (Burgess *et al.*, 1996), demonstrating the role of *Paraxis* in the formation and/or the maintenance of the epithelial architecture of the dermomyotome. *Paraxis* mutant embryos, however, activate *MRF* and *Pax* genes (see below), although their expression levels are reduced (Burgess *et al.*, 1996), establishing that epithelialization of somites is not required for the activation of somite regulatory genes.

---

[1] In these discussions, we use somite nomenclature of C. Ordhal (1993) in which the newest formed somite is designated as I.

5. Tissue Interactions and Signaling Mechanisms in Somite Myogenesis   173

Lbx1 is a vertebrate homologue of the *Drosophila* gene *ladybird*, and is expressed in somites of mouse and chick embryos (Jagla *et al.*, 1995; Dietrich *et al.*, 1998). *Lbx1* expression is specifically detected in migrating hypaxial muscle precursors of somites at the occipital, cervical, and limb levels of stage HH 14 and onward chick embryos (Fig. 1B) (Dietrich *et al.*, 1998). Therefore, it represents a unique marker for a subpopulation of hypaxial muscle cell progenitors. *C-met*, a tyrosine kinase receptor gene, is also expressed in hypaxial muscle progenitors (Daston *et al.*, 1996), and constitutes another hypaxial-specific marker.

C. *Pax3* Is an Essential Upstream Gene for Somite Myogenesis

Four *Pax* genes are expressed in somites, and these genes provide specific gene expression markers for the ventral sclerotome lineage and for the dorsal dermomyotome lineage. Two genes, *Pax1* and *Pax9*, are sclerotome-specific (Strachan and Read, 1994). In the avian embryos, *Pax1*, in contrast to *MyoD* and *Myf5*, is activated in the ventral somite in precise coordination with somite segmentation throughout development, indicating that the control of gene expression in the dorsal and ventral somite differs (Borycki *et al.*, 1997). Two additional *Pax* genes, *Pax3* and *Pax7*, are dermomyotome-specific (Fig. 1B) (Strachan and Read, 1994). In mouse embryos, *Pax3* is first transiently expressed throughout the newly formed, epithelial somite at E8.5 (Goulding *et al.*, 1991). Its expression then becomes rapidly restricted to the dermomyotome with higher expression in the lateral dermomyotome (Goulding *et al.*, 1994). Lateral somite cells, which undergo an epithelial to mesenchymal transition, detach from the lateral dermomyotome, and migrate into the limb bud, express *Pax3*, but not *MyoD* or *Myf5* (Fig. 1B) (Bober *et al.*, 1994a). *Pax3* function is essential for this cell migration, based on the observation that *Splotch* mice, which have a deficiency mutation of the *Pax3* locus, do not form limb muscles and are impaired in the migration of dermomyotomal cells into the limb (Daston *et al.*, 1996). *Pax3* function also is required for *c-met* expression, which encodes the receptor for the scatter factor Hepatocyte Growth Factor/Scatter Factor (HGF/SF) (Epstein *et al.*, 1996), which is essential for migration of somitic cells into the limb (Bladt *et al.*, 1995). In addition, *Splotch* mice exhibit dermomyotome-specific defects in epithelialization (Franz *et al.*, 1993) and disruptions in the spatial expression of *Myf5* and *MyoD* (Tajbakhsh *et al.*, 1997), and apoptosis is prevalent in the posterior somites (Borycki *et al.*, 1999), suggesting a more general functional role in somite integrity. Finally, mouse embryos that are double mutant for *Myf5* and *Pax3* lack all body muscles (Tajbakhsh *et al.*, 1997), indicating that *Pax3* has functions in the upstream regulation of *MyoD*. Interestingly, head muscle formation is unaffected in *Pax3/Myf5* double-mutant mouse embryos, revealing that *Pax3* functions are essential for somitic, but not head muscle specification. *Pax7* expression is first detected at E9.0 in the dermomyotome in a domain encompassing mostly the medial and central dermomyotome (Fig. 1B) (Jostes et al., 1991). Although Pax7 knockout mice do not exhibit muscle defects (Mansouri et al., 1996), Pax3/Pax7 double knockout mice display more severe somitic defects, confirming possible redundancy between Pax3 and Pax7 (A. Mansouri, personal communication, 1998). Furthermore, Pax7 is upregulated and its expression becomes expanded in somites and neural tube of Splotch embryos (Borycki et al., 1999a). Therefore, Pax3, in addition to its function in MyoD activation and inhibition of apoptosis, functions to repress Pax7 during neural tube and somite development.

IV. Tissue Interactions That Control Somite Myogenesis

A. Myogenic Determination Is Acquired during Somite Formation

Somite rotation and grafting experiments in avian and mammalian embryos show that the fates of cells of the unsegmented paraxial mesoderm are positionally determined, implying that somitic cells in newly formed somites are plastic and become determined in response to interactions with adjacent tissues (Aoyama and Asamoto, 1988; Ordahl and Le Douarin, 1992; Aoyama, 1993). Dorsal–ventral rotation of newly formed somites I–II from HH stage 14–15 quail embryos does not disturb normal development of the sclerotome and the myotome (Aoyama and Asamoto, 1988). Furthermore, when prospective dorsal dermomyotome cells from somite I of HH stage 12–13 quail embryos are grafted ventrally between somite I and neural tube in a chick embryo, these cells do not develop as epithelial dermomyotome, but instead are morphologically indistinguishable from the sclerotomal cells (Aoyama, 1993). Conversely, when prospective ventral sclerotome cells, are grafted to a dorsal position, between surface ectoderm and somite I in a chick embryo, these cells become determined to a myotomal fate within 2 days (Christ et al., 1992) and are incorporated into muscles 9 days after surgery (Aoyama, 1993). Similarly, when the lateral halves of somites I–II of a chick embryo are replaced with the medial halves of somites I–II of a HH stage 12–13 quail embryo, quail cells that would have formed myotomal muscles migrate instead to the limbs, where they form appendicular muscles (Ordahl and Le Douarin, 1992). Conversely, when medial halves of somites I–II of HH stage 12–13 chick embryos are replaced with the lateral halves of somites I–II of HH stage 12–13 quail embryos that would have formed limb muscles, quail cells form myotomal muscles (Ordahl and Le Douarin, 1992). Together, these experiments demonstrate that dorsal–ventral and medial–lateral cells of newborn somites are not determined along the anterior–posterior (AP) axis. These findings, furthermore, suggest that paraxial mesodermal cells and newly born somite cells are multipotential and require spatially directed environmental signals following segmentation and somite formation to become determined to the muscle lineage. Interestingly, Aoyama and Asamoto (1988) noted that the rotation of somite III of HH stage 14–15 quail embryos results in the formation of an ectopic sclerotome underneath the surface ectoderm without affecting the medial formation of the myotome, suggesting that the sclerotome lineage is determined in somites before the myotome lineage. This interpretation is consistent with more recent work using molecular markers, showing that myotome and sclerotome formation are regulated independently (see above) (Borycki et al., 1997).

B. Axial Tissues Control Epaxial Somite Myogenesis

The influence of axial tissues on somite maturation, and more specifically on cartilage induction, has been known for a long time (Grobstein and Holtzer, 1955; Kosher and Lash, 1975). More recently, the availability of muscle molecular markers and antibodies have made it possible to address the role of axial tissue interaction in muscle induction.

In HH stage 10–12 chick embryos, the separation of the neural tube/notochord complex from the segmental plate mesoderm and the somites or the ablation of the neural tube/notochord complex leads within 10–20 hr to a failure of somites to undergo muscle differentiation, as assayed by the lack of myosin positive cells in all, but the most anterior, somites (Rong et al., 1992). Seven days after the ablation, no epaxial vertebral muscles form, but hypaxial wing muscles form normally (Ordahl and Le Douarin, 1992; Rong et al., 1992), demonstrating that the neural tube/notochord complex is required for the differentiation of epaxial, but not hypaxial muscles. Subsequent studies of axial tissue interactions with presegmental mesoderm and somites in quail embryos established that this requirement for somite interactions with the neural tube/notochord complex acts on the earliest regulatory processes of epaxial myogenic determination, specifically on the activation of *MyoD* and *Myf5* at the time of somite formation (Pownall et al., 1996). These studies also established that, in the absence of neural tube and notochord interactions, newly formed somites turn over all *MyoD* and *Myf5* transcripts in less than 2 hr, indicating that *MRF* mRNAs in newly formed somites are very unstable and that the epaxial myogenic cells are not stably determined to the myogenic fate immediately following activation of *MRF* genes. As somites mature, however, *MyoD* and *Myf5* expression becomes stabilized and autonomous of neural tube and notochord interactions. In the absence of axial tissue interactions, somite cells also undergo apoptosis (Teillet et al., 1998). As will be discussed below, notochord signaling can replace axial tissue interactions for *MyoD* and *Myf5* activation and prevent apoptosis, although apoptosis appears to be a later response that does not specifically affect epaxial muscle cells (Teillet et al., 1998; Borycki et al., 1999b). In any case, these studies establish that axial tissue interactions are essential for *MRF* genes expression and somite cell survival.

The capacity of somites of avian embryos to respond to axial tissue interactions is developmentally regulated and stage specific, as determined by somite and segmental plate transplantation into embryos at early embryonic stages when *MRF* genes are not expressed (Borycki et al., 1997). Transplantation of HH stage 14 segmental plate or somites I–III, which express *MyoD* and *Myf5,* into a HH stage 10 embryo, which have no *MRF* expression, results in the loss of *MyoD* expression, indicating that in the early avian embryo, axial tissues are not competent to produce signals to promote myogenesis. Conversely, transplantation of HH stage 10 segmental plate or somites I–III to a HH stage 14 embryo does not lead to *MyoD* activation, indicating that presomitic mesoderm and somites of HH stage 10 embryos are not competent to respond to myogenesis-inducing signals until after HH stage 11. Therefore, the regulation of epaxial myogenesis involves regulatory processes that act at levels of both the production of myogenic signals from axial tissues and the capacity of somites to respond to these axial signals.

C. Notochord Provides Signals for the Initiation of Epaxial Somite Myogenesis

1. In Vivo Studies

Tissue grafting and genetic studies have provided recent evidence that the notochord has an essential role in myotome and sclerotome formation. Surgical ablation of the notochord and the floor plate or the notochord alone at the level of the posterior segmental plate of HH stage 10–12 chick or quail embryos prior to floor plate formation results in the failure of *Pax3* expression to become dorsalized in somites formed in the absence of notochord, and the failure of somites to activate both *MyoD* in the epaxial myotome progenitors and *Pax1* in the sclerotome progenitors (Pownall *et al.,* 1996; Dietrich *et al.,* 1997). This would indicate that the notochord has a negative regulatory role in dorsal–ventral patterning of *Pax3* expression and a positive regulatory role in the activation of *MyoD* and *Pax1* expression. Somites that form in the absence of notochord initially are morphologically normal (Hirano *et al.,* 1995) and continue to express genes such as *Pax3*. After longer incubation of embryos following notochord ablation, apoptosis is observed in some somite cells and other cells migrate beneath the neural tube (Teillet *et al.,* 1998). In cells within one somite diameter of the cut notochord, *MyoD* expression is observed (Goulding *et al.,* 1994), whereas larger notochord ablations lead to the complete loss of *MyoD* expression (Pownall *et al.,* 1996), suggesting that the notochord signal has action limited to the distance of a somite diameter. This interpretation is consistent with more recent molecular studies of the observed action of Sonic hedgehog, the notochord signal now known to be required for *MyoD* activation in somites (Fan *et al.,* 1995). The notochord in the absence of neural tube also is sufficient to rescue epaxial myogenesis (Rong *et al.,* 1992) and *MyoD* expression (Pownall *et al.,* 1996), establishing that the notochord is both essential and sufficient for *MRF* activation and epaxial muscle determination.

The response of somites to notochord grafts can be complex, depending on the positioning of the graft along the dorsal–ventral and medial–lateral axes of the somite. Ventral notochord grafts at the level of the segmental plate expand the domain of expression of *Pax1* and *Pax9* (Brand-Saberi *et al.*, 1993; Goulding *et al.*, 1994), establishing that the notochord functions in sclerotome formation. In contrast, dorsal notochord grafts give rise to variable responses, depending on the medial–lateral location of the graft and the response time. When the notochord is grafted in a dorsal medial position between the segmental plate (or the newest formed somites) and the neural tube, the dermomyotome becomes deepithelialized within 24 hr (Brand-Saberi *et al.*, 1993), *Pax3* expression is lost (Goulding *et al.*, 1993, 1994), and *Pax1* is expressed around the graft site (Brand-Saberi *et al.*, 1993; Dietrich *et al.*, 1997), consistent with the role of the notochord as a negative regulator of *Pax3* and positive regulator of *Pax1/Pax9* and sclerotome formation. However, *MyoD* is ectopically activated at a distance from the graft (Dietrich *et al.*, 1997), consistent with the capacity of medial notochord grafts to induce *MyoD* expression in the absence of neural tube (Pownall *et al.*, 1996). This demonstrates the requirement of notochord for *MyoD* activation during somite formation. After longer incubations of 2–5 days following a dorsal medial graft of notochord, ectopic cartilage is induced in the dorsal compartment of somites, myotome formation is inhibited, and an ectopic floor plate is induced in the dorsal neural tube (Brand-Saberi *et al.*, 1993; Pourquié *et al.*, 1993).

When a notochord is grafted in a dorsal lateral position at the level of the segmental plate in HH stage 10–12 avian embryos, within 6 hr, *Pax3* expression is lost in the dorsal lateral region of the somites and *MyoD* is upregulated (Dietrich *et al.*, 1997). The upregulation of *MyoD* is maintained for 24 hr (Bober *et al.*, 1994b; Goulding *et al.*, 1994; Dietrich *et al.*, 1997), after which *MyoD* transcripts are located more distantly from the graft, and gradually disappear. Ectopic *Pax1* transcripts then become predominant (Bober *et al.*, 1994b; Goulding *et al.*, 1994; Dietrich *et al.*, 1997). Two days following grafting, myogenesis is lost (Brand-Saberi *et al.*, 1993; Bober *et al.*, 1994b; Xue and Xue, 1996). Interestingly, lateral ventral notochord grafts do not activate ectopic *MyoD* expression, but instead activate ectopic cartilage formation (Brand-Saberi *et al.*, 1993; Dietrich *et al.*, 1997), indicating that dorsal, but not the ventral somite, is preferentially responsive to the myogenic activities of the notochord. The complex spatial responses of the somite to notochord grafts suggest that notochord signals are interacting with and affecting the functions of additional signals produced by surrounding neural tube, lateral plate mesoderm, and surface ectoderm tissues, which all are now known to produce positive and negative myogenic signals; and as will be discussed below, understanding of the interactions of these multiple signals can provide reasonable explanations of these observations.

2. *In Vitro* Studies

Explant culture methods have been utilized to investigate notochord functions in somite myogenesis in the chick and mouse embryo, and the results of these studies complement *in vivo* surgical and grafting studies, discussed above. Explants of segmental plate and somites I–III from HH stage 11–13 chick embryos, when cultured in isolation of axial tissues (neural tube and notochord) and surface ectoderm for 4 days, do not express *MRF* or *contractile protein* genes (Buffinger and Stockdale, 1994; Münsterberg and Lassar, 1995; Stern and Hauschka, 1995). However, coculture with notochord induces myogenesis in somites I–III but not in presegmental plate explants, indicating that presegmental mesoderm alone cannot respond to notochord signals (Buffinger and Stockdale, 1994; Stern and Hauschka, 1995). Segmental plate explants cultured with surface ectoderm and the lateral plate form epithelial somites that do not undergo myogenesis (Buffinger and Stockdale, 1994; Stern and Hauschka, 1995) unless cocultured with notochord (Buffinger and Stockdale, 1994). Consistent with the above *in vivo* studies, these findings provide evidence that tissue interactions between the surface ectoderm and/or the lateral plate are required for the notochord to have myogenic activity, as will be discussed later. A similar requirement for the notochord/neural tube complex has been demonstrated in mouse presomitic mesoderm explant studies (Cossu *et al.*, 1996b). The floor plate of the neural tube can substitute for the notochord in the activation of myogenesis in somites I–III of HH stage 11–13 chick embryos (Buffinger and Stockdale, 1995), suggesting that notochord and floor plate produce similar myogenic signals.

3. Genetic Studies

Notochord mutants in mouse embryos provide support for a role for the notochord in somite myogenesis. The mouse *Brachyury* mutant, *T/T*, has greatly reduced expression in the dermomyotomal expression of *Pax3* and embryos die at midgestation (Dietrich *et al.*, 1993). *HNF-3β* mutant mouse embryos also lack a notochord and die at midgestation (Ang and Rossant, 1994; Weinstein *et al.*, 1994). In absence of the notochord, somites fuse under the neural tube, and the dermomyotomal markers *Pax3* and Neural Cell Adhesion–Prostate Specific Antigen (NCAM-PSA) extend ventrally and closer to the neural tube (Ang and Rossant, 1994; Weinstein *et al.*, 1994), an observation also reported in notochordectomized avian embryos (Goulding *et al.*, 1994). Analysis of *MyoD* or *Myf5* expression in these mutant embryos would be informative. *Truncate* (*tc/tc*), *Danforth's short tail* (*Sd/Sd*), and *Pintail* (Pt/Pt) mutant embryos display partial loss, truncation, or degeneration of the notochord. The dermomyotomal gene, *Pax3,* is expressed ventrally and sclerotome is missing in these mutants (Dietrich *et al.*, 1993). More recent studies show that epaxial expression of the myotomal gene *Myf5* is lost where the notochord is disrupted, but lateral hypaxial expression of *MyoD* is normal (Asakura *et al.,* 1995), consistent with current evidence that the notochord and Shh control epaxial, but not hypaxial myogenesis (Borycki *et al.*, 1999b). Studies of *Danforth's short tail* mutant embryos also provide evidence that the notochord is required to prevent apoptosis in the dorsal dermomyotome and the dorsal ectoderm (Asakura and Tapscott, 1998).

Zebrafish notochord mutants also reveal the essential role of the notochord in somite myogenesis. *bozozok* (*boz^{tz}*) mutants completely lack a notochord and a

5. Tissue Interactions and Signaling Mechanisms in Somite Myogenesis   179

floor plate and do not express *MyoD* in adaxial cells that produce slow muscles (Stemple et al., 1996; Blagden et al., 1997). *no tail/brachyury* (*ntl*), *floating head/not-1* (*flh*), and *momo* (*mom*) do not form notochord or undergo notochord degeneration in the posterior region of the embryo, and somitic cells do not express *MyoD* or show myosin heavy chain at sites along the AP axis where notochord precursors are missing (Odenthal et al., 1996; Weinberg et al., 1996; Blagden et al., 1997).

Some caution is required in interpreting the phenotypes of these zebrafish and mouse notochord mutants until these mutant genes are identified and their tissue expression is determined. Some of these genes may have autonomous functions in somitic mesoderm as well as notochord, which would contribute to their somite phenotypes. However, these genetic studies in combination with the extensive surgical, grafting, and explant studies in the chick embryo provide compelling evidence for an essential role of the notochord in the initiation of epaxial myogenesis.

D. Neural Tube Provides Signals for Maintenance of Epaxial Somite Myogenesis

1. In Vivo Studies

The neural tube, like the notochord, forms well before somites and myogenesis initiate in vertebrate embryos. Later, it undergoes differentiation along its dorsoventral axis to form floor plate and a variety of motor and sensory neurons, and roof plate (Bronner-Fraser and Fraser, 1997; Sporle and Schughart, 1997). These dorsal and ventral domains are distinguished by their distinct cellular morphologies, their patterns of gene expression, and their production of signal molecules, including Sonic hedgehog (Shh), Bone Morphogenetic Proteins (BMPs), and Wnts. Floor plate and notochord have similar signaling activities that are required for the ventral differentiation of the neural tube (Placzek et al., 1990; Roelink et al., 1994). These ventral signaling activities are antagonistic to roof plate activities, which are required for the differentiation of cell types specific to the dorsal neural tube (Liem et al., 1995).

Floor plate and dorsal neural tube myogenic activities can be assessed by ablating the notochord/neural tube complex in chick embryos and replacing it by a graft of quail or mouse neural tube alone (Rong et al., 1992; Fontaine-Perus et al., 1997). These graft experiments show maintenance or restoration of myogenesis within 4–6 days after ablation (Rong et al., 1992; Fontaine-Perus et al., 1997). However, if, in addition to the notochord ablation, the floor plate is absent, no *MyoD* activation or myogenesis will occur (Pownall et al., 1996; Dietrich et al., 1997; Borycki et al., 1998), presumably because the floor plate is an alternative source of notochord signaling molecules such as Shh, which is required for the induction of *MyoD* expression (Borycki et al., 1998). Therefore, these experiments demonstrate that dorsal neural tube alone is insufficient for *MyoD* activation (Pownall et al., 1996; Dietrich et al., 1997; Borycki et al., 1998). However, the neural tube is required to maintain *MyoD* expression, once activated. In quail embryos, separating *MyoD*-expressing somites I to VII from the neural tube/notochord complex, which removes inductive signals, results in the rapid loss within 2 hr of *MyoD* expression (Pownall *et al.*, 1996). Ablation of the dorsal neural tube alone at the level of the segmental plate of chick embryos does not affect *Pax3* and *MyoD* expression in newly formed somites for an initial 6–9 hr period (Dietrich *et al.*, 1997), and does not disrupt the epithelial organization of the dermomyotome (Hirano *et al.*, 1995). These results indicate that the continuous presence of the notochord and its myogenic signals can maintain *MyoD* expression and somite integrity for an extended period. However, 18–20 hr after dorsal neural tube removal, the epithelial organization of the dorsal somite is lost (Spence *et al.*, 1996) as is expression of *Wnt 11*, a marker of the dorsal medial lip of the dermomyotome (Marcelle *et al.*, 1997). Forty-eight hours after ablation of the entire neural tube, *MyoD* expression is no longer maintained in the epaxial somite domain (Christ *et al.*, 1992; Bober *et al.*, 1994b; Spence *et al.*, 1996; Dietrich *et al.*, 1997), and epithelial architecture of the dorsomedial dermomyotome is lost (Hirano *et al.*, 1995). In contrast, abdominal and shoulder muscles, which derive from the lateral hypaxial myotome, retain *MyoD* expression in the absence of the neural tube (Bober *et al.*, 1994b), and the epithelial architecture of the ventrolateral dermomyotome is unaffected (Hirano *et al.*, 1995). These findings establish that maintenance of epaxial myogenesis requires the neural tube, whereas maintenance of hypaxial myogenesis is independent of neural tube, but likely requires other signals (Cossu *et al.*, 1996b). More mature somites, anterior to somite VII, are autonomous of neural tube for maintenance of *MyoD* expression (Pownall *et al.*, 1996), and for the formation and the maintenance of the medial dermomyotome (Spence *et al.*, 1996), confirming that older somites maintain *MRF* gene expression independent of neural tube signals.

In addition to its role in the maintenance of epaxial gene expression such as *MyoD*, the neural tube produces signals that counterbalance the inhibitory signals from the lateral plate. Competition between neural tube and lateral plate signals establishes the mediolateral patterning of the somite and the boundaries of the epaxial and hypaxial dermomyotome (see below) (Pourquié *et al.*, 1996; Hirsinger *et al.*, 1997). Neural tube also provides signals that are redundant with the surface ectoderm signals. These additional signals are required for the activation of the downstream components of the Shh signaling pathway, which is an essential step for somites to acquire competence to respond to the myogenic-inducing activities of Shh (see below) (Borycki *et al.*, 1998).

2. *In Vitro* Studies

In vitro explant studies generally support the conclusions of the above *in vivo* data. However, because of the multiple integrated functions of the neural tube, it is difficult to separate individual activities *in vitro*. As will become more evident in later discussions, the newly forming somite is exposed to multiple signal interac-

5. Tissue Interactions and Signaling Mechanisms in Somite Myogenesis

tions, both positive and inhibitory, from notochord, neural tube, surface ectoderm, and lateral plate mesoderm. These multiple interactions are perturbed and imbalanced in *in vitro* explant essays in which somites are allowed to interact with only one of these inductive tissues. As also will be discussed in more detail, epaxial and hypaxial myogenesis are controlled by multiple, independently operating, and redundant signaling systems that activate *Myf5* and *MyoD* (Cossu *et al.*, 1996b; Borycki *et al.*, 1998, 1999; Tajbakhsh *et al.*, 1998), which themselves are redundant for myogenic determination. Accordingly, *in vitro* assays, which usually are carried out over long periods of time (3–5 days), can blend the entire developmental complexity of epaxial and hypaxial myogenesis in bird and mouse embryos. Therefore, *in vitro* assays do not separate the temporal and spatial processes of epaxial and hypaxial myogenesis, as these occur *in vivo*. These considerations provide a cautionary basis for our evaluation of the often contradictory and complex responses of somites to neural tube interactions, as revealed by *in vitro* assays. However, with these considerations in mind, we also believe that *in vitro* assays will provide an experimental approach to dissect these complexities, as initiated in the recent work of Tajbakhsh *et al.* (1998).

Experiments utilizing chick and mouse presomitic mesoderm explants have established that neural tube can support myogenesis, as assayed by expression of differentiation markers or earlier *MRF* markers (Vivarelli and Cossu, 1986; Kenny-Mobbs and Thorogood, 1987; Buffinger and Stockdale, 1994, 1995; Gamel *et al.*, 1995; Münsterberg and Lassar, 1995; Stern *et al.*, 1995; Stern and Hauschka, 1995; Cossu *et al.*, 1996b). *In vitro*, as well as *in vivo* (Pownall *et al.*, 1996), more mature somites can undergo muscle differentiation autonomous of tissue interactions (Gamel *et al.*, 1995; Stern *et al.*, 1995; Stern and Hauschka, 1995), whereas the differentiation of presomitic mesoderm requires neural tube with associated floor plate, which presumably replaces the notochord requirement, discussed earlier (Buffinger and Stockdale, 1995; Gamel *et al.*, 1995; Münsterberg and Lassar, 1995; Stern *et al.*, 1995). Interestingly, in mouse embryos, neural tube/notochord from E9.5 mouse embryos appears to preferentially promote *Myf5* expression in presomitic mesoderm explants, whereas neural tube/notochord from E10.5 mouse embryos preferentially induces *MyoD* expression through a *Myf5*-independent pathway (Tajbakhsh *et al.*, 1998). This indicates that the maturation of axial tissues during embryonic development may account for the temporal separation of *Myf5*-dependent and *Myf5*-independent myogenesis (Tajbakhsh *et al.*, 1997). The dorsal neural tube alone also is capable of maintaining expression of *Pax3* and promoting expression of the dermomyotomal marker, *Pax7*, in explants of presomitic mesoderm from E9.5 mouse embryos (Fan and Tessier-Lavigne, 1994). Also, Stern *et al.* (1995) and Spence *et al.* (1996) have observed a high percentage of differentiation, as assayed by myosin-positive cells, in avian segmental plate explants cultured for 3–4 days with a dorsal half of the neural tube alone. Therefore, the dorsal neural tube can provide signals for the induction and/or maintenance of a *Pax3*-dependent myogenesis, which may preferentially participate in the function of the hypaxial lineage (Tajbakhsh *et al.*, 1997).

Although explant assays reveal *in vitro* myogenic activities of the neural tube, the data concerning the relative contributions of dorsal versus ventral neural tube (Buffinger and Stockdale, 1995; Stern *et al.*, 1995) are not entirely consistent with the *in vivo* data that the neural tube has maintenance functions, but is not required for the initiation of myogenesis, at least epaxial myogenesis. It is of interest that the "independent" myogenic activities of the dorsal neural tube are revealed in the absence of the ventral neural tube, which may be producing inhibitory as well as inductive factors for myogenesis, as also has been suggested by *in vivo* data (Pownall *et al.*, 1996).

3. Genetic Studies

The recently described *Open brain* (*Opb*) mutant provides evidence that the dorsal neural tube has a role in the maintenance of the dorsal dermomyotome and epaxial myotome in the mouse embryo (Gunther *et al.*, 1994). Mutant mouse embryos have severe neural defects, including an absence of dorsalization of the neural tube and failure of neural tube closure (Gunther *et al.*, 1994). As a consequence, epaxial musculature is severely impaired, while intercostal and limb (hypaxial) muscles are normal. Defects in the dorsal muscles are visible in E11.5 embryos, with a severe reduction of *Myogenin* expression in dorsal muscles of the cervical and upper thoracic region, and a complete loss of *Myogenin* expression in the dorsal muscles of the lumbar region (Sporle *et al.*, 1996). E10.5 embryos show normal expression of *Myf5* in the posterior somites, but a dramatic reduction of *Myf5* expression in the epaxial, dorsomedial myotome from the lumbar to the cervical region (Sporle *et al.*, 1996), suggesting that maintenance rather than activation of *Myf5* is affected in this more anterior region of these embryos. Surprisingly, in the posterior somites of E9.5 *Opb* embryos, *Myf5* activation in the epaxial, dorsomedial somite is impaired, but this defect is not evident in E10.5 *Opb* embryos, suggesting that alternative signaling systems may compensate to promote *Myf5* activation in these mutant embryos (Sporle *et al.*, 1996). Alternatively, *Opb* mutant embryos of earlier stages may overproduce ventralizing signals (Gunther *et al.*, 1994) that inhibit *Myf5* expression in somites, again raising the caution that interpretation of phenotypes in these mutant embryos will require identification of the *Opb* gene and analysis of its sites of expression in somites and other interacting tissues.

E. Surface Ectoderm Signals Control Somite Formation, Competence to Respond to Axial Signals, and Hypaxial Myogenesis

1. *In Vivo* Studies

A role for the surface ectoderm in somite maturation *in vivo* has been recognized for some time (Gallera, 1966). However, recently, with the availability of molecular markers, specific functions of the surface ectoderm in dermomyotome formation and muscle determination has begun to be elucidated. *Paraxis,* a bHLH transcription factor, is expressed in presomitic mesoderm and newly forming somites, and its expression becomes restricted to the dermomyotome as somites mature (Sosic *et al.,* 1997). Gene targeting of *Paraxis* in the mouse has shown that *Paraxis* is required for the formation of epithelial somites and for the formation of the dermomyotome (Burgess *et al.,* 1996). Experiments in avian embryos have demonstrated that redundant signals from the surface ectoderm and the neural tube control *Paraxis* expression (Sosic *et al.,* 1997). Removal of the surface ectoderm overlying the segmental plate in HH stage 10–14 avian embryos results in a delay in *Paraxis* expression, but in combination with the separation from the axial tissues, results in the loss of *Paraxis* expression (Sosic *et al.,* 1997), and the lack of epithelial somites (Borycki *et al.,* 1997; Dietrich *et al.,* 1997; Sosic *et al.,* 1997; Borycki *et al.,* 1998).

In absence of surface ectoderm, epaxial *MyoD* activation in newly forming somites also is delayed, but recovers after 24 hr (Borycki *et al.,* 1997; Dietrich *et al.,* 1997), suggesting that the initial delay can be rescued by redundant signals from the dorsal neural tube. This observation indicates that surface ectoderm signals are necessary for mediating the myogenic activation by notochord signals. Indeed, the expression of some *Gli* transcription factor genes, which are the downstream transcription factor effectors of Shh signaling, is controlled by surface ectoderm signals (see below, Section V,A) (Borycki *et al.,* 1998). The activation of the *Gli* transcription factor gene expression in coordination with somite formation is an essential step in the regulation of the competence of the paraxial mesoderm to respond to myogenic-inducing signals. Therefore, redundant signals from the neural tube and the surface ectoderm control the initiation of Shh-mediated signaling processes in myogenic determination.

Finally, *in vivo* studies involving ablation of surface ectoderm in chick embryos have provided evidence that surface ectoderm signals also participate to the determination of hypaxial muscle lineage (Dietrich *et al.,* 1998), which is controlled by different mechanisms than is epaxial muscle determination (Cossu *et al.,* 1996b). Separation of paraxial mesoderm from surface ectoderm in HH stage 10–13 chick embryos disrupts totally the hypaxial expression of *Lbx1* and *Pax3,* which are markers for hypaxial myogenesis (Dietrich *et al.,* 1998). Conversely, ventral graft of surface ectoderm between the segmental plate and the lateral plate results after 24 hr in the ectopic activation of *Lbx1* and *Pax3* (Dietrich *et al.,* 1998).

2. *In Vitro* Studies

In vitro explant studies support a role for surface ectoderm in somite myogenesis, and in agreement with *in vivo* data, reveal multiple functions for surface ectoderm signals in somite formation, in competence to respond to epaxial muscle-inducing signals, and in hypaxial muscle determination. For instance, culture of segmental plate explants without their overlying ectoderm blocks both segmentation of the paraxial mesoderm to form somites and activation of *Gli2* expression, which occurs in coordination with somite formation (Borycki *et al.*, 1998). Coculture of segmental plate and newly forming somites from chick embryos with surface ectoderm results (after 5–6 days) in extensive myogenic differentiation (Kenny-Mobbs and Thorogood, 1987), which likely reflects hypaxial myogenesis. In newly formed somites (IV–VI) from HH stage 10 chick embryos, activation of expression of *MyoD* and *Myf5*, as well as maintenance of *Pax3* and *Pax7*, requires surface ectoderm (Maroto *et al.*, 1997). In more mature somites (VII–IX), *MyoD* and *Myf5* are stably maintained in absence of surface ectoderm, whereas *Pax3* and *Pax7* are downregulated, indicating that these latter genes have a more sustained requirement for surface ectoderm signals for their maintenance (Reshef *et al.*, 1998).

Surface ectoderm from mouse embryos also functions to maintain *Pax3* expression and can function to induce *Pax7*, *MyoD*, and *Myf5* in presomitic mesoderm explants of E9.5 mouse embryos (Fan and Tessier-Lavigne, 1994; Cossu *et al.*, 1996b), demonstrating that surface ectoderm has conserved functions in avian and mammalian embryos. However, in avian embryos, surface ectoderm signals appear to be required to induce *Myf5* first, followed by *MyoD* expression (Maroto *et al.*, 1997), whereas in mouse embryos, the surface ectoderm appears to act first on *MyoD*, followed by *Myf5* (Cossu *et al.*, 1996b). Cossu *et al.* (1996b) demonstrated that surface ectoderm preferentially promotes myogenesis in the lateral hypaxial domain of somites, thereby providing further evidence for a role of surface ectoderm signals in hypaxial muscle determination. Infection of segmental plate mesoderm explants of avian embryos with a *Pax3*-expressing retrovirus can mimic the effects of surface ectoderm signaling in avian embryos, as assayed by the induction of *MyoD* and *Myf5* expression (Maroto *et al.*, 1997), suggesting that surface ectoderm signals support hypaxial myogenesis via a *Pax3*-dependent pathway. This conclusion is consistent with the findings that the *MyoD*-expressing lateral dermomyotome is disrupted in the *Pax3* mutant *Splotch* mice (Franz *et al.*, 1993), and mouse embryos that are double-mutant mice for *Myf5* and *Pax3* lack all trunk muscles (Tajbakhsh *et al.*, 1997), demonstrating that *Pax3* has an essential upstream function in *MyoD* regulation in hypaxial muscle formation.

F. Lateral Mesoderm Provides Inhibitory Signals That Spatially Restrict Epaxial Somite Myogenesis

Lateral plate mesoderm is required in avian embryos to restrict *MyoD* expression to the epaxial, dorsal medial domain of newly forming somites (Pourquié *et al.*, 1995, 1996; Pownall *et al.*, 1996). Grafts of lateral plate at the level of the segmental plate upregulate the expression of *Pax3* and *Sim1*, and downregulate the expression of the epaxial gene *Myf5* (Pourquié *et al.*, 1995, 1996), indicating that the lateral plate provides positive signals for lateral dermomyotome genes and inhibitory signals to restrict expression of epaxial myogenic genes to the medial somite. Moreover, the separation of the lateral plate from the segmental plate by a surgical incision or the insertion of an impermeable membrane in stage HH 12 avian embryos provokes the loss of *Pax3*, *Sim1*, and *Lbx1* in the lateral domain of somites, and the subsequent upregulation of *Myf5* and *MyoD* (Pourquié *et al.*, 1995, 1996; Pownall *et al.*, 1996; Dietrich *et al.*, 1998), confirming that lateral plate signals participate in determining the lateral specificity of the dermomyotome. Further surgical experiments demonstrated that the dorsal (somatopleura) and the ventral (splanchnopleura) lateral mesoderm, but not the Wolffian duct (intermediate mesoderm), is involved in this activity (Pourquié *et al.*, 1996). Interestingly, dorsal neural tube signals counteract the lateral plate signals, allowing the establishment in the dermomyotome of an epaxial, medial compartment expressing *MyoD* and *Myf5* and a hypaxial, lateral compartment expressing *Pax3*, *Sim1*, and *Lbx1* (Pourquié *et al.*, 1995, 1996; Pownall *et al.*, 1996; Dietrich *et al.*, 1998).

V. Signaling Molecules and Transduction Pathways Controlling Myogenesis in Somites

The development of molecular probes and *in vivo* and *in vitro* assays that have revealed the importance of tissue interactions in somite myogenesis has led to recent studies to identify the relevant signal molecules produced by surrounding tissues and the transduction pathways in responding somites (Fig. 1C). The candidate signal molecules identified to date are familiar players in early developmental signaling, including members of the Sonic hedgehog, Wnt, and BMP [transforming growth factor-β (TGFβ)] and fibroblast growth factor (FGF) families of signaling molecules. In this section, we review the basic signal transduction pathways of candidate signal molecules (Fig. 2) and the embryological studies that have led to the identification of specific signal molecules as regulators of somite myogenesis. We also summarize the current genetic information available on the phenotypes resulting from mutations in signaling molecule genes in mice (Table I).

A. Sonic Hedgehog Is an Essential and Sufficient Notochord Signal for Epaxial Myogenesis

1. Sonic Hedgehog Signal Transduction Mechanisms

Sonic hedgehog (Shh) is a member of the Hedgehog family of secreted proteins that comprises a single gene in *Drosophila* (*hedgehog*), three genes in mammals and avians (*Sonic hedgehog*, *Indian hedgehog*, and *Desert hedgehog*), and three

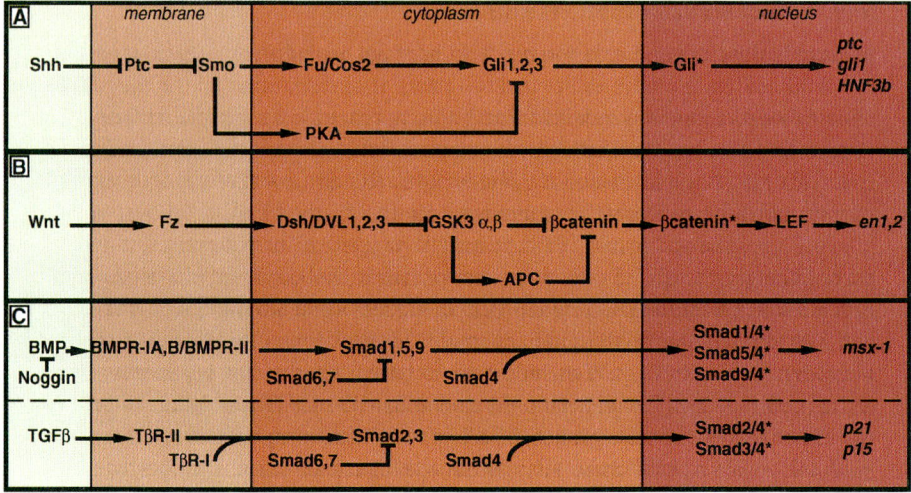

Figure 2 Signal transduction pathways of Shh (A), Wnt (B), and BMP/TGFβ (C) families. Asterisks (*) indicate functionally activated forms of proteins.

Table I Mouse Mutants in Developmental Signal Transduction Pathways

Genotype	Expression	Phenotype	References[a]
Tgf-β1 +/−	Notochord	Normal	1, 2
Tgf-β1 −/−		Newborns from +/− female die about 20 days after birth due to multifocal inflammatory disease (liver, lungs, heart skeletal muscles, . . .). Rescue during fetal and perinatal life by maternal TGFβ1. Newborns from −/− female die at 1 day of birth due to cardiac abnormalities.	
Bmp2 +/−	Myocardium at E9.5; precartilagenous mesenchyme at E12.5; hypertrophic cartilage at E15.5	Normal	3
Bmp2 −/−		Embryos die between E7.0 and E10.5. Defect in the formation of the amnion and chorion due to the lack of closure of the proamniotic canal. Abnormal development of the heart in the exocoelomic cavity possibly due to defects in interactions between mesodermal and ectodermal/endodermal cells. Somites smaller	
Bmp4 +/−	Lateral mesoderm, limb, dorsal neural tube, surface ectoderm, somites	Normal	4
Bmp4 −/−		Most embryos die at egg cylinder stage (E7.25). No *Brachyury* (*T*) expression, no mesoderm formation.	

Table I *Continued*

Genotype	Expression	Phenotype	References[a]
		Some embryos survive until neural fold/early somite stage or beating heart. No *Cdx-4* expression in posterior primitive streak. Possible roles: proliferation and survival of epiblast cells, proliferation and differentiation of posterior primitive streak mesoderm.	
Bmp7 +/−	Notochord, surface	Normal	5, 6
Bmp7 −/−	ectoderm, limbs	Embryos die perinatally. Eye (bilateral or unilateral micro- or anophtalmia) and kidney (bilateral renal dysplasia) defects. Early inductive epithelial–mesenchymal events are normal in developing kidney, but subsequent proliferation and survival is impaired. Primary induction of lens placode and optic vesicle is normal, but eye degeneration occurs by E13. In some cases, unilateral hind limb polydactyly (7–50%, 82%). Also, skull and rib cage abnormalities (rib fusion, impaired rib attachment).	
Bmpr-I +/−	Ubiquitous from E7.5	Normal	7
Bmpr-I −/−		Embryos die at egg cylinder stage (E7.0–E8.5), no mesoderm formation. No expression of *Brachyury, Lim-1, HNF3β*, and *Goosecoid* at E7.0. limited proliferation of epiblast cells. Possible role: required for the shortening of the length of the cell cycle of epiblast cells.	
Wnt1 +/−	Dorsal neural tube	Normal	8–10
Wnt1 −/−	from E8.5	Embryos die within 24 hr of birth. Caudal 2/3 of midbrain, midbrain–metencephalon junction, rostral metencephalon are missing. No *En-1* expression from the five somite stage and cell death occurs at stage 20–24 somite, in midbrain–hindbrain junction.	
Wnt3a +/−	Primitive streak at E7.5,	Normal	11, 12
Wnt3a −/−	dorsal neural tube from E8.5	Embryos die between E10.5 and E12.5. Lack of somites caudal to the first seven or nine somites, and formation of a ventral ectopic neural tube with full polarity. Extensive cell death in dorsal mesodermal cells in caudal region. No *Mox1* expression in tail region.	

continued

Table I *Continued*

Genotype	Expression	Phenotype	References[a]
Wnt1/Wnt3a −/−		Possible role: determination of the paraxial mesoderm fate. Normal activation of *Myf5* in cervical somites but reduced level of expression at E9.5. Loss of dorsal medial lip gene-specific expression (*Wnt11, Noggin,* etc) and epithelial dermomyotome reduced in its medial compartment. Disorganization of *MyoD* and *Myf5* expressing cells at E11.0.	13
Wnt4 +/− Wnt4 −/−	Dorsal neural tube, kidney mesenchyme	Normal Embryos die within 24 hr of birth. Kidney defects due to the failure in the formation of epithelial tubular derivatives from kidney mesenchyme. No *Pax8* induction in kidneys. Possible role: regulation of mesenchymal to epithelial transition.	14
Wnt7a +/− Wnt7a −/−	Dorsal ectoderm from E8.75, dorsal limb ectoderm from E9.25 (forelimb) and E9.75 (hindlimb)	Normal Viable but sterile and limb abnormalities. Loss of dorsal limb structures and ventralization of mutant limbs. More severe phenotype in distal than proximal limb, and in posterior than anterior structures. Decrease in *Shh* and *Bmp2* expression in the ZPA, and *Fgf4* in the AER.	15
Shh +/− Shh −/−	Primitive streak, notochord, limbs	Normal Cyclopia, cranofacial defects, lack of vertebral bone and ribs; defects in limbs, heart, lung, kidneys. Reduced expression of *Myf5* and *Pax1* in somites at E9.5.	16
Ptc +/− Ptc −/−	Ubiquitous at low level, neural tube, limb, somites	Development defects in neural tube, polydactyly, syndactyly; cancers including medulloblastomas (8/10), rhabdomyosarcoma (1/10). Die E9.0–E10.5; CNS (overgrowth of neural tube and brain) and heart defects.	17
Gli2 +/− Gli2 −/−, Dominant hemimelia mutants (*Dh*)	Neural tube, somites, limbs	No defect; poly/oligodactyly, visceral abnormalities (spleen). Die between E18.5 and birth; craniofacial abnormalities, lack of intervertebral disks, extra ribs.	18 19

5. Tissue Interactions and Signaling Mechanisms in Somite Myogenesis

Table I Continued

Genotype	Expression	Phenotype	References[a]
Gli3 +/−	Neural tube, somites, limbs	Craniofacial abnormalities, polydactyly.	18
Gli3 −/− natural deletions or insertions (*extra-toe, anterior digit deformity*)		Die at birth; severe craniofacial defects, polydactyly, unfused sternum, fusion of neural arches.	19
Fgf-3 +/−	Primitive streak, rhombomeres 5 and 6, cerebellum, retina, teeth, inner ear	Normal	20
Fgf-3 −/−		22% of homozygous survive to weaning, 78% die within 3 weeks of birth. Short curly or kinked tail visible at E12.5. Partially penetrant inner ear abnormalities. Otic vesicles form normally but defects in endolymphatic duct formation.	
Fgf-4 +/−	Morula, ICM (blastocyst stage), ectoderm (egg cylinder stage), muscle, tooth mesenchyme, ...	Normal	21
Fgf-4 −/−		Embryos die between E5.5 and E6.5. Normal uterine implantation, but no postimplantation development due to growth failure of the inner cell mass (ICM).	
Fgf-5 +/−	Embryonic ectoderm, myotome, spinal cord, and hyppocampus of adults, hair follicles	Normal	22
Fgf-5 −/− Angora (*go*) mutants		Viable, but hair 50% longer starting at 3 weeks after birth. Increase in the length of anagen VI, the phase of hair growth.	
Fgf-6 +/−	Myotome	Normal	23
Fgf-6 −/−		Normal; 2.5-fold increase in the number of muscle satellite cells. Defect in muscle regeneration after crush injury. Reduced expression of *MyoD* in injured muscles. Possible role: regulation of entry of satellite cells into the cell cycle.	
Fgfr-1 +/−	Ectoderm (egg cylinder stage)	Normal	24, 25
Fgfr-1 −/−		Embryos die between E7.5 and E9.5. Embryos are smaller, lack somites, have abnormal heart tube, and broader midline structure in the tail. Formation and survival of paraxial mesoderm impaired, no limb bud, abnormal axial mesoderm (notochord, neural tube). Axial and paraxial phenotype may result from abnormal migration of cells from the primitive streak, rather than from a defect in cell differentiation. It also has role in cell growth.	

continued

Table I *Continued*

Genotype	Expression	Phenotype	References[a]
Noggin +/− Noggin −/−	Notochord, dorsal neural tube, somites	Normal Embryos die at birth. Defects in neural tube closure, loss of caudal vertebrae. In somites, defects are mostly found between E8.5 and E9.5 in the posterior embryo; shortening of dermomyotome, reduction of medial *MyoD* and *Pax1* expression, expansion of lateral *Pax3* and *Sim1* into dorsal medial dermomyotome.	26

[a] Key to references:
(1). Shull, M. M., Ormsby, I., Kier, A. B., Pawlowski, S., Diebold, R. J., Yin, M., Allen, R., Sidman, C., Proetzel, G., Calvin, D., Annunziata, N., and Doetschman, T. (1992). Targeted disruption of the mouse transforming growth factor-beta 1 gene results in multifocal inflammatory disease. *Nature* **359**, 693–699; (2). Letterio, J. J., Geiser, A. G., Kulkarni, A. B., Roche, N. S., Sporn, M. B., and Roberts, A. B. (1994). Maternal rescue of transforming growth factor-beta 1 null mice. *Science* **264**, 1936–1938; (3). Zhang, H., and Bradley, A. (1996). Mice deficient for BMP2 are nonviable and have defects in amnion/chorion and cardiac development. *Development* **122**, 2977–2986; (4). Winnier, G., Blessing, M., Labosky, P., and Hogan, B. L. M. (1995). Bone morphogenetic protein-4 is required for mesoderm formation and patterning in the mouse. *Genes Dev.* **9**, 2105–2116; (5). Luo, G., Hofmann, C., Bronckers, A. L., Sohocki, M., Bradley, A., and Karsenty, G. (1995). BMP-7 is an inducer of nephrogenesis, and is also required for eye development and skeletal patterning. *Genes Dev.* **9**, 2808–2820; (6). Dudley, A. T., Lyons, K. M., and Robertson, E. J. (1995). A requirement for bone morphogenetic protein-u during development of the mammalian kidney and eye. *Genes Dev.* **9**, 2795–2807; (7). Mishina, Y., Suzuki, A., Ueno, N., and Behringer, R. R. (1995). Bmpr encodes a type I bone morphogenetic protein receptor that is essential for gastrulation during mouse embryogenesis. *Genes Dev.* **9**, 3027–3037; (8). Serbedzija, G. N., Dickinson, M., and McMahon, A. P. (1996). Cell death in the CNS of the Wnt-1 mutant mouse. *J. Neurobiol.* **31**, 275–282; (9). McMahon, A. P., and Bradley, A. (1990). The Wnt-1 (int-1) proto-oncogene is required for development of a large region of the mouse brain. *Cell* **62**, 1073–1085; (10). McMahon, A. P., Joyner, A. L., Bradley, A., and McMahon, J. A. (1992). The midbrain–hindbrain phenotype of Wnt-1-/Wnt-1- mice results from stepwise deletion of engrailed-expressing cells by 9.5 days postcoitum. *Cell* **69**, 581–595; (11). Takada, S., Stark, K. L., Shea, M. J., Vassileva, G., McMahon, J. A., and McMahon, A. P. (1994). Wnt-3a regulates somite and tailbud formation in the mouse embryo. *Genes Dev.* **8**, 174–189; (12). Yoshikawa, Y., Fujimori, T., McMahon, A. P., and Takada, S. (1997). Evidence that absence of Wnt-3a signaling promotes neuralization instead of paraxial mesoderm development in the mouse. *Dev. Biol.* **183**, 234–242; (13). Ikeya, M., and Takada, S. (1998). Wnt signaling from the dorsal neural tube is required for the formation of the medial dermomyotome. *Development* **125**, 4969–4976; (14). Stark, K., Vainio, S., Vassileva, G., and McMahon, A. P. (1994). Epithelial transformation of metanephric mesenchyme in the developing kidney regulated by Wnt-4. *Nature* **372**, 679–683; (15). Parr, B. A., and McMahon, A. P. (1995). Dorsalizing signal Wnt-7a required for normal polarity of D–V and A–P axes of mouse limb. *Nature* **374**, 350–353; (16). Chiang, C., Litingtung, Y., Lee, E., Young, K. E., Corden, J. L., Westphal, H., and Beachy, P. A. (1996). Cyclopia and defective axial patterning in mice lacking sonic hedgehog gene function. *Nature* **383**, 407–413; (17). Goodrich, L. V., Milenkovic, L., Higgins, K. M., and Scott, M. P. (1997). Altered neural cell fates and medulloblastoma in mouse patched mutants. *Science* **277**, 1109–1113; (18). Mo, R., Freer, A. M., Zinyk, D. L., Crackower, M. A., Michaud, J., Heng, H. H., Chik, K. W., Shi, X. M., Tsui, L. C., Cheng, S. H., Joyner, A. L.,

5. Tissue Interactions and Signaling Mechanisms in Somite Myogenesis

Table I *Continued*

and Hui, C. (1997). Specific and redundant functions of Gli2 and Gli3 zinc finger genes in skeletal patterning and development. *Development* **124**, 113–123; (19). Motoyama, J., Liu, J., Mo, R., Ding, Q., Post, M., and Hui, C. C. (1998). Essential function of Gli2 and Gli3 in the formation of lung, trachea and oesophagus. *Nat. Genet.* **20**, 54–57; (20). Mansour, S. L., Goddard, J. M., and Capecchi, M. R. (1993). Mice homozygous for a targeted disruption of the proto-oncogene int-2 have developmental defects in the tail and inner ear. *Development* **117**, 13–28; (21). Feldman, B., Poueymirou, W., Papaioannou, V. E., DeChiara, T. M., and Goldfarb, M. (1995). Requirement of FGF-4 for postimplantation mouse development. *Science* **267**, 246–249; (22). Hebert, J. M., Rosenquist, T., Gotz, J., and Martin, G. R. (1994). FGF5 as a regulator of the hair growth cycle: Evidence from targeted and spontaneous mutations. *Cell* **78**, 1017–1025; (23). Floss, T., Arnold, H. H., and Braun, T. (1997). A role for GFG-6 in skeletal muscle regeneration. *Genes Dev.* **11**, 2040–2051; (24). Deng, C. X., Wynshaw-Boris, A., Shen, M. M., Daugherty, C., Ornitz, D. M., and Leder, P. (1994). Murine FGFR-1 is required for early postimplantation growth and axial organization. *Genes Dev.* **8**, 3045–3057; (25). Yamaguchi, T. P., Harpal, K., Henkemeyer, M., and Rossant, J. (1994). fgfr-1 is required for embryonic growth and mesodermal patterning during mouse gastrulation. *Genes Dev.* **8**, 3032–3044; (26). McMahon, J. A., Takada, S., Zimmerman, L. G., Fan, C. M., Harland, R. M., and McMahon, A. P. (1998). Noggin-mediated antagonism of BMP signaling is required for growth and patterning of the neural tube and somite. *Genes Dev.* **12**, 1438–1452.

genes in zebrafish (*Sonic hedgehog, Echidna hedgehog,* and *Tiggy winkle hedgehog*) (Hammerschmidt *et al.,* 1997). The biochemistry of Shh processing, secretion, and diffusion in the embryo is complex and still under investigation. Shh is synthesized as a precursor protein that undergoes an autoproteolytic cleavage leading to a 19-kDa N-terminal peptide and a 26- to 28-kDa C-terminal peptide (Lee *et al.,* 1994). N-terminal peptides, that have been shown to carry both the short- and long-range signaling activities of Shh (Porter *et al.,* 1995), are bound to the cell surface through cholesterol groups, creating locally high concentration of signaling molecules (Porter *et al.,* 1996). In fact, immunodetection of N-Shh in embryos reveals very poor diffusion of the peptide away from the cells producing it (Marti *et al.,* 1995a), leaving open the question of the mechanism of diffusion of Shh in the embryo. In *Drosophila,* the diffusion of Hedgehog in embryonic tissues is regulated by *Tout velu,* which has mammalian homologues (*Ext*), consistent with a role for Hedgehog as a morphogen that regulates spatial patterning of gene expression in the embryo (Bellaiche *et al.,* 1998). In addition, the crystal structure of N-Shh reveals the presence of a zinc cation in the N-terminal domain of the protein, reminiscent of zinc hydrolase proteins (Hall *et al.,* 1995). This raises the possibility of additional enzymatic activity for Shh. C-terminal Shh peptides are released in the intercellular space *in vivo* and *in vitro,* but have not been implicated in the signaling activity of Shh (Marti *et al.,* 1995b; Porter *et al.,* 1995). However, recent studies indicate that C-Shh shares sequence homologies with the self-splicing protein domain of yeast and bacterial inteins (Hall *et al.,* 1997), a discovery that may bring light to the mechanism of autoproteolytic cleavage used by Shh.

In addition to its complex biochemistry, Shh signaling involves multiple components and levels of control (Fig. 2A), which have been elucidated in *Drosophila*. Shh binds and inactivates a transmembrane protein, Patched (Ptc) (Marigo *et al.*, 1996a; Stone *et al.*, 1996), releasing the inhibition of Smoothened (Smo), another transmembrane protein, by Ptc (Alcedo *et al.*, 1996; van den Heuvel and Ingham, 1996). Therefore, in absence of Shh, Ptc represses the constitutively active downstream signaling pathway of Shh. Downstream components of this signaling pathway in flies include Fused (Fu), a serine/threonine kinase, Costal 2 (Cos2), a kinesin-related protein recently shown to be associated with microtubules, and Cubitus interruptus (Ci), a zinc finger transcription factor (Ruiz i Atalba, 1997). In the current model, in cells subject to low concentrations of Hedgehog (Hh) or in absence of Hh, Ci is associated into a complex with Cos2 and Fu to the microtubules (Robbins *et al.*, 1997; Sisson *et al.*, 1997). In this cytoplasmic complex, Ci undergoes a cleavage generating a shorter N-terminal 75-kDa peptide, which then translocates to the nucleus where it acts as a repressor of the Hh target genes, *ptc, wingless* (*wg*), and *decapentaplegic* (*dpp*) (Aza-Blanc *et al.*, 1997). In cells subject to high concentrations of Hh, the complex Cos2/Fu/Ci dissociates from the microtubules and the cleavage of Ci into a repressive form is prevented. It is postulated that the 155-kDa form of Ci undergoes posttranslational modifications allowing its translocation to the nucleus where it now acts as an activator of Hh target genes. In vertebrates, the existence of three homologues of Ci (Gli 1, Gli 2, and Gli 3) suggest that the dual repressor/activator function of Ci in flies may be achieved through different proteins in vertebrates (Ruiz i Atalba, 1997). In fact, recent data argue for Gli 1 being a positive regulator, and Gli 3, possibly redundant with Gli 2 (Mo *et al.*, 1997), being a negative regulator of Shh target genes in vertebrates (Marigo *et al.*, 1996b; Lee *et al.*, 1997; Marine *et al.*, 1997; Sasaki *et al.*, 1997).

2. Sonic Hedgehog Signaling and Transduction in Embryos

In vertebrate embryos, Shh is synthesized at diverse sites overlapping with organizing centers, such as the notochord, the floor plate, and the zone of polarizing activity (ZPA) (Fig. 1C) (Marti *et al.*, 1995a). Because of its notochord expression, Shh is a logical candidate as a notochord inducer of somite determination, as has now been demonstrated by extensive lines of experimentation. Initial work demonstrated that Shh can induce expression of *Pax 1* in explants of mouse presomitic mesoderm (Fan and Tessier-Lavigne, 1994) and chick segmental plate explants (Münsterberg *et al.*, 1995). In these assays, Shh represses *Pax3* expression over a long distance, indicating that Shh can diffuse (Fan and Tessier-Lavigne, 1994). Since *Pax3* expression occurs in the dorsal somite and is required for skeletal myogenesis (Bober *et al.*, 1994a; Goulding *et al.*, 1994), this result initially suggested that Shh activity is antagonistic to myogenesis. However, Shh also promotes myogenic gene expression in the segmental plate mesoderm assays (Münsterberg

5. Tissue Interactions and Signaling Mechanisms in Somite Myogenesis

et al., 1995). Also, ectopic expression of Shh in HH stage 10 chick embryos, via retrovirus infection of the segmental plate mesoderm, induces ectopic expression of both *MyoD* and *Pax1*, as well as the loss of *Pax3* expression (Johnson *et al.,* 1994). These findings suggested that Shh is required for both *Pax1* and *MyoD* expression in somite progenitors, but that additional factors are required for their dorsal and ventral patterning of expression. *In vitro* explant experiments provided additional evidence that, in presence of the dorsal neural tube, Shh can substitute for the notochord in *MyoD*-inducing activity (Münsterberg *et al.,* 1995). Finally, implantation of beads of N-Shh recombinant protein in the ventral neural tube in place of the notochord promotes *MyoD*, *Myf5*, and *Pax1* activation in newly forming somites of quail embryos (Borycki *et al.,* 1998), and engraftment of Shh-expressing cells between the lateral plate and the segmental plate mesoderm in chick embryos leads to the upregulation of *MyoD* and *Pax1* in the lateral somite (Hirsinger *et al.,* 1997). In complementary experiments, inhibition of Shh production by antisense oligonucleotide treatment inhibits *MyoD*, *Myf5*, and *Pax1* activation in newly forming somites (Borycki *et al.,* 1998). These studies, therefore, establish that Shh is a necessary and sufficient notochord factor for both epaxial myotome and sclerotome gene activation in newly forming somites of avian embryos and that Shh alone can replace the notochord in these activities.

In mouse embryos, targeted mutation of the *Shh* gene disrupts, but does not eliminate somite myogenesis (Chiang *et al.,* 1996) (Table I). Initial observations reported that *Shh* mutant embryos express reduced levels of *Myf5* transcripts, but that *MyoD* expression is normally activated (Chiang *et al.,* 1996). However, further analysis of the expression of *MRF* and other somitic genes in *Shh* mutant mice reveals that *Myf5* and *MyoD* fail to become activated in the medial epaxial somite, whereas their activation in the lateral hypaxial somite occurs normally (Borycki *et al.,* 1999b). These results demonstrate the requirement for Shh in early regulatory processes of murine epaxial myogenesis. Thus, Shh functions similarly in epaxial myogenesis in mouse and avian embryos.

In zebrafish, Hh family proteins also have lineage-specific myogenic functions. In the zebrafish embryo, the myotome encompasses most of the somitic mesoderm, but *MyoD* expression is activated prior to somite formation in the adaxial cells, which are the mesodermal cells located immediately adjacent to the notochord, a subset of which gives rise to the muscle pioneer cells (Devoto *et al.,* 1996). A family of three Hh homologues are known in zebrafish: *sonic hedgehog* (*shh*), which is expressed in the notochord and the floor plate, *echidna hedgehog* (*ehh*), which is expressed in the notochord only, and *tiggy winkle hedgehog* (*twhh*), which is expressed in the floor plate only (Krauss *et al.,* 1993; Ekker *et al.,* 1995; Currie and Ingham, 1996). Interestingly, two *Hh* genes, *shh* and *ehh*, contribute to muscle formation. *shh* is required for adaxial expression of *MyoD*, and more specifically for the formation of slow muscle, as shown by ectopic expression studies in which *shh* mRNA is injected into wild-type and *floating head* (*flh*) or *bozozok* (*boz*) notochord mutants (Weinberg *et al.,* 1996; Blagden *et al.,* 1997). Injected *shh* induces ectopic *myoD* expression and slow muscle myosin expression throughout the somitic mesoderm (Weinberg *et al.*, 1996; Blagden *et al.*, 1997). *ehh* controls the formation of muscle pioneer cells (Currie and Ingham, 1996). Notably, injection of *ehh* mRNA in *no tail* (*ntl*) and *flh* mutants restores muscle pioneer cells, and injection of *shh* and *ehh* mRNA induces supernumerary muscle pioneer cells (Currie and Ingham, 1996). Recently, it has been shown that *sonic-you* (*syu*) is a *shh* mutation (Schauerte *et al.*, 1998). *syu* mutants have reduced expression of *myoD* in adaxial cells and muscle pioneer cells do not form (van Eeden *et al.*, 1996). These studies, therefore, establish that *shh* has an essential and sufficient role in pioneer muscle cell formation in the zebrafish. The availability of a zebrafish *shh* mutant provides a genetic tool to examine the specific functions of *shh* in *myoD* activation and muscle pioneer determination.

3. The Trophic Effects of Sonic Hedgehog

In avian embryos, the ablation of the notochord, as well as the neural tube, results in the rapid decay in *MyoD* expression (Pownall *et al.*, 1996; Dietrich *et al.*, 1997; Teillet *et al.*, 1998), and within 24 hr in massive apoptosis in somitic cells (Teillet *et al.*, 1998). Similarly, *Danforth's short tail* (*Sd*) mutant mouse embryos, which undergo progressive notochord degeneration, display massive apoptosis in somites by E10.5, specifically in the dorsal part of somites where myogenesis takes place (Asakura and Tapscott, 1998). These observations suggest that the notochord produces factors essential for the survival of the somitic mesoderm. Because implantation of beads or cells expressing Sonic hedgehog can rescue the loss of *MyoD* expression and prevent cell death, it has been proposed that Shh is involved in somite cell survival (Teillet *et al.*, 1998). Whether loss of *MyoD* expression in notochordectomized embryos is the result of the loss of the lineage-specific trophic effect of Shh or the consequence of the lack of inductive signals for cell determination is an issue that can be resolved only by simultaneous examination of cell death and gene expression to determine which comes first. Evidence that trophic functions of Shh are independent of its inductive functions has been obtained in studies of *Shh* mutant mouse embryos. In E9.5 *Shh* mutant embryos, epaxial *Myf5* expression is undetectable, but no apoptosis is detected in the dorsal somite, which is the site of *Myf5* activation and epaxial myogenesis (Borycki *et al.*, 1999b). Thus, loss of *Myf5* activation in *Shh* null mouse embryos is not the consequence of apoptosis. However, Shh has lineage-specific functions in cell survival, selectively in the ventral compartment of somite and in the ventral and dorsal neural tube (Borycki *et al.*, 1999b). Therefore, the extensive apoptosis observed in somites of notochordectomized avian embryos and *Danforth's short tail* mouse embryos is likely due to the loss of other notochord cell survival molecules, that can be replaced by local, high levels of Shh. Shh also has activity as a lineage-specific mitogen (Fan *et al.*, 1995; Duprez *et al.*, 1998). However, in *Shh* mutant embryos, no specific defect in cell proliferation was observed in the epaxial domain of somites that lack *Myf5* expression (Borycki *et al.*, 1999b), although the smaller size of *Shh*

5. Tissue Interactions and Signaling Mechanisms in Somite Myogenesis 195

null embryos suggests that Shh may have general proliferative function. Therefore, Shh is multifunctional and provides signals for cell proliferation, cell survival, and cell determination in the development of somites. Future studies to identify the direct target genes of Shh signaling are required to define the specific functions of Shh in the processes of somitogenesis.

4. The Downstream Effectors of Shh Signaling Are Key Players in the Control of Myogenesis

The expression of Shh in the developing notochord precedes somite formation and the activation somite regulatory genes, which are precisely regulated and coordinated with somite formation. Shh protein expression is first detected in the node and node-derived cells in E8.0 mouse embryos and HH stage 5–6 in chick embryos, then in the notochord and ventral midbrain, and finally in the floor plate (Marti et al., 1995a), while *Myf5* and *MyoD* transcripts are not detected in mouse before E8.5 and E10.5, respectively; and in avian embryos, *MRF* genes are not activated until HH stage 11 (Charles de la Brousse and Emerson, 1990; Buckingham, 1992; Pownall and Emerson, 1992; Borycki et al., 1997). Shh proteins are expressed in the notochord throughout the anteroposterior axis of the embryo while *Myf5* and *MyoD* are activated in coordination with somite formation (Borycki et al., 1997). Finally, unsegmented paraxial mesoderm and newly formed somites prior to HH stage 11 in quail embryos are not competent to respond to *MyoD*-inducing signals (Borycki et al., 1997). Therefore, although Shh is an essential signal for somite myogenesis, production of the Shh signal cannot play a direct regulatory role in the control of somite determination. These findings suggested that genes in the downstream Shh signal transduction pathway in somites have a regulatory role.

The signal transduction genes in the Shh pathway include *Ptc*, which encodes the Shh receptor, and the *Gli* genes, *Gli1*, *Gli2*, and *Gli3*, which are the transcription factor effectors of Shh signaling (Fig. 2A). Analysis of the developmental expression of *Ptc*, and *Gli1* and *Gli2* established that activation of these genes in quail somites is precisely coordinated with somite formation, and occurs prior to activation of myogenic and sclerotome genes that are under the control of Shh signaling (Borycki et al., 1998). Similar results have been obtained for expression of *Gli3* (Borycki and Emerson, 1999). These findings, therefore, provide evidence for the regulatory functions of genes in the Shh transduction pathway in controlling the timing of activation of somite genes. In addition, *Ptc*, *Gli1*, and *Gli2* also have dynamic spatial patterns of expression during somite maturation and are differentially expressed in the ventral and dorsal, epaxial and sclerotomal compartments of somites, respectively (Borycki et al., 1998). The spatially regulated expression of *Gli1* and *Gli2* provides a potential mechanism for the differential transduction of Shh signaling to the somite through differential transcriptional activities of the *Gli* transcription factors. In quail embryos, *Gli1* is expressed in the ventral somite as early as stage HH 8, when *Pax1* expression is being activated in the sclerotome, whereas dorsal expression of *Gli2* is not detected before stage HH 11 (Borycki *et al.*, 1998), when *Myf5* and *MyoD* are activated in the epaxial myogenic lineage (Borycki *et al.*, 1997). It remains to be determined whether *Pax1* and *MyoD* and *Myf5* are direct targets of *gli* binding or are activated by upstream Gli-independent transcription factors. Finally, *Gli1* and *Gli2* activation in somites is controlled by different mechanisms. The mechanisms that activate the expression of *Ptc* and the *Gli* genes are currently under investigation. At this time, there is evidence that *Gli1* and *Ptc* require Shh for activation in the somite (Borycki *et al.*, 1998) as well as in neural tube (Lee *et al.*, 1997) and the limb bud (Marigo *et al.*, 1996b). In contrast, activation of *Gli2* (and also *Gli3*) during somite formation is upstream of *Gli1* and *Ptc*, as these genes are activated independent of Shh and axial tissues and their activation is controlled by signals from the surface ectoderm and the dorsal neural tube (Borycki *et al.*, 1998). These observations provide a basis for the requirement of surface ectoderm and/or the dorsal neural tube, in combination with Shh, for *MyoD* expression in segmental plate explants, as discussed previously.

In *Xenopus*, *Gli4*, which is closely related to mouse and avian *Gli2*, is specifically expressed in myotomal cells and may act as a transcriptional repressor (Marine *et al.*, 1997). In zebrafish, *you-too* (*yot*) mutations affect the formation of the myoseptum and *myoD* expression (van Eeden *et al.*, 1996). Epistatic studies recently showed that *yot* is downstream of *shh* in the Shh signaling pathway (van Eeden *et al.*, 1996; Schauerte *et al.*, 1998), and linkage analyses indicate that the *yot* mutation is likely in the *Gli2* locus (Karlstrom *et al.*, 1999), demonstrating the importance of Shh and the downstream components of its signaling cascade in *myoD* activation. Mouse mutants for *Ptc*, *Smo*, *Gli1*, *Gli2*, and *Gli3* have been generated by gene targeting and have been utilized in studies of tumorigenesis and bone formation (see Table I), but studies of somite myogenesis in these mutants remain to be performed.

B. Surface Ectoderm Wnts Are Signals for Activation of Hypaxial Myogenesis, and Neural Tube Wnts Are Signals for Maintenance of Epaxial Myogenesis

1. Wnt Signal Transduction Mechanisms

Wnt proteins have functions in embryonic patterning in *Drosophila* (Russell *et al.*, 1992; Klingensmith and Nusse, 1994; Graba *et al.*, 1995). Twenty genes have been isolated in vertebrates (Cadigan and Nusse, 1997). They encode cysteine-rich secreted glycoproteins of 37–48 kDa that have been shown to remain associated largely with the cell surface (Nusse and Varmus, 1992). The biochemistry of Wnt proteins has been hampered because methods are not yet available to isolate biologically active, soluble forms of Wnt. Nevertheless, recent studies in *Drosophila* and *Caenorhabditis elegans* suggest that synthesis and processing of active Wnt proteins requires the functions of accessory proteins such as Porcupine, an endoplasmic reticulum-associated protein (Kadowaki et al., 1996). In *Drosophila* where antibodies against Wingless (wg), the Wnt1 homologue, are available, the protein can be detected at a long distance from its secretion site. This is consistent with its proposed function as a morphogen (Lawrence et al., 1996), although morphogen activity of Wnt remains to be demonstrated in vertebrates.

A Wnt signal transduction pathway has been elucidated in *Drosophila* and vertebrates (Fig. 2B). Wnt proteins bind to a family of seven transmembrane receptor proteins encoded by the *frizzled* (*fz*) genes (Bhanot et al., 1996), specifically through a cysteine-rich extracellular domain of the receptor (Bhanot et al., 1996). Signaling through Frizzled activates and recruits Dishevelled, a cytoplasmic phosphoprotein, to the plasma membrane (Yang-Snyder et al., 1996), leading to the inhibition of a serine/threonine kinase, Glycogen Synthase Kinase 3 (GSK3) (see Fig. 2). GSK3 normally functions by promoting the degradation of β-catenin, and Wnt signaling results in the stabilization of β-catenin and its accumulation in the nucleus (Yost et al., 1996), where it interacts with High Motility Groups (HMG)-box transcription factors such as LEF1 to stimulate transcription of target genes (Behrens et al., 1996). In addition to this complex signaling pathway, a family of extracellular proteins, called Frzb, sharing sequence homology with the cysteine-rich domain of Frizzled was recently isolated, and shown to bind and sequester Wnt proteins from their receptor (Moon et al., 1997). Frzb is thought to regulate the spatial activities of Wnt proteins in the embryo, analogous to the inhibitory activities of Noggin on the spatial activities of BMP proteins, as will be discussed below. Interestingly, FrzB proteins are themselves spatially regulated (Hoang et al., 1998), and also have differential affinities for different Wnts (Wang et al., 1997), suggesting multiple levels of regulation of the Wnt signaling pathways.

In vertebrates, there are two classes of Wnt families, based on their domains of expression in the developing embryo. The first class includes Wnt 1, 3, 3a, 4, 5a, 7a, and 7b, which are expressed in the neural tube, whereas the second class includes Wnt 4, 6, 7a, and 7b, which are expressed in the surface ectoderm (Fig. 1C) (Dealy et al., 1993; Parr et al., 1993; Hollyday et al., 1995). Within the neural tube, three subclasses can be defined: the first includes Wnts that are expressed in the dorsal neural tube and roof plate, including Wnt 1, 3, and 3a; the second class includes Wnts that are expressed in the ventral neural tube, including Wnt 5a, 7a, and 7b; and the third class includes Wnt4, which is expressed in a broader ventral domain, including the floor plate (Parr et al., 1993; Hollyday et al., 1995). This distinctive spatial expression of the Wnt proteins, both in the neural tube and the surface ectoderm, makes them attractive candidates for mediating the redundant neural tube and/or surface ectoderm signaling involved in somite myogenesis in avian and mouse embryos.

2. Wnt Signaling and Transduction in Embryos

The initial studies of Wnt function in avian somites showed that neural tube Wnts, Wnt1 or Wnt3, and to a lesser extent Wnt4, are capable of maintaining *MyoD* expression when Wnt expressing cells are cocultured with explants from somite I–IV of stage HH 10–11 chick embryos, while the surface ectoderm Wnts, Wnt7a and Wnt7b, lacked activity (Münsterberg et al., 1995). However, segmental plate mesoderm explants cultured under similar conditions in presence of Wnt 1, 3, or 4 expressing cells did not activate MyoD (Münsterberg et al., 1995). These results indicate that the Wnt proteins are involved in maintenance rather than activation of MyoD expression. In support of this interpretation, culture of segmental plate mesoderm explants with Shh and Wnt1- or Wnt3-expressing cells results in the activation of MyoD (Münsterberg et al., 1995). In these assays, Shh is required only for a short time, whereas Wnt is required continuously over the 3–5 days of culture, indicating that Shh provide an essential activating signal for MyoD expression and Wnts provide a maintenance signal (Münsterberg et al., 1995). Interestingly, Wnt signaling also promotes increased cell proliferation in segmental plate mesoderm cultures (Münsterberg et al., 1995), suggesting that Wnts also have mitotic activities. In vivo experiments also show that Wnt1 has a maintenance, rather than activating activity for MyoD expression in somites, as assayed by the ability of Wnt1-expressing cells to rescue MyoD expression in somites of chick embryos following neural tube ablation (Hirsinger et al., 1997). In addition, injection of Wnt1-expressing retroviruses in the presegmental plate mesoderm of stage HH 10 chick embryos does not induce ectopic MyoD expression in somites, although it results in the expansion of Pax3 transcripts in the dermomyotome and the downregulation of Pax1 (Capdevila et al., 1998), suggesting that Wnt1 does not influence MyoD expression directly, but acts indirectly on dermomyotome specification. In Xenopus, ectopic Wnt gene expression can specify dorsal or ventral mesoderm, depending on the stage of mRNA injection, and injection of a dominant-negative Wnt or Frzb induces MyoD expression, suggesting that Wnts can have both positive and negative regulatory functions (Hoppler et al., 1996; Leyns et al., 1997). Interestingly, expression of the dorsal neural tube-specific Wnts, Wnt 1, 3a, and 4, is activated in the neural tube in chick embryos at the time of somite formation when somite myogenesis initiates, but are not expressed earlier in the segmental plate mesoderm (Marcelle et al., 1997), suggesting the existence of a mechanism to coordinate Wnt expression with somite formation and myogenesis.

Further support for a role of Wnt proteins in somite myogenesis comes from in vitro studies on chick and mouse presomitic mesoderm explants (Stern et al., 1995; Fan et al., 1997). Coculture of somites I–III or presomitic mesoderm of stage HH 9–12 chick embryos with Wnt1-expressing cells results in high levels of myosin heavy chain expression after a 3 day culture period, and myosin expression is dose-dependent on the number of Wnt1-expressing cells (Stern et al., 1995). Similarly, Wnt1- and Wnt3a-expressing cells, and to a lesser extent Wnt4- and Wnt6-expressing cells, are sufficient to maintain the expression of Pax3 and Sim1, two dermomyotomal genes, and to induce the expression of Pax7 in mouse (Fan et al., 1997) and chick (Maroto et al., 1997) presomitic mesoderm explants.

5. Tissue Interactions and Signaling Mechanisms in Somite Myogenesis 199

Wnt1-expressing cells also activate ectopic expression of *Wnt11* when injected *in vivo* in the paraxial mesoderm of chick embryos (Marcelle *et al.*, 1997). As *Wnt11* is normally expressed in the dorsal medial lip of the dermomyotome, this suggests that the neural tube Wnts could act indirectly on dermomyotome specification, through their regulation of other Wnts.

In the mouse embryo, dorsal neural tube, surface ectoderm, and different Wnts have been implicated in the signaling required for epaxial and hypaxial somite myogenesis (Tajbakhsh *et al.*, 1998). In explants of presomitic mesoderm from E8.5 mouse embryos, Shh synergizes with Wnt1 or Wnt7a to induce myogenesis, as assayed by Myosin light chain 3F (MLC3F) expression (Tajbakhsh *et al.*, 1998), consistent with the induction of *MyoD* expression in avian embryos (Münsterberg *et al.*, 1995) and in mouse embryos (Borycki *et al.*, 1999b) indicating that Shh and Wnts participate in the activation and maintenance of *MRF* genes during epaxial myogenesis. Of particular interest are the findings that presomitic mesoderm explants from older embryos (E9.5) activate *MRF* genes in response to Wnt-expressing cells alone, without Shh (Tajbakhsh *et al.*, 1998). This finding indicates that, in addition to the Shh-dependent signaling pathway involved in epaxial myogenesis (Borycki *et al.*, 1999b), Shh-independent signaling pathways involving Wnts are primary candidates for activation of *MRF* genes in hypaxial myogenesis and myogenesis at other sites in the embryo (Tajbakhsh *et al.*, 1998). In further support of this possibility, Wnt1, which is produced by dorsal neural tube, differentially promotes *Myf5* activation in mouse presomitic mesoderm explant assays, whereas Wnt7a, which is produced by the surface ectoderm, differentially promotes *MyoD* activation (Tajbakhsh *et al.*, 1998). These findings, therefore, provide evidence that the somite can differentially express *MRF* genes in response to multiple, Shh-independent signals produced by neural tube and surface ectoderm and involving different Wnt family members. The basis for this complexity of signaling is hypothesized to be related to the diversity of myogenic pathways in the vertebrate embryo, including epaxial and hypaxial lineages, and *Myf5*-dependent and *Myf5*-independent lineages (Cossu *et al.*, 1996b; Tajbakhsh *et al.*, 1997).

Studies of Wnt signaling in mouse embryos to date have been limited to *in vitro* explant experiments. Genetic studies of mouse embryos with targeted mutations in four *Wnt* genes, *Wnt1, 3a, 4,* and *7a,* have not produced strong myogenic phenotypes, consistent with the finding that multiple *Wnt* genes are expressed in the neural tube and surface ectoderm that likely have redundant functions (Table I). Of those mutants, *Wnt3a* mutants lack posterior somites, suggesting that *Wnt3a* is required for processes involved in the specification of the paraxial mesoderm (Takada *et al.*, 1994; Yoshikawa *et al.*, 1997). In contrast, *Wnt1/Wnt3a* double-mutant embryos display somite patterning defects, confirming that Wnt proteins have redundant functions. *Wnt1/Wnt3a* double mutants lack a recognizable epithelial dermomyotome in the medial compartment of somites, but retain the lateral dermomyotome (Ikeya and Takada, 1998). This phenotype is associated with the loss of dorsomedial lip markers such as *Wnt11, Noggin,* and *Notch2* (Ikeya and

Takada, 1998). *Myf5* and *MyoD* are activated at the right time in these embryos, however levels of *Myf5* expression are reduced at E9.5 and *MyoD*-expressing cells appear to be highly disorganized (Ikeya and Takada, 1998), indicating that Wnt1 and Wnt3a together control the formation of the medial dermomyotome rather than *MRFs* activation. TdT-mediated dUTP nick end labeling (TUNEL) and bromodeoxyuridine incorporation analysis of these mutants did not reveal defects in cell survival or cell proliferation in the medial dermomyotome (Ikeya and Takada, 1998), leaving open the question regarding the fate of the dorsomedial lip cells in these mutants. Additional mouse genetic studies, involving dominant negative and null mutations of *Wnt* genes, *FrzB* genes, the natural Wnt inhibitor, and genes in the Wnt signal transduction pathway, including *Frizzled* receptor genes and *beta-Catenin/LEF* transcription factors, are required to define the specific functions of Wnts in somite myogenesis *in vivo*.

C. BMP4 and the Bone Morphogenetic Protein Antagonist, Noggin, Regulate Mediolateral Boundaries of the Epaxial and Hypaxial Somite Domains

1. BMP Signal Transduction Mechanisms

The BMP gene family consists of about 10 family members that are a subclass of the TGFβ-like protein family, which is involved in many aspects of embryonic development from gastrulation to organogenesis (Hogan, 1996). BMPs and other members of the TGFβ family are synthesized as a precursor that undergoes a proteolytic cleavage, dimerization, and secretion. Although not demonstrated yet for BMPs, studies on TGFβ processing suggest that the mature protein remains inactive at the cell surface, noncovalently bound to the cleaved product, until further proteolytic cleavage releases the active protein (Hogan, 1996). In addition, BMPs can form homo- or heterodimers, which increases the possibilities of diversity of action of these factors, and in some cases, heterodimerization appears to enhance and/or modify their activity (Suzuki *et al.*, 1997). BMPs transduce their signal by binding to a dimeric receptor constituted of type I and type II BMP receptor (Fig. 2C). Binding induces transphosphorylation of BMP-RI by BMP-RII, and the activation downstream of Smad1, which binds to Smad4 and translocates to the nucleus (Kretzschmar *et al.*, 1997). The existence of a common signaling pathway for all members of the BMP family suggests that their developmental specificity of action may in part result from variations in their differential timing and spatial domains of expression, and their ability to form heterodimeric BMP complexes. In addition, BMP7 homodimers can efficiently bind to receptors formed of the type II activin receptor and the type I BMP receptor (Yamashita *et al.*, 1995), suggesting that various receptor combinations on receiving cells also may transduce different signal responses. Finally, three BMP binding proteins, Noggin, Chordin,

5. Tissue Interactions and Signaling Mechanisms in Somite Myogenesis

and Follistatin, have been identified and been shown to bind BMP4 and prevent the interaction with its receptor (Fig. 2C) (Piccolo et al., 1996; Zimmerman et al., 1996; Fainsod et al., 1997). As will be discussed, these binding proteins likely play a key role in regulating BMP functions in specific spatial domains of the developing embryo.

2. BMP Signaling and Transduction in Embryos

BMPs are expressed at diverse sites in the embryo, including the surface ectoderm, the notochord, and the lateral mesoderm (Fig. 1C). More specifically, BMP4 is strongly expressed in the lateral mesoderm of avian embryos (Pourquié et al., 1996), although mesodermal expression in mouse embryos has been only described in the most posterior part at E8.5 (Jones et al., 1991). BMP4 is also expressed in the dorsal neural tube and surface ectoderm (Pourquié et al., 1996; Watanabe and Le Douarin, 1996). BMP7 is expressed in the notochord and the surface ectoderm between E8.5 and E10.5 in the mouse embryo (Lyons et al., 1995). BMP2 is also expressed in the notochord (Lyons et al., 1995). In contrast, the BMP-binding proteins are found in opposing sites of BMPs expression. For instance, the avian homologues of *Noggin* and *Follistatin* (Amthor et al., 1996; Patel et al., 1996; Connolly et al., 1997) are expressed in somites beginning in the lateral dorsal quadrant and then becoming more restricted to the dorsal medial quadrant in a domain overlapping epaxial *MyoD* expression (Fig. 1C) (Amthor et al., 1996; Hirsinger et al., 1997; Marcelle et al., 1997; Reshef et al., 1998). In mouse embryos, *Noggin* is found in the notochord and also in the dorsal medial lip of somites (McMahon et al., 1998). Both *Noggin* and *Follistatin* are activated in somites in response to axial signals from the neural tube and notochord (Amthor et al., 1996; Hirsinger et al., 1997; Marcelle et al., 1997), and more specifically in response to Wnt1, in the case of *Noggin* (Hirsinger et al., 1997; Marcelle et al., 1997; Reshef et al., 1998).

These opposing patterns of expression reflect the antagonistic functions of BMPs and Noggin during embryonic development (Piccolo et al., 1996; Zimmerman et al., 1996; Fainsod et al., 1997), and more specifically during somitogenesis. In avian embryos, there is a balance between BMP4 levels in the lateral plate and Noggin levels in the somite. These opposing activities define the mediolateral axis of newly formed somites, and the identity of lateral plate prior to somite formation. In support of this interpretation, implantation of BMP4-expressing cells into the dorsal medial segmental plate of HH stage 12 chick embryos ectopically activates *Cytokeratin*, which is normally a lateral plate-specific gene, in the somitic region (Tonegawa et al., 1997). Conversely, implantation of Noggin beads in the lateral plate results in the formation of additional somites (Tonegawa et al., 1997), demonstrating that a gradient of BMP4 activity, controlled by the level of Noggin proteins, specifies the somites and the lateral plate from the paraxial mesoderm. In addition, the gradient of BMP4 activity also determines the mediolateral patterning within somites, and disruption of this gradient by modifying the levels of BMP4 or Noggin proteins results in the expansion of the lateral or the medial domain of somites. Therefore, BMP4 positively regulates expression of lateral dermomyotomal genes such as *Pax3* and *Sim1*, and inhibits expression of medial genes such as *MyoD* and *Myf5* (Pourquié *et al.*, 1995, 1996). Dilutions of BMP4-expressing cells induce differential effects on dermomyotomal gene expression, such that high concentrations of BMP4 completely abolish *Pax3* and *Sim1* expression in somites as well as inhibit the formation of epithelial dermomyotome (Watanabe and Le Douarin, 1996; Tonegawa *et al.*, 1997). In contrast, low concentrations of BMP4 promote expansion of *Pax3* and *Sim1* in the medial dermomyotome (Pourquié *et al.*, 1996; Tonegawa *et al.*, 1997). These findings have been confirmed in *in vitro* explant cultures of somites, where BMP4 at high concentration (100 ng/ml) inhibits *Pax3*, *MyoD*, and *Myf5* activation by the surface ectoderm, by the neural tube/notochord complex, and by Shh and Wnt1 (Reshef *et al.*, 1998). In contrast, a 2-fold decrease in BMP4 concentration (50 ng/ml) restores *Pax3*, but not *MyoD* expression, in explants of somites IV–VI cocultured with surface ectoderm, whereas a 10-fold decrease in BMP4 concentration restores both *Pax3* and *MyoD* expression (Reshef *et al.*, 1998). BMP4 and BMP2, which are the vertebrate homologues of the *Drosophila dpp* gene, share this lateralizing activity, whereas BMP7, the homologue of the *Drosophila 60A* gene, does not have lateral plate mesoderm-inducing activity (Tonegawa *et al.*, 1997).

The effects of BMP4 on mediolateral patterning of somites also may explain why BMPs were originally described as antagonists of myogenesis. For instance, BMP2 inhibits the differentiation of C2C12 and L6 myoblasts *in vitro* (Murray *et al.*, 1993; Katagiri *et al.*, 1994), and in micromass cultures of chick limb buds, BMP2 also inhibits muscle differentiation and promotes cartilage formation (Duprez *et al.*, 1996). *In vivo*, when BMP4-expressing cells are grafted medially between the neural tube and the segmental plate at HH stage 12 chick embryos, *MyoD* activation in somites is inhibited, mimicking the repressive effect of a medial graft of lateral plate mesoderm on epaxial expression of *MRF*s (Pourquié *et al.*, 1996). The mechanism by which BMP4 inhibits *MyoD* remains unknown although it is likely that the inhibitory effects of BMPs on myogenic differentiation are the result of BMP inhibition of *MRF* transcription (Murray *et al.*, 1993; Pourquié *et al.*, 1996; Katagiri *et al.*, 1997). Conversely, lateral grafts of *Noggin* expressing cells in the segmental plate mesoderm of HH stage 11–12 chick embryos leads to the lateral expansion of *MyoD* and the inhibition of *Pax3* and *Sim1* expression in the lateral dermomyotome (Hirsinger *et al.*, 1997; Reshef *et al.*, 1998), mimicking the effects observed following removal of lateral plate as a source of BMP4 (Pourquié *et al.*, 1996; Pownall *et al.*, 1996). During later embryonic stages (HH stage 20), the regulated expression of BMP4 and Noggin appears to control the balance between *Pax3* and *MyoD* expression in order to maintain a pool of proliferating and differentiating myogenic progenitor cells in the epaxial domain of chicken somites (Amthor *et al.*, 1999).

5. Tissue Interactions and Signaling Mechanisms in Somite Myogenesis

These data indicate that BMP4 can act as a morphogen to determine medial–lateral specification of the dorsal somite, and that Noggin functions *in vivo* to counteract BMP4 signaling by generating a functional gradient of BMP4 activity along the lateral to medial axis of the dorsal somite. In zebrafish embryos, dorsal–ventral somite patterning is analogous to the medial–lateral patterning in avian embryos. The identification of the BMP2 mutant, *swirl*, in the zebrafish confirms the involvement of the *dpp*-like members of the BMP family in the medial–lateral patterning (Kishimoto *et al.*, 1997). *swirl* mutants display a dramatic expansion of the somitic mesoderm and *myoD* expression such that somites encircle the embryo (Mullins *et al.*, 1996), an observation reminiscent of the expansion of *myoD* expression in avian embryos following lateral plate separation. Examination of homozygous mouse mutants of the BMP family genes or BMPR-I receptor has not provided genetic evidence for a role of BMPs in myogenesis, likely because BMPs have early functions in development that result in early embryonic lethality well before myogenesis initiates (see Table I). In contrast to BMP mutants, mouse embryos with a targeted mutation in the *Noggin* gene display somitic defects (McMahon *et al.*, 1998) (Table I). Interestingly, in these mutants, somites form normally (McMahon *et al.*, 1998), indicating that *Noggin* is not required for somite formation. However, defects in somite patterning are visible in caudal and interlimb regions, although rostral somites appear normal. In particular, *MyoD* expression is reduced in the dorsal medial somite of E10.5 embryos, although expression in the ventral lateral somite is normal. This reduction of *MyoD* expression is accompanied by the expansion of *Pax3* into the dorsomedial dermomyotome (McMahon *et al.*, 1998). These observations are consistent with avian results demonstrating that *BMP4/Noggin* are not involved in the direct activation of the *MRF* genes, but control their domains of expression along the mediolateral axis of somites. Absence of somite phenotype in the rostral embryo, as well as the transient nature of the somite phenotype, may be due to compensation by *Follistatin*, which becomes upregulated from day 8.5 in the dorsal medial somite of *Noggin* mutant embryos.

D. Transforming Growth Factors β Have Positive and Negative Functions in Myogenesis

1. TGFβ Signal Transduction Mechanisms

Three TGFβ factors have been isolated in mammals (TGFβ1, TGFβ2, and TGFβ3) and birds (TGFβ2, TGFβ3, and TGFβ4) (Massague, 1996), TGFβ RNAs and proteins are broadly distributed in embryos and are expressed in the notochord and the neural tube in early stage mouse and chicken embryos (Fig. 1C) (Pelton *et al.*, 1991; Jakowlew *et al.*, 1992, 1994; Roelen *et al.*, 1994; Sanders *et al.*, 1994; Unsicker *et al.*, 1996). TGFβ family members transduce their signal to the nucleus by binding to a serine/threonine kinase receptor, TGFβ receptor type II (TGFβ-RII), and subsequently recruiting the TGFβ receptor type I (TGFβ-RI) into a heterodimeric receptor complex (Fig. 2C) (Massague, 1996). Transphosphorylation of the type I receptor by type II receptor induces a cytoplasmic signaling cascade that leads to the phosphorylation of Smad3, its association with Smad4, and translocation to the nucleus where it can activate the transcription of target genes (Fig. 2C) (Lagna et al., 1996; Derynck and Feng, 1997; Kretzschmar et al., 1997). TGFβ-RII, which mediates TGFβ signaling, is expressed at diverse sites of the mouse embryo including the floor plate of the neural tube and the mesenchymal cells of somites (Roelen et al., 1994; Wang et al., 1995), indicating that cells are competent to respond to TGFβ signaling.

2. TGFβ Signaling and Transduction in Embryos

Involvement of TGFβ in the proliferation and differentiation of muscle cell lines is well known (Olson et al., 1986; Brennan et al., 1991; Zentella and Massague, 1992; Cusella-De Angelis et al., 1994; Filvaroff et al., 1994). Specifically, TGFβ induces muscle cell differentiation in some culture conditions (Zentella and Massague, 1992), and truncated, dominant negative type II TGFβ mutant receptor can partially block myogenic differentiation (Filvaroff et al., 1994). Although TGFβ also can inhibit muscle differentiation of some cultured myoblast cell lines (Massague et al., 1986; Schofield and Wolpert, 1990; Brennan et al., 1991), this inhibitory activity may be cell specific, as suggested by the finding that TGFβ inhibits differentiation of primary fetal myoblasts, but not embryonic myoblasts (Cusella-De Angelis et al., 1994) or embryonic stem (ES) cells (Slager et al., 1993).

In segmental plate and somite explants, TGFβ1 promotes Myosin heavy chain (MHC) expression of HH stage 13 chick embryos, but not in explants of earlier HH stage 10–11 embryos (Stern et al., 1997). Interestingly, bFGF and TGFβ1 act synergistically to promote myogenesis in segmental plate explants (Stern et al., 1997), an observation also reported for *Xenopus* embryos (Kimelman and Kirschner, 1987). Moreover, in these studies, exposure to TGFβ is required for only the first 12 hr to achieve full muscle activation, while bFGF is required throughout the culture period (Stern et al., 1997), indicating that a functional hierarchy exists in the actions of these factors. Blocking antibodies against TGFβ also block the muscle-inducing activity of dorsal neural tube in a segmental plate explant assay (Stern et al., 1997), demonstrating that, in addition to Wnts, TGFβ family members may participate in neural tube signaling of myogenesis. Finally, TGFβ proteins are also expressed in developing muscle cells throughout the period of secondary myogenesis and muscle maturation (Pelton et al., 1991; Jakowlew et al., 1992; Unsicker et al., 1996), suggesting that they play an additional role in later stages of muscle differentiation. Targeted mutations of *TGFβ1* do not have a myogenic phenotype (see Table I), although it is possible that other TGFs provide compensatory functions. However, embryos with targeted mutations of *GDF8*, a new member of the TGFβ family, have 30% increases in muscle mass resulting from hyperplasia of the muscle fibers (McPherron *et al.*, 1997), providing support for the negative regulatory role of some TGFβ factors in the regulation of myogenic cell proliferation and differentiation.

E. Fibroblast Growth Factors Promote Mesoderm Formation and Have Later Maintenance Functions in Myogenesis

1. FGF Signal Transduction Mechanisms

The FGF family comprises 15 members that display broad expression during embryonic development (Fig. 1C) (Yamaguchi and Rossant, 1995). Signaling through FGF requires the participation of three transmembrane proteins, a FGF tyrosine kinase receptor, a cysteine-rich receptor, and a heparan sulfate proteoglycan (Olwin *et al.*, 1994).

2. FGF Signaling and Transduction in Embryos

Somitic cells express predominantly the high affinity FGF-R1 (Peters *et al.*, 1992; Patstone *et al.*, 1993), but also FGF-R2 (Orr-Urtreger *et al.*, 1991) and FREK (Marcelle *et al.*, 1994), indicating that somites can respond to FGF signals. In vertebrate embryos, bFGF is expressed in the neural tube (Kalcheim and Neufeld, 1990; Savage and Fallon, 1995; Yamaguchi and Rossant, 1995). Neural tube FGFs have a functional role in somite myogenesis, as indicated by the finding that anti-bFGF antibodies inhibit the muscle-inducing activity of dorsal neural tube on explant cultures of newly formed somites I–III of HH stage 12–13 chicken embryos (Stern *et al.*, 1997). Explants of segmental plate mesoderm cultured in presence of bFGF and TGFβ required bFGF only 12 hr after the initial activation of myogenesis in the culture (Stern *et al.*, 1997), indicating that FGF does not function in the early phase of the myogenic response, but rather promotes proliferation and facilitates cell survival. Consistent with this interpretation, bFGF promotes myoblast proliferation and is a potent inhibitor of myoblast differentiation in cultured myoblast cell lines (Lathrop *et al.*, 1985; Seed and Hauschka, 1988), and mouse embryos mutant for *FGF-R1* display growth retardation defects (Deng *et al.*, 1994), and *FGF6* mutant mice display defects in the capacity for muscle regeneration (Floss *et al.*, 1997) (see Table I). However, targeted gene mutations of either *FGF-R1* or *SHP-2*, a downstream transduction component in the FGF signaling pathway, result in mouse embryos with defects in paraxial mesoderm formation, consistent with a role of FGF in mesoderm determination (Yamaguchi *et al.*, 1994; Ciruna *et al.*, 1997; Saxton *et al.*, 1997). This possibility is supported by the discovery of a notochord-expressed FGF, eFGF, in *Xenopus* that can function in mesoderm determination (Isaacs *et al.*, 1995). Avian and mammalian homologues of eFGFs have not yet been reported. FGF also can function synergistically with TGFβ, as revealed by experiments in which bFGF can potentiate the activity of Activin to induce mesodermal genes in *Xenopus* vegetal cap assays (Cornell *et al.*, 1995), and by experiments on the promotion of myogenesis in segmental plate mesoderm (Stern *et al.*, 1997). Finally, many FGF family members, including aFGF, bFGF, FGF4, FGF5, FGF6, FGF7, and FGF8, are expressed in myotomal muscle, indicating that these FGFs have developmental functions in muscle differentiation, growth, or survival (Alterio *et al.*, 1990; deLapeyriere *et al.*, 1993; Drucker and Goldfarb, 1993; Han and Martin, 1993; Coulier *et al.*, 1994; Mason *et al.*, 1994; Savage and Fallon, 1995; Grass *et al.*, 1996; Hannon *et al.*, 1996; Floss *et al.*, 1997).

F. The Notch/Delta Signaling Pathway May Control the Maintenance of the Epithelial Organization and Myogenesis in Somites

1. Notch/Delta Signal Transduction Mechanisms

Numerous transmembrane proteins from the DSL family (Delta, Serrate, Lag2) constitute the ligands of the Notch receptors (Nye and Kopan, 1995). In mammals, four transmembrane receptors have been characterized, called Notch1, Notch2, Notch3, and Notch4 (Lardelli *et al.*, 1995). Because both DSL proteins and Notch proteins are transmembrane proteins, the Notch signaling pathway likely mediates cell–cell signaling interactions. Binding of a DSL protein to Notch activates a downstream signal transduction pathway which has been investigated in detail in *C. elegans* and *Drosophila* (Artavanis-Tsakonas *et al.*, 1996; Greenwald, 1998). Vertebrate homologues of proteins in the Notch transduction pathway also have been identified. In this pathway, activation of the transcription factor CBF1/Su(H)/LAG-1 following ligand binding to Notch results in the induction of different transcriptional targets such as the transcription factor *E(spl)* in some tissues, or *vestigial* and *wingless* in the developing wing margin (Neumann and Cohen, 1996).

2. Notch Signaling and Transduction in Embryos

Although analysis of *Notch* expression during embryogenesis is still incomplete, *Notch1* and *Notch2* are expressed in the paraxial mesoderm (Del Amo *et al.*, 1992; Reaume *et al.*, 1992; Weinmaster *et al.*, 1992; Williams *et al.*, 1995). *Notch1*, however, is activated in the presomitic mesoderm, and its expression is downregulated soon after somite formation (Del Amo *et al.*, 1992; Reaume *et al.*, 1992), ruling out a direct role during myogenesis. *Delta1* is expressed in cells immediately adjacent to *Notch1*-expressing cells, leading to the hypothesis that *Notch1* may be involved in segmentation and somite epithelialization (Gossler and Hrabe de Angelis, 1998), and as predicted, targeted mutation of *Notch1* in mouse embryos results in defects in somite segmentation (Conlon *et al.*, 1995). In contrast, *Notch2* is activated specifically in the dorsal medial lip of the dermomyotome (Weinmaster *et al.*, 1992; Williams *et al.*, 1995), indicating that it plays a more direct role in myogenesis. In *Drosophila*, genetic evidence has established a role of Notch in myogenesis, specifically as an inhibitor that functions to select a small group of muscle progenitors that are maintained in an undifferentiated state until they are properly positioned in the embryo (Corbin *et al.*, 1991; Bate *et al.*, 1993). Notch has a similar function in the selection of neural progenitors for neurogenesis (Doe *et al.*, 1998). In vertebrates, Notch signaling also is proposed to have a negative regulatory function in myogenesis. A constitutively active Notch receptor lacking the extracellular domain represses the myogenic conversion of cultured fibroblasts by *MyoD* (Jarriault *et al.*, 1995). This repression appears to occur through a CBF1/HES-1 pathway, in which HES-1 blocks the functional activity of MyoD (Sasai *et al.*, 1992; Kopan *et al.*, 1994; Jarriault *et al.*, 1995). Additional experiments in embryos are required to confirm the inhibitory role of Notch in myogenesis. As in *Drosophila*, one can hypothesize that Notch may function to maintain myogenic progenitors in an undifferentiated state in order to regulate the initiation of differentiation. Alternatively, Notch could maintain the epithelial state of the dorsal medial lip, a structure that is required for the production of a pool of muscle progenitors, as has been proposed in *Drosophila* embryos in which cell differentiation is frequently associated with epithelial delamination (Hartenstein *et al.*, 1992).

VI. Overview and Future Perspectives

Current research in the field of signaling mechanisms and somite myogenesis has been propelled by the discovery of the *MRF* genes, *MyoD* and *Myf5*. These genes are activated during the early regulatory processes that commit somitic cells to myogenesis, and therefore, have provided specific molecular markers for embryological studies of tissue interactions and signal molecules that influence the development of somites, in the embryo and in culture explants. These myogenic regulatory genes also have opened up these studies to embryological analyses in birds and frogs, and genetic analysis in mice and zebrafish, through examination of mutants of encoding signaling molecules and their signal transduction components.

Much of the signaling research in avian and mammalian embryos has been focused on the formation of epaxial muscles, which arise first in the vertebrate embryo. However, recent genetic and explant studies have begun to define signaling components of hypaxial myogenesis. We now know that the regulation of epaxial myogenesis is under the control of tightly balanced, positive and negative signals from the tissues surrounding the somite (Fig. 3). It is necessary to keep in mind that this process occurs following somite formation, and that it requires only a few

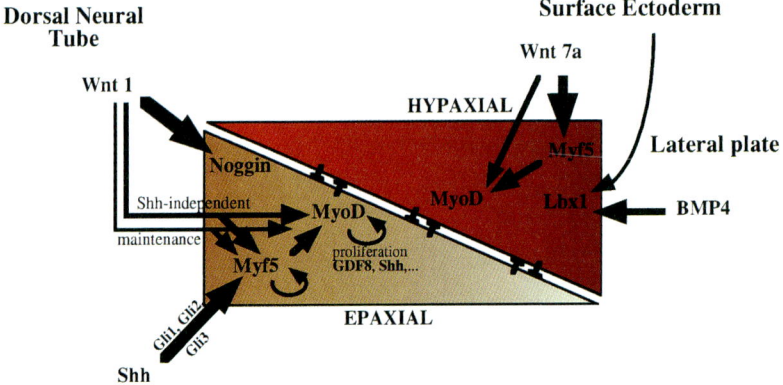

Figure 3   Model for epaxial and hypaxial myogenesis and medial–lateral patterning of the dermomyotome in the avian and mouse embryo. Epaxial myogenesis is controlled by Shh signals from the ventral notochord that induce *Myf5*, which in turn induces *MyoD*. Signaling of somitic mesoderm by Shh requires prior induction of the *Gli* genes, the transcription factor effectors of Shh signaling, by surface ectoderm signals. Neural tube signals (Wnt1) also can induce *Myf5* and *MyoD* and maintain their expression in the newly formed somites. Additional factors, under Shh control, promote the proliferation of the muscle cell progenitors and TGFβ family members such as GDF8 are likely negative regulators of myoblast proliferation. Hypaxial myogenesis is controlled by dorsal surface ectoderm signals (Wnt7a) that induce *Myf5* as well as *MyoD* expression. A subset of the lateral hypaxial muscle progenitors induce *Lbx1* expression under the control of lateral plate and surface ectoderm signals, which are currently unknown. Domains of epaxial and hypaxial gene expression within the dorsal somite are tightly controlled along the mediolateral axis by opposing signals from Noggin (yellow), which is induced in the dorsomedial dermomyotome by Wnt1, and from BMP4 (red) in the lateral plate. Shh also contributes to the establishment of epaxial and hypaxial boundaries through repression of hypaxial gene expression in the ventral medial somite.

hours to be completed. Therefore, these signals function to coordinate both the temporal activation and the spatial localization of *Myf5* and *MyoD* expression in the newly formed somite. The notochord produces Shh, which is essential for *Myf5/MyoD* activation and epaxial myogenesis following somite formation. The surface ectoderm and dorsal neural tube produce signals that act in concert with somite formation to make somites competent to respond to Shh signals via the activation of the Shh effector transcription factors genes, *Gli2* and *Gli3*. The ventral neural tube and notochord also produce negative signals to dorsalize the expression of *Myf5/MyoD*, allowing the establishment of the sclerotome lineage in the ventral somite. Dorsal neural tube signals, Wnt1/Wnt3a, establish the dorsal medial lip and likely have a function to maintain *Myf5* and *MyoD* expression prior to the establishment of their autonomous expression. To counter the medial Shh regulatory system for epaxial myogenesis, the lateral plate mesoderm produces BMP4, which is a positive regulator of hypaxial genes such as *Lbx1*, but is a negative regulator that prevents the expansion of epaxial *Myf5* and *MyoD* expression into the lateral hypaxial somite domain. The medial diffusion of BMP4 is limited by Noggin, whose expression in the medial, epaxial somite domain of *Myf5/MyoD* expression is controlled by neural tube signals, thereby establishing the medial–lateral patterning of epaxial and hypaxial domains in the dermomyotome.

In contrast to epaxial myogenesis, which is coordinated with the early processes of somite segmentation, hypaxial myogenesis occurs later, after somites have a well-formed dermomyotome and ventral lateral lip. It is notable that, in the mouse embryo, *Myf5* is the initial target of activation by the epaxial and hypaxial signaling systems. *MyoD* is activated initially by *Myf5*-dependent mechanisms, but in *Myf5* mutant embryos, *MyoD* can be activated by *Myf5*-independent mechanisms in both epaxial and hypaxial domains. Therefore, *Myf5* and *MyoD* can respond to signals independently. Hypaxial myogenesis is initiated by surface ectoderm signals, apparently Wnts, that activate *Myf5/MyoD* completely independent of the epaxial Shh signaling mechanism (Fig. 3). The upstream regulation of these genes, however, does diverge, as *MyoD* is dependent on *Pax3*, and *Myf5* is *Pax3*-independent. Combinatorial signals from the surface ectoderm and the lateral mesoderm (BMP4) likely regulate the upstream genes of hypaxial myogenesis, such as *Pax3*, as well as the expression of *Lbx1*, which defines a subset of hypaxial muscle progenitors.

Finally, there are additional regional differences in the regulation of *Myf5/MyoD* genes in the limb, tongue, diaphragm, branchial arch, and other head myogenic lineages, where tissue interactions and signaling systems for myogenesis appear to be divergent. The anatomical dispersion of myogenesis in the vertebrate embryo provides additional complexities that will require future experimentation to resolve. Molecular evidence for the regional complexity of signaling regulation of myogenesis in the mouse embryo comes from the discovery of multiple, distinct transcriptional regulatory elements that control *Myf5* expression at different anatomical sites in the embryo (Yoon *et al.*, 1997; M. Buckingham and P. Rigby, personal communication, 1999). This observation suggests that *Myf5* activation at these sites is regulated by fundamentally distinct upstream control systems. In contrast, *MyoD* activation in all skeletal myogenic progenitor cells in the mouse embryo is regulated by a single core enhancer (Goldhamer *et al.*, 1995), although maintenance and later expression of *MyoD* in these embryonic lineages requires a second regulatory element (Tapscott *et al.*, 1992; Asakura *et al.*, 1995). These considerations support the view that *MyoD* is subject to regulatory control downstream of *Myf5*, which, at least in the epaxial and hypaxial somite in the mouse embryo, is activated first and, therefore, is the primary target of epaxial and hypaxial signal transduction systems.

The adaptive value for the embryo of redundant signaling systems and transcription factors (Myf5 and MyoD) for myogenesis remains to be elucidated. One possibility is that this complexity is an evolutionary mechanism that has been exploited by the embryo to promote the formation of new muscle groups at sites throughout the embryo through the capture of local developmental signals to activate *Myf5* and *MyoD* in somitic progenitors that migrate into these signal fields. The validity of this signal capture hypothesis will be tested as future studies define, in molecular detail, the signals and signal transduction systems that regulate the activation of *Myf5* and *MyoD* at different anatomical sites in a diversity of vertebrate embryos, as well as define the response elements and interacting transfactors that control the transcription of the *Myf5* and *MyoD* regulatory genes.

Among the most intriguing and fundamentally important questions for future investigation in the field of signaling and somite myogenesis are (1) the molecular mechanisms by which developmental signals and their signal transduction pathways interact with one another to control the precise temporal activation and spatial restriction of expression of the *Myf5* and *MyoD* genes, and (2) the molecular mechanisms by which the expression of *Myf5* and *MyoD* commit somite cells to myogenic progenitors and mediate their differentiation. In our view, these questions can be simply framed and directly addressed by experiments that undertake to define the target genes that are activated and repressed by the transcription factor effectors of these signaling pathways and the target genes of *Myf5* and *MyoD*, which are required to commit somitic cells to myogenesis.

The developmental signal transduction pathways that contribute to somite myogenesis are not unique to myogenesis, but participate in and regulate many lineage determination processes in the embryo. Of particular significance for somite myogenesis are the Shh, Wnt, and BMP signaling pathways. Each of these transduction pathways activates distinct transcription factor effectors, the Gli, β Catenin/ lymphoid enhancing factor (LEF), and smothers against dcca pentaplegic (SMAD) groups, that bind DNA at specific sequences on the regulatory elements of target genes in cells that are responsive to these signals. Therefore, the identification of the target genes that respond to these signal effectors in somites will provide insights into and define the functions of these different signaling pathways in the processes of myogenesis. In the case of epaxial myogenesis, there are currently two divergent views of the functions of Shh in the control of *Myf5* and *MyoD* and myogenesis that will be clarified by a molecular understanding of the target genes regulated by Shh. One possibility is that Shh is a trophic signal that specifically regulates the proliferation and survival of a subpopulation of somite cells that are committed to proceed with myogenesis as a default state (Pourquié *et al.*, 1993; George-Weinstein *et al.*, 1996). In support of this maintenance model are the findings that Shh can promote/maintain cell proliferation and cell survival in newly formed somites (Teillet *et al.*, 1998; Borycki *et al.*, 1999b). Furthermore, epiblast cells from early avian embryos give rise in majority to muscle cells when dissociated and cultured at high density in a minimal medium (George-Weinstein *et al.*, 1996), indicating that, in epiblast cells, myogenic regulatory genes are already activated at a low subthreshold level but are subject to repression, possibly by signaling components of the extracellular matrix, including the cadherins (George-Weinstein *et al.*, 1997). An alternative, but not mutually exclusive possibility is

5. Tissue Interactions and Signaling Mechanisms in Somite Myogenesis 211

that Shh (and Wnts) has inductive functions that, in addition to controlling genes important for somite cell proliferation and cell survival, regulate genes that function to convert somite cells from a multipotential state into a determined lineage of myogenic cells, through the activation of *Myf5* and *MyoD*. According to this inductive model, these inductive functions could be mediated either through the direct interaction of Gli (β Catenin/LEF) transcription factors with *Myf5/MyoD* regulatory elements, or through Gli-mediated activation of upstream genes that encode *Myf5/MyoD* regulatory factors.

Future experiments will define the target genes regulated by Gli transcription factors, and will examine whether these genes regulate cell division and apoptosis, or regulate *Myf5* and *MyoD* activation. Our prediction is that Shh and Wnt transcription factor effectors will have activator and repressor functions and, through the regulation of multiple target genes, they will control both maintenance and inductive processes during epaxial and hypaxial myogenesis in the somite.

A second area of future research on somite signaling will be molecular investigations of the mechanisms by which different signaling pathways intersect and interact to mediate somite myogenesis, particularly to mediate the precise spatial patterning of gene expression during epaxial and hypaxial myogenesis to the dorsal medial and ventral lateral somite. Shh and Wnt signaling have synergistic functions (Münsterberg *et al.*, 1995), and surface ectoderm signals regulate Gli transcription and somite formation, as a key regulatory step in the establishment of the Shh signal transduction pathway for epaxial myogenesis (Borycki *et al.*, 1998). Additionally, Shh is a potent inhibitor of BMP4 signaling and lateral patterning of somites (Pourquié *et al.*, 1996; Hirsinger *et al.*, 1997). Our prediction is that these pathways are interlocked in feedback loops mediated by their transcription factor effectors, which activate and/or repress specific target genes that encode critical functional components (receptors, kinases, and transcription factors) of the signal transduction pathways with which they synergyze or compete. We also expect that the interactions of these signals and transduction pathways function to coordinate the precise spatial and temporal regulation of myotome formation with sclerotome and dermatome formation, which also are controlled by these same signals and signal transduction pathways. This suggests that these signal molecules have graded functions in different somite compartments, perhaps acting as morphogens. Alternatively, these signals may be part of complex, combinatorial controls and feedback regulatory networks that modify their transduction outputs, spatially and temporally, within the somite.

A third area of research in somite signaling will be identification of target genes regulated by *Myf5* and *MyoD*. These genes are essential for the establishment of myogenic progenitor cells and muscle differentiation, and therefore, must have functionally significant targets. Myf5 and MyoD are members of the bHLH transcription factor family, which bind E-box recognition sequences (Weintraub, 1993). Some target genes for these myogenic regulators have been identified and include downstream transcription factor genes, such as *Myogenin,* which regulates the initiation of myogenic differentiation in progenitor cells (Yee and Rigby, 1993). We are predicting that Myf5 and MyoD will have additional functions in the activation and/or repression of target genes in myogenic progenitor cells that determine somite cells to the myogenic lineage. Such target genes will likely be involved in the control of myogenic progenitor cell proliferation and migration, as well as in the initiation of differentiation in spatially restricted sites of myogenesis in the embryo. Such studies may also reveal that Myf5 and MyoD may not be entirely redundant, but may have overlapping and distinct target genes in different myogenic lineages.

There are many challenges ahead to understand the tissue interactions and signaling mechanisms that regulate somite myogenesis in vertebrates. Clearly, this field is in its infancy, but recent progress has provided important new molecular markers, genetic tools, including mouse and zebrafish mutants in somite signaling pathways, and well-defined models to be tested. These advances lay a clear path for future experiments to define the specific molecular mechanisms that underlie somite signaling, both in the establishment of myogenic lineages in the vertebrate embryo, and in the temporal and spatial coordination of myogenesis within the dynamic context of the development of the vertebrate body plan in the developing embryo. We also expect that future findings in this field will have important impact on our general understanding of fundamental developmental mechanisms that regulate cell lineage determination and cell differentiation.

Acknowledgments

We thank Mark Fortini and Drew Noden for critical reading of aspects of the manuscript.

References

Alcedo, J., Ayzenzon, M., Von Ohlen, T., Noll, M., and Hooper, J. E. (1996). The *Drosophila* smoothened gene encodes a seven-pass membrane protein, a putative receptor for the hedgehog signal. *Cell* **86,** 221–232.

Alterio, J., Courtois, Y., Robelin, J., Bechet, D., and Martelly, I. (1990). Acidic and basic fibroblast growth factor mRNAs are expressed by skeletal muscle satellite cells. *Biochem. Biophys. Res. Commun.* **166,** 1205–1212.

Amthor, H., Connolly, D., Patel, K., Brand-Saberi, B., Wilkinson, D. G., Cooke, J., and Christ, B. (1996). The expression and regulation of follistatin and a follistatin-like gene during avian somite compartmentalization and myogenesis. *Dev. Biol.* **178,** 343–362.

Amthor, H., Christ, B., and Patel, K. (1999). A molecular mechanism enabling continuous embryonic muscle growth—a balance between proliferation and differentiation. *Development* **126,** 1041–1053.

Ang, S. L., and Rossant, J. (1994). HNF-3 beta is essential for node and notochord formation in mouse development. *Cell* **78,** 561–574.

Aoyama, H. (1993). Developmental plasticity of the prospective dermatome and the prospective sclerotome region of avian somite. *Dev. Growth Differ.* **35,** 507–519.

5. Tissue Interactions and Signaling Mechanisms in Somite Myogenesis

Aoyama, H., and Asamoto, K. (1988). Determination of somite cells: Independence of cell differentiation and morphogenesis. *Development* **104**, 15–28.
Artavanis-Tsakonas, S., Matsuno, K., and Fortini, M. E. (1996). Notch signaling. *Science* **268**, 225–232.
Asakura, A., and Tapscott, S. J. (1998). Apoptosis of epaxial myotome in Danforth's short-tail (Sd) mice in somites that form following notochord degeneration. *Dev. Biol.* **203**, 276–289.
Asakura, A., Lyons, G. E., and Tapscott, S. J. (1995). The regulation of myoD gene-expression—conserved elements mediate expression in embryonic axial muscle. *Dev. Biol.* **171**, 386–398.
Aza-Blanc, P., Ramirez-Weber, F. A., Laget, M. P., Schwartz, C., and Kornberg, T. B. (1997). Proteolysis that is inhibited by hedgehog targets Cubitus interruptus protein to the nucleus and converts it to a repressor. *Cell* **89**, 1043–1053.
Bate, M., Rushton, E., and Frasch, M. (1993). A dual requirement for neurogenic genes in *Drosophila* myogenesis. *Development Suppl.*, 149–161.
Behrens, J., von Kries, J. P., Kuhl, M., Bruhn, L., Wedlich, D., Grosschedl, R., and Birchmeier, W. (1996). Functional interaction of beta-catenin with the transcription factor LEF-1. *Nature* **382**, 638–642.
Bellaiche, Y., The, I., and Perrimon, N. (1998). Tout-velu is a *Drosophila* homologue of the putative tumour suppressor EXT-1 and is needed for Hh diffusion. *Nature* **394**, 85–88.
Bhanot, P., Brink, M., Samos, C. H., Hsieh, J. C., Wang, Y., Macke, J. P., Andrew, D., Nathans, J., and Nusse, R. (1996). A new member of the frizzled family from *Drosophila* functions as a Wingless receptor. *Nature* **382**, 225–230.
Bladt, F., Riethmacher, D., Isenmann, S., Aguzzi, A., and Birchmeier, C. (1995). Essential role for the c-met receptor in the migration of myogenic precursor cells into the limb bud [see comments]. *Nature* **376**, 768–771.
Blagden, C. S., Currie, P. D., Ingham, P. W., and Hughes, S. M. (1997). Notochord induction of zebrafish slow muscle mediated by Sonic hedgehog. *Genes Dev.* **11**, 2163–2175.
Bober, E., Franz, T., Arnold, H. H., Gruss, P., and Tremblay, P. (1994a). Pax-3 is required for the development of limb muscles: A possible role for the migration of dermomyotomal muscle progenitor cells. *Development* **120**, 603–612.
Bober, E., Brand-Saberi, B., Ebensperger, C., Wilting, J., Balling, R., Paterson, B. M., Arnold, H. H., and Christ, B. (1994b). Initial steps of myogenesis in somites are independent of influence from axial structures. *Development* **120**, 3073–3082.
Borycki, A. G., Strunk, K., Savary, R., and Emerson, C. P., Jr. (1997). Distinct signal/response mechanisms regulate pax1 and QmyoD activation in sclerotomal and myotomal lineages of quail somites. *Dev. Biol.* **185**, 185–200.
Borycki, A.-G., Mendham, L., and Emerson, C. P., Jr. (1998). Control of somite patterning by Sonic hedgehog and its downstream signal response genes. *Development* **125**, 777–790.
Borycki, A.-G., Li, J., Jin, F., Emerson, C. P., and Epstein, J. (1999a). Pax3 functions in cell survival and in *Pax7* regulation. *Development* **126**, 1665–1674.
Borycki, A.-G., Brunk, B., Tajbakhsh, S., Buckingham, M., Chiang, C., and Emerson, C. P., Jr. (1999b). Sonic hedgehog controls epaxial muscle determination through *Myf5* activation. *Development*, in press.
Brand-Saberi, B., Ebensperger, C., Wilting, J., Balling, R., and Christ, B. (1993). The ventralizing effect of the notochord on somite differentiation in chick embryos. *Anat. Embryol.* **188**, 239–245.
Braun, T., Rudnicki, M. A., Arnold, H. H., and Jaenisch, R. (1992). Targeted inactivation of the muscle regulatory gene Myf-5 results in abnormal rib development and perinatal death. *Cell* **71**, 369–382.
Braun, T., Bober, E., Rudnicki, M. A., Jaenisch, R., and Arnold, H. H. (1994). MyoD expression marks the onset of skeletal myogenesis in Myf-5 mutant mice. *Development* **120**, 3083–3092.
Brennan, T. J., Edmondson, D. G., Li, L., and Olson, E. N. (1991). Transforming growth factor beta represses the actions of myogenin through a mechanism independent of DNA binding. *Proc. Natl. Acad. Sci. U.S.A.* **88**, 3822–3826.

Bronner-Fraser, M., and Fraser, S. E. (1997). Differentiation of the vertebrate neural tube. *Curr. Opin. Cell Biol.* **9,** 885–891.
Buckingham, M. (1992). Making muscle in mammals. *Trends Genet.* **8,** 144–148.
Buffinger, N., and Stockdale, F. E. (1994). Myogenic specification in somites: Induction by axial structures. *Development* **120,** 1443–1452.
Buffinger, N., and Stockdale, F. E. (1995). Myogenic specification of somites is mediated by diffusible factors. *Dev. Biol.* **169,** 96–108.
Burgess, R., Cserjesi, P., Ligon, K. L., and Olson, E. N. (1995). Paraxis: a basic helix-loop-helix protein expressed in paraxial mesoderm and developing somites. *Dev. Biol.* **168,** 296–306.
Burgess, R., Rawls, A., Brown, D., Bradley, A., and Olson, E. N. (1996). Requirement of the paraxis gene for somite formation and musculoskeletal patterning. *Nature* **384,** 570–573.
Cadigan, K. M., and Nusse, R. (1997). Wnt signaling: A common theme in animal development. *Genes Dev.* **11,** 3286–3305.
Capdevila, J., Tabin, C., and Johnson, R. L. (1998). Control of dorsoventral somite patterning by Wnt-1 and b-catenin. *Dev. Biol.* **193,** 182–194.
Charles de la Brousse, F., and Emerson, C. P., Jr. (1990). Localized expression of a myogenic regulatory gene, *qmf1*, in the somite dermatome of avian embryos. *Genes Dev.* **4,** 567–581.
Chiang, C., Litingtung, Y., Lee, E., Young, K. E., Corden, J. L., Westphal, H., and Beachy, P. A. (1996). Cyclopia and defective axial patterning in mice lacking sonic hedgehog gene function. *Nature* **383,** 407–413.
Christ, B., and Ordahl, C. P. (1995). Early stages of chick somite development. *Anat. Embryol.* **191,** 381–396.
Christ, B., and Wilting, J. (1992). From somites to vertebral column. *Anat. Anz.* **174,** 23–32.
Christ, B., Brand-Saberi, B., Grim, M., and Wilting, J. (1992). Local signalling in dermomyotomal cell type specification. *Anat. Embryol.* **186,** 505–510.
Ciruna, B. G., Schwartz, L., Harpal, K., Yamaguchi, T. P., and Rossant, J. (1997). Chimeric analysis of fibroblast growth factor receptor-1 (Fgfr1) function: A role for FGFR1 in morphogenetic movement through the primitive streak. *Development* **124,** 2829–2841.
Conlon, R. A., Reaume, A. G., and Rossant, J. (1995). Notch1 is required for the coordinate segmentation of somites. *Development* **121,** 1533–1545.
Connolly, D. J., Patel, K., and Cooke, J. (1997). Chick noggin is expressed in the organizer and neural plate during axial development, but offers no evidence of involvement in primary axis formation. *J. Dev. Biol.* **41,** 389–396.
Corbin, V., Michelson, A. M., Abmayr, S. M., Neel, V., Alcamo, E., Maniatis, T., and Young, M. W. (1991). A role for the *Drosophila* neurogenic genes in mesoderm differentiation. *Cell* **67,** 311–323.
Cornell, R. A., Musci, T. J., and Kimelman, D. (1995). FGF is a prospective competence factor for early activin-type signals in *Xenopus* mesoderm induction. *Development* **121,** 2429–2437.
Cossu, G., Tajbakhsh, S., and Buckingham, M. (1996a). How is myogenesis initiated in the embryo? *Trends Genet.* **12,** 218–223.
Cossu, G., Kelly, R., Tajbakhsh, S., Di Donna, S., Vivarelli, E., and Buckingham, M. (1996b). Activation of different myogenic pathways: myf-5 is induced by the neural tube and MyoD by the dorsal ectoderm in mouse paraxial mesoderm. *Development* **122,** 429–437.
Coulier, F., Pizette, S., Ollendorff, V., deLapeyriere, O., and Birnbaum, D. (1994). The human and mouse fibroblast growth factor 6 (FGF6) genes and their products: Possible implication in muscle development. *Prog. Growth Factor Res.* **5,** 1–14.
Currie, P. D., and Ingham, P. W. (1996). Induction of a specific muscle cell type by a hedgehog-like protein in zebrafish. *Nature* **382,** 452–455.
Cusella-De Angelis, M. G., Molinari, S., Le Donne, A., Coletta, M., Vivarelli, E., Bouche, M., Molinaro, M., Ferrari, S., and Cossu, G. (1994). Differential response of embryonic and fetal myoblasts to TGF beta: A possible regulatory mechanism of skeletal muscle histogenesis. *Development* **120,** 925–933.

5. Tissue Interactions and Signaling Mechanisms in Somite Myogenesis

Daston, G., Lamar, E., Olivier, M., and Goulding, M. (1996). Pax3 is necessary for migration but not differentiation of limb muscle precursor in the mouse. *Development* **122,** 1017–1027.

Dealy, C. N., Roth, A., Ferrari, D., Brown, A. M., and Kosher, R. A. (1993). Wnt-5a and Wnt-7a are expressed in the developing chick limb bud in a manner suggesting roles in pattern formation along the proximodistal and dorsoventral axes. *Mech. Dev.* **43,** 175–186.

Del Amo, F. F., Smith, D. E., Swiatek, P. J., Gendron-Maguire, M., Greenspan, R. J., McMahon, A. P., and Gridley, T. (1992). Expression pattern of Motch, a mouse homolog of *Drosophila* Notch, suggests an important role in early postimplantation mouse development. *Development* **115,** 737–744.

deLapeyriere, O., Ollendorff, V., Planche, J., Ott, M. O., Pizette, S., Coulier, F., and Birnbaum, D. (1993). Expression of the *Fgf6* gene is restricted to developing skeletal muscle in the mouse embryo. *Development* **118,** 601–611.

Deng, C. X., Wynshaw-Boris, A., Shen, M. M., Daughterty, C., Ornitz, D. M., and Leder, P. (1994). Murine FGFR-1 is required for early postimplantation growth and axial organization. *Genes Dev.* **8,** 3045–3057.

Derynck, R., and Feng, X. H. (1997). TGF-beta receptor signaling. *Biochim. Biophys. Acta* **1333,** F105–F150.

Devoto, S. H., Melancon, E., Eisen, J. S., and Westerfield, M. (1996). Identification of separate slow and fast muscle precursor cells *in vivo*, prior to somite formation. *Development* **122,** 3371–3380.

Dietrich, S., Schubert, F. R., and Gruss, P. (1993). Altered Pax gene expression in murine notochord mutants: The notochord is required to initiate and maintain ventral identity in the somite. *Mech. Dev.* **44,** 189–207.

Dietrich, S., Schubert, F. R., and Lumsden, A. (1997). Control of dorsoventral pattern in the chick paraxial mesoderm. *Development* **124,** 3895–3908.

Dietrich, S., Schubert, F. R., Healy, C., Sharpe, P. T., and Lumsden, A. (1998). Specification of the hypaxial musculature. *Development* **125,** 2235–2249.

Doe, C. Q., Fuerstenberg, S., and Peng, C. Y. (1998). Neural stem cells: From fly to vertebrates. *J. Neurobiol.* **36,** 111–127.

Drucker, B. J., and Goldfarb, M. (1993). Murine FGF-4 gene expression is spatially restricted within embryonic skeletal muscle and other tissues. *Mech. Dev.* **40,** 155–163.

Duprez, D. M., Coltey, M., Amthor, H., Brickell, P. M., and Tickle, C. (1996). Bone morphogenetic protein-2 (BMP-2) inhibits muscle development and promotes cartilage formation in chick limb bud cultures. *Dev. Biol.* **174,** 448–452.

Duprez, D. M., Fournier-Thibault, C., and Le Douarin, N. (1998). Sonic hedgehog induces proliferation of committed skeletal muscle cells in the chick limb. *Development* **125,** 495–505.

Ekker, S. C., McGrew, L. L., Lai, C. J., Lee, J. J., von Kessler, D. P., Moon, R. T., and Beachy, P. A. (1995). Distinct expression and shared activities of members of the hedgehog gene family of *Xenopus laevis*. *Development* **121,** 2337–2347.

Epstein, J. A., Shapiro, D. N., Cheng, J., Lam, P. Y., and Maas, R. L. (1996). Pax3 modulates expression of the c-Met receptor during limb muscle development. *Proc. Natl. Acad. Sci. U.S.A.* **93,** 4213–4218.

Fainsod, A., Deissler, K., Yelin, R., Marom, K., Epstein, M., Pillemer, G., Steinbeisser, H., and Blum, M. (1997). The dorsalizing and neural inducing gene follistatin is an antagonist of BMP-4. *Mech. Dev.* **63,** 39–50.

Fan, C. M., and Tessier-Lavigne, M. (1994). Patterning of mammalian somites by surface ectoderm and notochord: Evidence for sclerotome induction by a hedgehog homolog. *Cell* **79,** 1175–1186.

Fan, C. M., Porter, J. A., Chiang, C., Chang, D. T., Beachy, P. A., and Tessier-Lavigne, M. (1995). Long-range sclerotome induction by sonic hedgehog: Direct role of the amino-terminal cleavage product and modulation by the cyclic AMP signaling pathway. *Cell* **81,** 457–465.

Fan, C. M., Kuwana, E., Bulfone, A., Fletcher, C. F., Copeland, N. G., Jenkins, N. A., Crews, S., Martinez, S., Puelles, L., Rubenstein, J. L., and Tessier-Lavigne, M. (1996). Expression patterns of two murine homologs of *Drosophila* single-minded suggest possible roles in embryonic patterning and in the pathogenesis of Down syndrome. *Mol. Cell. Neurosci.* **7,** 1–16.

Fan, C. M., Lee, C. S., and Tessier-Lavigne, M. (1997). A role for WNT proteins in induction of dermomyotome. *Dev. Biol.* **191,** 160–165.
Filvaroff, E. H., Ebner, R., and Derynck, R. (1994). Inhibition of myogenic differentiation in myoblasts expressing a truncated type II TGF-beta receptor. *Development* **120,** 1085–1095.
Floss, T., Arnold, H. H., and Braun, T. (1997). A role for FGF-6 in skeletal muscle regeneration. *Genes Dev.* **11,** 2040–2051.
Fontaine-Perus, J., Halgand, P., Cheraud, Y., Rouaud, T., Velasco, M., Cifuentes Diaz, C., and Rieger, F. (1997). Mouse–chick chimera: A developmental model of murine neurogenic cells. *Development* **124,** 3025–3036.
Franz, T., Kothary, R., Surani, M. A., Halata, Z., and Grim, M. (1993). The Splotch mutation interferes with muscle development in the limbs. *Anat. Embryol.* **187,** 153–160.
Gallera, J. (1966). Mise en evidence du role de l'ectoblaste dans la differenciation des somites chez les Oiseaux. *Rev. Suisse Zool.* **73,** 492–503.
Gamel, A. J., Brand-Saberi, B., and Christ, B. (1995). Halves of epithelial somites and segmental plate show distinct muscle differentiation behavior *in vitro* compared to entire somites and segmental plate. *Dev. Biol.* **172,** 625–639.
George-Weinstein, M., Gerhart, J., Reed, R., Flynn, J., Callihan, B., Mattiacci, M., Miehle, C., Foti, G., Lash, J. W., and Weintraub, H. (1996). Skeletal myogenesis: The preferred pathway of chick embryo epiblast cells *in vitro*. *Dev. Biol.* **173,** 279–291.
George-Weinstein, M., Gerhart, J., Blitz, J., Simak, E., and Knudsen, K. A. (1997). N-cadherin promotes the commitment and differentiation of skeletal muscle precursor cells. *Dev. Biol.* **185,** 14–24.
Goldhamer, D. J., Faerman, A., Shani, M., and Emerson, C. P., Jr. (1992). Regulatory elements that control the lineage-specific expression of myoD. *Science* **256,** 538–542.
Goldhamer, D. J., Brunk, B. P., Faerman, A., King, A., Shani, M., and Emerson, C. P., Jr. (1995). Embryonic activation of the *myoD* gene is regulated by a highly conserved distal control element. *Development* **121,** 637–649.
Gossler, A., and Hrabe de Angelis, M. (1998). Somitogenesis. *Curr. Top. Dev. Biol.* **38,** 225–287.
Goulding, M. D., Chalepakis, G., Deutsch, U., Erselius, J. R., and Gruss, P. (1991). Pax-3, a novel murine DNA binding protein expressed during early neurogenesis. *EMBO J.* **10,** 1135–1147.
Goulding, M. D., Lumsden, A., and Gruss, P. (1993). Signals from the notochord and floor plate regulate the region-specific expression of two Pax genes in the developing spinal cord. *Development* **117,** 1001–1016.
Goulding, M., Lumsden, A., and Paquette, A. J. (1994). Regulation of Pax-3 expression in the dermomyotome and its role in muscle development. *Development* **120,** 957–971.
Graba, Y., Gieseler, K., Aragnol, D., Laurenti, P., Mariol, M. C., Berenger, H., Sagnier, T., and Pradel, J. (1995). DWnt-4, a novel *Drosophila Wnt* gene acts downstream of homeotic complex genes in the visceral mesoderm. *Development* **121,** 209–218.
Grass, S., Arnold, H. H., and Braun, T. (1996). Alterations in somite patterning of Myf-5-deficient mice: A possible role for FGF-4 and FGF-6. *Development* **122,** 141–150.
Greenwald, I. (1998). LIN-12/Notch signaling: Lessons from worms and flies. *Genes Dev.* **12,** 1751–1762.
Grobstein, C., and Holtzer, H. (1955). *In vitro* studies of cartilage induction in mouse somite mesoderm. *J. Exp. Zool.* **128,** 333–356.
Gunther, T., Struwe, M., Aguzzi, A., and Schughart, K. (1994). *Open brain*, a new mouse mutant with severe neural tube defects, shows altered gene expression patterns in the developing spinal cord. *Development* **120,** 3119–3130.
Hacker, A., and Guthrie, S. (1998). A distinct developmental programme for the cranial paraxial mesoderm in the chick embryo. *Development* **125,** 3461–3472.
Hall, T. M., Porter, J. A., Beachy, P. A., and Leahy, D. J. (1995). A potential catalytic site revealed by the 1.7-A crystal structure of the amino-terminal signalling domain of Sonic hedgehog. *Nature* **378,** 212–216.
Hall, T. M., Porter, J. A., Young, K. E., Koonin, E. V., Beachy, P. A., and Leahy, D. J. (1997). Crystal

structure of a Hedgehog autoprocessing domain: Homology between Hedgehog and self-splicing proteins. *Cell* **91,** 85–97.
Hammerschmidt, M., Brook, A., and McMahon, A. P. (1997). The world according to hedgehog. *Trends Genet.* **13,** 14–21.
Han, J. K., and Martin, G. R. (1993). Embryonic expression of Fgf-6 is restricted to the skeletal muscle lineage. *Dev. Biol.* **158,** 549–554.
Hannon, K., Kudla, A. J., McAvoy, M. J., Clase, K. L., and Olwin, B. B. (1996). Differentially expressed fibroblast growth factors regulate skeletal muscle development through autocrine and paracrine mechanisms. *J. Cell Biol.* **132,** 1151–1159.
Hartenstein, A. Y., Rugendorff, A., Tepass, U., and Hartenstein, V. (1992). The function of the neurogenic genes during epithelial development in the *Drosophila* embryo. *Development* **116,** 1203–1220.
Hatada, Y., and Stern, C. D. (1994). A fate map of the epiblast of the early chick embryo. *Development* **120,** 2879–2889.
Hinterberger, T. J., Sassoon, D. A., Rhodes, S. J., and Konieczny, S. F. (1991). Expression of the muscle regulatory factor MRF4 during somite and skeletal myofiber development. *Dev. Biol.* **147,** 144–156.
Hirano, S., Hirako, R., Kajita, N., and Norita, M. (1995). Morphological analysis of the role of the neural tube and notochord in the development of somites. *Anat. Embryol.* **192,** 445–457.
Hirsinger, E., Duprez, D., Jouve, C., Malapert, P., Cooke, J., and Pourquié, O. (1997). Noggin acts downstream of Wnt and Sonic Hedgehog to antagonize BMP4 in avian somite patterning. *Development* **124,** 4605–4614.
Hoang, B. H., Thomas, J. T., Abdul-Karim, F. W., Correia, K. M., Conlon, R. A., Luyten, F. P., and Ballock, R. T. (1998). Expression pattern of two Frizzled-related genes, *Frzb-1* and *Sfrp-1,* during mouse embryogenesis suggests a role for modulating action of Wnt family members. *Dev. Dyn.* **212,** 364–374.
Hogan, B. L. (1996). Bone morphogenetic proteins: Multifunctional regulators of vertebrate development. *Genes Dev.* **10,** 1580–1594.
Hollyday, M., McMahon, J. A., and McMahon, A. P. (1995). Wnt expression patterns in chick embryo nervous system. *Mech. Dev.* **52,** 9–25.
Hoppler, S., Brown, J. D., and Moon, R. T. (1996). Expression of a dominant–negative Wnt blocks induction of MyoD in *Xenopus* embryos. *Genes Dev.* **10,** 2805–2817.
Hopwood, N. D., Pluck, A., and Gurdon, J. B. (1989). MyoD expression in the forming somites is an early response to mesoderm induction in *Xenopus* embryos. *EMBO J.* **8,** 3409–3417.
Hopwood, N. D., Pluck, A., and Gurdon, J. B. (1991). *Xenopus* Myf-5 marks early muscle cells and can activate muscle genes ectopically in early embryos. *Development* **111,** 551–560.
Ikeya, M., and Takada, S. (1998). Wnt signaling from the dorsal neural tube is required for the formation of the medial dermomyotome. *Development* **125,** 4969–4976.
Isaacs, H. V., Pownall, M. E., and Slack, J. M. (1995). eFGF is expressed in the dorsal midline of *Xenopus laevis. Dev. Biol.* **39,** 575–579.
Jacob, M., Jacob, J. H., and Christ, B. (1975). The early differentiation of the perinotochordal connective tissue. A scanning and transmission electron microscopic study on chick embryos. *Experientia* **31,** 1083–1086.
Jagla, K., Dolle, P., Mattei, M. G., Jagla, T., Schuhbaur, B., Dretzen, G., Bellard, F., and Bellard, M. (1995). Mouse Lbx1 and human LBX1 define a novel mammalian homeobox gene family related to the *Drosophila* lady bird genes. *Mech. Dev.* **53,** 345–356.
Jakowlew, S. B., Ciment, G., Tuan, R. S., Sporn, M. B., and Roberts, A. B. (1992). Pattern of expression of transforming growth factor-beta 4 mRNA and protein in the developing chicken embryo. *Dev. Dyn.* **195,** 276–289.
Jakowlew, S. B., Ciment, G., Tuan, R. S., Sporn, M. B., and Roberts, A. B. (1994). Expression of transforming growth factor-beta 2 and beta 3 mRNAs and proteins in the developing chicken embryo. *Differentiation* **55,** 105–118.

Jarriault, S., Brou, C., Logeat, F., Schroeter, E. H., Kopan, R., and Israel, A. (1995). Signalling downstream of activated mammalian Notch. *Nature* **377**, 355–358.

Johnson, R. L., Laufer, E., Riddle, R. D., and Tabin, C. (1994). Ectopic expression of Sonic hedgehog alters dorsal–ventral patterning of somites. *Cell* **79**, 1165–1173.

Jones, C. M., Lyons, K. M., and Hogan, B. L. (1991). Involvement of Bone Morphogenetic Protein-4 (BMP-4) and Vgr-1 in morphogenesis and neurogenesis in the mouse. *Development* **111**, 531–542.

Jostes, B., Walther, C., and Gruss, P. (1991). The murine paired box gene, *Pax7*, is expressed specifically during the development of the nervous and muscular system. *Mech. Dev.* **33**, 27–38.

Kadowaki, T., Wilder, E., Klingensmith, J., Zachary, K., and Perrimon, N. (1996). The segment polarity gene *porcupine* encodes a putative multitransmembrane protein involved in Wingless processing. *Genes Dev.* **10**, 3116–3128.

Kalcheim, C., and Neufeld, G. (1990). Expression of basic fibroblast growth factor in the nervous system of early avian embryos. *Development* **109**, 203–215.

Karlstrom, R. O., Talbot, W. S., and Schier, A. F. (1999). The zebrafish *you-too* locus encodes the hedgehog target gli2 and affects patterning and axon guidance in the ventral forebrain. *Genes Dev.* **13**, 388–393.

Katagiri, T., Yamaguchi, A., Komaki, M., Abe, E., Takahashi, N., Ikeda, T., Rosen, V., Wozney, J. M., Fujisawa-Sehara, A., and Suda, T. (1994). Bone morphogenetic protein-2 converts the differentiation pathway of C2C12 myoblasts into the osteoblast lineage. *J. Cell Biol.* **127**, 1755–1766.

Katagiri, T., Akiyama, S., Namiki, M., Komaki, M., Yamaguchi, A., Rosen, V., Wozney, J. M., Fujisawa-Sehara, A., and Suda, T. (1997). Bone morphogenetic protein-2 inhibits terminal differentiation of myogenic cells by suppressing the transcriptional activity of MyoD and myogenin. *Exp. Cell Res.* **230**, 342–351.

Kenny-Mobbs, T., and Thorogood, P. (1987). Autonomy of differentiation in avian branchial somites and the influence of adjacent tissues. *Development* **100**, 449–462.

Kimelman, D., and Kirschner, M. (1987). Synergistic induction of mesoderm by FGF and TGF-beta and the identification of an mRNA coding for FGF in the early *Xenopus* embryo. *Cell* **51**, 869–877.

Kishimoto, Y., Lee, K. H., Zon, L., Hammerschmidt, M., and Schulte-Merker, S. (1997). The molecular nature of zebrafish swirl: BMP2 function is essential during early dorsoventral patterning. *Development* **124**, 4457–4466.

Klingensmith, J., and Nusse, R. (1994). Signaling by wingless in *Drosophila*. *Dev. Biol.* **166**, 396–414.

Kopan, R., Nye, J. S., and Weintraub, H. (1994). The intracellular domain of mouse Notch: A constitutively activated repressor of myogenesis directed at the basic helix-loop-helix region of MyoD. *Development* **120**, 2385–2396.

Kosher, R. A., and Lash, J. W. (1975). Notochordal stimulation of *in vitro* somite chondrogenesis before and after enzymatic removal of perinotochordal materials. *Dev. Biol.* **42**, 362–378.

Krauss, S., Concordet, J. P., and Ingham, P. W. (1993). A functionally conserved homolog of the *Drosophila* segment polarity gene *hh* is expressed in tissues with polarizing activity in zebrafish embryos. *Cell* **75**, 1431–1444.

Kretzschmar, M., Liu, F., Hata, A., Doody, J., and J., M. (1997). The TGF-beta family mediator Smad1 is phosphorylated directly and activated functionally by the BMP receptor kinase. *Genes Dev.* **11**, 984–995.

Lagna, G., Hata, A., Hemmati-Brivanlou, A., and Massague, J. (1996). Partnership between DPC4 and SMAD proteins in TGF-beta signalling pathways. *Nature* **383**, 832–836.

Lardelli, M., Williams, R., and Lendahl, U. (1995). Notch-related genes in animal development. *Int. J. Dev. Biol.* **39**, 769–780.

Lathrop, B., Thomas, K., and Glaser, L. (1985). Control of myogenic differentiation by fibroblast growth factor is mediated by position in the G1 phase of the cell cycle. *J. Cell Biol.* **101**, 2194–2198.

Lawrence, P. A., Sanson, B., and Vincent, J. P. (1996). Compartments, wingless and engrailed: Patterning the ventral epidermis of *Drosophila* embryos. *Development* **122**, 4095–4103.

5. Tissue Interactions and Signaling Mechanisms in Somite Myogenesis

Le Douarin, N. M. (1973). A Feulgen-positive nucleolus. *Exp. Cell Res.* **77,** 459–468.
Lee, J. J., Ekker, S. C., von Kessler, D. P., Porter, J. A., Sun. B. I., and Beachy, P. A. (1994). Autoproteolysis in hedgehog protein biogenesis. *Science* **266,** 1528–1537.
Lee, J., Platt, K. A., Censullo, P., and Ruiz i Altaba, A. (1997). Gli1 is a target of Sonic hedgehog that induces ventral neural tube development. *Development* **124,** 2537–2552.
Leyns, L., Bouwmeester, T., Kim, S. H., Piccolo, S., and De Robertis, E. M. (1997). Frzb-1 is a secreted antagonist of Wnt signaling expressed in the Spemann organizer. *Cell* **88,** 747–756.
Liem, K. F., Jr., Tremml, G., Roelink, H., and Jessell, T. M. (1995). Dorsal differentiation of neural plate cells induced by BMP-mediated signals from epidermal ectoderm. *Cell* **82,** 969–979.
Lyons, K. M., Hogan, B. L., and Robertson, E. J. (1995). Colocalization of BMP 7 and BMP 2 RNAs suggests that these factors cooperatively mediate tissue interactions during murine development. *Mech. Dev.* **50,** 71–83.
McMahon, J. A., Takada, S., Zimmerman, L. B., Fan, C. M., Harland, R. M., and McMahon, A. P. (1998). Noggin-mediated antagonism of BMP signaling is required for growth and patterning of the neural tube and somite. *Genes Dev.* **12,** 1438–1452.
McPherron, A. C., Lawler, A. M., and Lee, S. J. (1997). Regulation of skeletal muscle mass in mice by a new TGF-beta superfamily member. *Nature* **387,** 83–90.
Mansouri, A., Stoykova, A., Torres, M., and Gruss, P. (1996). Dysgenesis of cephalic neural crest derivatives in Pax7−/− mutant mice. *Development* **122,** 831–838.
Marcelle, C., Eichmann, A., Halevy, O., Breant, C., and Le Douarin, N. M. (1994). Distinct developmental expression of a new avian fibroblast growth factor receptor. *Development* **120,** 683–694.
Marcelle, C., Stark, M. R., and Bronner-Fraser, M. (1997). Coordinate actions of MPBs, Shh and Noggin mediate patterning of the dorsal somite. *Development* **124,** 3955–3963.
Marigo, V., Davey, R. A., Zuo, Y., Cunningham, J. M., and Tabin, C. J. (1996a). Biochemical evidence that patched is the Hedgehog receptor. *Nature* **384,** 176–179.
Marigo, V., Johnson, R. L., Vortkamp, A., and Tabin, C. J. (1996b). Sonic hedgehog differentially regulates expression of GLI and GLI3 during limb development. *Dev. Biol.* **180,** 273–283.
Marine, J.-C., Bellefoid, E. J., Pendeville, H., Martial, J. A., and Pieler, T. (1997). A role for *Xenopus* Gli-type zinc-finger proteins in the early embryonic patterning of the mesoderm and neuroectoderm. *Mech. Dev.* **63,** 211–225.
Maroto, M., Reshef, R., Münsterberg, A. E., Koester, S., Goulding, M., and Lassar, A. B. (1997). Ectopic Pax-3 activates MyoD and Myf-5 expression in embryonic mesoderm and neural tissue. *Cell* **89,** 139–148.
Marti, E., Takada, R., Bumcrot, D. A., Sasaki, H., and McMahon, A. P. (1995a). Distribution of Sonic hedgehog peptides in the developing chick and mouse embryo. *Development* **121,** 2537–2547.
Marti, E., Bumcrot, D. A., Takada, R., and McMahon, A. P. (1995b). Requirement of 19K form of Sonic hedgehog for induction of distinct ventral cell types in CNS explants [see comments]. *Nature* **375,** 322–325.
Mason, I. J., Fuller-Pace, F., Smith, R., and Dickson, C. (1994). FGF-7 (keratinocyte growth factor) expression during mouse development suggests roles in myogenesis, forebrain regionalisation and epithelial–mesenchymal interactions. *Mech. Dev.* **45,** 15–30.
Massague, J. (1996). TGFbeta signaling: Receptors, transducers, and Mad proteins. *Cell* **85,** 947–950.
Massague, J., Cheifetz, S., Endo, T., and Nadal-Ginard, B. (1986). Type beta transforming growth factor is an inhibitor of myogenic differentiation. *Proc. Natl. Acad. Sci. U.S.A.* **83,** 8206–8210.
Michaud, J. L., Rosenquist, T., May, N. R., and Fan, C. M. (1998). Development of neuroendocrine lineages requires the bHLH-PAS transcription factor SIM1. *Genes Dev.* **12,** 3264–3275.
Mo, R., Freer, A. M., Zinyk, D. L., Crackower, M. A., Michaud, J., Heng, H. H., Chik, K. W., Shi, X. M., Tsui, L. C., Cheng, S. H., Joyner, A. L., and Hui, C. (1997). Specific and redundant functions of Gli2 and Gli3 zinc finger genes in skeletal patterning and development. *Development* **124,** 113–123.
Moon, R. T., Brown, J. D., Yang-Snyder, J. A., and Miller, J. R. (1997). Structurally related receptors and antagonists compete for secreted Wnt ligands. *Cell* **88,** 725–728.
Mullins, M. C., Hammerschmidt, M., Kane, D. A., Odenthal, J., Brand, M., van Eeden, F. J., Furutani-

Seiki, M., Granato, M., Haffter, P., Heisenberg, C. P., Jiang, Y. J., Kelsh, R. N., and Nusslein-Volhard, C. (1996). Genes establishing dorsoventral pattern formation in the zebrafish embryo: The ventral specifying genes. *Development* **123**, 81–93.

Münsterberg, A. E., and Lassar, A. B. (1995). Combinatorial signals from the neural tube, floor plate and notochord induce myogenic bHLH gene expression in the somite. *Development* **121**, 651–660.

Münsterberg, A. E., Kitajewski, J., Bumcrot, D. A., McMahon, A. P., and Lassar, A. B. (1995). Combinatorial signaling by sonic hedgehog and wnt family members induces myogenic bHLH gene expression in the somite. *Genes Dev.* **9**, 2911–2922.

Murray, S. S., Murray, E. J., Glackin, C. A., and Urist, M. R. (1993). Bone morphogenetic protein inhibits differentiation and affects expression of helix-loop-helix regulatory molecules in myoblastic cells. *J. Cell. Biochem.* **53**, 51–60.

Neumann, C. J., and Cohen, S. M. (1996). A hierarchy of cross-regulation involving Notch, wingless, vestigial and cut organizes the dorsal/ventral axis of the *Drosophila* wing. *Development* **122**, 3477–3485.

Noden, D. M. (1983). The embryonic origins of avian cephalic and cervical muscles and associated connective tissues. *Am. J. Anat.* **168**, 257–276.

Noden, D. M. (1986). Patterning of avian craniofacial muscles. *Dev. Biol.* **116**, 347–356.

Noden, D. M. (1988). Interactions and fates of avian craniofacial mesenchyme. *Development* **103**, 121–140.

Nusse, R., and Varmus, H. E. (1992). Wnt genes. *Cell* **69**, 1073–1087.

Nye, J. S., and Kopan, R. (1995). Developmental signaling. Vertebrate ligands for Notch. *Curr. Biol.* **5**, 966–969.

Odenthal, J., Haffter, P., Vogelsang, E., Brand, M., van Eeden, F. J., Furutani-Seiki, M., Granatao, M., Hammerschmidt, M., Heisenberg, C. P., Jiang, Y. J., Kane, D. A., Kelsh, R. N., Mullins, M. C., Warga, R. M., Allende, M. L., Weinberg, E. S., and Nusslein-Volhard, C. (1996). Mutations affecting the formation of the notochord in the zebrafish, *Danio rerio*. *Development* **123**, 103–115.

Olson, E. N., Sternberg, E., Hu, J. S., Spizz, G., and Wilcox, C. (1986). Regulation of myogenic differentiation by type beta transforming growth factor. *J. Cell Biol.* **103**, 1799–1805.

Olwin, B. B., Arthur, K., Hannon, K., Hein, P., McFall, A., Riley, B., Szebenyi, G., Zhou, Z., Zuber, M. E., Rapraeger, A. C., Fallon, J. F., and Kudla, A. J. (1994). Role of FGFs in skeletal muscle and limb development. *Mol. Reprod. Dev.* **39**, 90–100.

Ordahl, C. P. (1993). "Myogenic Lineages within the Developing Somite," *In* Molecular Basis of Morphogenesis (M. Bernfield, Ed.), pp. 165–176. Alan R. Liss, New York.

Ordahl, C. P., and Le Douarin, N. M. (1992). Two myogenic lineages within the developing somite. *Development* **114**, 339–353.

Orr-Urtreger, A., Givol, D., Yayon, A., Yarden, Y., and Lonai, P. (1991). Developmental expression of two murine fibroblast growth factor receptors, flg and bek. *Development* **113**, 1419–1434.

Ott, M. O., Bober, E., Lyons, G., Arnold, H., and Buckingham, M. (1991). Early expression of the myogenic regulatory gene, *myf-5*, in precursor cells of skeletal muscle in the mouse embryo. *Development* **111**, 1097–1107.

Parr, B. A., Shea, M. J., Vassileva, G., and McMahon, A. P. (1993). Mouse Wnt genes exhibit discrete domains of expression in the early embryonic CNS and limb buds. *Development* **119**, 247–261.

Patel, K., Connolly, D. J., Amthor, H., Nose, K., and Cooke, J. (1996). Cloning and early dorsal axial expression of *Flik*, a chick follistatin-related gene: Evidence for involvement in dorsalization/neural induction. *Dev. Biol.* **178**, 327–342.

Patstone, G., Pasquale, E. B., and Maher, P. A. (1993). Different members of the fibroblast growth factor receptor family are specific to distinct cell types in the developing chicken embryo. *Dev. Biol.* **155**, 107–123.

Pelton, R. W., Saxena, B., Jones, M., Moses, H. L., and Gold, L. I. (1991). Immunohistochemical localization of TGF beta 1, TGF beta 2, and TGF beta 3 in the mouse embryo: Expression patterns suggest multiple roles during embryonic development. *J. Cell Biol.* **115**, 1091–1105.

Peters, K. G., Werner, S., Chen, G., and Williams, L. T. (1992). Two FGF receptor genes are differ-

5. Tissue Interactions and Signaling Mechanisms in Somite Myogenesis 221

entially expressed in epithelial and mesenchymal tissues during limb formation and organogenesis in the mouse. *Development* **114**, 233–243.

Piccolo, S., Sasai, Y., Lu, B., and De Robertis, E. M. (1996). Dorsoventral patterning in *Xenopus*: Inhibition of ventral signals by direct binding of chordin to BMP-4. *Cell* **86**, 589–598.

Pinney, D. F., de la Brousse, F. C., Faerman, A., Shani, M., Maruyama, K., and Emerson, C. P., Jr. (1995). Quail myoD is regulated by a complex array of cis-acting control sequences. *Dev. Biol.* **170**, 21–38.

Placzek, M., Tessier-Lavigne, M., Yamada, T., Jessell, T., and Dodd, J. (1990). Mesodermal control of neural cell identity: Floor plate induction by the notochord. *Science* **250**, 985–988.

Porter, J. A., von Kessler, D. P., Ekker, S. C., Young, K. E., Lee, J. J., Moses, K., and Beachy, P. A. (1995). The product of hedgehog autoproteolytic cleavage active in local and long-range signaling. *Nature* **374**, 363–366.

Porter, J. A., Young, K. E., and Beachy, P. A. (1996). Cholesterol modification of hedgehog signaling proteins in animal development. *Science* **274**, 255–259.

Pourquié, O., Coltey, M., Teillet, M. A., Ordahl, C., and Le Douarin, N. M. (1993). Control of dorso-ventral patterning of somitic derivatives by notochord and floor plate. *Proc. Natl. Acad. Sci. U.S.A.* **90**, 5242–5246.

Pourquié, O., Coltey, M., Breant, C., and Le Douarin, N. M. (1995). Control of somite patterning by signals from the lateral plate. *Proc. Natl. Acad. Sci. U.S.A.* **92**, 3219–3223.

Pourquié, O., Fan, C. M., Coltey, M., Hirsinger, E., Watanabe, Y., Breant, C., Francis-West, P., Brickell, P., Tessier-Lavigne, M., and Le Douarin, N. M. (1996). Lateral and axial signals involved in avian somite patterning: A role for BMP4. *Cell* **84**, 461–471.

Pownall, M. E., and Emerson, C. P., Jr. (1992). Sequential activation of three myogenic regulatory genes during somite morphogenesis in quail embryos. *Dev. Biol.* **151**, 67–79.

Pownall, M. E., Strunk, K. E., and Emerson, C. P., Jr. (1996). Notochord signals control the transcriptional cascade of myogenic bHLH genes in somites of quail embryos. *Development* **122**, 1475–1488.

Reaume, A. G., Conlon, R. A., Zirngibl, R., Yamaguchi, T. P., and Rossant, J. (1992). Expression analysis of a Notch homologue in the mouse embryo. *Dev. Biol.* **154**, 377–387.

Reshef, R., Maroto, M., and Lassar, A. B. (1998). Regulation of dorsal somitic cell fates: BMPs and Noggin control the timing and pattern of myogenic regulator expression. *Genes Dev.* **12**, 290–303.

Robbins, D. J., Nybakken, K. E., Kobayashi, R., Sisson, J. C., Bishop, J. M., and Therond, P. P. (1997). Hedgehog elicits signal transduction by means of a large complex containing the kinesin-related protein costal2. *Cell* **90**, 225–234.

Roelen, B. A., Lin, H. Y., Knezevic, V., Freund, E., and Mummery, C. L. (1994). Expression of TGF-betas and their receptors during implantation and organogenesis of the mouse embryo. *Dev. Biol.* **166**, 716–728.

Roelink, H., Augsburger, A., Heemskerk, J., Korzh, V., Norlin, S., Ruiz i Altaba, A., Tanabe, Y., Placzek, M., Edlund, T., Jessell, T. M., and Dodd, J. (1994). Floor plate and motor neuron induction by vhh-1, a vertebrate homolog of hedgehog expressed by the notochord. *Cell* **76**, 761–775.

Rong, P. M., Teillet, M. A., Ziller, C., and Le Douarin, N. M. (1992). The neural tube/notochord complex is necessary for vertebral but not limb and body wall striated muscle differentiation. *Development* **115**, 657–672.

Rudnicki, M. A., Braun, T., Hinuma, S., and Jaenisch, R. (1992). Inactivation of MyoD in mice leads to up-regulation of the myogenic HLH gene *Myf-5* and results in apparently normal muscle development. *Cell* **71**, 383–390.

Rudnicki, M. A., Schnegelsberg, P. N., Stead, R. H., Braun, T., Arnold, H. H., and Jaenisch, R. (1993). MyoD or Myf-5 is required for the formation of skeletal muscle. *Cell* **75**, 1351–1359.

Ruiz i Atalba, A. (1997). Catching a Gli-mpse of hedgehog. *Cell* **90**, 193–196.

Russell, J., Gennissen, A., and Nusse, R. (1992). Isolation and expression of two novel Wnt/wingless gene homologues in *Drosophila*. *Development* **115**, 475–485.

Sanders, E. J., Khare, M. K., Ooi, V. C., and Bellairs, R. (1986). An experimental and morphological analysis of the tail bud mesenchyme of the chick embryo. *Anat. Embryol.* **174,** 179–185.

Sanders, E. J., Hu, N., and Wride, M. A. (1994). Expression of TGF beta 1/beta 3 during early chick embryo development. *Anat. Rec.* **238,** 397–406.

Sasai, Y., Kageyama, R., Tagawa, Y., Shigemoto, R., and Nakanishi, S. (1992). Two mammalian helix-loop-helix factors structurally related to *Drosophila* hairy and Enhancer of split. *Genes Dev.* **6,** 2620–2634.

Sasaki, H., Hui, C., Nakafuku, M., and Kondoh, H. (1997). A binding site for Gli proteins is essential for HNF-3beta floor plate enhancer activity in transgenics and can respond to Shh *in vitro. Development* **124,** 1313–1322.

Sassoon, D., Lyons, G., Wright, W. E., Lin, V., Lassar, A., Weintraub, H., and Buckingham, M. (1989). Expression of two myogenic regulatory factors myogenin and MyoD1 during mouse embryogenesis. *Nature* **341,** 303–307.

Savage, M. P., and Fallon, J. F. (1995). FGF-2 mRNA and its antisense message are expressed in a developmentally specific manner in the chick limb bud and mesonephros. *Dev. Dyn.* **202,** 343–353.

Saxton, T. M., Henkemeyer, M., Gasca, S., Shen, R., Rossi, D. J., Shalaby, F., Feng, G., and Pawson, T. (1997). Abnormal mesoderm patterning in mouse embryos mutant for the SH2 tyrosine phosphatase Shp-2. *EMBO J.* **16,** 2352–2364.

Schauerte, H. E., van Eeden, F. J., Fricke, C., Odenthal, J., Strahle, U., and Haffter, P. (1998). Sonic hedgehog is not required for the induction of medial floor plate cells in the zebrafish. *Development* **125,** 2983–2993.

Schofield, J. N., and Wolpert, L. (1990). Effect of TGF-beta 1, TGF-beta 2, and bFGF on chick cartilage and muscle cell differentiation. *Exp. Cell Res.* **191,** 144–148.

Seed, J., and Hauschka, S. D. (1988). Clonal analysis of vertebrate myogenesis. VIII. Fibroblasts growth factor (FGF)-dependent and FGF-independent muscle colony types during chick wing development. *Dev. Biol.* **128,** 40–49.

Selleck, M. A., and Stern, C. D. (1991). Fate mapping and cell lineage analysis of Hensen's node in the chick embryo. *Development* **112,** 615–626.

Sisson, J. C., Ho, K. S., Suyama, K., and Scott, M. P. (1997). Costal2, a novel kinesin-related protein in the Hedgehog signaling pathway. *Cell* **90,** 235–245.

Slager, H. G., Van Inzen, W., Freund, E., Van den Eijnden-Van Raaij, A. J., and Mummery, C. L. (1993). Transforming growth factor-beta in the early mouse embryo: Implications for the regulation of muscle formation and implantation. *Dev. Genet.* **14,** 212–224.

Sosic, D., Brand-Saberi, B., Schmidt, C., B., C., and Olson, E. N. (1997). Regulation of paraxis expression and somite formation by ectoderm- and neural tube-derived signals. *Dev. Biol.* **185,** 229–243.

Spence, M. S., Yip, J., and Erickson, C. A. (1996). The dorsal neural tube organizes the dermomyotome and induces axial myocytes in the avian embryo. *Development* **122,** 231–241.

Sporle, R., and Schughart, K. (1997). Neural tube morphogenesis. *Curr. Opin. Genet. Dev.* **7,** 507–512.

Sporle, R., Gunther, T., Struwe, M., and Schughart, K. (1996). Severe defects in the formation of epaxial musculature in open brain (opb) mutant mouse embryos. *Development* **122,** 79–86.

Spratt, N. T. (1955). Analysis of the organizer center in the early chick embryo. I. Localization of the prospective notochord and somite cells. *J. Exp. Zool.* **128,** 121–164.

Stemple, D. L., Solnica-Krezel, L., Zwartkruis, F., Neuhauss, S. C., Schier, A. F., Malicki, J., Stainier, D. Y., Abdelilah, S., Rangini, Z., Mountcastle-Shah, E., and Driever, W. (1996). Mutations affecting development of the notochord in zebrafish. *Development* **123,** 117–128.

Stern, H. M., and Hauschka, S. D. (1995). Neural tube and notochord promote *in vitro* myogenesis in single somite explants. *Dev. Biol.* **167,** 87–103.

Stern, H. M., Brown, A. M. C., and Hauschka, S. D. (1995). Myogenesis in paraxial mesoderm: Preferential induction by dorsal neural tube and by cells expressing Wnt-1. *Development* **121,** 3675–3686.

5. Tissue Interactions and Signaling Mechanisms in Somite Myogenesis 223

Stern, H. M., Lin-Jones, J., and Hauschka, S. D. (1997). Synergistic interactions between bFGF and a TGF-beta family member may mediate myogenic signals from the neural tube. *Development* **124,** 3511–3523.

Stone, D. M., Hynes, M., Armanini, M., Swanson, T. A., Gu, Q., Johnson, R. L., Scott, M. P., Pennica, D., Goddard, A., Phillips, H., Noll, M., Hooper, J. E., de Sauvage, F., and Rosenthal, A. (1996). The tumour-suppressor gene patched encodes a candidate receptor for Sonic hedgehog. *Nature* **384,** 129–134.

Strachan, T., and Read, A. P. (1994). PAX genes. *Curr. Opin. Genet. Dev.* **4,** 427–438.

Suzuki, A., Kaneko, E., Maeda, J., and Ueno, N. (1997). Mesoderm induction by BMP-4 and -7 heterodimers. *Biochem. Biophys. Res. Commun.* **232,** 153–156.

Tajbakhsh, S., Bober, E., Babinet, C., Pournin, S., Arnold, H., and Buckingham, M. (1996a). Gene targeting the myf-5 locus with nlacZ reveals expression of this myogenic factor in mature skeletal muscle fibres as well as early embryonic muscle. *Dev. Dyn.* **206,** 291–300.

Tajbakhsh, S., Rocancourt, D., and Buckingham, M. (1996b). Muscle progenitor cells failing to respond to positional cues adopt non-myogenic fates in myf5 null mice. *Nature* **384,** 266–270.

Tajbakhsh, S., Rocancourt, D., Cossu, G., and Buckingham, M. (1997). Redefining the genetic hierarchies controlling skeletal myogenesis: Pax-3 and Myf-5 act upstream of MyoD. *Cell* **89,** 127–138.

Tajbakhsh, S., Borello, U., Vivarelli, E., Kelly, R., Papkoff, J., Duprez, D., Buckingham, M., and Cossu, G. (1998). Differential activation of Myf5 and MyoD by different Wnts in explants of mouse paraxial mesoderm and the later activation of myogenesis in the absence of Myf5. *Development* **125,** 4155–4162.

Takada, S., Stark, K. L., Shea, M. J., Vassileva, G., McMahon, J. A., and McMahon, A. P. (1994). Wnt-3a regulates somite and tailbud formation in the mouse embryo. *Genes Dev.* **8,** 174–189.

Tapscott, S. J., Lassar, A. B., and Weintraub, H. (1992). A novel myoblast enhancer element mediates MyoD transcription. *Mol. Cell. Biol.* **12,** 4994–5003.

Teillet, M.-A., Watanabe, Y., Jeffs, P., Duprez, D., Lapointe, F., and Le Douarin, N. M. (1998). Sonic hedgehog is required for survival of both myogenic and chondrogenic somitic lineages. *Development* **125,** 2019–2030.

Tonegawa, A., Funayama, N., Ueno, N., and Takahashi, Y. (1997). Mesodermal subdivision along the mediolateral axis in chicken controlled by different concentrations of BMP-4. *Development* **124,** 1975–1984.

Unsicker, K., Meier, C., Krieglstein, K., Sartor, B. M., and Flanders, K. C. (1996). Expression, localization, and function of transforming growth factor-beta s in embryonic chick spinal cord, hindbrain, and dorsal root ganglia. *J. Neurobiol.* **29,** 262–276.

van den Heuvel, M., and Ingham, P. W. (1996). *smoothened* encodes a receptor-like serpentine protein required for hedgehog signalling. *Nature* **382,** 547–551.

van Eeden, F. J., Granato, M., Schach, U., Brand, M., Furutani-Seiki, M., Haffter, P., Hammerschmidt, M., Heisenberg, C. P., Jiang, Y. J., Kane, D., Kelsh, R. N., Mullins, M. C., Odenthal, J., Warga, R. M., Allende, M. L., Weinberg, E. S., and C. N. V. (1996). Mutations affecting somite formation and patterning in the zebrafish, *Danio rerio*. *Development* **123,** 153–164.

Vivarelli, E., and Cossu, G. (1986). Neural control of early myogenic differentiation in cultures of mouse somites. *Dev. Biol.* **117,** 319–325.

Wachtler, F., and Christ, B. (1992). The basic embryology of skeletal muscle formation in vertebrates: The avian model. *Semin. Dev. Biol.* **3,** 217–227.

Wang, S., Krinks, M., and Moos, M. J. (1997). Frzb-1, an antagonist of Wnt-1 and Wnt-8, does not block signaling by Wnts-3A, -5A, or -11. *Biochem. Biophys. Res. Commun.* **236,** 502–504.

Wang, Y. Q., Sizeland, A., Wang, X. F., and Sassoon, D. (1995). Restricted expression of type-II TGF beta receptor in murine embryonic development suggests a central role in tissue modeling and CNS patterning. *Mech. Dev.* **52,** 275–289.

Watanabe, Y., and Le Douarin, N. M. (1996). A role for BMP-4 in the development of subcutaneous cartilage. *Mech. Dev.* **57,** 69–78.

Weinberg, E. S., Allende, M. L., Kelly, C. S., Abdelhamid, A., Murakami, T., Andermann, P., Doerre, O. G., Grunwald, D. J., and Riggleman, B. (1996). Developmental regulation of zebrafish MyoD in wild-type, no tail and spadetail embryos. *Development* **122,** 271–280.

Weinmaster, G., Roberts, V. J., and Lemke, G. (1992). Notch2: A second mammalian Notch gene. *Development* **116,** 931–941.

Weinstein, D. C., Ruiz i Altaba, A., Chen, W. S., Hoodless, P., Prezioso, V. R., Jessell, T. M., and Darnell, J. J. (1994). The winged-helix transcription factor HNF-3 beta is required for notochord development in the mouse embryo. *Cell* **78,** 575–588.

Weintraub, H. (1993). The MyoD family and myogenesis: Redundancy, networks, and thresholds. *Cell* **75,** 1241–1244.

Weintraub, H., Davis, R., Tapscott, S., Thayer, M., Krause, M., Benezra, R., Blackwell, T. K., Turner, D., Rupp, R., Hollenberg, S., Zhuang, Y., and Lasser, A. (1991). The myod gene family: Nodal point during specification of the muscle cell lineage. *Science* **251,** 761–766.

Williams, R., Lendahl, U., and Lardelli, M. (1995). Complementary and combinatorial patterns of Notch gene family expression during early mouse development. *Mech. Dev.* **53,** 357–368.

Xue, X. J., and Xue, Z. G. (1996). Spatial and temporal effects of axial structures on myogenesis of developing somites. *Mech. Dev.* **60,** 73–82.

Yamaguchi, T. P., and Rossant, J. (1995). Fibroblast growth factors in mammalian development. *Curr. Opin. Genet. Dev.* **5,** 485–491.

Yamaguchi, T. P., Harpal, K., Henkemeyer, M., and Rossant, J. (1994). fgfr-1 is required for embryonic growth and mesodermal patterning during mouse gastrulation. *Genes Dev.* **8,** 3032–3044.

Yamashita, H., ten Dijke, P., Huylebroeck, D., Sampath, T. K., Andries, M., Smith, J. C., Heldin, C. H., and Miyazono, K. (1995). Osteogenic protein-1 binds to activin type II receptors and induces certain activin-like effects. *J. Cell Biol.* **130,** 217–226.

Yang-Snyder, J., Miller, J. R., Brown, J. D., Lai, C. J., and Moon, R. T. (1996). A frizzled homolog functions in a vertebrate Wnt signaling pathway. *Curr. Biol.* **6,** 1302–1306.

Yee, S. P., and Rigby, P. W. (1993). The regulation of myogenin gene expression during the embryonic development of the mouse. *Genes Dev.* **7,** 1277–1289.

Yoon, J. K., Olson, E. N., Arnold, H. H., and Wold, B. J. (1997). Different MRF4 knockout alleles differentially disrupt Myf-5 expression: cis-regulatory interactions at the MRF4/Myf-5 locus. *Dev. Biol.* **188,** 349–362.

Yoshikawa, Y., Fujimori, T., McMahon, A. P., and Takada, S. (1997). Evidence that absence of Wnt-3a signaling promotes neuralization instead of paraxial mesoderm development in the mouse. *Dev. Biol.* **183,** 234–242.

Yost, C., Torres, M., Miller, J. R., Huang, E., Kimelman, D., and Moon, R. T. (1996). The axis-inducing activity, stability, and subcellular distribution of beta-catenin is regulated in *Xenopus* embryos by glycogen synthase kinase 3. *Genes Dev.* **10,** 3639–3650.

Zentella, A., and Massague, J. (1992). Transforming growth factor beta induces myoblast differentiation in the presence of mitogens. *Proc. Natl. Acad. Sci. U.S.A.* **89,** 5176–5180.

Zimmerman, L. B., De Jesus-Escobar, J. M., and Harland, R. M. (1996). The Spemann organizer signal noggin binds and inactivates bone morphogenetic protein 4. *Cell* **86,** 599–606.

6
The Birth of Muscle Progenitor Cells in the Mouse: Spatiotemporal Considerations

Shahragim Tajbakhsh and Margaret Buckingham
CNRS URA 1947
Department of Molecular Biology
Pasteur Institute
25 rue du Dr. Roux
75724 Paris Cedex 15, France

I. Introduction
II. Origins of Skeletal Muscle in Vertebrates
III. Domains of the Dermomyotome
   A. Epaxial and Hypaxial Domains
   B. The Central Dermomyotome Domain
   C. Nonmuscle Dermomyotomal Derivatives
IV. Differences between Somites at Different Axial Levels
V. Myogenic Regulatory Factor Expression Patterns in the Somite and Myotome Heterogeneity
VI. Extrinsic Factors Govern Somite Differentiation and the Activation of Myf5 and MyoD
VII. Do Myf5 and MyoD Define Different Subpopulations of Muscle Cells?
VIII. The Roles of Myf5 and MyoD in Muscle Progenitor Cell Determination
   A. Myf5 Initiates the Early Mouse Myotome
   B. Aberrant Rib Formation in *Myf5* Null Mice
   C. Differences between Myogenic Regulatory Factors and Threshold Levels. Are All MRFs Functionally Alike?
   D. Myf5 and Pax3 Act Genetically Upstream of MyoD
IX. Conclusions
   References

I. Introduction

Since the identification of the MyoD family of myogenic regulatory factors (MRFs), MyoD, Myf5, myogenin, and MRF4 (Weintraub *et al.*, 1991), their critical role in the formation of the muscle precursor cell population as well as in the differentiation of these cells in the embryo was demonstrated by gene knockout experiments performed in the mouse at the beginning of the 1990s. Myf5 and MyoD were identified as "upstream" determination factors, whereas myogenin is required for most muscle cell differentiation (see Molkentin and Olson, 1996; Yun and Wold, 1996; Tajbakhsh and Cossu, 1997). While it was clear at the time that

Myf5 is activated before *MyoD* in the mouse embryo (see Sassoon *et al.*, 1989; Ott *et al.*, 1991; Buckingham, 1992), a finer appreciation of the spatial and temporal features which characterize the onset of myogenesis has only emerged more recently. This detailed anatomy, accompanied by the expression patterns of key genes in normal and mutant mice has led to changes in the conceptual framework which influences our current thinking about myogenesis. In particular the importance of the localization of progenitor cell populations in relation to signaling molecules produced by surrounding tissues has become a major issue. The role of other gene families in this context has already modified our view of the genetic hierarchy that regulates myogenesis. These considerations will be the focus of this chapter, which will deal primarily with myogenic progenitor cells in the mouse somite, with reference to other species where appropriate.

II. Origins of Skeletal Muscle in Vertebrates

Most of our understanding of the origins of various skeletal muscles in the body and head comes from experimental manipulations carried out in avians (see Wachtler and Christ, 1992; Christ and Ordahl, 1995). Chick–quail chimeras have served as the paradigm permitting the identification of tissues derived from the transplant after further development *in ovo*. For the most part, the origins of skeletal muscles and the migrations of their precursor cells to distal sites have been conserved between avians and mice (Milaire, 1976). Some difficulties arise, however, when comparing neck and craniofacial muscles due to the anatomical differences in these regions between these species.

Somites give rise to all skeletal muscles of the body and some head muscles, whereas the remainder of head muscles originate from paraxial head mesoderm (located anterior to the first formed somite), and most anteriorly from prechordal mesoderm (see Wachtler and Christ, 1992; Christ and Ordahl, 1995). To avoid confusion between paraxial mesoderm (giving rise to somites) and paraxial head mesoderm, we employ the term "presomitic mesoderm" (PSM; also referred to as unsegmented paraxial mesoderm in mice, or segmental plate in avians) as the paraxial mesoderm that gives rise to somites. To understand how early skeletal muscle develops both spatially and temporally, it is necessary to review how somites develop and differentiate. In the mouse, the first somite forms between embryonic day (E) 7.5 and 8. During gastrulation, future mesenchymal somite precursor cells feed into the caudal PSM as they move rostrally, they begin to compact to form the somite unit which exits the PSM as an epithelial sphere surrounding a mesenchymal core. Somites are produced continuously in pairs on either side of the neuraxis until midgestation following a rostrocaudal developmental gradient (see Christ and Ordahl, 1995; Gossler and Hrabe de Angelis, 1997).

During the course of somite maturation, at the epithelial somite stage, the ventral compartment (adjacent to the notochord) disperses first into mesenchymal sclerotomal cells, which will contribute to the cartilage and bone of the vertebrae and ribs, while the dorsal somite, underneath the surface ectoderm, retains a dermomyotome epithelium which later contributes to skeletal muscle, distal ribs, and dorsal dermis. This dermomyotome serves, in part, as a source of muscle progenitor cells (MPCs) over several days, both for the underlying myotome, the first skeletal muscle mass to form, and for migration to more distal muscle masses.

III. Domains of the Dermomyotome

A. Epaxial and Hypaxial Domains

As the somite differentiates into mesenchymal cells ventrally, the dermomyotomal epithelium initially extends between the prospective mediolateral and dorsoventral body axis. This epithelium then elongates as the embryo grows and later becomes polarized along the dorsoventral body axis (Figs. 1 and 2; see also Tajbakhsh and Cossu, 1997). During the course of this positional shift, myotomal progenitor cells leave the dermomyotome, form a compartment of cells which are mainly postmitotic, and differentiate into skeletal muscle directly underneath this epithelium. A basal lamina separates the myotomal and sclerotomal compartments. Separate epaxial (dorsal) and hypaxial (ventral) myotomes are generated, respectively, primarily from the extremities of this epithelium. The dermomyotome edge adjacent to the neural tube is characterized by an epithelium folding onto itself, referred to as the epaxial dermomyotome lip. The dermomyotome adjacent to the lateral plate mesoderm is characterised by a crescent shaped epithelium, referred to as the hypaxial somitic bud (see Fig. 2). As the epaxial and hypaxial myotomes develop, they merge to form the continuous myotome layer (Figs. 1, 2, and 3A). They themselves can then no longer be distinguished as separate domains either morphologically or with currently available molecular markers; they can be distinguished by their juxtaposition to surrounding structures. Muscles derived from the epaxial myotomes are innervated by the dorsal rami of the spinal nerves, whereas the derivatives of the hypaxial myotomes are innervated by the ventral rami of the spinal nerves, thus further distinguishing these muscle domains (see Ordahl, 1993). At early stages of development (prior to E12.5) the term "hypaxial" (underneath the axis—notochord) is used for both the ventral myotome, which is the muscle mass forming underneath the dermomyotome epithelium, and hypaxial muscles located either adjacent to the somite (e.g., shoulder muscles or body wall muscles) or at more distant sites such as those in the limb (appendicular muscles). The latter groups are distinct anatomically and appear to form independently of the hypaxial myotome, although their progenitor cells may be responding to similar signaling pathways. In some cases, such as for limb muscles, the requirements for certain myogenic factors may be different from myotomal muscle (see Section VIII, C).

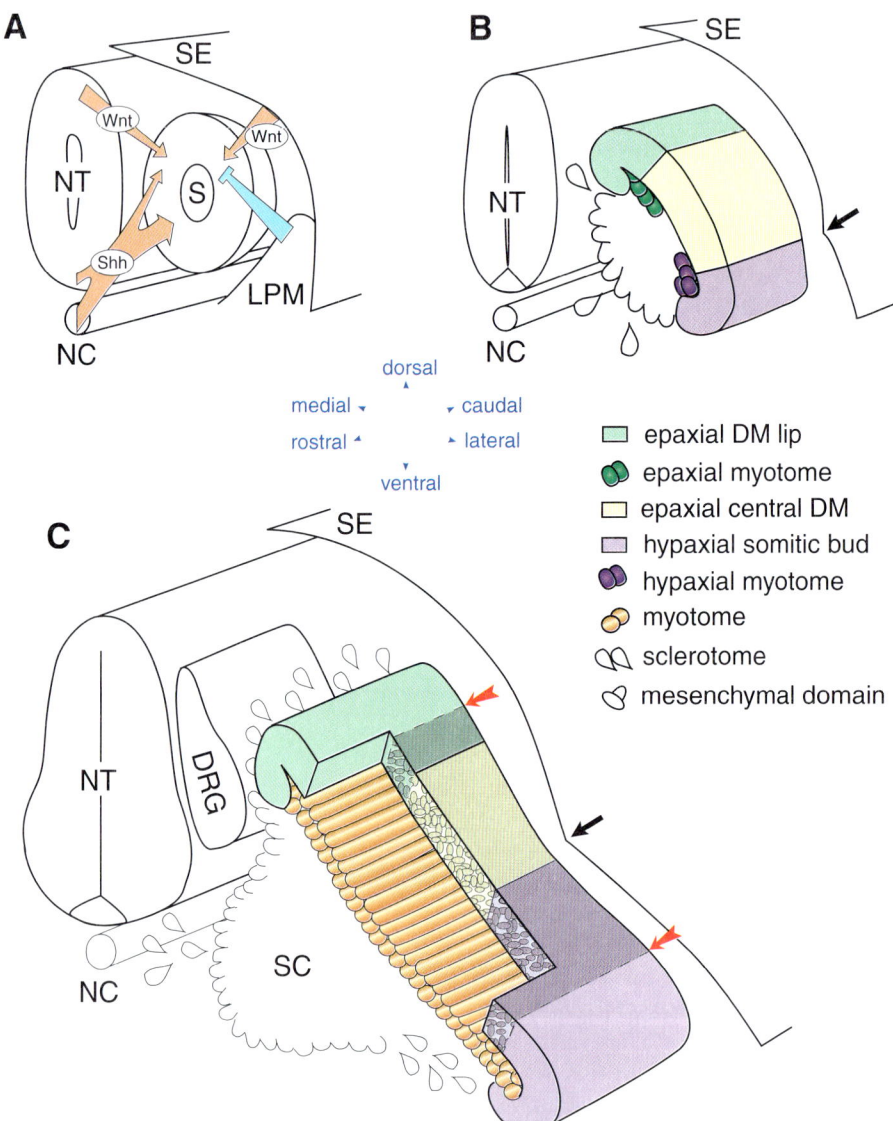

Figure 1. Schema of a transverse section through a maturing interlimb (thoracic) mouse somite. (A) In an epithelial somite (S), myogenesis initiates along the medial aspect of the dorsomedial quarter of the dermomyotome, in response to positive signals (arrows) from axial structures (neural tube and notochord) such as Shh and Wnts and negative signals (bar) from the lateral plate mesoderm. Muscle progenitor cells in the epaxial domain will later form epaxial (deep back) muscles. (B) The ventral somite is composed of mesenchymal sclerotomal cells, and a dermomyotome epithelium is retained dorsally, from which an epaxial (dorsal) and hypaxial (ventral) myotome are established. These can be dis-

6. Muscle Progenitor Cells: Spatiotemporal Considerations   229

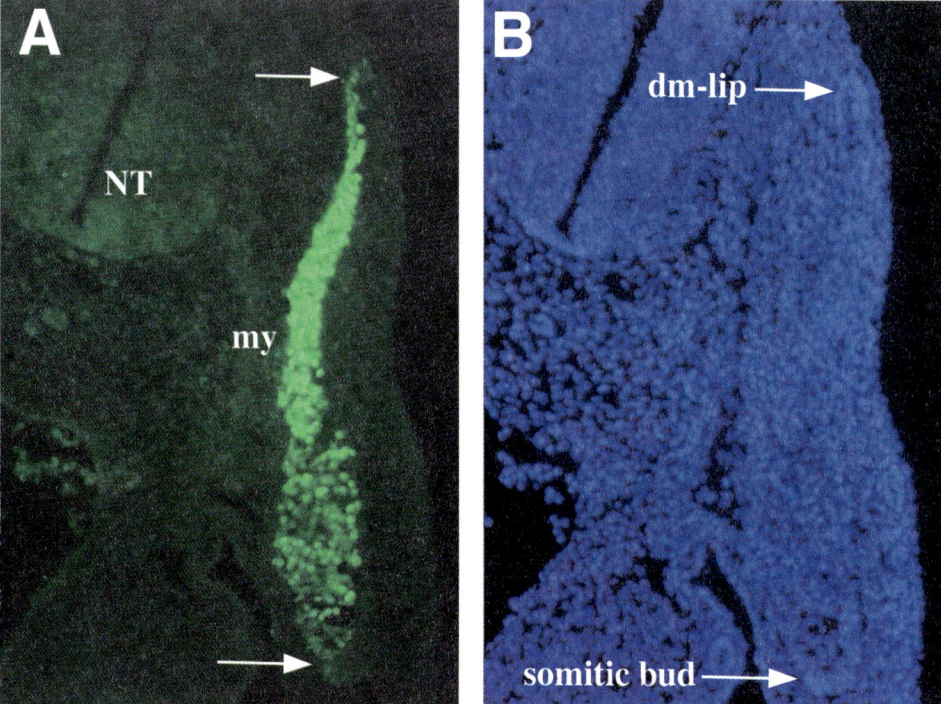

Figure 2. The epaxial and hypaxial dermomyotome lips persist as epithelial structures until later stages as the central dermomyotome becomes mesenchymal. (A) An E11 interlimb somite from a *Myf5-nlacZ* embryo was stained with an anti-β-galactosidase polyclonal antibody to reveal *Myf5* expressing cells in the myotome (my). (B) Hoechst staining of the section in A indicating the epaxial dermomyotome lip (upper arrow; dm-lip) and the hypaxial somitic bud (lower arrow) as epithelial structures while the central dermomyotome has undergone an epithelial–mesenchymal transition (see Fig. 1).

---

tinguished as independent domains only transiently with molecular markers. (C) The epaxial and hypaxial myotomes then become continuous and can no longer be distinguished by conventional muscle markers by E10.5. A basal lamina (not shown) separates the myotome from the sclerotome. As the embryo grows, the dermomyotome rotates into a dorsoventral position by E11; at this stage it has an epithelial medial and lateral edge, and a mesenchymal center (see Fig. 2). The lateral quarter of the dermomyotome curves ventrally to form the somitic bud (whereas the medial folds onto itself); it produces muscle precursors as it penetrates the flank during embryonic growth. Dermis overlying the back and somite originates from the somite. The epaxial and hypaxial dermomyotomal extremities remain epithelial longest, while the central dermomyotome becomes mesenchymal (delimited by red arrows). Black arrow indicates indentation in the body wall, a morphological landmark. DRG, dorsal root ganglion; LPM, lateral plate mesoderm (lateral + intermediate mesoderm); NC, notochord; NT, neural tube; S, somite; SE, surface ectoderm. (Modified from Tajbakhsh and Spörle, 1998.)

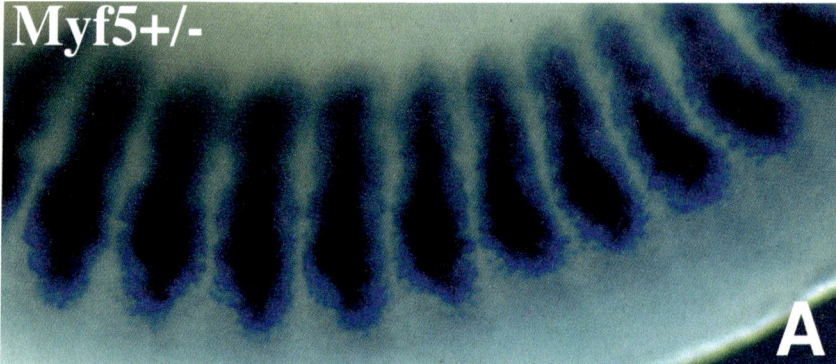
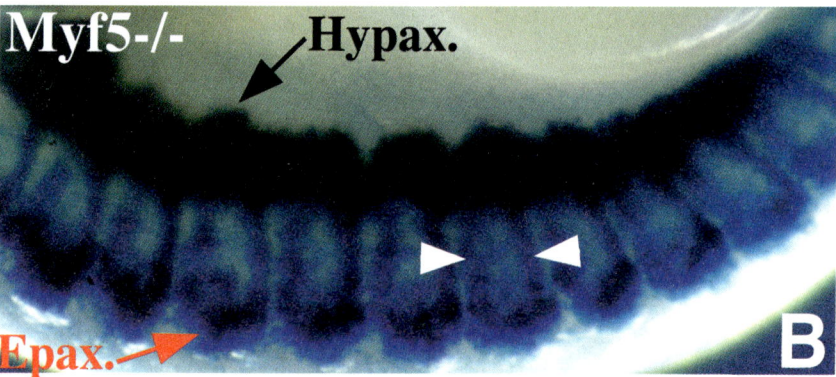

Figure 3. Tracking muscle progenitor cells in *Myf5-nlacZ* embryos. Interlimb somites of E10.5 embryos are shown. (A) Heterozygous *Myf5-nlacZ* embryo showing a developing myotome as marked by *Myf5* expression. (B) Homozygous *Myf5-nlacZ* embryo showing myogenic progenitor cells accumulated along the dorsal (epaxial), ventral (hypaxial), rostral (left arrowhead), and caudal (right arrowhead) edges of the somite. The majority of these progenitors are located along the dorsal and ventral edges suggesting that the bulk of muscle progenitors are born along these edges of the dermomyotome.

The use of the term hypaxial points to their site of origin within the dermomyotome. Indeed, muscles derived from the epaxial and hypaxial myotomes originate from different subpopulations of progenitor cells located within the medial and lateral epithelial somite, respectively (Ordahl and Le Douarin, 1992).

The migratory routes of muscle progenitor cells from the dermomyotome epithelium to form the myotome have been described extensively and different models have been proposed; at the present time, this issue remains unresolved. Williams (1910) suggested that MPCs originate along the epaxial (dorsomedial) dermomyotome lip and migrate from this site, under the lip, to form the myotome. By contrast, also based on morphological criteria, and using desmin as a marker, Kaehn and colleagues (Kaehn *et al.*, 1988) proposed that MPCs do not originate from the epaxial dermomyotome lip but, rather, that this migration is initiated at the rostromedial (craniomedial) corner of the dermomyotome. Consistent with this observation, Myf5 and myogenin protein accumulation were first detected in the rostral (anterior) myotomal compartment (Smith *et al.*, 1994). *In situ* hybridization experiments subsequently pointed to an initial accumulation of Myf5 (and in avians MyoD) transcripts along the medial edge of the somite, and this is particularly evident in *Myf5-nlacZ* mutant mice with an *nlacZ* reporter gene targeted into the *Myf5* locus (Tajbakhsh and Buckingham, 1994; Cossu *et al.*, 1996a). This would not be expected if all MPCs originated exclusively from the rostral (cranial) edge of the somite. Based on these morphological observations using *Myf5-nlacZ* heterozygous embryos, Cossu and colleagues (Cossu *et al.*, 1996a) proposed that, at least in the mouse, the majority of epaxial muscle precursors are born along the medial edge of the dermomyotome lip. At early stages, however, greater numbers of β-galactosidase positive cells do accumulate along the rostral compared to the caudal edge. *Myf5-nlacZ* homozygous embryos have also proved to be informative. In these mutants, MPCs undergo an epithelial to mesenchymal transition and leave the dermomyotome, but they remain developmentally arrested, not entering the myogenic program until the activation of *MyoD* at later stages. This "freeze-frame" image of blocked MPCs (Fig. 3B) confirmed the observations made with the heterozygous mutant; a substantial number of MPCs accumulate along the entire medial dermomyotome edge by E10.5–E11.5. Similarly, in these *Myf5-nlacZ* null embryos, hypaxial myotomal precursors accumulate along the entire edge of the hypaxial somitic bud (Fig. 3B), consistent with observations made with *Myf5-nlacZ* heterogyzous embryos, and with *in situ* studies, that MPCs activate *Myf5* independently in this domain of the somite (see Section V). In addition to the medial and lateral edges of the dermomyotome, a minor contribution from the rostral and caudal edges of the somite is evidenced by the accumulation of MPCs along both of these edges of the dermomyotome in the *Myf5-nlacZ* homozygous embryo (Fig. 3B). Therefore, these observations suggest that the bulk of MPCs are born along the medial and lateral dermomyotome edges. Prior to somite formation, *Myf5* expression is detected in the PSM of *Myf5-nlacZ* heterozygote embryos (see Cossu *et al.*, 1996a), and Myf5 transcripts have been detected by RT–PCR in the PSM (Kopan *et al.*, 1994). However, it is not yet clear if this expression occurs within cells that will later express *Myf5* in the somite, thus marking future muscle progenitors, or if it is functionally significant.

An elegant series of lineage tracing experiments in the chick, using DiI labeled cells, demonstrated that the medial lip is indeed the source of the majority of muscle precursors of the epaxial myotome (Denetclaw *et al.*, 1997). Their subsequent migratory route prior to their entry into the myotome, however, remains a controversial issue (Denetclaw *et al.*, 1997; Kahane *et al.*, 1998a). Studies by Kahane and colleagues (Kahane *et al.*, 1998a), also using DiI as a cell lineage marker, suggested that although MPCs arise along the medial edge of the dermomyotome, they enter the myotome predominantly from the rostral and caudal edges. Indeed, experiments by K. Tosney have shown that all dermomyotome edges, even those experimentally induced via an incision within the dermomyotome sheet, will produce myocytes suggesting that the entire dermomyotome epithelium is potentially a source of myogenic progenitor cells (see Tajbakhsh and Spörle, 1998). Both rostral and caudal dermomyotome lips have been reported to be positive for MRF gene expression (Smith *et al.*, 1994). More experiments are necessary to resolve the extent of MPC contribution from these different regions, and their subsequent migratory routes. Within the myotome itself, individual myocytes span the width of the somite (Fig. 1C), and how this orientation is achieved is also a consideration. Migration from the rostral (see Kaehn *et al.*, 1988) or caudal edges, for example, achieves this without reorientation of the cells, however, it is not clear if this is a critical requirement for constructing the myotome.

B. The Central Dermomyotome Domain

Recently, a central dermomyotome moiety was identified as a third territory, potentially contributing to the myotome. It can be identified by markers such as *En1* and *Sim1* (vertebrate homologues of *Drosophila engrailed* and *single minded* genes, respectively) during embryonic development as the epaxial and hypaxial myotomes are being established (see Tajbakhsh and Spörle, 1998). Sim1 transcripts have been described as a marker of the lateral somite in the chick at early stages (Pourquié *et al.*, 1996). At later stages in the mouse, however, *Sim1* expression is strongest in the central dermomyotome territory which is intercalated between the epaxialmost and hypaxial dermomyotomes. This *Sim1* expression appears adjacent to the lateral plate mesoderm at early stages (e.g., E10.5) in the occipital, cervical, and hind limb level somites where the hypaxial myotome is not prominent or not yet detectable. In the interlimb somites, however, it bisects the somite along the dorsoventral body axis revealing a prominent hypaxial (ventral) myotome. Within this intercalated dermomyotome territory, *En1* is expressed more dorsally, and *Sim1* more ventrally along the dorsoventral body axis. The intercalated domain defined by these markers is localized primarily to the ventral moiety of the epaxial dermomyotome, which is delimited by the morphological indentation of the body wall (R. Spörle, personal communication) (see Fig. 1). Indeed, it is interesting if *En1* expression marks this epaxial boundary, given its role in the establishment of boundaries in *Drosophila* and vertebrates (see Siegfried and Perrimon, 1994; Wurst *et al.*, 1994). This centrally located intercalated domain is also revealed by *Pax3* expression in *Myf5* null mutants (compare Figs. 3B and 4B). *Pax3*, which encodes a paired/homeodomain transcription factor of the *Pax* gene family (see Tremblay and Gruss, 1994), is first expressed in the presomitic mesoderm. As somites form, this expression covers the entire somite and subsequently becomes confined to the dermomyotome as the sclerotome forms. Pax3 transcripts are present at low levels throughout the dermomyotome, but are more concentrated in the epaxial and caudal dermomyotome lips, and are most abundant in the

6. Muscle Progenitor Cells: Spatiotemporal Considerations 233

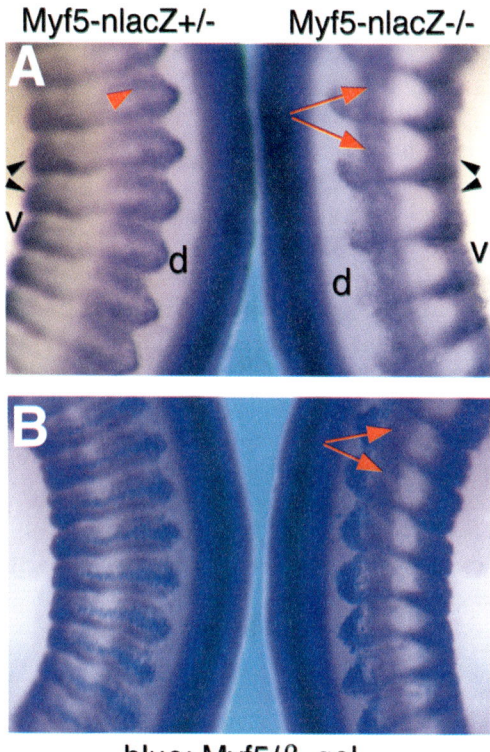

Figure 4. *Pax3* and *Myf5* expression reveal compartments within the dorsal somite. (A, B) Whole mount *in situ* hybridizations using an antisense Pax3 probe (purple) on E11.5 unstained (A) and briefly X-gal stained (B, blue) *Myf5-nlacZ* +/− and *Myf5-nlacZ* −/− embryos (interlimb region shown). Ventral (v) staining of Pax-3 corresponds to the epithelial somitic bud in the distalmost dermomyotome in the heterozygote (double arrowheads), while dorsal (d) staining in the cytoplasm of myotomal cells (red arrowhead) surrounds the β-galactosidase$^+$ nuclei. Note that *Pax3* expression in the *Myf5* null reveals an intercalated compartment in the somite (red arrows). In the ventral dermomyotome of the *Myf5* null embryo, *Myf5* and *Pax3* are coexpressed. A continuous intense staining in the neural tube can also be observed with the Pax3 probe (Anterior is to the top). (From Tajbakhsh *et al.*, 1997).

hypaxial somitic bud. As MPCs leave the dermomyotome to form the myotome, *Pax3* expression is rapidly downregulated. A reactivation of *Pax3* then occurs in the mature myotome. This later activation can be observed in normal *Myf5-nlacZ* +/− embyros (Fig. 4A; red arrowhead). In *Myf5-nlacZ* null embryos, the early myotome does not form, the epaxial and hypaxial myotomal precursors are displaced and myotomal *Pax3* expression is missing (see Section VIII,A). The remaining *Pax3* expression in the center of the somite clearly marks the intercalated domain of the dermomyotome (Fig. 4; red arrows).

Evidence for a muscle cell population which is distinct from that of the epaxial and hypaxial myotomal cells comes from studies with the zebrafish (see Kimmel et al., 1995). In fish and amphibians, unlike amniotes (birds and mammals), the bulk of the early somite is composed of skeletal muscle, and this compartment differentiates first, even before segmentation. So-called adaxial muscle cells, expressing *MyoD* (Weinberg et al., 1996) and marked by slow myosin expression, form within the paraxial mesoderm directly adjacent to the notochord even before somitogenesis (Devoto et al., 1996). These large cuboidal-shaped cells subsequently migrate radially from a medial position adjacent to the notochord, across the surface of the myotome to occupy a more lateral position. A subset of adaxial cells, misleadingly named muscle pioneers since they are not equivalent to those described in the grasshopper (Ho et al., 1983), are marked by the expression of *engrailed* (Hatta et al., 1991) and remain more central, later localizing to the myoseptum. The epaxial and hypaxial myotomes which comprise the bulk of skeletal muscle in the fish, are derived from the somites and form later. These muscle groups can be distinguished from the adaxial cells, which they replace medially, by antibodies to fast myosin isoforms (Devoto et al., 1996; Blagden et al., 1997; Du et al., 1997). The temporal pattern of myotome development between fish and amniote vertebrates is therefore clearly different, and it is not clear to what extent equivalent MPCs within these different subdomains of the somite have been conserved.

In amniotes, an added complexity in somite maturation and differentiation is the regionalized de-epithelialization of the central dermomyotome, from about E11 in the mouse. This region includes the ventral moiety of the epaxial domain and the dorsal moiety of the hypaxial domain and encompasses the intercalated $En1^+$ domain (Fig. 1C). As the central dermomyotome becomes mesenchymal, the epaxial dermomyotome lip and the hypaxial somitic bud remain as epithelial structures, continuing to produce MPCs for at least another 12 hr. Although it remains to be proved that the intercalated $En1^+$ domain in the amniote somite will contribute to the myotome, this has been suggested to be the case when the central dermomyotome becomes mesenchymal (Fig. 1) (see Tajbakhsh and Spörle, 1998). Indeed, $En1^+$ myocytes have been detected in the myotome (Gardner and Barald, 1992) suggesting that the intercalated dermomyotome moiety, defined by *En1* and *Sim1* expression, may give rise to myocytes independently of the epaxial and hypaxial myotomes and thereby contribute in part to the forming myotome. Experiments carried out with *Pax3* (*splotch*) mutant mice suggest that Pax3 is not necessary for *Sim1* expression in the intercalated cell population (S. Tajbakhsh and M. Buckingham, unpublished).

C. Nonmuscle Dermomyotomal Derivatives

Other cell derivatives of the dermomyotome include the dermis of the back, connective tissue, and blood vessels. Experiments in avians have suggested that

6. Muscle Progenitor Cells: Spatiotemporal Considerations 235

somite-derived connective tissue patterns the myotomal muscle. This contrasts with the limb where connective tissue derived from the lateral plate mesoderm patterns the forming muscle masses which are of somitic origin (see Christ and Ordahl, 1995). Differentiation of the dermis (Brill *et al.*, 1995; Kanzler *et al.*, 1997), like that of the connective tissue and endothelial cells (see Christ and Ordahl, 1995), has received less attention due partly to a lack of specific early markers for these different cell types. Although it is well established that the dermomyotome gives rise to these different cell types, an important recent finding has forced a reevaluation of the origins of certain cartilage(s) and bones. It has been generally accepted that ribs arise from the sclerotome and that the dermomyotome does not contribute to their formation. However, by performing quail–chick dermomyotome grafts, Kato and Aoyama (1998) showed that the distal and sternal parts of the ribs may indeed have a dermomyotomal origin. The precise location of these nonmuscle progenitor cells within the dermomyotome and the time they leave it remain to be defined.

The dermomyotome is a site of epithelial to mesenchymal transitions, giving rise to multiple cell types as the somite differentiates. A major mechanism for generating different cell types from pluripotential progenitors involves asymmetric cell division, first described in *Drosophila* neurogenesis, through the action of genes such as *inscuteable* and *numb* (see Jan and Jan, 1998). These findings may be relevant to mammalian myogenesis since recent studies indicate that this asymmetric apparatus also acts in muscle progenitor cells in *Drosophila* (Ruiz Gomez and Bate, 1997; Carmena *et al.*, 1998; Lu *et al.*, 1998).

IV. Differences between Somites at Different Axial Levels

Somites at different levels along the rostrocaudal body axis have different morphological characteristics and this has resulted in some confusion in the interpretation of gene expression data, in particular with the myogenic regulatory factors (see Section V). To identify somites located at different axial levels, two numbering schemes are used: the classic scheme where, for example, in a 15-somite embryo, the most mature (anterior) somite is indicated as somite 1 (S-1) whereas the newly formed (posterior) somite is referred to as S 15, and a more recently introduced scheme (Ordahl, 1993) in which somite I (S-I) refers to the posterior-most newly formed somite and S-XV to the oldest somite. Since at least at later stages, the first one or two somites are not readily detectable with markers, the latter scheme presents some advantages. Another somite identification scheme based on the morphological characteristics of both the spinal ganglia and somite (derivatives) permits accurate individual identification of somites at any embryonic stage and level along the rostrocaudal body axis (Spörle and Schughart, 1997).

Since the first one or two occipital somites (S-1 and S-2) rapidly dissociate, contributing to head muscles, they are difficult to distinguish as segmented structures by conventional muscle markers at later stages ($>$ E9.5). Myotomes in the remaining occipital somites can be readily distinguished although the myotome of the third somite (S-3) is considerably reduced in size. Cervical somites extend to the posterior forelimb level. In occipital, cervical, and some lumbal somites (opposite the hind limb), the hypaxial myotome is not as readily distinguishable as it is in interlimb somites. Furthermore, occipital and cervical somites contribute to the mesenchymal hypoglossal chord which gives rise to hypaxial muscles of the tongue, pharynx, and shoulder. This mesenchymal chord develops early, and is located adjacent to the dermomyotome in the lateral hypaxial body wall (see Franz *et al.*, 1993). In contrast, it is in somites at the level of the posterior forelimb and interlimb region that the hypaxial myotome can be clearly observed as it grows under this dermomyotome epithelium (see Franz *et al.*, 1993; Smith *et al.*, 1994; Spörle *et al.*, 1996; Tajbakhsh *et al.*, 1996a; Tajbakhsh and Cossu, 1997). The morphological distinction between epaxial and hypaxial myotomes can be made only at early developmental stages (prior to E10.5) and with a limited number of markers (e.g., Myf5 and MyoD). *En1* and *Sim1* expression patterns in the somite also demonstrate that the hypaxial myotome is more readily distinguishable at interlimb levels (R. Spörle, personal communication).

Most strikingly, somites opposite the limbs differ from interlimb level somites in that individual mesenchymal cells which express genes such as *Pax3, c-met, Mox2,* and *Lbx1,* leave the lateral dermomyotome and migrate into the limb field (Bober *et al.*, 1994; Goulding *et al.*, 1994; Williams and Ordahl, 1994; Jagla *et al.*, 1995; Yang *et al.*, 1996; Mankoo *et al.*, 1999) (see Section VIII,D). These muscle progenitor cells have similar characteristics to those which form muscles of the hypoglossal chord and diaphragm, for example, in that they require Pax3 and c-met for migration to their sites of permanent residence. The lateral dermomyotome of somites at limb levels is reduced in comparison to that of interlimb somites; somite transplant studies have suggested that this lateral compartment undergoes apoptosis induced by the limb bud environment (Tosney, 1994).

Caudal to the hind limb, somites continue to form until E16 in the mouse, and these have been less extensively studied. This is due, in part, to the fact that somite maturation proceeds more rapidly at later developmental stages, thus rendering their analysis more difficult.

Evidence for the differential programming of muscles derived from somites at different axial levels was provided recently by the finding that in *Pax3(splotch)/Myf5* double mutant embryos, muscle masses do not form in the trunk, tail, and limbs, whereas muscles derived from the anterior-most occipital somites are present and contribute to head muscles (see Tajbakhsh *et al.*, 1997) (see Section VIII,D). Although these somites, unlike the head paraxial and prechordal mesoderm, do express *Pax3,* the specification and subsequent determination of myogenic progenitor cells located here can take place in the absence of *Pax3* and *Myf5.* By contrast, MPCs born in somites located in the trunk do require Pax3 and Myf5 to form muscle. Therefore, it appears that the anterior-most occipital somites, located in the head environment, are programmed differently from other somites.

As a result of experimental manipulation with avian embryos, it was initially thought that when early undifferentiated somites/presomitic mesoderm are transplanted to different axial positions, they would then acquire the appropriate characteristics. For example, when PSM corresponding to non-limb level somites is transplanted into an embryo at a position that will yield limb level somites, the resulting somites in their ectopic position will give rise to muscle progenitor cells which will migrate into the limb bud (see Christ and Ordahl, 1995). More recently, transplantation of trunk mesoderm into the paraxial head mesoderm region resulted in the downregulation of *Pax3* (Hacker and Guthrie, 1998) thus underscoring the importance of environmental influences. In spite of these findings, it is also clear that positional information exists from an early stage, since transplantation of thoracic level PSM to the cervical region results in rib formation in this ectopic location (Kieny *et al.*, 1972). Interestingly, dermal derivatives also show early positional identity. When cervical PSM is grafted to the thoracic/lumbar level in chick embryos, cervical-type plummage develops (Mauger, 1972). Additional evidence comes from experiments where the rostrocaudal inversion of the PSM results in a reversed polarity of neural crest migration, suggesting that the rostrocaudal somite axis is already established in the presomitic mesoderm (see Keynes and Stern, 1988). It remains unclear if MPCs in the somite already have positional identity, due to the lack of specific markers defining muscles at different axial levels. A transgenic mouse line, however, provided evidence of intrinsic differences between muscle cells from different somites. The myosin light chain 1F CAT transgenic line showed higher expression levels in the myotomes of more posterior somites in the thoracic region (Donoghue *et al.*, 1992; Grieshammer *et al.*, 1992), and this was maintained in clonal cultures of these cells (Donoghue *et al.*, 1992). In another study, the transplantation of muscle cells from the masseter in the cat, to the hind limb, resulted in the formation of muscle fibers which expressed masseter specific myosin at this ectopic position instead of the fast myosin isoforms characteristic of limb muscles (Hoh and Hughes, 1988). These experiments point to the acquisition of intrinsic positional information by muscle precursor cells at different axial levels. The time when this identity is established and how this is effected are important considerations. Hox genes known to have characteristic expression patterns in mesoderm at different anterior/posterior levels are obvious candidate effectors (see Krumlauf, 1994), and indeed Hox genes have been implicated in conferring positional identity to skin/feather dermal precursors (Kanzler *et al.*, 1997). However, whereas alterations in the Hox code clearly affect the positional identity of motorneurons (Tiret *et al.*, 1998), this remains to be shown for skeletal muscle.

V. Myogenic Regulatory Factor Expression Patterns in the Somite and Myotome Heterogeneity

Considerable information has accumulated over the last decade on the role of MRFs as transcriptional activators of muscle gene regulatory elements in cell lines, either alone or in cooperation with positive and negative effector molecules (see Olson et al., 1995; Molkentin and Olson, 1996; Yun and Wold, 1996). A few studies in the embryo have also pointed to the importance of particular sequence motifs (e.g., CANNTG) for reporter gene activation, probably via the action of the MRFs (Cheng et al., 1993; Yee and Rigby, 1993). Much has been learned about MRF function in activating genes in differentiating muscle cells, however, information is lacking for direct downstream targets of Myf5 and MyoD in precursor myoblasts. Consequently, limited domain specific markers are available for the investigation of early myotome formation in the embryo and indeed, the molecular mechanisms by which Myf5 and MyoD confer this early myogenic identity, prior to *myogenin* and *MRF4* activation, remain unclear.

The observation that skeletal muscles and myoblasts are absent in *MyoD/Myf5* double null embryos (Rudnicki et al., 1993) demonstrated that there is an absolute requirement for Myf5 or MyoD to initiate epaxial and hypaxial myotome formation. Therefore, to understand the spatiotemporal complexities of myotome formation it is essential to assess precisely when and where *Myf5* and *MyoD* are expressed in somites.

In the chick (see Hacker and Guthrie, 1998), as in the mouse (Ott et al., 1991; see Buckingham, 1992), *Myf5* is expressed first in the embryo and *MyoD* expression generally follows within a few hours. In most somites of the mouse embryo, however, *MyoD* is expressed about 2 days after *Myf5* (Sassoon et al., 1989; see Buckingham, 1992). Somitic expression of *Myf5* first occurs in a few cells in the medial epithelial somite adjacent to the neural tube and subsequently *Myf5* is expressed in the epaxial dermomyotome lip as the somite matures. These latter Myf5[+] cells will establish the epaxial myotome, first in the occipital somites and then following the rostrocaudal developmental gradient, progressively in more caudal somites. Therefore, early *Myf5* expression initiates and programs the epaxial myotome in the mouse (see Section VIII,A). *MyoD* expression is first detectable, with sensitive *in situ* hybridization techniques, on whole embryos in the epaxial myotome at E9.75 (about 27 somites) and progresses in a rostrocaudal gradient, to become clearly evident in most somites by E10.5 when the myotome is already well established (Sassoon et al., 1989; Buckingham, 1992; S. Tajbakhsh and M. Buckingham, unpublished).

A second and distinct *MyoD* expression is detected in the hypaxial myotome in somites opposite the posterior forelimb bud, at E9.75 (about 27 somites; Goldhamer et al., 1992; Faerman et al., 1995; S. Tajbakhsh and M. Buckingham, unpublished), although a few dispersed MyoD[+] cells have been detected on sections as early as 20 somites in outbred CD1 mouse embryos (Smith et al., 1994). The main wave of hypaxial *MyoD* expression occurs simultaneously, or just slightly after the epaxial myotome expression is detected in the rostral (occipital) somites. This distinct site of expression also progresses in a rostrocaudal gradient as clearly observed in the hypaxial myotomes of the interlimb somites. *Myf5* is also activated early in this domain (Spörle et al., 1996; Tajbakhsh et al., 1996a), in fact before *MyoD* (see below), however its expression had tended to be missed probably because, very rapidly (within 12 hr), MyoD transcript and protein accumulation in the hypaxial myotome predominates. The accumulation of Myf5 transcripts (Spörle et al., 1996; S. Tajbakhsh and M. Buckingham, unpublished), and of β-galactosidase positive MPCs in the hypaxial somitic bud of *Myf5-nlacZ* heterozygous, and notably homozygous embryos (Tajbakhsh et al., 1996a), clearly points to *Myf5* expression in the forming hypaxial myotome. Additional evidence that *Myf5* expression precedes that of *MyoD* in the hypaxial as well as the epaxial myotome comes from examination of various mutant mice. An examination of *Myf5-nlacZ* null embryos has demonstrated that myotome formation does not occur at the time when *MyoD* is normally activated (E9.75–E10.5). Notably, a delay in both epaxial and hypaxial myotome formation (about 1 day) corresponds to a delay in the activation of *MyoD* in *Myf5-nlacZ* null embryos (Tajbakhsh et al., 1997). This observation was more unexpected for hypaxial myogenesis since, as discussed above, *MyoD* expression was thought to precede that of *Myf5* (see Cossu et al., 1996b). These findings strongly implicate Myf5 in the initial activation of *MyoD* in the hypaxial myotome. The delay in *MyoD* activation can be fully appreciated by examining the expression patterns of these genes by whole mount *in situ* hybridization. Indeed, previous studies (Braun et al., 1994) relied on *in situ* hybridization on sections, and this delay in *MyoD* activation was missed. At other sites in the mouse embryo such as the limbs and head, *MyoD* expression follows that of *Myf5* within a few hours. One exception, however, is a group of shoulder muscles which, curiously, activate *MyoD* first, and these muscles do not show the perturbations that are observed later in the remainder of the trunk muscles in *Myf5-nlacZ* null embryos (Fig. 5). These caveats have to be taken into consideration when examining somites located at different axial levels, and arise because of the anatomical complexity of muscle development.

Heterogeneity in the onset of *MRF* gene expression within the myotome is also seen for *Myogenin* and *MRF4*. Their expression patterns follow that of *Myf5;* Myogenin transcripts are first detected at E8.5 and MRF4 mRNA transitorily from E9 to E11.5 in the myotome; a second major wave of *MRF4* expression occurs during fetal stages (Sassoon et al., 1989; Bober et al., 1991; Hinterberger et al., 1991). The localization of transcripts coding for differentiation markers such as myosins and actins to the central part of the myotome when these genes are first expressed (Lyons et al., 1990, 1991) indicates that these cells are more mature. This pattern of expression is also seen with transgenic mice (Kelly et al., 1995; Biben et al., 1996). Indeed, Myogenin transcript accumulation is most prominent in a central subset of myotomal cells. Furthermore, it is in this central domain extending along the dorsoventral body axis (oriented from the neural tube to the lateral plate mesoderm) of the maturing myotome where *Myf5* transcription is downregulated (S. Tajbakhsh and M. Buckingham, unpublished; R. Sporle, personal communication). In *Myf5-nlacZ* embryos at the same stage, however, the β-galactosidase staining persists due to the greater stability of this bacterial enzyme (Fig. 3A; Tajbakhsh et al., 1996a, b).

In looking at gene expression, the sensitivity of detection obviously influences

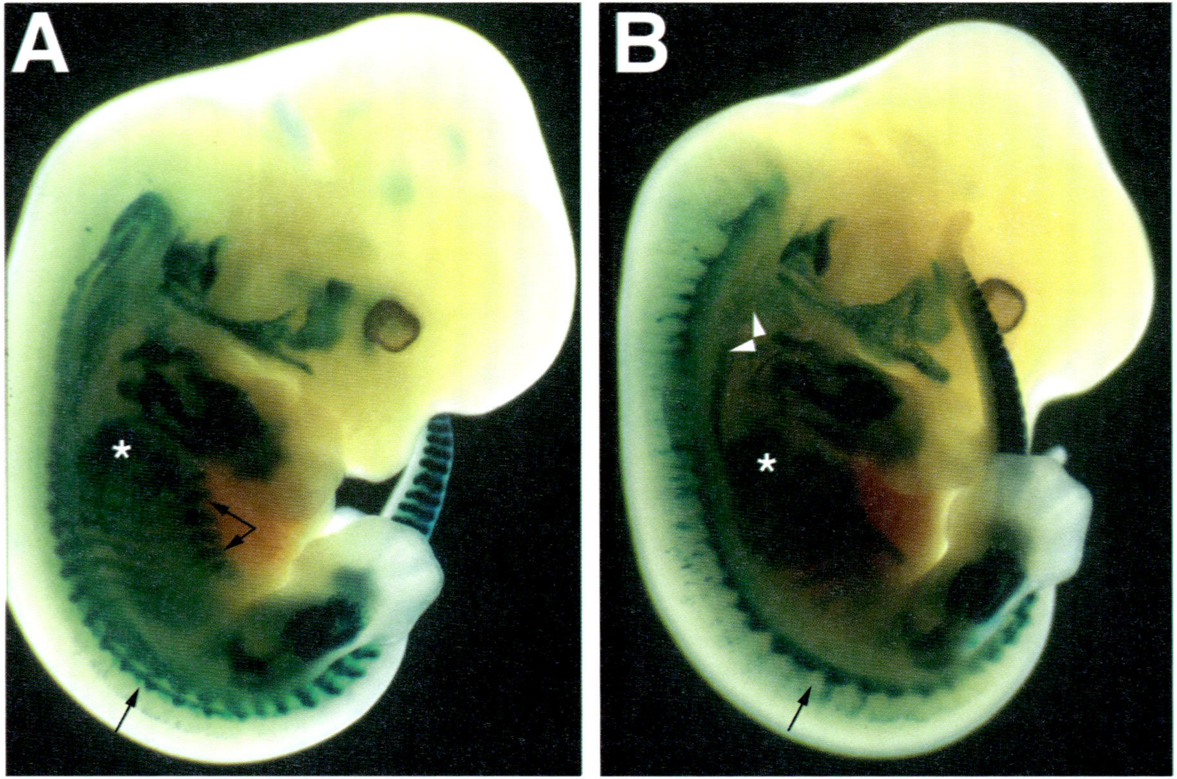

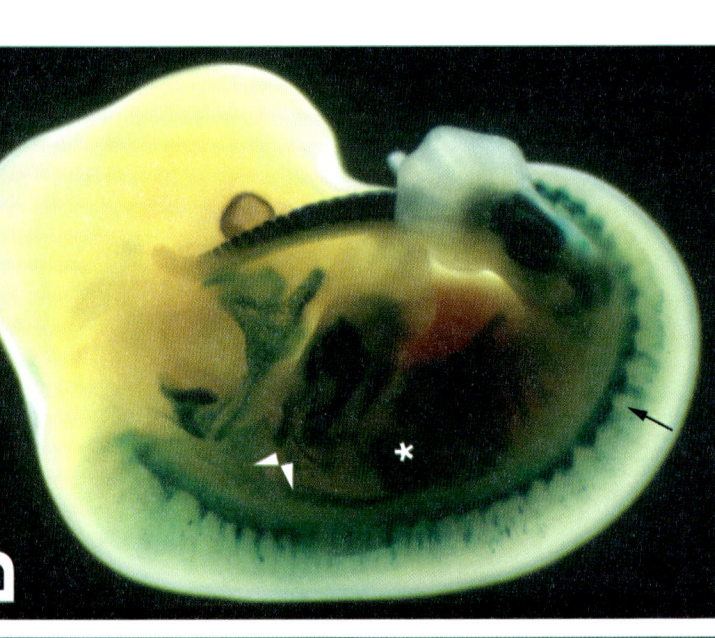

Figure 5. Some skeletal muscles escape perturbations in the trunk of *Myf5-nlacZ* null embryos. E12.5 embryos were analyzed by X-gal staining *in toto*. (A) *Myf5-nlacZ+/−* embryo. The forming M. latissimus dorsi (asterisk), deep back (arrow) and intercostal (double arrows) muscles are indicated. (B) *Myf5-nlacZ−/−* embryo showing deep back (arrow) and shoulder muscle perturbations. Note that a group of shoulder muscles opposite and anterior to the forelimb (arrowheads), which activates *MyoD* relatively early, does not show the pertur-

the result, as illustrated for *Myf5* expression in the hypaxial myotome. Some of the heterogeneity reported for proteins in the myotome may also reflect this limitation. Myosin heavy chain positive myotomal cells, which were reported to be negative with MRF antibodies when cultured *in vitro* (Smith *et al.*, 1993), were found to be MRF positive with more sensitive immunodetection methods (Smith *et al.*, 1994). The level of protein that is functionally significant is another consideration (see Section VIII,C). In the case of myogenin, Cusella de Angelis and colleagues (1992) reported transcript but not protein accumulation in the myotome prior to E10.5, although myogenin protein has been detected in myotomal cells from E9 (Smith *et al.*, 1994; Patapoutian *et al.*, 1995). The phenotype of *myogenin* null embryos, in which the myotome is not perturbed initially, is consistent with a role for this MRF only at later developmental stages (see Buckingham, 1994). A striking example of gene transcription, which does not reflect gene activity, is provided by *Myf5* transcription in the central nervous system, detected at the transcript level and as β-galactosidase$^+$ cells in *Myf5-nlacZ* embryos (Tajbakhsh and Buckingham, 1995). However, no Myf5 protein has been detected in subdomains of the brain (P. Daubas, S. Tajbakhsh, M. Primig, and M. Buckingham, unpublished), in keeping with the lack of myogenic conversion *in vitro*.

Targeted disruption of the *MRF4* gene might suggest that this MRF, rather than myogenin, is important for early myotome differentiation (Patapoutian *et al.*, 1995). However, interpretation of the *MRF4* knockout phenotypes is complicated by interference with expression of the nearby *Myf5* gene (see Olson *et al.*, 1996; Yoon *et al.*, 1997). In light of the delay in myogenin protein accumulation discussed above, it is important to examine differentiation in the early myotome (prior to E10) in a situation in which only *MRF4* gene function in perturbed, to evaluate the respective roles of these MRFs in early myotome differentiation. That MRF4 can substitute for myogenin during myogenesis was shown by expression of an *MRF4* transgene in *myogenin* null mice (Zhu and Miller, 1997).

In addition to observations on the expression of *MRF* genes in normal and mutant embryos, morphological observations have also pointed to a heterogeneity in the developing myotome. Kahane and colleagues (1998b) suggested that a "pioneer," postmitotic, myotome acts as a scaffold for the subsequent migration of myogenic precursors from the dermomyotome onto the surface of the myotome immediately adjacent to the sclerotome. Marcelle and colleagues (1995) reported that myotomal cells expressing the FGF receptor FREK in the chick also become localized to this surface of the myotome. These cells may represent recent dermomyotomal cells which have not yet exited the cell cycle, or quiescent cells poised to proliferate and expand the myotome at later stages in response to FGFs, such as FGF4 and FGF6, which are produced by the myotomal cells themselves (see Grass *et al.*, 1996). Entry of new myotomal cells directly from the dermomyotome to the myotome may also take place as the central dermomyotome becomes mesenchymal. It remains possible that the En1$^+$ cells present in this domain may integrate directly into the myotome without transiting via the edges of the dermomyotome (see Section III,B). If this is the case, evaluating how this migratory route also contributes to myotome heterogeneity, both morphologically and in terms of gene expression, will be a challenging task. As development procedes, the myotome greatly increases in size. The epaxial dermomyotomal lip and hypaxial somitic bud, which persist for longer, probably represent a continuing source of myogenic progenitor cells. However, the extent to which "myoblasts," such as the FREK positive cells in the chick, are present in the myotome and contribute to its growth is unclear. It has become important to identify markers that can reveal such a potential heterogeneity in the myotome.

VI. Extrinsic Factors Govern Somite Differentiation and the Activation of Myf5 and MyoD

Cells in the presomitic mesoderm/immature somite are multipotent, and it is the surrounding tissues which influence their fate. This important concept was first demonstrated by classic embryological experiments, mainly with avian embryos (see Christ and Ordahl, 1995). In studies on the initiation of myogenesis, most experiments *in ovo* with avian embryos have been carried out at early stages at the level of the forelimb (wing) level, where epaxial myogenesis predominates. This depends on the presence of the axial structures (neural tube and notochord) and has been characterized by the expression of *MyoD*. Experiments with avian and mouse explants have confirmed that the presence of axial structures results in the onset of myogenesis in explants of presomitic mesoderm/immature somites and have also indicated that the surface ectoderm can play this role. The presence of lateral plate mesoderm inhibits or delays activation (reviewed in Cossu *et al.*, 1996a).

Models for the onset of myogenesis in mouse embryos are based on explant experiments where somite geometry is perturbed. However, the data, mainly with material from the interlimb level somites, suggest that the axial structures are required for epaxial myogenesis, characterized in the mouse embryo by the early activation of *Myf5*, whereas surface ectoderm alone leads to preferential activation of *MyoD* (Cossu *et al.*, 1996b). The *in vivo* onset of myogenesis in the hypaxial domain in response to the overlying surface ectoderm occurs later, consistent with a delaying effect of the adjacent lateral plate mesoderm.

A number of candidate signaling molecules produced by the surrounding tissues have been proposed as positive effectors of myogenesis. These include the Wnts which are present in the dorsal neural tube and in the surface ectoderm and Sonic hedgehog produced by the notochord and floor plate of the neural tube (Münsterberg *et al.*, 1995; Stern *et al.*, 1995; Borycki *et al.*, 1998; Tajbakhsh *et al.*, 1998). In mouse explants, coculture of presomitic mesoderm with cells producing Wnts demonstrates that different Wnts have different effects on the expression of *Myf5* and *MyoD*. Wnt1 producing cells lead to preferential expression of *Myf5*, consis-

6. Muscle Progenitor Cells: Spatiotemporal Considerations   243

tent with a role for this signaling molecule, produced by the dorsal neural tube, in epaxial myogenesis, whereas Wnt7a, present in surface ectoderm more ventrally, promotes preferential expression of *MyoD* (Tajbakhsh *et al.*, 1998). Interestingly Wnt1 and Wnt7a act through different intracellular signaling pathways, the former being β-catenin dependent, whereas the latter is β-catenin independent (Kengaku *et al.*, 1998). The notochord, which produces Sonic hedgehog, is clearly important as a ventralizing influence on the somite, leading to sclerotome differentiation (Brand-Saberi *et al.*, 1993; Pourquié *et al.*, 1993). Its effects on the dorsal somite and particularly myogenesis have been controversial. However, Sonic hedgehog can have a positive myogenic influence *in ovo* (Borycki *et al.*, 1998). This has also been observed in explant experiments and a cooperative effect with different Wnts has been noted (Münsterberg *et al.*, 1995; Tajbakhsh *et al.*, 1998). In the case of the Wnts a complicating factor is the production, possibly as a "relay," of Wnts by the dorsal somite itself as well as by the adjacent tissues; Wnt11, for example, is present in the epaxial dermomyotome in chick (Marcelle *et al.*, 1995) and mouse (Christiansen *et al.*, 1995) embryos.

One of the difficulties with experiments aimed at identifying the role of signaling tissues/molecules is that of quantitation. *In vivo* there are probably threshold levels that determine activation/inhibition effects, and this may well explain some of the apparently contradictory results with notochord transplantation experiments (see Dietrich *et al.*, 1997). Indeed the complexity of quantitative fine tuning is evident in the case of the Wnts (see Wodarz and Nusse, 1998); not only are there different Wnts and different Wnt receptors of the frizzled family but also dominant negative forms of the latter, the Frzbs (see Zorn, 1997), expressed in and around the somite. The quantities and relative binding affinities of all these components are not yet understood.

Another complicating factor is that the same families of signaling molecules are frequently involved in cell specification and/or morphogenesis at different sites in the embryo, raising the possibility of interference. The transcription of the *Myf5* gene in the brain in a region adjacent to sites of Sonic hedgehog and Wnt1 production is probably an example of this; in this case the Myf5 protein does not accumulate and consequently there is no myogenesis (P. Daubas, S. Tajbakhsh, M. Primig, and M. Buckingham, unpublished). An example which concerns the somite is that of the BMPs, a family of signaling molecules which also illustrate the critical role of concentration on their effect as activators or inhibitors (Tonegawa *et al.*, 1997; Reshef *et al.*, 1998). In chick embryos, BMP4 will inhibit *MyoD* activation in the somite (Pourquié *et al.*, 1996) and since BMP4 is produced by lateral plate mesoderm this may explain the inhibitory action of this tissue on hypaxial myogenesis. BMPs, including BMP4, are also produced by the dorsal neural tube (Monsoro-Burq *et al.*, 1996), and therefore would potentially also antagonize epaxial myogenesis. However, noggin, a BMP inhibitor, is transcribed in the epaxial domain of the chick somite (Hirsinger *et al.*, 1997; Marcelle *et al.*, 1997). It is also present in the epaxial somite of the mouse, and the inactivation of this gene leads to a disruption of epaxial myogenesis (McMahon *et al.*, 1998). Noggin expression in the absence of the neural tube, can be rescued by Wnt1 (Hirsinger et al., 1997), suggesting that Wnt1 may also play an indirect role in potentiating myogenesis.

In addition to spatial considerations, temporal differences in signaling systems are also important. Initially in the mouse, *Myf5* is activated in the epaxial domain under the influence of the neural tube. However later, *MyoD* is also expressed epaxially from about E10, and in *Myf5* null embryos *MyoD* activation, although delayed (Tajbakhsh et al., 1997), takes place (Braun et al., 1994). This is not due to extension of *MyoD* activation from the hypaxial domain, under the influence of surface ectoderm, but rather to independent activation of *MyoD* in the epaxial domain by the older neural tube, as shown by explant experiments (Tajbakhsh et al., 1998). Whether signaling molecules implicated in *MyoD* activation, and produced by the older neural tube ($<$ E10), are distinct from those acting at earlier stages is not known. It is also not clear what signaling systems operate as the central dermomyotome disintegrates when some of these cells may participate in later myotome formation. In addition to phenomena of this kind which affect the mature somite, it is clear that the presomitic mesoderm/immature somites present more caudally in older embryos are developmentally more advanced than those in younger embryos. This is indicated by the effect of Sonic hedgehog on myogenesis (Münsterberg et al., 1995), which is much more evident with mouse explants from younger (E8.5 rather than E9.5) embryos (Tajbakhsh et al., 1998). Further evidence is provided by the observation that myogenic markers are activated more rapidly in immature somites of older embryos.

Mouse mutants for genes encoding components of signaling systems provide an increasingly valuable resource. Mutants for a single *Wnt* gene, for example, may not have a myogenic phenotype due to the overlap of expression of related Wnts, but double *Wnt* mutants are becoming available as are other mutations such as those in components of the intracellular signaling pathway A recent example is provided by the *Wnt1/Wnt3a* double mutant in which the epaxial dermomyotome is perturbed and *Myf5* expression in this domain is much reduced (Ikeya and Takada, 1998). This is consistent with a role for Wnt1 (and Wnt3a) in epaxial myogenesis and is in keeping with the expression of these Wnts in the dorsal neural tube. Indeed, the *opb* (open brain) mutant in which formation of the dorsal pole of the neural tube remains incomplete and the expression of Wnt1 and Wnt3a are partially lost, exhibits a similar myogenic phenotype (Spörle et al., 1996; Nagai et al., 1997). In the case of *Sonic hedgehog* null mice (Chiang et al., 1996), re-examination of the muscle phenotype shows a clear-cut ablation of epaxial myogenesis, characterized by the absence of *Myf5* expression in this domain (Borycki et al., 1999). Hypaxial myogenesis, on the other hand, takes place. This is in keeping with the importance of surface ectoderm as a positive signaling source for the latter, and of axial structures (neural tube and notochord) for the former. However, another difficulty in dealing with signaling molecules that influence myogenesis is distinguishing effects on the proliferation/survival of precursor cells as distinct from activation of *Myf5* or *MyoD* genes per se. Many of the signaling molecules implicated, such as Shh (Duprez et al., 1998; Teillet et al., 1998), can potentially play both roles. In the *Shh* mutant embryos somite growth is affected. Initially however, *Pax3* expression, which marks the dorsal somite, is normal in the epaxial domain at a time when *Myf5* should be activated, suggesting that an initial effect of SHH is on *Myf5* expression. Furthermore, apoptosis, which is evident particularly in the neural tube and ventral somite at later stages, is not detectable. In the *Shh* mutant, therefore, the informative role of this signaling molecule can be distinguished from a trophic effect. In the case of the *Wnt1/Wnt3a* double mutant, an increase in apoptosis was not observed (Ikeya and Takada, 1998), however, the potential effects of these molecules on somite cell proliferation remain an issue.

In embryos which have lacked a key signaling molecule throughout their development, it is difficult to eliminate indirect effects. Conditional mutations introduced once the embryonic tissues have been established will help to deal with this problem. Such mutants will also be useful in the context of initiation versus maintenance of myogenesis. Based on manipulations in the avian embryo it has been suggested that initial signals (perhaps Sonic hedgehog) are necessary to activate *MyoD* in the avian somite (Borycki *et al.,* 1998), but that this is not maintained if the axial structures are not present (Pownall *et al.,* 1996). Subsequently in more mature somites, once myogenesis is established, such signals are no longer essential. Indeed the morphogenesis and differentiation of the somite in general depends initially on a balance between signals. For example, a number of studies have shown that continuous signaling, mainly from surface ectoderm, is required to maintain the dermomyotome as an epithelium (see Tajbakhsh and Spörle, 1998), and to prolong *Pax3* expression (Fan and Tessier-Lavigne, 1994), potentially via Wnt signaling (Fan *et al.,* 1997). Indeed *Pax3* and and its orthologue *Pax7,* which is also expressed in the dermomyotome, play an integral part in the establishment of this epithelium since embryos lacking both genes have a much reduced dermomyotome epithelium (A. Mansouri and P. Gruss, personal communication). Curiously, however, in *paraxis* null embryos where an epithelial somite does not form, muscle markers are nevertheless expressed in the myotome (Burgess *et al.,* 1995), suggesting that an epithelial–mesenchymal transition is not critical for the formation of myotomal cells (see also Section VIII,A).

The preceding discussion has been centered on the activation of myogenesis in the somite, under the influence of signals from surrounding tissues. However, derepression may also be an important factor, and the Notch pathway is a candidate in this respect (see Cossu *et al.,* 1996a; Weinmaster, 1997). A number of observations suggest that a myogenic potential can be revealed in nonmuscle cells when these are removed from their tissue context in the embryo (George-Weinstein *et al.,* 1996; Ferrari *et al.,* 1998). Members of the Notch family and its ligands are expressed by presomitic mesoderm and some somitic cells, and in cultured muscle cell lines manipulation of the Notch signaling pathway results in inhibition of *MyoD* activation and myogenesis. Furthermore in *Drosophila, Notch* mutants affect myogenesis (Corbin *et al.,* 1991; see Baylies *et al.,* 1998) and indeed it has been proposed that wingless (*Drosophila* homologue of vertebrate Wnt) may be a ligand for the Notch receptor, antagonizing repression by its ligand, Delta (Couso and Martinez Arias, 1994). It remains to be demonstrated that the Notch pathway affects the onset of myogenesis during normal vertebrate development *in vivo*, but repression as well as activation may be important, potentially operating at the level of the myogenic genes (*Myf5* and *MyoD*), the proliferation/survival of muscle progenitor cells, and the restriction of cell fate to specific lineages.

VII. Do Myf5 and MyoD Define Different Subpopulations of Muscle Cells?

While it is clear from the mutant results that *Myf5* and *MyoD* genes are activated independently in the embryo, and indeed differential activation by Wnt1 and Wnt7a in explants is also indicative of this, it is not clear that this takes place in different cell populations *in vivo*. Although epaxial and hypaxial (and the central, intercalated) myotomal domains can be regarded as arising from separate lineages in that they are derived from differently positioned mesodermal precursors in the embryo (Selleck and Stern, 1991; Psychoyos and Stern, 1996; J. F. Nicolas, personal communication) there is no evidence that within these domains, separate muscle cell "lineages" identified as Myf5$^+$ or MyoD$^+$ exist. Therefore, the term "subpopulations" instead of the misnomer "lineages" is preferable when referring to cells within subdomains of the somite which may exhibit differences.

This issue was investigated by Braun and Arnold (1996) using primarily an *in vitro* assay. One allele of *Myf5* was targeted with *lacZ* and the other with the herpes simplex virus *thymidine kinase* (HSV-TK) gene, in embryonic stem cells. The cells were then allowed to differentiate into skeletal muscle in the presence of the nucleotide analogue FIAU (toxic; incorporated into DNA by HSV-TK) to eliminate *Myf5*-expressing cells undergoing mitosis. In the presence of the drug, MyoD-positive cells remained, leading Braun and Arnold (1996) to conclude that *MyoD* and *Myf5* are expressed in independent cells and to propose, by extrapolating to the embryo, that muscle in the *MyoD* null mutant is reconstituted by a *Myf5* dependent lineage and vice versa for *Myf5* null embryos. Whereas this study underlines the independent activation of *Myf5* and *MyoD*, the interpretation of these observations and their extrapolation to the *in vivo* situation require discussion. First, HSV-TK is toxic to dividing cells, and therefore those cells expressing *Myf5* but not in rapid proliferation would not necessarily be eliminated. Second, Myf5 (and MyoD) protein is subject to cell cycle degradation/resynthesis (Kitzmann *et al.,* 1998; Lindon *et al.,* 1998) at least in cell culture. Whereas Myf5 appears to peak in G0 and decreases during G1, MyoD protein is absent in G0 and peaks in mid-G1 (Kitzmann *et al.,* 1998), and this may also obscure the detection of cells that coexpress both proteins. Third, the coexpression of *Myf5* and *MyoD* has been demonstrated *in vivo* (Fig. 6) (Cossu *et al.,* 1996b), in explants of embryos (Cossu

6. Muscle Progenitor Cells: Spatiotemporal Considerations

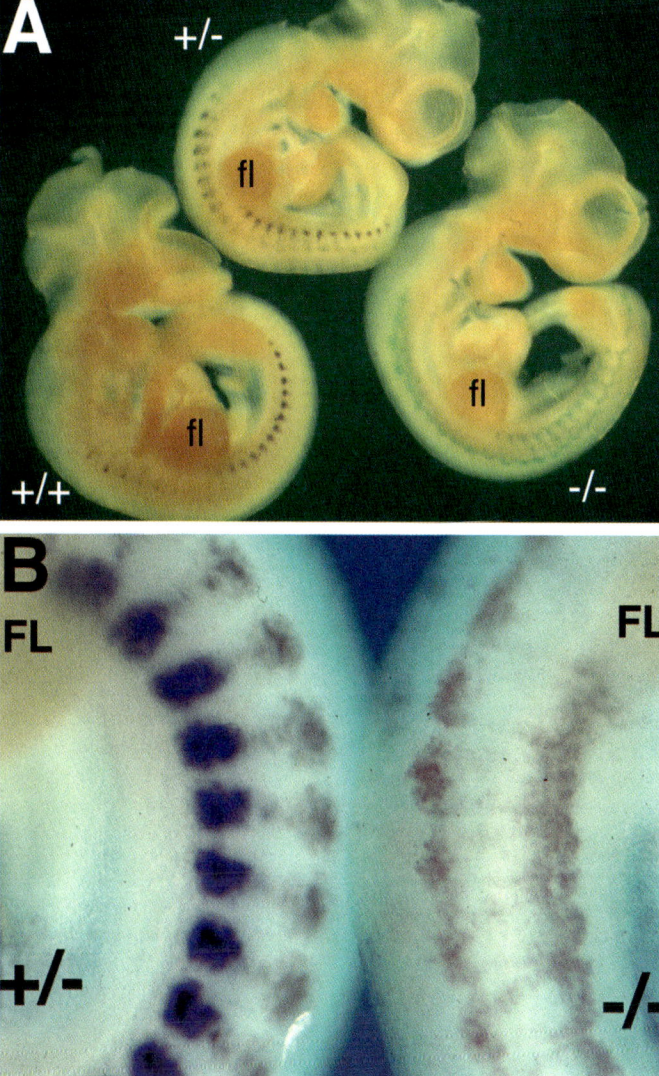

Figure 6. *MyoD* activation is delayed in *Myf5-nlacZ−/−* embryos. (A) Wild-type, *Myf5-nlacZ+/−*, and *Myf5-nlacZ−/−* embryos from E10.75 (39 somite) outbred mice were stained briefly with X-gal (blue), then subjected to whole mount *in situ* hybridization with an antisense MyoD probe (purple). Both the wild-type and *Myf5-nlacZ+/−* outbred embryos reveal MyoD transcript accumulation in the ventral domain of interlimb somites and in the dorsal domain in more anterior somites. (B) Enlargement of the interlimb region of heterozygous and homozygous mutant embryos obtained from inbred C57Bl6 mice (E10.75; 39 somites). The Salmon-gal labeling (pink) distinguishes the heterozygous (left) from the homozygous (right) embryo. Note the absence of MyoD transcript accumulation (purple) in the homozygous mutant indicating that the initial activation of MyoD is dependent on Myf5. FL, fl, forelimb. (Modified from Tajbakhsh *et al.*, 1997.)

et al., 1996b), and in cell lines (Kitzmann *et al.*, 1998; Lindon *et al.*, 1998; Yoshida *et al.*, 1998). When the coexpression of *Myf5* and *MyoD* is not detected, the relative sensitivity of the reagents used must be taken into consideration. As muscle cells mature, *Myf5* tends to be downregulated. This is seen in myotubes in culture (Montarras *et al.*, 1991; Mangiacapra *et al.*, 1992) and in the embryo from E11 when *MyoD* is widely expressed (Ott *et al.*, 1991; see Buckingham, 1992). The use of *Myf5-nlacZ* embryos has been informative in permitting the identification of cells that had expressed *Myf5*, or continue to express it at a low level, due to the detection of the relatively stable β-galactosidase protein. Analysis of heterozygous *Myf5-nlacZ* embryos has demonstrated that most muscle cells in the embryo express both genes once *MyoD* has been activated and, furthermore, that the nuclei in muscle fibers at later stages of development remain β-galactosidase positive indicating that *Myf5* was activated (Fig. 6) (Cossu *et al.*, 1996b; Tajbakhsh *et al.*, 1996b). The phenotype of the *MyoD* null mouse would also argue against the importance of a "MyoD only" cell population since the expected muscle groups form (Rudnicki *et al.*, 1992).

Myf5-nlacZ null embryos are also informative in this respect. An *in situ* hybridization analysis of these embryos demonstrates that even under conditions where a robust MyoD signal is observed in normal *Myf5-nlacZ+/−* embryos at E10.5, no MyoD transcripts are detectable in the *Myf5-nlacZ* null embryo at this stage (Fig. 6) (Tajbakhsh *et al.*, 1997) (see Section VIII.A), with the exception of some shoulder muscles (S. Tajbakhsh and M. Buckingham, unpublished) (see Fig. 5). This delay in *MyoD* expression in the null embryo suggests that (**1**) *MyoD* is normally activated within cells having previously activated *Myf5* and lack of Myf5 protein provokes a delay in *MyoD* activation within these cells; and/or (**2**) *MyoD* is activated in an independent cell population and, in the *Myf5* null, aberrantly blocked MPCs impede the exit of these MyoD-MPCs that would normally activate *MyoD* first; and/or (**3**) some MPCs do activate *MyoD* first but they require nonautonomous cell signaling from Myf5 programmed MPCs. Evaluating these possibilities requires further experimentation, however, we do know that *MyoD* is indeed activated later (after the delay indicated) within β-galactosidase positive developmentally arrested MPCs in *Myf5-nlacZ* null embryos (S. Tajbakhsh and M. Buckingham, unpublished). This also suggests that the bulk of the developmentally arrested MPCs in the *Myf5* null are recruited into myogenesis. This is consistent with the observation that by E14.5, most of the skeletal muscle deficits in the *Myf5* null have been overcome (Fig. 7A, B; S. Tajbakhsh and M. Buckingham, unpublished). Experiments with explants of murine presomitic mesoderm cultured with surface ectoderm which showed activation of *MyoD* prior to *Myf5* (Cossu *et al.*, 1996b) appear to be in contradiction with these genetic results. However, other adjacent tissues such as lateral mesoderm may well modulate this effect *in vivo*, whereas in explants this geometry may be lost. Since early *Myf5* expression is very low compared to *MyoD* in the hypaxial myotome, there may also be a sensitivity issue. In the case of single satellite cells obtained from isolated fibers and

6. Muscle Progenitor Cells: Spatiotemporal Considerations 249

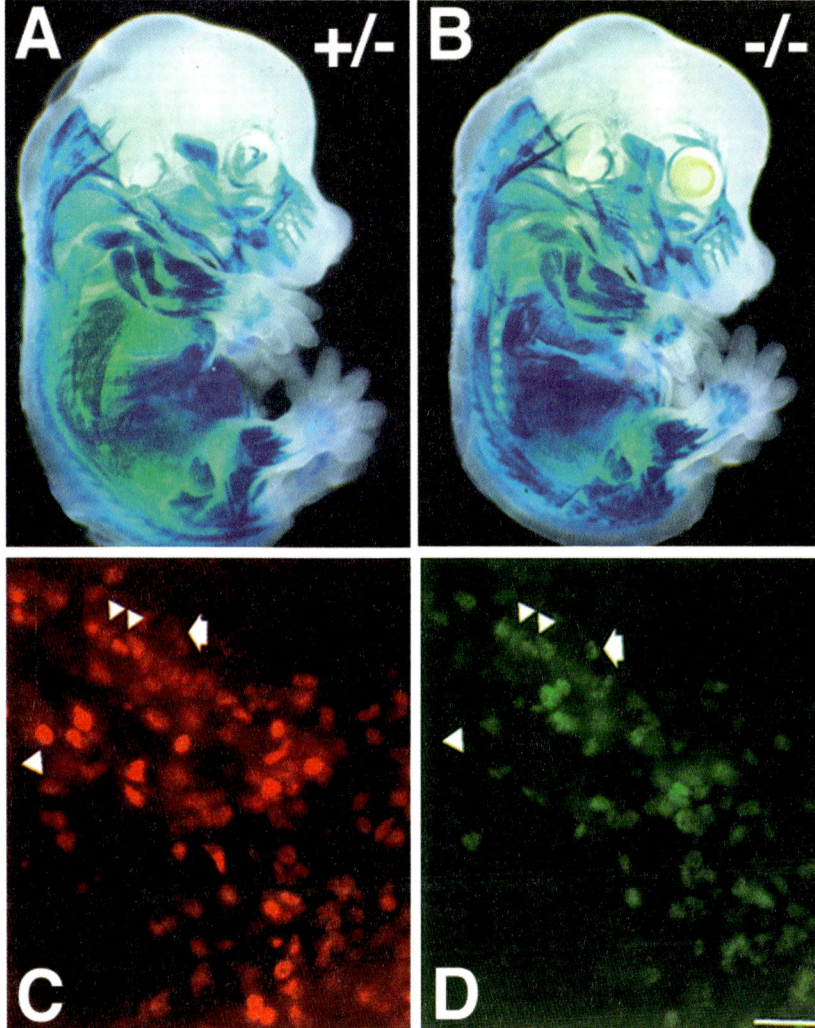

Figure 7. *Myf5* and *MyoD* are coexpressed in the same cell, and most skeletal muscle defects in the *Myf5-nlacZ* null are reconstituted by fetal stages. E14.5 heterozygous (A) and homozygous (B) mutant embryos were stained with X-gal to reveal *Myf5* expression in the majority of skeletal muscles, notably in the null mutant, indicating that *Myf5* and *MyoD* are coexpressed in essentially all muscles. Transverse cryostat section of a forelimb level somite from an E10.5 *Myf5-nlacZ* embryo was reacted with anti-β-galactosidase monoclonal (C) and anti-MyoD polyclonal (D) antibodies (reprinted from Cossu *et al.*, 1996b). Arrowhead, Myf5$^+$/MyoD$^-$; arrow, Myf5$^-$/MyoD$^+$; double arrowheads, Myf5$^+$/MyoD$^+$ which is the case for the great majority of cells. Bar = 40 μm.

assayed by RT-PCR, MyoD or Myf5 positive cells are initially detected, although subsequently, these genes are coexpressed (Cornelison and Wold, 1997; see also Cooper *et al.*, 1999). Therefore, at least in this adult context, two subpopulations can be detected initially.

In conclusion, there are a number of reasons to suppose that independent activation of *Myf5* and *MyoD* does not take place in distinct cell populations to a major extent, within the somite. Although this remains to be resolved directly, for example, by cell ablation experiments in the embryo, it is inappropriate at present to refer to separate *Myf5* and *MyoD* muscle cell lineages during embryonic development.

VIII. The Roles of Myf5 and MyoD in Muscle Progenitor Cell Determination

All four *MRF* genes have been mutated in the mouse and in some cases, double mutants have been analyzed; some of the results are summarized in Table 1. These have led to the general classification of the MRFs into two groups: the "determination" factors Myf5 and MyoD, and the "differentiation" factors myogenin and MRF4. *Myf5/MyoD* double mutant embryos lack skeletal muscle fibers and myoblasts (Rudnicki *et al.*, 1993). In keeping with the observation that *Myf5*, but not *MyoD*, is expressed in the early myotome, *Myf5* null embryos lack this structure

Table 1 Summary of Phenotypes Obtained from Gene Inactivation Studies of the Myogenic Regulatory Factors

Gene(s) mutated	Myoblasts	Muscle fibers	Description[a]
Myf5	+	+	Lethal at birth due to rib truncations; ~2.5 day delay in myotome formation; early muscle perturbations in trunk and tail
MyoD	+	+	Viable and fertile; ~30% reduction in size; muscle fiber regeneration defects; limb muscles delayed
myogenin	+	−	Lethal at birth; most skeletal muscle fibers absent; myotome and some myotomal derivatives present
MRF4	+	+	Mildest allele is viable and fertile with some mild rib defects, possible interference with *Myf5* activity
Myf5/MyoD	−	−	Lethal at birth; total absence of muscle fibres and precursor myoblasts

[a] See text for references.

6. Muscle Progenitor Cells: Spatiotemporal Considerations 251

initially. They do not survive after birth due to respiratory problems related to rib defects which probably reflect repercussions in the delay in muscle formation (Braun *et al.*, 1992; Tajbakhsh *et al.*, 1996a). *MyoD* −/− mice are viable (Rudnicki *et al.*, 1992) but these animals exhibit an impairment of the satellite cell population, which affects muscle growth, and strikingly, muscle regeneration (Megeney *et al.*, 1996). They also display some defects in hypaxially derived musculature (Kablar *et al.*, 1997), consistent with the fact that although Myf5 initially programs myogenesis in this domain (see Section VIII,A), *MyoD* expression subsequently predominates here. In contrast, as expected from the expression pattern, *Myf5* null embryos have predominantly epaxial muscle defects (Kablar *et al.*, 1997; Tajbakhsh *et al.*, 1997; S. Tajbakhsh and M. Buckingham, unpublished). Early muscle differentiation is observed in *myogenin* −/− embryos (Nabeshima *et al.*, 1993; Venuti *et al.*, 1995) at a time when *MRF4* is expressed in the myotome (see Buckingham, 1994). Later, most differentiated skeletal muscle is lacking (Hasty *et al.*, 1993; Nabeshima *et al.*, 1993). The *MRF4* mutant which minimally disturbs *Myf5* gene activity (*MRF4* and *Myf5* are separated by about 7 kbp) yields viable and fertile mice (Zhang *et al.*, 1995; Yoon *et al.*, 1997). As is often the case, revisiting these mutants and examining them in more detail has been both necessary and illuminating. A discussion of these more recent developments follows in the context of the spatiotemporal programming of muscle progenitor cells.

A. Myf5 Initiates the Early Mouse Myotome

Given that the myotome does not form until *MyoD* is expressed in *Myf5* null embryos (Braun *et al.*, 1992, 1994; Tajbakhsh *et al.*, 1996a, 1997), a study of these mutants before *MyoD* activation allows investigation of the role that a single myogenic factor, Myf5, plays in establishing myogenic identity. Such an analysis was carried out with embryos in which both alleles of *Myf5* had been targeted with an *nlacZ* reporter gene resulting in disruption of the *Myf5* coding sequence. This permits the visualization of cells (β-galactosidase$^+$) in which the *Myf5* gene has been activated.

First, in *Myf5-nlacZ* null embryos, MPCs are developmentally arrested, remaining as mesenchymal cells along all edges of the dermomyotome epithelium (see Fig. 3B). In the absence of Myf5 protein, these β-galactosidase$^+$ MPCs exit the epithelium, but remain predominantly in the immediate viscinity of the epaxial lip and hypaxial somitic bud, whereas in normal embryos they migrate underneath the dermomyotome, toward the central domain of the somite (Tajbakhsh *et al.*, 1996a). Therefore, *Myf5* gene function in not necessary for the epithelial to mesenchymal transition of MPCs from the dermomyotome, a morphological change that is often associated with the restriction of cell fate (Hay, 1993). Furthermore, the numbers of MPCs in heterozyous and homozygous mutant embryos are similar, suggesting that at least in these early stages of myotome formation, Myf5 does not seem to influence MPC proliferation. Myf5 is, however, necessary for the correct localization of MPCs to the central domain of the somite where the myotome will form. A functional cell surface molecule is probably necessary to effect this displacement and the activation of such a gene appears to be at least one of the early roles of Myf5. Platelet-derived growth factor (PDGF) signaling may be involved since the ligand PDGF-A appears to be one of the downstream targets of Myf5 (P. Soriano, personal communication). The PDGFα receptor is expressed in the dermomyotome, as well as the sclerotome, and in its absence, the myotome does not form normally (Soriano, 1997).

Second in *Myf5-nlacZ* null embyros, β-galactosidase$^+$ cells are found in ectopic locations, where MPCs are not normally present (Tajbakhsh *et al.,* 1996a). In the dorsal somite, β-galactosidase$^+$ cells are found under the ectoderm and transcribe a dermal (nonmuscle) marker, Dermo1. In the ventral somite, β-galactosidase$^+$ cells are observed in the condensing mesenchymal cells of the ribs, transcribe a cartilage (nonmuscle) marker, scleraxis, and later are found in ribs (also observed in *Sp/Myf5* double mutants, see Fig. 8). In normal embryos, a basal lamina forms between the sclerotome and the myotome, presumably laid down by the myotomal cells themselves after the myotome is initiated (Tosney *et al.,* 1994), thus confining these cells to this compartment. In *Myf5* null embryos, this basal lamina is not detectable early (Tajbakhsh *et al.,* 1996a); this may contribute to, but is probably not the primary cause of, the aberrant location of MPCs in the somite. The continuous exit of MPCs from the dermomyotome and their subsequent accumulation along the edges of the dermomyotome probably provokes the passive movement of some of these MPCs into adjacent microenvironments as is observed most dramatically for the anterior (occipital and cervical) somites. The bulk of these developmentally arrested MPCs, however, appear to be recruited into myogenesis since *MyoD* is activated within these cells (see Section VII). Developmentally arrested β-galactosidase$^+$ MPCs accumulating in the hypaxial domain of *Myf5-nlacZ* null embryos continue to express *Pax3* (Fig. 4), which is normally downregulated as the myotome forms. This finding provides evidence that these MPCs originate from the dermomyotome and are not adjacent mesenchymal cells which have aberrantly activated the *nlacZ* reporter due to loss of *Myf5* function. Indeed, in an extrasomitic environment, in double *Myf5-nlacZ/MyoD* mutant embryos, this phenomenon is also observed in the limb where β-galactosidase$^+$ MPCs become incorporated into condensing cartilage (S. Tajbakhsh and M. Rudnicki, unpublished). This is also revealed by a *MyoD-lacZ* transgene in this mutant background (Kablar *et al.,* 1999). Furthermore, in *Sp/Myf5-nlacZ* double mutants (see Section VIII,D) where limb muscles do not form, β-galactosidase$^+$ MPCs are also detected in cartilage (Fig. 8). We conclude that in the absence of Myf5 protein, multipotential MPCs are capable of changing fate, according to their local environment, demonstrating that Myf5 (and by inference MyoD) is a determination factor confering myogenic identity. The *Myf5* null illustrates the distinction between myogenic "determination," i.e., identity, which depends on the presence of Myf5

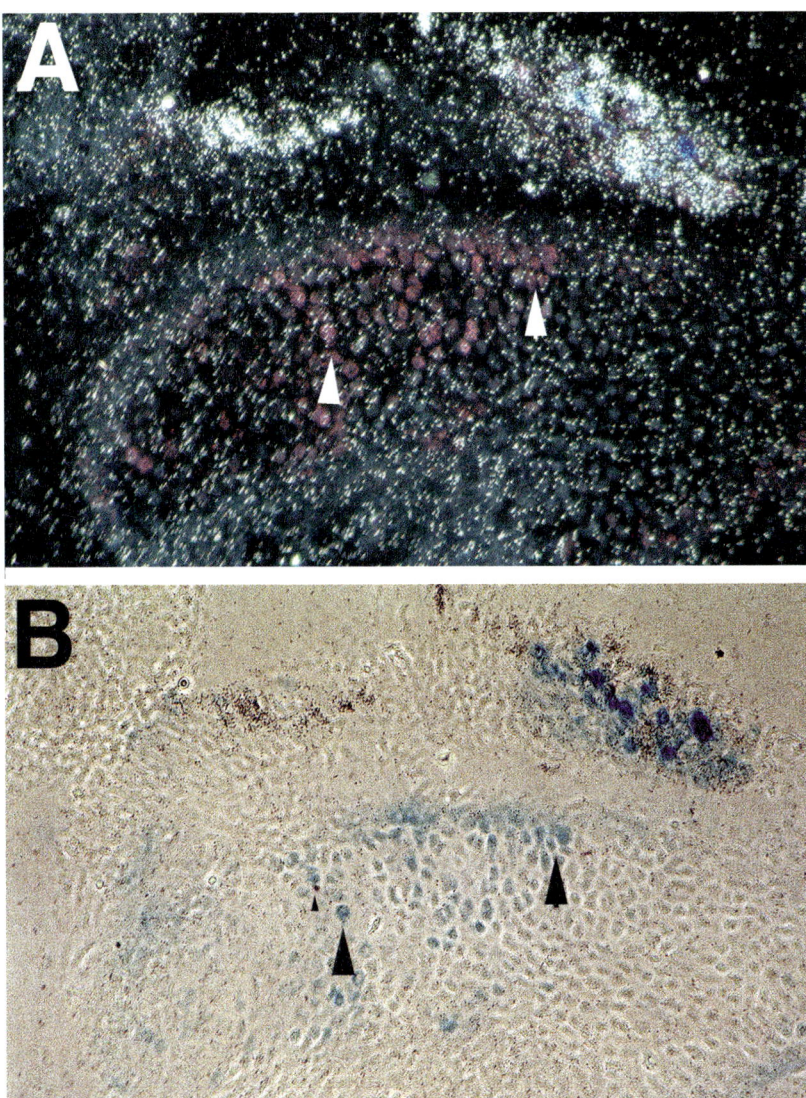

Figure 8. Muscle progenitor cells located in the ribs of $Sp/Myf5\text{-}nlacZ-/-$ double mutants do not express myogenin *in situ*. E14.5 paraffin section hybridized with an ^{35}S-antisense myogenin probe. (A) Dark-field image showing that β-galactosidase$^+$ cells (which appear as pink in dark field) in these double mutants do not express the muscle marker myogenin which marks adjacent neck muscles (see Tajbakhsh *et al.*, 1997), indicating that although they express *Myf5*, they have not entered the myogenic program in this ectopic location. Greater numbers of β-galactosidase$^+$ MPCs are located in the ribs of these double mutants as compared to the *Myf5-nlacZ* null (Tajbakhsh *et al.*, 1996a). (B) Bright-field image of A showing that β-galactosidase$^+$ cells are not colocalizing with the silver grains of the myogenin antisense probe.

protein, and myogenic "specification" which depends on environmental signals necessary for *Myf5* activation (see Tajbakhsh and Cossu, 1997). The latter leads to the identification of myogenic progenitor cells; whether these MPCs make muscle depends on functional Myf5 (or MyoD).

B. Aberrant Rib Formation in *Myf5* Null Mice

A hallmark of *Myf5* null mice is distal rib perturbations (Braun et al., 1992; Tajbakhsh et al., 1996a). This is an unexpected finding given that *Myf5* is not expressed in rib precursor cells in normal embryos (Tajbakhsh et al., 1996a). Consequently, *Myf5* null embryos provide an important model for studying non-cell autonomous interactions between myotomal and rib precursor cells. One model (Grass et al., 1996) suggests that growth factors and signaling molecules such as FGF4 and FGF6 (IGFs and PDGFs are also candidates) which are present in the myotome may signal to and promote the growth of condensing mesenchymal rib precursors in the sclerotome. Indeed, in a cell culture assay, TGFβ-2 (found in the dermomyotome) was found to synergize with FGFs to promote chondrogenesis (Grass et al., 1996). Therefore the absence of an early myotome may provoke perturbations in rib formation.

Another more mechanistic model was proposed after examining aberrant MPC accumulation and perturbations in muscle formation which persist even after *MyoD* activation, in the *Myf5* null embryo (Tajbakhsh et al., 1996a). Distal ribs have been thought to originate from precursors in the lateral somite. In normal embryos, the cells in the leading edge of the ventral compartment of the distal ribs appear to condense, as the ventral-most hypaxial myotomal edge forms the intercostal muscles which become intercalated between the nascent ribs. In *Myf5-nlacZ* null embryos MPCs accumulate aberrantly and enter into the adjacent compartment where rib precursors condense. It is not clear if, in normal embryos, the leading distal tip of the rib recruits new mesenchymal cells as it grows ventrally, or if a pre-existing bed of mesenchymal cells condense and pave the way to the lateral mesoderm-derived sternum. In either case, this lateral compartment is disrupted by the abnormal presence of MPCs which accumulate in front of the distal tip of the immature rib (S. Tajbakhsh and M. Buckingham, unpublished). Muscles in this domain fuse as do those in the abdominal compartment, due to the absence of ribs, and this may impair further rib growth. The incorporation of some β-galactosidase$^+$ MPCs into the ribs may also contribute to the rib defects, although these cells rapidly assume a cartilagenous morphology, do not express the muscle marker myogenin (Fig. 8), and express the cartilage marker gene *scleraxis* (Tajbakhsh et al., 1996a). It is not yet clear whether the absence of growth factors produced by the myotome or mechanical perturbations, or both, are responsible for the truncated rib phenotype observed in these mutants. In support of the latter model, recent studies (Kato and Aoyama, 1998) demonstrated that distal

6. Muscle Progenitor Cells: Spatiotemporal Considerations   255

and sternal rib progenitors originate from the dermomyotome. Their exit from the dermomyotome and subsequent ventral migration may be impeded by the aberrant accumulation of MPCs in this compartment. It seems unlikely that the absence of Myf5 interferes with cell fate decisions within the dermomyotome, since as indicated, MPCs exit this epithelium as expected in *Myf5* null embryos.

C. Differences between Myogenic Regulatory Factors and Threshold Levels. Are All MRFs Functionally Alike?

Since Myf5 and MyoD appear to perform similar roles in muscle cell determination, it is relevant to ask whether they would have identical roles in the embryo if they exhibited identical expression levels, and patterns of expression. Within the same vertebrate class, Myf5 and MyoD are highly divergent outside the bHLH domain. However, from an evolutionary standpoint, there is a striking conservation between *Xenopus* and human Myf5 (or MyoD) outside the bHLH region, which is itself highly conserved among all MRF family members and across lower and higher vertebrates. Such a strong evolutionary pressure across species, suggests that distinct functional roles have evolved for these myogenic factors. To date, no knock-in experiment has been reported for this pair of sequences, and therefore the question of whether Myf5 and MyoD activate the same genes with the same efficiency remains unresolved. Introduction of a *myogenin* coding sequence in place of *Myf5* resulted in rescue of the truncated rib phenotype and viable mice (Wang *et al.*, 1996); however, it would be important to examine skeletal muscle formation in these mutants, in light of the complexities of spatiotemporal muscle development detailed above. Further, it remains to be determined, as with all experiments of this kind, whether myogenin protein levels generated by the *Myf5* locus are equivalent to the normal levels of Myf5 protein in wild-type embryos. In *Myf5/MyoD* double null embryos carrying this *myogenin* knock-in allele(s) skeletal myogenesis was not rescued (Wang and Jaenisch, 1997), confirming the differences noted already between the determination class and differentiation class of MRFs. That compensation was only partial is not unexpected in view of the demonstration that a region outside the bHLH domain which is conserved between Myf5 and MyoD, and which is not present in myogenin or MRF4, promotes transcriptional activation by rendering chromatin accessible to the basal transcriptional apparatus (Gerber *et al.*, 1997). Remarkably, this property is confered on the myogenin protein when this motif is incorporated into *myogenin* coding sequences. The observation that all MRFs can effect myogenic conversion in non-muscle cells *in vitro* (Weintraub *et al.*, 1991) does not therefore imply that they are equivalent. Overexpression studies in cultured muscle cells force the system, and important regulatory subtleties are lost.

Myf5 and *MyoD* have distinct expression patterns implying that their regulatory sequences are different and that they probably respond to different signaling molecules (see Section VI). During myotome formation, if Myf5 and MyoD program predominantly epaxial and hypaxial myogenesis, respectively, as has been suggested (Cossu *et al.*, 1996b; Kablar *et al.*, 1997), it is possible that distinct roles have evolved for these MRFs in different regions of the embryo. However, at present there is no evidence that different muscle genes are expressed in "Myf5" versus "MyoD" muscles.

Differences in their responses to signaling molecules may mean that *Myf5* and *MyoD* are expressed at different levels; initial *Myf5* expression in the hypaxial domain is low, for example, compared to that seen for *MyoD*. This raises the question of threshold levels necessary to initiate myogenesis. On a *MyoD* null background, *Myf5* is haploinsufficient, two functional alleles being more efficient in programming myogenesis than one. The question of threshold levels has now begun to be addressed in greater detail by examining different intercrosses between myogenic factor mutant mice. Notably, *MyoD/MRF4* double mutants have severe deficiencies in skeletal muscle cell differentiation suggesting that *myogenin* alone does not suffice to drive differentiation (Rawls *et al.*, 1998). Even in the presence of two functional *Myf5* alleles, limb myogenesis is delayed in *MyoD* null embryos until E13.5 as assayed by the absence of desmin and myosin heavy chain protein expression (Kablar *et al.*, 1997). Normally *Myf5* is activated in limb muscle precursors prior to *MyoD* and it was expected that Myf5, and myogenin would program limb myogenesis even in the absence of MyoD. How myogenesis eventually procedes in the limbs of *MyoD* null embryos remains to be determined. These experiments provide compelling evidence that Myf5 and MyoD are not functionally redundant transcription factors, but that they can compensate for each others functions. In the myotome, *MRF4* is expressed together with *Myf5* and muscle forms, whereas in the limb, *MRF4* expression is not detected by *in situ* hybridization until fetal stages. Presumably in the early limb in the absence of MyoD, myogenin (and Myf5) protein has to accumulate to a sufficient threshold level for the activation of downstream muscle genes. Curiously, however, *desmin* is activated in the myotome prior to myogenin and MRF4 protein accumulation, presumably by *Myf5*. This underlines the differences between myotomal and limb myoblasts. In the older myotome when myogenin protein accumulates and MRF4 levels decline, MyoD is present.

That MyoD may play a role in muscle differentiation as well as determination is suggested by observations on cultured muscle cells. As these differentiate, MyoD levels increase whereas *Myf5* is downregulated. Furthermore, myoblasts from *myogenin* null embryos, when cultured *in vitro*, will differentiate (Nabeshima *et al.*, 1993). *MRF4* is not upregulated and MyoD is therefore the candidate "differentiation" MRF in this situation (see Buckingham, 1994). Its potential to function in differentiating muscle thus appears to distinguish MyoD from Myf5, although this again may reflect differences in the response of gene regulatory sequences to cell cycle withdrawal. However, it appears that both proteins are differentially regulated during the cell cycle (Kitzmann *et al.*, 1998). Their susceptibility to proteolysis, and phosphorylation, or other post-transcriptional modi-

D. Myf5 and Pax3 Act Genetically Upstream of MyoD

Based on the observation that skeletal muscles are present in both *Myf5* and *MyoD* null mice and the report that *MyoD* activation takes place normally in *Myf5* null mice (Braun *et al.*, 1994) it had been concluded that *MyoD* activation is independent of *Myf5*, and that *Myf5* and *MyoD* act in parallel genetic pathways. Recent experiments, however, using *Pax3* (*splotch;* Sp) and *Myf5* single and double mutant mice have necessitated a reinterpretation of this view (Tajbakhsh *et al.*, 1997). *Pax3* is expressed in neuroectoderm and neural crest cells, as well as in mesoderm, well before myogenesis is initiated (Goulding *et al.*, 1991). As the somite matures, Pax3 transcripts in the dermomyotome are mostly concentrated in the hypaxial somitic bud. *Pax3* expression also marks muscle progenitor cells migrating into the limb, diaphragm, and hypoglossal chord (Franz *et al.*, 1993; Bober *et al.*, 1994; Goulding *et al.*, 1994; Tajbakhsh *et al.*, 1997). Consistent with this expression pattern, *splotch* mutants exhibit truncation of the characteristically curved hypaxial somitic bud and muscle defects, seen primarily in hypaxially derived skeletal muscle masses. It would therefore have been predicted that *Sp/Myf5* double-mutant embryos would exhibit a delay in myotome function characteristic of *Myf5* null embryos and subsequently, once *MyoD* was activated, a *splotch* type phenotype. However, strikingly, in *Sp/Myf5* double mutants, body muscles are absent. Unexpectedly, MyoD cannot rescue these muscles in the absence of functional Pax3 and Myf5. Anterior muscles are present including head muscles formed from paraxial head and prechordal mesoderm (Tajbakhsh *et al.*, 1997). Neither *Pax3* nor the orthologue *Pax7* (Mansouri *et al.*, 1996) are expressed in the head mesoderm at early stages. However, surprisingly, anterior muscles derived also from occipital somites are present. The regulatory genes that substitute for *Pax3* in this anterior region of the embryo remain to be identified.

Although the absence of hypaxially derived muscles is not surprising in the *Sp/Myf5-nlacZ* double mutant, the absence of epaxial myogenesis was unexpected. However, *Pax3* is also expressed in the epaxial and caudal dermomyotome lips and skeletal muscle perturbations have been observed in epaxial muscles of *splotch* mice (Franz *et al.*, 1993; Tajbakhsh *et al.*, 1997). In the *Sp/Myf5-nlacZ* double mutant, epaxial muscles are not rescued by MyoD. Furthermore, since *Sp* mutants have some muscles in their hind limbs, it was interesting to determine if MPCs arriving in the hind limb of *Sp/Myf5* double mutants would be rescued by MyoD. This is not the case since β-galactosidase$^+$ MPCs in the hind limb of these double mutants are located in and around the cartilage of the femur and do not form muscle (Fig. 9; see also Fig. 8).

Although skeletal muscles do not form caudal to the head in *Sp/Myf5-nlacZ* double mutants, traces of *MyoD* expression and skeletal muscle myogenesis,

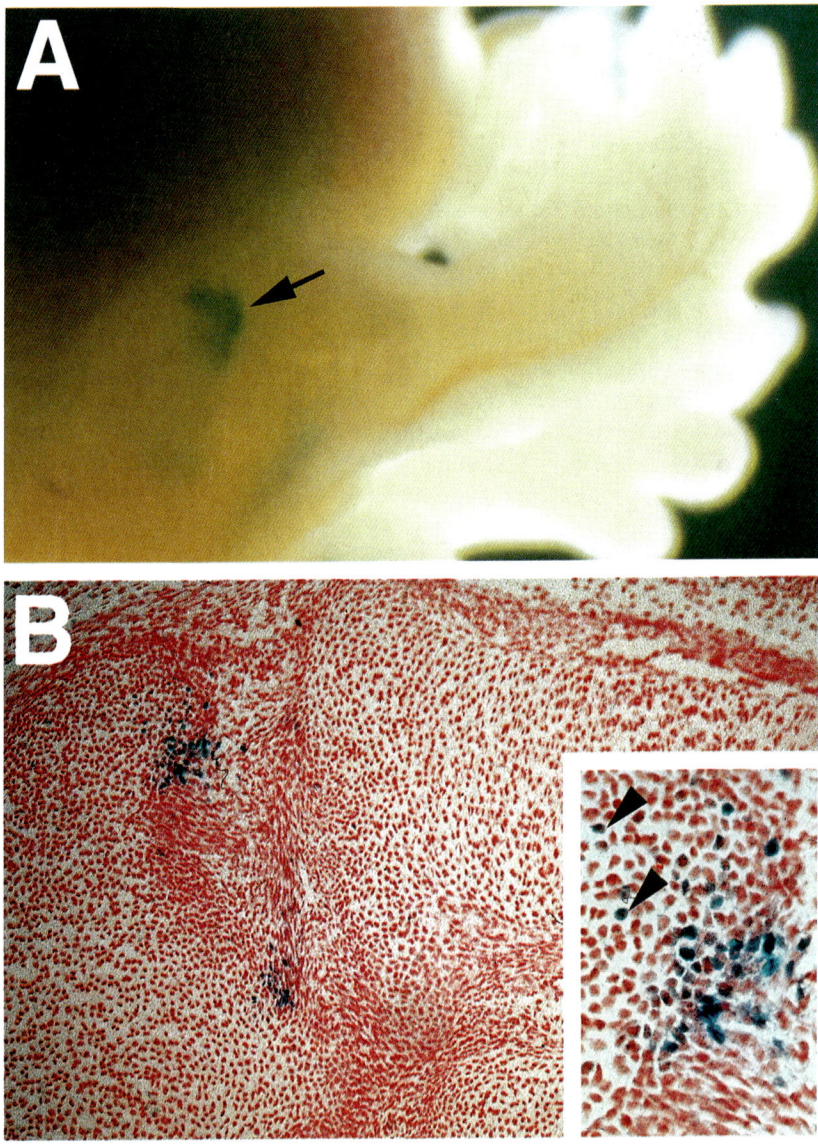

Figure 9. Muscle progenitor cells located in the hind limb of $Sp/myf5$-$nlacZ^{-/-}$ double homozygous embryos remain deficient in their ability to make muscle. Whole mount X-gal staining (A) and section (B) of an E14.5 $Sp/myf5$-$nlacZ^{-/-}$ embryo showing β-galactosidase$^+$ cells located in and surrounding the femur in the hind limb of these double mutants. These cells are not rescued by MyoD and therefore do not form muscle.

nevertheless, are observed in the body (Tajbakhsh et al., 1997; S. Tajbakhsh and M. Buckingham, unpublished). It is not yet clear from which domain(s) of the somite this residual myogenesis originates and how this molecular activation of *MyoD* takes place. One possibility is that the *splotch* allele does not reflect a true null. Indeed, 45 amino acids of the Pax3 protein are deleted in *splotch*, encompassing part of the paired domain and the octapeptide motif, leaving the rest of the Pax3 molecule intact (Goulding et al., 1993). Thus some *MyoD* activation may take place from a partially active Pax3 protein in this case. Another possiblility is that this residual myogenesis is an evolutionary carryover of what is observed in *Drosophila* or *Caenorhabditis elegans*, where a null mutation in the single myogenic factor gene does not result in the total absence of skeletal muscle in these organisms (Bate, 1992; Harfe et al., 1998).

At present, it is not clear precisely what role Pax3 plays in myogenesis. One suggestion is that Pax3 is necessary for the proliferation of dermomyotomal cells. Consistent with this notion, embryos doubly mutant for *Pax3* and *Pax7* exhibit a much reduced dermomyotome epithelium (A. Mansouri and P. Gruss, personal communication). *Pax7* null embryos, however, do not exhibit any obvious dermomyotomal defects (Mansouri et al., 1996). Using a genetic approach, it was demonstrated that Pax3 is not necessary for the activation of the *Myf5* gene (Tajbakhsh et al., 1997), although it has been suggested that Pax3 is necessary for the activation of *MyoD* (Maroto et al., 1997; Tajbakhsh et al., 1997), perhaps directly (Maroto et al., 1997). However, an examination of *Mox2* null embryos revealed that although *Pax3* expression was severely reduced in limb muscle precursors, this did not result in a downregulation of the *MyoD* gene, suggesting that *Pax3* activation of *MyoD* may be indirect (Mankoo et al., 1999).

As in the case of Myf5, Pax3 may play a key role in the correct positioning of myogenic progenitor cells. Defects similar to those seen in *splotch* embryos were observed in embryos null for the *c-met* tyrosine kinase receptor gene (Bladt et al., 1995; Maina et al., 1996) and its ligand, hepatocyte growth factor/scatter factor (HGF/SF) (Schmidt et al., 1995; Uehara et al., 1995). *c-met* is expressed on the surface of epithelial cells in the hypaxial somitic bud, and of mesenchymal cells migrating from this site; it is also expressed in the medial dermomyotome (Sonnenberg et al., 1993; Yang et al., 1996). HGF/SF is expressed along the pathway of migrating muscle progenitor cells (diaphragm, hypoglossal chord) and in the limb bud mesenchyme (Dietrich et al., 1999). Interaction of HGF/SF and c-met is necessary for cells to undergo an epithelial–mesenchymal transition but it remains unclear if the c-met signaling pathway is also necessary for the subsequent migration of these cells to more distal sites (Brand-Saberi et al., 1996; Heymann et al., 1996). It has been suggested that *c-met* is a downstream target of Pax3, since Pax3 binding sites are located in the *c-met* promoter (Epstein et al., 1996). Although *c-met* expression was not detected in *splotch* embryos (Yang et al., 1996) in keeping with the notion that it acts downstream of Pax3, more recent studies have demonstrated that some c-met transcripts are indeed present in the somites of *splotch* embryos (Mennerich et al., 1998). The role that *c-met* plays in the myogenic regulatory network was investigated by examining *c-met*/*Myf5* double mutants. Preliminary analysis reveals that although skeletal muscle defects are evident in the trunk and limbs of these double mutants, some skeletal muscle masses are present in the trunk suggesting that c-met does not mediate all the skeletal muscle functions of Pax3 (S. Tajbakhsh, C. Ponzetto, and M. Buckingham, unpublished).

IX. Conclusions

This review has been concerned primarily with the spatiotemporal dynamics of gene expression patterns in the somite, in muscle progenitor cells and their derivatives, skeletal muscles. The analysis of this molecular anatomy in embryos, with mutations in genes encoding key myogenic regulatory factors and signaling molecules which influence their expression, will continue to extend our understanding of how myogenesis is initiated. The cell biology of the system is only beginning to be understood *in vivo;* lineage studies and imaging analysis with vital dyes should provide more insight into the expression and function of regulatory genes in individual myogenic cells, as well as the fate of such cells in the embryo. With the discovery of molecules such as numb that intervene in cell fate decisions resulting from the asymmetric cell division of progenitor cells, it is now pertinent to examine how decisions are made in the dermomyotome with respect to cell proliferation, or entry into the muscle, dermal, or cartilage (rib) cell lineages. Once a progenitor cell has activated a key myogenic regulatory gene, such as *Myf5* or *Pax3*, identification of the immediate targets, prior to the onset of differentiation, remains an important question. In the case of these genes, their role in the correct positioning of precursor cells as they leave the dermomyotome is a key issue, as it is for the epithelial to mesenchymal transition itself, and the control of proliferation in the progenitor cell population. Finally, the developmental evolution of myogenesis can be very informative. Even for an organism such as the mouse where skeletal muscle formation very largely depends on the MyoD family of myogenic regulators, vestiges of other myogenic strategies that have assumed much more importance in invertebrates such as *Drosophila* or *C. elegans* can be detected. Important insights into the interplay between the environment and myogenic cells, both at the level of the initial specification of these cells and their subsequent organization within a muscle mass, come from these organisms and other vertebrate models such as the zebrafish.

Acknowledgments

We are grateful to our colleague G. Cossu for comments and D. Rocancourt for artwork. We also thank R. Spörle for his contribution both for discussions and for permission to quote unpublished data. The laboratory is supported by grants from the Pasteur Institute, CNRS, AFM, and a grant to ST from the HFSPO.

References

Bate, M. (1992). Mechanisms of muscle patterning in *Drosophila*. *Semin. Dev. Biol.* **3,** 267–275.
Baylies, M. K., Bate, M., and Ruiz Gomez, M. (1998). Myogenesis: A view from *Drosophila*. *Cell* **93,** 921–927.
Biben, C., Hadchouel, J., Tajbakhsh, S. and Buckingham, M. (1996). Developmental and tissue-specific regulation of the murine cardiac actin gene *in vivo* depends on distinct skeletal and cardiac muscle-specific enhancer elements in addition to the proximal promoter. *Dev. Biol.* **173,** 200–212.
Bladt, F., Riethmacher, D., Isenmann, S., Aguzzi, A., and Birchmeier, C. (1995). Essential role for the c-met receptor in the migration of myogenic precursor cells into the limb bud. *Nature* **376,** 768–771.
Blagden, C. S., Currie, P. D., Ingham, P. W. and Hughes, S. M. (1997). Notochord induction of zebrafish slow muscle mediated by sonic hedgehog. *Genes Dev.* **11,** 2163–2175.
Bober, E., Lyons, G. E., Braun, T., Cossu, G., Buckingham, M., and Arnold, H. H. (1991). The muscle regulatory gene, *Myf-6*, has a biphasic pattern of expression during early mouse development. *J. Cell Biol.* **113,** 1255–1265.
Bober, E., Franz, T., Arnold, H. H., Gruss, P., and Tremblay, P. (1994). Pax-3 is required for the development of limb muscles: A possible role for the migration of dermomyotomal muscle progenitor cells. *Development* **120,** 603–612.
Borycki, A. G., Mendham, L., and Emerson, C. P., Jr. (1998). Control of somite patterning by Sonic hedgehog and its downstream signal response genes. *Development* **125,** 777–790.
Borycki, A. G., Brunk, B., Tajbakhsh, S., Buckingham, M., Chiang, C., and Emerson, C. P. Jr. (1999). Sonic hedgehog controls epaxial muscle determination through *Myf5* activation. *Development,* in press.
Brand-Saberi, B., Ebensperger, C., Wilting, J., Balling, R., and Christ, B. (1993). The ventralizing effect of the notochord on somite differentiation in chick embryos. *Anat. Embryol.* **188,** 239–245.
Brand-Saberi, B., Muller, T. S., Wilting, J., Christ, B. and Birchmeier, C. (1996). Scatter factor/hepatocyte growth factor (SF/HGF) induces emigration of myogenic cells at interlimb level *in vivo*. *Dev. Biol.* **179,** 303–308.
Braun, T., and Arnold, H.-H. (1996). *myf-5* and *myoD* genes are activated in distinct mesenchymal stem cells and determine different skeletal muscle cell lineages. *EMBO J.* **15,** 310–318.
Braun, T., Rudnicki, M. A., Arnold, H. H., and Jaenisch, R. (1992). Targeted inactivation of the muscle regulatory gene Myf-5 results in abnormal rib development and perinatal death. *Cell* **71,** 369–382.
Braun, T., Bober, E., Rudnicki, M. A., Jaenisch, R., and Arnold, H. H. (1994). MyoD expression marks the onset of skeletal myogenesis in Myf-5 mutant mice. *Development* **120,** 3083–3092.
Brill, G., Kahane, N., Carmeli, C., von Schack, D., Barde, Y. A., and Kalcheim, C. (1995). Epithelial–mesenchymal conversion of dermatome progenitors requires neural tube-derived signals: Characterization of the role of Neurotrophin-3. *Development* **121,** 2583–2594
Buckingham, M. (1992). Making muscle in mammals. *Trends Genet.* **8,** 144–149.
Buckingham, M. (1994). Which myogenic factors make muscle? *Curr. Biol.* **4,** 61–63.
Burgess, R., Cserjesi, P., Ligon, K. L., and Olson, E. N. (1995). Paraxis: A basic helix-loop-helix protein expressed in paraxial mesoderm and developing somites. *Dev. Biol.* **168,** 296–306.
Carmena, A., Murugasu-Oei, B., Menon, D., Jimenez, F., and Chia, W. (1998). Inscuteable and numb mediate asymmetric muscle progenitor cell divisions during *Drosophila* myogenesis [published erratum appears in *Genes Dev.* (1998) **Apr 15;** 12(8):1241]. *Genes Dev* **12,** 304–315.
Cheng, T. C., Wallace, M. C., Merlie, J. P. and Olson, E. N. (1993). Separable regulatory elements governing myogenin transcription in mouse embryogenesis. *Science* **261,** 215–218.
Chiang, C., Ying, L. T. T., Lee, E., Young, K. E., Corden, J. L., Westphal, H., and Beachy, P. A. (1996). Cyclopia and defective axial patterning in mice lacking sonic hedgehog gene function. *Nature* **383,** 407–413.

Christ, B., and Ordahl, C. P. (1995). Early stages of chick somite development. *Anat. Embryol.* **191**, 381–396.

Christiansen, J. H., Dennis, C. L., Wicking, C. A., Monkley, S. J., Wilkinson, D. G. and Wainwright, B. J. (1995). Murine Wnt-11 and Wnt-12 have temporally and spatially restricted expression patterns during embryonic development. *Mech. Dev.* **51**, 341–350.

Cooper, R. N., Tajbakhsh, S., Mouly, V., Cossu, G., Buckingham, M., and Butler-Browne, G. S. (1999). Two distinct pathways of *in vivo* satellite cell activation in regenerating mouse skeletal muscle. *J. Cell Sci.* in press.

Corbin, V., Michelson, A. M., Abmayr, S. M., Neel, V., Alcamo, E., Maniatis, T., and Young, M. W. (1991). A role for the *Drosophila* neurogenic genes in mesoderm differentiation. *Cell* **67**, 311–323.

Cornelison, D. D. W., and Wold, B. J. (1997). Single-cell analysis of regulatory gene expression in quiescent and activated mouse skeletal muscle satellite cells. *Dev. Biol.* **191**, 270–283.

Couso, J. P., and Martinez Arias, A. (1994). Notch is required for wingless signaling in the epidermis of *Drosophila*. *Cell* **79**, 259–272.

Cossu, G., Tajbakhsh, S., and Buckingham, M. (1996a). How is myogenesis initiated in the embryo? *Trends Genet.* **12**, 218–223.

Cossu, G., Kelly, R., Tajbakhsh, S., Donna, S. D., Vivarelli, E., and Buckingham, M. (1996b). Activation of different myogenic pathways: myf-5 is induced by the neural tube and MyoD by the dorsal ectoderm in mouse paraxial mesoderm. *Development* **122**, 429–437.

Cusella de Angelis, M. G., Lyons, G., Sonnino, C., de Angelis, L., Vivarelli, E., Farmer, K., Wright, W. E., Molinaro, M., Bouche, M., Buckingham, M., and Cossu, G. (1992). MyoD, myogenin independent differentiation of primordial myoblasts in mouse somites. *J. Cell Biol.* **116**, 1243–1255.

Denetclaw, W. F., Christ, B., and Ordahl, C. P. (1997). Location and growth of epaxial myotome precursor cells. *Development* **124**, 1601–1610.

Devoto, S. H., Melancon, E., Eisen, J. S., and Westerfield, M. (1996). Identification of separate slow and fast muscle precursor cells *in vivo*, prior to somite formation. *Development* **122**, 3371–3380.

Dietrich, S., Schubert, F. R., and Lumsden, A. (1997). Control of dorsoventral pattern in the chick paraxial mesoderm. *Development* **124**, 3895–3908.

Dietrich, S., Abou-Rebyeh, F., Brohmann, H., Bladt, F., Sonnenberg-Riethmacher, E., Yamaai, T., Lumsden, A., Brand-Saberi, B., and Birchmeier, C. (1999). The role of SF/HGF and c-met in the development of skeletal muscle. *Development* **126**, 1621–1629.

Donoghue, M. J., Morris-Valero, R., Johnson, Y. R., Merlie, J. P., and Sanes, J. R. (1992). Mammalian muscle cells bear a cell-autonomous, heritable memory of their rostrocaudal position. *Cell* **69**, 67–77.

Du, J. S., Devoto, S. H., Westerfield, M., and Moon, R. T. (1997). Positive and negative regulation of muscle cell identity by members of the hedgehog and Tgfβ gene families. *J. Cell Biol.* **139**, 145–156.

Duprez, D., Fournierthibault, C., and Le Douarin, N. (1998). Sonic hedgehog induces proliferation of committed skeletal muscle cells in the chick limb. *Development* **125**, 495–505.

Epstein, J. A., Shapiro, D. N., Cheng, J., Lam, P. Y., and Maas, R. L. (1996). Pax3 modulates expression of the c-Met receptor during limb muscle development. *Proc. Natl. Acad. Sci. U.S.A.* **93**, 4213–4218.

Faerman, A., Goldhamer, D. J., Puzis, R., Emerson, C. P., Jr., and Shani, M. (1995). The distal human myoD enhancer sequences direct unique muscle-specific patterns of lacZ expression during mouse development. *Dev. Biol.* **171**, 27–38.

Fan, C.-M., and Tessier-Lavigne, M. (1994). Patterning of mammalian somites by surface ectoderm and notochord: Evidence for sclerotome induction by a *hedgehog* homolog. *Cell* **79**, 1175–1186.

Fan, C.-M., Lee, C. S., and Tessier-Lavigne, M. (1997). A role for Wnt proteins in induction of dermomyotome. *Dev. Biol.* **191**, 160–165.

Ferrari, G., Cusella de Angelis, G., Coletta, M., Paolucci, E., Stornaiuolo, A., Cossu, G., and Mavilio, F. (1998). Muscle regeneration by bone marrow-derived myogenic progenitors [see com-

ments] [published erratum appears in *Science* (1998) **Aug. 14;** 281(5379):923]. *Science* **279,** 1528–1530.
Franz, T., Kothary, R., Surani, M. A., Halata, Z., and Grim, M. (1993). The Splotch mutation interferes with muscle development in the limbs. *Anat. Embryol.* **187,** 153–160.
Gardner, C. A., and Barald, K. F. (1992). Expression patterns of engrailed-like proteins in the chick embryo. *Dev. Dyn.* **193,** 370–388.
George-Weinstein, M., Gerhart, J., Reed, R., Flynn, J., Callihan, B., Mattiacci, M., Miehle, C., Foti, G., Lash, J. W., and Weintraub, H. (1996). Skeletal myogenesis: The preferred pathway of chick embryo epiblast cells *in vitro*. *Dev. Biol.* **173,** 279–291.
Gerber, A. N., Klesert, T. R., Bergstrom, D. A., and Tapscott, S. J. (1997). Two domains of myoD mediate transcriptional activation of genes in repressive chromatin—A mechanism for lineage determination in myogenesis. *Genes Dev.* **11,** 436–450.
Goldhamer, D. J., Faerman, A., Shani, M., and Emerson, C., Jr. (1992). Regulatory elements that control the lineage-specific expression of myoD. *Science* **256,** 538–542.
Gossler, A., and Hrabe de Angelis, M. (1997). Somitogenesis. *Curr. Top. Dev. Biol.,* **38,** 225–287.
Goulding, M. D., Chalepakis, G., Deutsch, U., Erselius, J. R., and Gruss, P. (1991). Pax-3, a novel murine DNA binding protein expressed during early neurogenesis. *EMBO J.* **10,** 1135–1147.
Goulding, M., Sterrer, S., Fleming, J., Balling, R., Nadeau, J., Moore, K. J., Brown, S. D., Steel, K. P., and Gruss, P. (1993). Analysis of the *Pax-3* gene in the mouse mutant splotch. *Genomics* **17,** 355–363.
Goulding, M., Lumsden, A., and Paquette, A. J. (1994). Regulation of Pax-3 expression in the dermomyotome and its role in muscle development. *Development* **120,** 957–971.
Grass, S., Arnold, H. H., and Braun, T. (1996). Alterations in somite patterning of Myf-5 deficient mice: A possible role for FGF-4 and FGF-6. *Development* **122,** 141–150.
Grieshammer, U., Sassoon, D., and Rosenthal, N. (1992). A transgene target for positional regulators marks early rostrocaudal specification of myogenic lineages. *Cell* **69,** 79–93.
Hacker, A., and Guthrie, S. (1998). A distinct developmental programme for the cranial paraxial mesoderm in the chick embryo. *Development* **125,** 3461–3472.
Harfe, B. D., Branda, C. S., Krause, M., Stern, M. J., and Fire, A. (1998). MyoD and the specification of muscle and non-muscle fates during postembryonic development of the *C. elegans* mesoderm. *Development* **125,** 2479–2488.
Hasty, P., Bradley, A., Morris, J. H., Edmondson, D. G., Venuti, J. M., Olson, E. N., and Klein, W. H. (1993). Muscle deficiency and neonatal death in mice with a targeted mutation in the myogenin gene. *Nature* **364,** 501–506.
Hatta, K., BreMiller, R. A., Westerfield, M., and Kimmel, C. B. (1991). Diversity of expression of *engrailed* homeoproteins in zebrafish. *Development* **112,** 821–832.
Hay, E. D. (1993). Extracellular matrix alters epithelial differentiation. *Curr. Opin. Cell Biol.* **5,** 1029–1035.
Heymann, S., Koudrova, M., Arnold, H., Koster, M., and Braun, T. (1996). Regulation and function of SF/HGF during migration of limb muscle precursor cells in chicken. *Dev. Biol.* **180,** 566–578.
Hinterberger, T. J., Sassoon, D. A., Rhodes, S. J., and Konieczny, S. F. (1991). Expression of the muscle regulatory factor MRF4 during somite and skeletal myofiber development. *Dev. Biol.* **147,** 144–156.
Hirsinger, E., Duprez, D., Jouve, C., Malapert, P., Cooke, J., and Pourquié, O. (1997). Noggin acts downstream of Wnt and Sonic Hedgehog to antagonize BMP4 in avian somite patterning. *Development* **124,** 4605–4614.
Ho, R. K., Ball, E. E., and Goodman, C. S. (1983). Muscle pioneers: Large mesodermal cells that erect a scaffold for developing muscles and motoneurones in grasshopper embryos. *Nature* **301,** 66–69.
Hoh, J. F. Y., and Hughes, S. (1988). Myogenic and neurogenic regulation of myosin gene expression in cat jaw-closing muscles. *J. Muscle Res. Cell Motil.* **9,** 59–72.
Ikeya, M. and Takada, S. (1998). Wnt signaling from the dorsal neural tube is required for the formation of the medial dermomyotome. *Development* **125,** 4969–4976.

Jagla, K., Dollé, P., Mattei, M.-G., Jagla, T., Schuhbaur, B., Dretzen, G., Bellard, F., and Bellard, M. (1995). Mouse *Lbx1* and human *LBX1* define the novel mammalian homeobox gene family related to the *Drosophila lady bird* genes. *Mech. Dev.* **53**, 345–356.

Jan, Y. N., and Jan, L. Y. (1998). Asymmetric cell division. *Nature* **392**, 775–778.

Kablar, B., Krastel, K., Ying, C., Asakura, A., Tapscott, S. J., and Rudnicki, M. A. (1997). MyoD and Myf-5 differentially regulate the development of limb versus trunk skeletal muscle. *Development* **124**, 4729–4738.

Kablar, B., Krastel, K., Ying, C., Tapscott, S. J., Goldhamer, D. J., Rudnicki, M. A. (1999). Myogenic determination occurs independently in somites and limb buds. *Dev. Biol.* **206**, 219–231.

Kaehn, K., Jacob, H. J., Christ, B., Hinrichsen, K., and Poelmann, R. E. (1988). The onset of myotome formation in the chick. *Anat. Embryol.* **177**, 191–201.

Kahane, N., Cinnamon, Y., and Kalcheim, C. (1998a). The cellular mechanism by which the dermomyotome contributes to the second wave of myotome development. *Development* **125**, 4259–4271.

Kahane, N., Cinnamon, Y., and Kalcheim, C. (1998b). The origin and fate of pioneer myotomal cells in the avian embryo. *Mech. Dev.* **74**, 59–73.

Kanzler, B., Prin, F., Thelu, J., and Dhouailly, D. (1997). CHOXC-8 and CHOXD-13 expression in embryonic chick skin and cutaneous appendage specification. *Dev. Dyn.* **210**, 274–287.

Kato, N., and Aoyama, H. (1998). Dermomyotomal origin of the ribs as revealed by extirpation and transplantation experiments in chick and quail embryos. *Development* **125**, 3437–3443.

Kelly, R., Alonso, S., Tajbakhsh, S., Cossu, G., and Buckingham, M. (1995). Myosin light chain 3F regulatory sequences confer regionalised cardiac and skeletal muscle reporter gene expression in transgenic mice. *J. Cell Biol.* **129**, 383–396.

Kengaku, M., Capdevila, J., Rodriguez-Esteban, C., De La Pena, J., Johnson, R. L., Belmonte, J. C. I., and Tabin, C. J. (1998). Distinct WNT pathways regulating AER formation and dorsoventral polarity in the chick limb bud. *Science* **280**, 1274–1277.

Keynes, R. J., and Stern, C. D. (1988). Mechanisms of vertebrate segmentation. *Development* **103**, 413–429.

Kieny, M., Mauger, A., and Sengel, P. (1972). Early regionalization of somitic mesoderm as studied by the development of axial skeleton of the chick embryo. *Dev. Biol.* **28**, 142–161.

Kimmel, C. B., Ballard, W. W., Kimmel, S. R., Ullmann, B., and Schilling, T. F. (1995). Stages of embryonic development of the zebrafish. *Dev. Dyn.* **203**, 253–310.

Kitzmann, M., Carnac, G., Vandromme, M., Primig, M., Lamb, N. J. C., and Fernandez, A. (1998). The muscle regulatory factors MyoD and Myf-5 undergo distinct cell cycle-specific expression in muscle cells. *J. Cell Biol.* **142**, 1447–1459.

Kopan, R., Nye, J. S., and Weintraub, H. (1994). The intracellular domain of mouse Notch: A constitutively activated repressor of myogenesis directed at the basic helix-loop-helix region of MyoD. *Development* **120**, 2385–2396.

Krumlauf, R. (1994). Hox genes in vertebrate development. *Cell* **78**, 191–201.

Lindon, C., Montarras, D., and Pinset, C. (1998). Cell cycle-regulated expression of the muscle determination factor Myf5 in proliferating myoblasts. *J. Cell Biol.* **140**, 111–118.

Lu, B., Rothenberg, M., Jan, L. Y., and Jan, Y. N. (1998). Partner of Numb colocalizes with Numb during mitosis and directs Numb asymmetric localization in *Drosophila* neural and muscle progenitors. *Cell* **95**, 225–235.

Lyons, G. E., Ontell, M., Cox, R., Sassoon, D., and Buckingham, M. (1990). The expression of myosin genes in developing skeletal muscle in the mouse embryo. *J. Cell Biol.* **111**, 1465–1476.

Lyons, G. E., Buckingham, M. E., and Mannherz, H. G. (1991). alpha-Actin proteins and gene transcripts are colocalized in embryonic mouse muscle. *Development* **111**, 451–454.

McMahon, J. A., Takada, S., Zimmerman, L. B., Fan, C. M., Harland, R. M., and McMahon, A. P. (1998). Noggin-mediated antagonism of BMP signaling is required for growth and patterning of the neural tube and somite. *Genes Dev.* **12**, 1438–1452.

Maina, F., Casagranda, F., Audero, E., Simeone, A., Comoglio, P. M., Klein, R., and Ponzetto, C.

6. Muscle Progenitor Cells: Spatiotemporal Considerations

(1996). Uncoupling of Grb2 from the Met Receptor *in vivo* reveals complex roles in muscle development. *Cell* **87,** 531–542.

Mangiacapra, F. J., Roof, S. L., Ewton, D. Z., and Florini, J. R. (1992). Paradoxical decrease in myf-5 messenger RNA levels during induction of myogenic differentiation by insulin-like growth factors. *Mol. Endocrinol.* **6,** 2038–2044.

Mankoo, B. S., Collins, N. S., Ashby, P., Grigoleva, E., Pevny, L. H., Candia, A., Wright, C. V., Rigby, P. W., and Pachnis, V. (1999). Mox2 is a component of the genetic hierarchy controlling limb muscle development. *Nature* **400,** 69–73.

Mansouri, A., Stoykova, A., Torres, M., and Gruss, P. (1996). Dysgenesis of cephalic neural crest derivatives in Pax7−/− mutant mice. *Development* **122,** 831–838.

Marcelle, C., Wolf, J., and Bronner-Fraser, M. (1995). The *in vivo* expression of the FGF receptor FREK mRNA in avian myoblasts suggests a role in muscle growth and differentiation. *Dev. Biol.* **172,** 100–114.

Marcelle, C., Stark, M. R., and Bronner-Fraser, M. (1997). Coordinate actions of BMPs, Wnts, Shh and noggin mediate patterning of the dorsal somite. *Development* **124,** 3955–3963.

Maroto, M., Reshef, R., Munsterberg, A. E., Koester, S., Goulding, M., and Lassar, A. B. (1997). Ectopic Pax-3 activates MyoD and Myf-5 expression in embryonic mesoderm and neural tissue. *Cell* **89,** 139–148.

Mauger, A. (1972). Rôle du mésoderme somitique dans le développement du plumage dorsal chez l'embryon de poulet. II. Régionalisation du mésoderme plumigène. *J. Embryol. Exp. Morphol.* **28,** 343–366.

Megeney, L. A., Kablar, B., Garrett, K., Anderson, J. E., and Rudnicki, M. A. (1996). MyoD is required for myogenic stem cell function in adult skeletal muscle. *Genes Dev.* **10,** 1173–1183.

Mennerich, D., Schafer, K., and Braun, T. (1998). Pax-3 is necessary but not sufficient for lbx1 expression in myogenic precursor cells of the limb. *Mech. Dev.* **73,** 147–158.

Milaire, J. (1976). Contribution cellulaire des somites à la genèse des bourgeons de membres postérieurs chez la souris. *Arch. Biol.* **87,** 315–343.

Molkentin, J. D., and Olson, E. N. (1996). Defining the regulatory networks for muscle development. *Curr. Opin. Genet. Dev.* **6,** 445–453.

Monsoro-Burq, A. H., Duprez, D., Watanabe, Y., Bontoux, M., Vincent, C., Brickell, P., and Le Douarin, N. (1996). The role of bone morphogenetic proteins in vertebral development. *Development* **122,** 3607–3616.

Montarras, D., Chelly, J., Bober, E., Arnold, H., Ott, M. O., Gros, F., and Pinset, C. (1991). Developmental patterns in the expression of Myf5, MyoD, myogenin, and MRF4 during myogenesis. *New Biol.* **3,** 592–600.

Münsterberg, A. E., Kitajewski, J., Bumcrot, D. A., McMahon, A. P., and Lassar, A. B. (1995). Combinatorial signaling by Sonic hedgehog and Wnt family members induces myogenic bHLH gene expression in the somite. *Genes Dev.* **9,** 2911–2922.

Nabeshima, Y., Hanaoka, K., Hayasaka, M., Esumi, E., Li, S., Nonaka, I., and Nabeshima, Y. (1993). Myogenin gene disruption results in perinatal lethality because of severe muscle defect. *Nature* **364,** 532–535.

Nagai, T., Aruga, J., Takada, S., Gunther, T., Sporle, R., Schughart, K., and Mikoshiba, K. (1997). The expression of the mouse *Zic1*, *Zic2*, and *Zic3* gene suggests an essential role for *Zic* genes in body pattern formation. *Dev. Biol.* **182,** 299–313.

Olson, E. N., Perry, M., and Schulz, R. A. (1995). Regulation of muscle differentiation by the MEF2 family of MADS box transcription factors. *Dev. Biol.* **172,** 2–14.

Olson, E. N., Arnold, H.-H., Rigby, P. W. J., and Wold, B. J. (1996). Know your neighbors: Three phenotypes in null mutants of the myogenic bHLH gene *MRF4*. *Cell* **85,** 1–4.

Ordahl, C. P. (1993). Myogenic lineages within the developing somite. *In* "Molecular Basis of Morphogenesis" (M. Bernfield, Ed.), pp. 165–176. Wiley-Liss, New York.

Ordahl, C. P., and Le Douarin, N. M. (1992). Two myogenic lineages within the developing somite. *Development* **114,** 339–353.

Ott, M.-O., Bober, E., Lyons, G., Arnold, H., and Buckingham, M. (1991). Early expression of the myogenic regulatory gene, *myf-5,* in precursor cells of skeletal muscle in the mouse embryo. *Development* **111,** 1097–1107.
Patapoutian, A., Yoon, J. K., Miner, J. H., Wang, S., Stark, K., and Wold, B. (1995). Disruption of the mouse *MRF4* gene identifies multiple waves of myogenesis in the myotome. *Development* **121,** 3347–3358.
Pourquié, O., Coltey, M., Teillet, M.-A., Ordahl, C., and Le Douarin, N. M. (1993). Control of dorsoventral patterning of somitic derivatives by notocord and floor plate. *Proc. Natl. Acad. Sci. U.S.A.* **90,** 5242–5246.
Pourquié, O., Fan, C.-M., Coltey, M., Hirsinger, E., Wanatabe, Y., Bréant, C., Francis-West, P., Brickell, P., Tessier-Lavigne, M., and Le Douarin, N. (1996). Lateral and axial signals involved in avian somite patterning: A role for BMP4. *Cell* **84,** 461–471.
Pownall, M. E., Strunk, K. E., and Emerson, C. P., Jr. (1996). Notochord signals control the transcriptional cascade of myogenic bHLH genes in somites of quail embryos. *Development* **122,** 1475–1488.
Psychoyos, D., and Stern, C. D. (1996). Fates and migratory routes of primitive streak cells in the chick embryo. *Development* **122,** 1523–1534.
Rawls, A., Valdez, M. R., Zhang, W., Richardson, J., Klein, W. H., and Olson, E. N. (1998). Overlapping functions of the myogenic bHLH genes MRF4 and MyoD revealed in double mutant mice. *Development* **125,** 2349–2358.
Reshef, R., Maroto, M., and Lassar, A. B. (1998). Regulation of dorsal somitic cell fates: BMPs and Noggin control the timing and pattern of myogenic regulator expression. *Genes Dev.* **12,** 290–303.
Rudnicki, M. A., Braun, T., Hinuma, S., and Jaenisch, R. (1992). Inactivation of *MyoD* in mice leads to up-regulation of the myogenic HLH gene *Myf-5* and results in apparently normal muscle development. *Cell* **71,** 383–390.
Rudnicki, M. A., Schneglesberg, P. N. J., Stead, R. H., Braun, T., Arnold, H.-H., and Jaenisch, R. (1993). Myod or myf-5 is required for the formation of skeletal muscle. *Cell* **75,** 1351–1359.
Ruiz Gomez, M., and Bate, M. (1997). Segregation of myogenic lineages in *Drosophila* requires numb. *Development* **124,** 4857–4866.
Sassoon, D., Lyons, G., Wright, W. E., Lin, V., Lassar, A., Weintraub, H., and Buckingham, M. (1989). Expression of two myogenic regulatory factors myogenin and MyoD1 during mouse embryogenesis. *Nature* **341,** 303–307.
Schmidt, C., Bladt, F., Goedecke, S., Brinkmann, V., Zschiesche, W., Sharpe, M., Gherardi, E., and Birchmeier, C. (1995). Scatter factor/hepatocyte growth factor is essential for liver development. *Nature* **373,** 699–702.
Selleck, M. A., and Stern, C. D. (1991). Fate mapping and cell lineage analysis of Hensen's node in the chick embryo. *Development* **112,** 615–626.
Siegfried, E., and Perrimon, N. (1994). *Drosophila* wingless: A paradigm for the function and mechanism of Wnt signaling. *BioEssays* **16,** 395–404.
Smith, T. H., Block, N. E., Rhodes, S. J., Konieczny, S. F., and Miller, J. B. (1993). A unique pattern of expression of the four muscle regulatory factor proteins distinguishes somitic from embryonic, fetal and newborn mouse myogenic cells. *Development* **117,** 1125–1133.
Smith, T. H., Kachinsky, A. M., and Miller, J. B. (1994). Somite subdomains, muscle cell origins, and the four muscle regulatory factor proteins. *J. Cell Biol.* **127,** 95–105.
Sonnenberg, E., Meyer, D., Weidner, K. M., and Birchmeier, C. (1993). Scatter factor/hepatocyte growth factor and its receptor, the c-met tyrosine kinase, can mediate a signal exchange between mesenchyme and epithelia during mouse development. *J. Cell Biol.* **123,** 223–235.
Soriano, P. (1997). The PDGF alpha receptor is required for neural crest cell development and for normal patterning of the somites. *Development* **124,** 2691–2700.
Spörle, R., and Schughart, K. (1997). System to identify individual somites and their derivatives in the developing mouse embryo. *Dev. Dyn.* **210,** 216–226.

6. Muscle Progenitor Cells: Spatiotemporal Considerations 267

Spörle, R., Günther, T., Struwe, M. and Schughart, K. (1996). Severe defects in the formation of epaxial musculature in *open brain* (*opb*) mutant mouse embryos. *Development* **122,** 79–86.

Stern, H. M., Brown, A. M., and Hauschka, S. D. (1995). Myogenesis in paraxial mesoderm: Preferential induction by dorsal neural tube and by cells expressing Wnt-1. *Development* **121,** 3675–3686.

Tajbakhsh, S., and Buckingham, M. E. (1994). Mouse limb muscle is determined in the absence of the earliest myogenic factor myf-5. *Proc. Natl. Acad. Sci. U.S.A.* **91,** 747–751.

Tajbakhsh, S., and Buckingham, M. E. (1995). Lineage restriction of the myogenic conversion factor myf-5 in the brain. *Development* **121,** 4077–4083.

Tajbakhsh, S., and Cossu, G. (1997). Establishing myogenic identity during somitogenesis. *Curr. Opin. Genet. Dev.* **7,** 634–641.

Tajbakhsh, S., and Spörle, R. (1998). Somite development: Constructing the vertebrate body. *Cell* **92,** 9–16.

Tajbakhsh, S., Rocancourt, D., and Buckingham, M. (1996a). Muscle progenitor cells failing to respond to positional cues adopt non-myogenic fates in *myf-5* null mice. *Nature* **384,** 266–270.

Tajbakhsh, S., Bober, E., Babinet, C., Pournin, S., Arnold, H., and Buckingham, M. (1996b). Gene targeting the myf-5 locus with LacZ reveals expression of this myogenic factor in mature skeletal muscle fibres as well as early embryonic muscle. *Dev. Dyn.* **206,** 291–300.

Tajbakhsh, S., Rocancourt, D., Cossu, G., and Buckingham, M. (1997). Redefining the genetic hierarchies controlling skeletal myogenesis: *Pax-3* and *Myf-5* act upstream of *MyoD*. *Cell* **89,** 127–138.

Tajbakhsh, S., Vivarelli, E., Kelly, R., Papkoff, J., Duprez, D., Buckingham, M., and Cossu, G. (1998). Differential activation of Myf5 and MyoD by different Wnts in explants of mouse paraxial mesoderm and the later activation of myogenesis in the absence of Myf5. *Development* **125,** 4155–4162.

Teillet, M. A., Watanabe, Y., Jeffs, P., Duprez, D., Lapointe, F., and Ledouarin, N. M. (1998). Sonic hedgehog is required for survival of both myogenic and chondrogenic somitic lineages. *Development* **125,** 2019–2030.

Tiret, L., Le Mouellic, H., Maury, M., and Brulet, P. (1998). Increased apoptosis of motoneurons and altered somatotopic maps in the brachial spinal cord of Hoxc-8-deficient mice. *Development* **125,** 279–291.

Tonegawa, A., Funayama, N., Ueno, N., and Takahashi, Y. (1997). Mesodermal subdivision along the mediolateral axis in chicken controlled by different concentrations of BMP-4. *Development* **124,** 1975–1984.

Tosney, K. (1994). Programmed death of axial muscle primordia in somites. *J. Cell. Biochem.* **18D,** 462.

Tosney, K. W., Dehnbostel, D. B., and Erickson, C. A. (1994). Neural crest cells prefer the myotome's basal lamina over the sclerotome as a substratum. *Dev. Biol.* **163,** 389–406.

Tremblay, P., and Gruss, P. (1994). Pax: Genes for mice and men. *Pharmacol. Ther.* **61,** 205–226.

Uehara, Y., Minowa, O., Mori, C., Shiota, K., Kuno, J., Noda, T., and Kitamura, N. (1995). Placental defect and embryonic lethality in mice lacking hepatocyte growth factor/scatter factor. *Nature* **373,** 702–705.

Venuti, J. M., Morris, J. H., Vivian, J. L., Olson, E. N., and Klein, W. H. (1995). Myogenin is required for late but not early aspects of myogenesis during mouse development. *J. Cell Biol.* **128,** 563–576.

Wachtler, F., and Christ, B. (1992). The basic embryology of skeletal muscle formation in vertebrates: The avian model. *Semin. Dev. Biol.* **3,** 217–227.

Wang, Y., and Jaenisch, R. (1997). Myogenin can substitute for Myf5 in promoting myogenesis but less efficiently. *Development* **124,** 2507–2513.

Wang, Y., Schnegelsberg, P. N., Dausman, J., and Jaenisch, R. (1996). Functional redundancy of the muscle-specific transcription factors Myf5 and myogenin. *Nature* **379,** 823–825.

Weinberg, E. S., Allende, M. L., Kelly, C. S., Abdelhamid, A., Murakami, T., Andermann, P., Doerre, O. G., Grunwald, D. J., and Riggleman, B. (1996). Developmental regulation of zebrafish MyoD in wild-type, no tail and spadetail embryos. *Development* **122,** 271–280.

Weinmaster, G. (1997). The ins and outs of notch signaling. *Mol. Cell Neurosci.* **9,** 91–102.

Weintraub, H., Davis, R., Tapscott, S., Thayer, M., Krause, M., Benezra, R., Blackwell, T. K., Turner, D., Rupp, R., Hollenberg, S., Zhuang, Y., and Lassar, A. (1991). The MyoD gene family: Nodal point during specification of the muscle cell lineage. *Science* **251,** 761–766.

Williams, B. A., and Ordahl, C. P. (1994). Pax-3 expression in segmental mesoderm marks early stages in myogenic cell specification. *Development* **120,** 785–793.

Williams, L. W. (1910). The somites of the chick. *Am. J. Anat.* **2,** 55–100.

Wodarz, A., and Nusse, R. (1998). Mechanisms of Wnt signalling in development. *Annu. Rev. Cell. Dev. Biol.* **14,** 59–88.

Wurst, W., Auerbach, A. B., and Joyner, A. L. (1994). Multiple developmental defects in Engrailed-1 mutant mice: An early mid-hindbrain deletion and patterning defects in forelimbs and sternum. *Development* **120,** 2065–2075.

Yang, X. M., Vogan, K., Gros, P., and Park, M. (1996). Expression of the met receptor tyrosine kinase in muscle progenitor cells in somites and limbs is absent in Splotch mice. *Development* **122,** 2163–2171.

Yee, S.-P., and Rigby, P. W. J. (1993). The regulation of myogenin gene expression during the embryonic development of the mouse. *Genes Dev.* **7,** 1277–1289.

Yoon, J. K., Olson, E. N., Arnold, H. H., and Wold, B. J. (1997). Different MRF4 knockout alleles differentially disrupt Myf-5 expression: cis-regulatory interactions at the MRF4/Myf-5 locus. *Dev. Biol.* **188,** 349–362.

Yoshida, N., Yoshida, S., Koishi, K., Masuda, K., and Nabeshima, Y. (1998). Cell heterogeneity upon myogenic differentiation: Down-regulation of MyoD and Myf-5 generates 'reserve cells.' *J. Cell Sci.* **111,** 769–779.

Yun, K. S., and Wold, B. (1996). Skeletal muscle determination and differentiation—Story of a core regulatory network and its context. *Curr. Opin. Cell Biol.* **8,** 877–889.

Zhang, W., Behringer, R. R., and Olson, E. N. (1995). Inactivation of the myogenic bHLH *MRF4* gene results in up-regulation of myogenin and rib anomalies. *Genes Dev.* **9,** 1388–1399.

Zhu, Z., and Miller, J. B. (1997). MRF4 can substitute for myogenin during early stages of myogenesis. *Dev. Dyn.* **209,** 233–241.

Zorn, A. M. (1997). Cell–cell signalling: Frog frizbees. *Curr. Biol.* **7,** R501–R504.

7
Mouse–Chick Chimera: An Experimental System for Study of Somite Development

Josiane Fontaine-Pérus
CNRS EP1593
Faculté des Sciences et des Techniques
44322 Nantes Cedex 03, France

I. Introduction
II. Technical Aspects
   A. Creating Chimeras
   B. Cell Markers
III. Somitic Lineage into Somite Chimera
   A. Dermal Lineage
   B. Sclerotomal Lineage
   C. Myogenic Lineage
IV. Epaxial Myogenic Lineage into Neural Tube Chimera
V. What Is the Benefit of the Somite–Neural Tube Chimera?
VI. Concluding Remarks
   References

As the mammalian embryo is implanted in the uterus and not readily accessible to direct observation or manipulation, much of our understanding of mammalian somite development is based on findings in lower vertebrates. One means of overcoming the difficulties raised by intrauterine development is to engraft mouse tissue *in ovo*. The experiments described in this chapter relate to the unilateral replacement of somites in chick embryo with those from mouse fetus. Mouse somites differentiate *in ovo* in dermis, cartilage, and skeletal muscle and are able to migrate into chick host limb. A LacZ transgenic mouse strain was used to ascertain the role of the implanted somites in forming epaxial and hypaxial muscle in the chick embryo. Myogenesis occurred normally in *in ovo* developing mouse somites, and muscle cells from mouse myotome formed neuromuscular contacts with chick motor axons. After fragments of fetal mouse neural primordium were transplanted into chick embryo, mouse neural tube contributed to the mechanism maintaining myogenesis in the somites of the host embryo. A recently developed double-grafting procedure involving neural tube and somites from knockout mouse strains should elucidate the molecular events involved in early somitogenesis.

I. Introduction

Experimental production of chimeras has provided biologists with material to pose and investigate a range of questions of importance to developmental biology:

What is the fate and potential of particular embryonic cells? What is the nature of their determination? Can cell differentiation be reversed? How do cells interact during development? How is gene expression controlled?

The technical aspects of creating chimeras are directly dependent on the type of embryos used. Bird embryos have several advantages over those of other vertebrates, making some approaches interesting and feasible. The greatest advantage is continual accessibility to experimentation throughout the developmental period. Another is the ease with which the different rudiments can be delineated and thus removed and replaced with extreme precision. Avian chimeric embryos have been useful in studying the fate of various derivatives. Quail–chick chimeras have provided valuable information about the mechanisms of somite development. Orthotopical replacement of somite halves and quarters between quail and chick species has indicated which structures are derived from different parts of the somites because the distinctive nucleolar marker of the quail nuclei (Le Douarin 1969, 1973) allows host cells to be distinguished from donor cells. Experiments have shown that the dorsal half becomes the dermomyotome, that is, the precursor of striated muscle and dermis, and that the ventral half is the source of the sclerotome which gives rise to axial skeleton (Christ *et al.,* 1978). Subsequently, Ordahl and Le Douarin (1992) demonstrated that epaxial muscle originates from the dorsomedial quarter, whereas hypaxial muscle is derived from the dorsolateral quarter. Endothelial cells arise from all parts of the somites (Wilting *et al.,* 1997).

Although the avian embryo is a practical model perfectly suitable for tissue graft experiments after the incubated egg is opened, it is difficult to undertake this type of investigation in the mammalian fetus *in utero.* Consequently, much of our understanding of mammalian development is still based on findings in avian embryogenesis. Yet various questions about the origin and developmental potential of many derivatives in mammals remain unanswered. To help identify these derivatives, several authors have labeled mouse embryonic cells with a solution of DiI (1.1′,dioctadecyl-3,3,3′3′,-tetramethylindo-carbocyanine perchlorate, Molecular Probes, OR) or DiO (3,3′-dioctadecyloxa-carbocyanine perchlorate, Molecular Probes, OR), fluorescent dyes (Serbedzija *et al.,* 1990; Lee and Sze, 1993; Trainor and Tam, 1995, Tam *et al.,* 1997), or a lysinated rhodamine dextran (Beddington and Martin, 1989; Tam and Tan, 1992; Lee and Sze, 1993; Serbedzija *et al.,* 1994; Sze *et al.,* 1995) prior to culturing them. As embryo cultures can be maintained for only 40 hr, the use of this technique is too restrictive for efficient study of the developmental mechanisms in mammalian embryogenesis. This difficulty can be overcome through a technique developed in our laboratory, which consists of implanting xenografts of fetal mouse tissue into the avian embryo. This procedure is of particular interest because the murine fetal cell can be studied by the classic methods of experimental embryology in terms of what happens to avian embryo cells.

Our objective was to determine whether the development of somitic derivatives in mammals occurs according to the pattern of events in birds. Accordingly, we

7. Mouse–Chick Chimera: Studying Somite Development 271

created a chimeric mouse–chick model in which mouse somites or mouse neural tube or both are implanted into a chick host.

II. Technical Aspects

A. Creating Chimeras

1. Preparation of Donor Somites and Neural Tubes from 8.5- and 9-Day Postcoitum Mouse Fetus

After pregnant mice are killed by decapitation and their bodies washed with 70% ethanol, abdominal skin is cut off and the uterus rinsed in Mac Even's solution. Mouse embryos are then isolated and stored at 4°C in Mac Even's solution. The truncal part of the embryo is dissected out and immersed in pancreatin (GIBCO, Grand Island, New York), diluted 1:10 in Tyrode's solution for 5 min on ice before dissociation. The ectoderm is then mechanically extirpated using very thin stainless steel needles, and the somites and neural tube are isolated from surrounding tissues (i.e., lateral and intermediate plates, mesenchyme, notochord, and endoderm). These structures are transferred into Tyrode's solution to which fetal calf serum is added.

2. Preparation of Donor Dermomyotomes from 10-Day Postcoitum Mouse Fetus

The mouse dermomyotome transplantation procedure was carried out according to a previously described method used in avian embryos (Auda-Boucher and Fontaine-Pérus, 1994). The axial portions of mouse embryos are cut out at 10 days postcoitum. Pancreatin is used to facilitate dermomyotome removal, and midportions of the first 10 cranial dermomyotomes are then excised and stored in Tyrode's solution and fetal calf serum.

3. Preparation of 2-Day-Old Chick Host Embryo

Albumen (0.5 ml) is extirpated from the chick embryo by aspiration through a small hole pierced at the top of the egg, which is then obturated with a drop of sealing wax. This allows the upper part of the shell to be cut off without damaging the embryo developing at the bottom of the yolk sac. After the shell is opened, the embryonic area is rinsed with a few drops of Tyrode's solution. A mixture of Tyrode's solution and drawing ink is then microinjected beneath the embryonic area to allow its vizualization. The vitelline membrane is removed, and a few drops of Tyrode's solution are placed on the embryo to prevent dehydration.

  a. Somite Ablation.   A slit is made in the ectoderm between the neural tube and the somite area to be extirpated (generally the last five somites), which is then

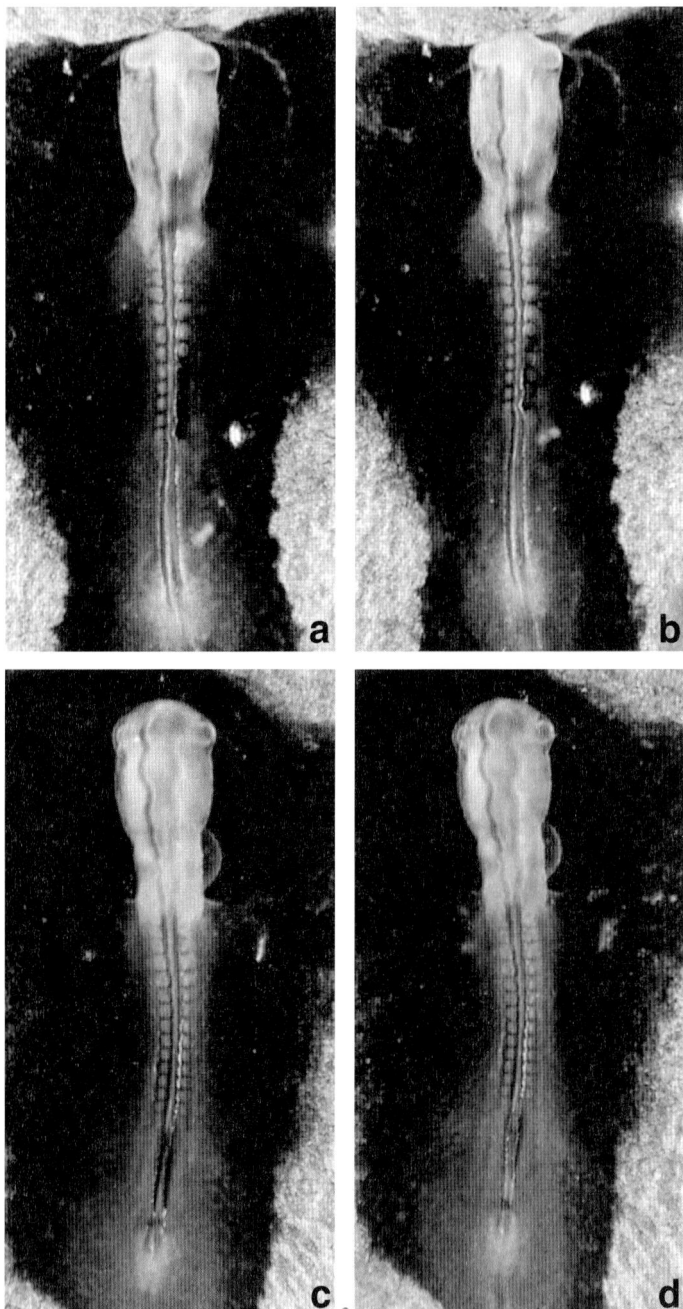

Figure 1 Somite replacement. In (a) the area appears as a black zone where the somites have been removed in the chick recipient embryo. In (b) the mouse donor somites have been grafted (magnification: ×8). Neural tube replacement. In (c) the naked notochord appears as a white line between the unsegmental plates where the tube has been removed in the chick recipient embryo. In (d) the mouse neural tube fragment has been inserted (magnification: ×8).

pushed aside. Using a micropipette, a drop of pancreatin diluted 1:2 in Tyrode's solution is laid on. Host somites are then mechanically extirpated, and the surgical zone is rapidly washed with a few drops of Tyrode's solution to which 20% fetal calf serum is added (Fig. 1a).

b. Somite Replacement. Mouse donor somites are transferred by micropipette into chick embryos. The orientation of the transplanted somites is determined with respect to that of adventitial tissues (Fig. 1b).

c. Dermomyotome Replacement. The dorsal halves of the last formed somites, together with the overlying ectoderm, are removed unilaterally in the chick host. Mid-portions of mouse dermomyotomes are then grafted into the space left by the ablation. In most cases, four to six dorsal halves of host somites are replaced with four to six dermomyotome explants. The dorsoventral positioning of the grafted tissue is not disturbed.

d. Neural Tube Ablation. The neural tube region to be isolated is delimited in the chick host by making two longitudinal slits between the neural tube and the somites and two transverse slits at the anterior and posterior limits of the graft level. The tube is then mechanically extirpated from surrounding tissues (Fig. 1c).

e. Neural Tube Replacement. Mouse neural tube fragments are transferred by micropipette into chick embryos. The dorsolateral and anteroposterior orientation is respected (Fig. 1d). The chick host is rinsed with a few drops of Tyrode's solution.

f. Neural Tube and Somite Replacement. Host neural tube and somites are first ablated, as described above, and then replaced with mouse neural tube and somites.

4. Implantation Level

Mouse somites are extirpated from 8.5 (8–15 somites) and 9 (15–18 somites) dpc embryos (day of vaginal plug: day 0 of gestation). In mouse embryos to the 10-somite stage, somites 2 to 10 are retained for transplantation into the 15- to 18-somite region of chick host. In mouse embryos between the 15- and 18-somite stage, only the last 5 somites are kept and then grafted into the last 5-somite level of the 15- to 18-somite chick host.

Mouse neural tubes arise from the caudalmost part of 8.5–9 dpc embryos, extending the posteriormost somites of 12- to 18-somite embryos. They are implanted into the 15- to 18-somite chick embryo over a length corresponding to the last 5 somites. The latter conditions are retained when mouse neural tube and somites are transplanted.

Mouse dermomyotomes are isolated from 10-dpc mouse embryos and grafted into chick host embryos at least at the 15- but no more than 20-somite stage. They are grafted into the last 5-somite level of the chick host.

B. Cell Markers

1. Cytological Marker

Interphase nuclei of mouse cells show several highly compacted heterochromatic masses which stain purple after application of the technique of Feulgen and Rossenbeck (1924), whereas chick cells do not show such masses and stain pink. This difference in staining serves as a useful cell marker when chick and mouse cells are associated to form chimeras (Fig. 2a,b). The difference in chromatin staining ability between the chick and mouse is also quite readily apparent after acridine orange (Fontaine-Pérus *et al.*, 1985) and bisbenzimide staining (Fontaine-Pérus *et al.*, 1997) (Fig. 2c). As this cytological marker is detected in virtually all nucleated cell types, most cells in a section of chimeric tissue can be usually identified as either chick or mouse. The marker has thus proved extremely useful in studies of the cell migration process.

2. Genetic Marker

Techniques allowing the manipulation of certain genes in embryonic cells have provided new insight into their role in the development of the organism. In the mouse embryo, the analysis of cell lineages has developed widely owing to a labeling method based on genetic recombination. A LacZ gene coding for β-galactosidase (β-gal) is inserted into a cell, and the descendants of this modified cell can be identified histochemically. With this method, the myotomal lineage in mouse somite-derived cells can be analyzed using LacZ gene driven by a promoter that confers an expression specifically on cells of this compartment. In addition, the LacZ desmin transgenic (Des 1-n LacZ) mouse line makes it possible to study the myogenic fate of *in ovo* grafted mouse somites (Li *et al.*, 1993). A sequence of 228 base pairs (bp) situated upstream from the transcription initiation site is sufficient to confer a low level of specifically muscular expression on the desmin gene, even in myoblasts. The high level of expression of this gene depends on a muscle-specific enhancing sequence of 280 bp located between -693 and -973 bp upstream from the transcription initiation site. The construction contains all of these sequences coupled to the β-gal enzyme gene targeted on the nuclei. The product of the β-gal gene is easily detected after addition of X-gal substrate, and desmin gene activation can then be visualized even in a single cell (Sanes *et al.*, 1986).

III. Somitic Lineage into Somite Chimera

In the mouse, somites begin to form at 8 dpc and develop caudally over a period of several days (Tam, 1981). Initially, they take the form of epithelial balls of cells

7. Mouse-Chick Chimera: Studying Somite Development

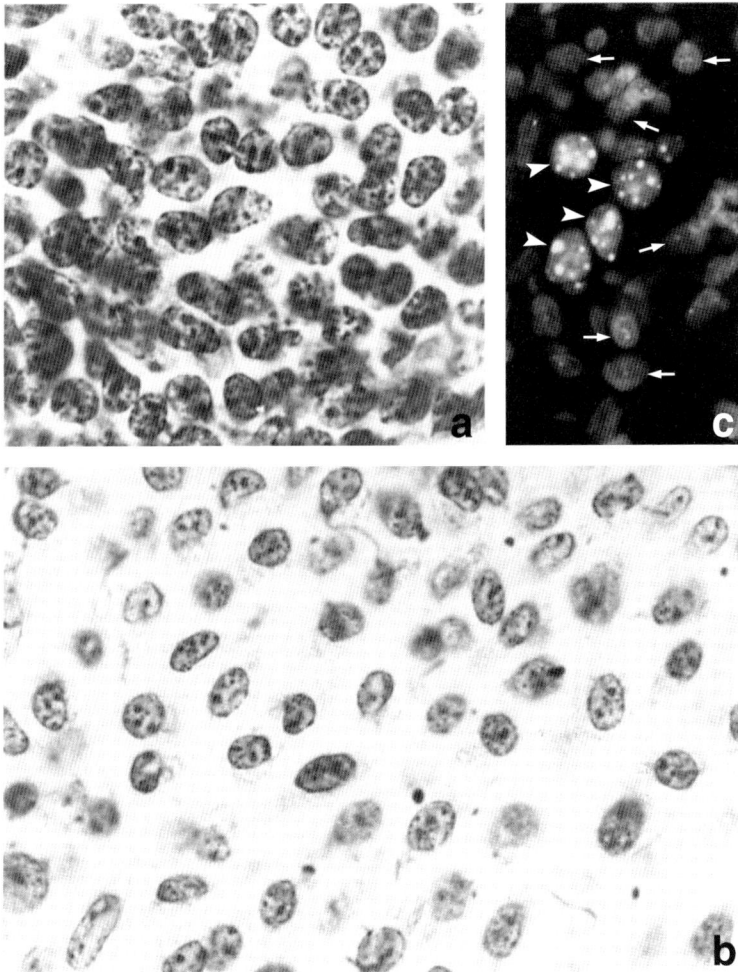

Figure 2 Distinction of mouse and chick cells. Feulgen–Rossenbeck staining of mouse myotomal cells in chick chimera at day 5 of incubation. Mouse (a) and chick (b) nuclei are easy to distinguish because of differences in DNA staining (magnification: ×960). (c) Mouse nuclei (arrowheads) appear more brightly fluorescent than chick nuclei (arrows) after Hoechst staining (magnification: ×900).

with no discernible morphological heterogeneity (Rugh, 1968; Tam and Beddington, 1986). Early morphological change occurs when cells in the ventromedial part of the somite lose their epithelial arrangement and form a loosely structured mesenchyme from which sclerotome cells are derived. The sclerotome ultimately participates in axial structures such as vertebrae and ribs. Cells in the dorsal part of the somite retain their epithelial organization and form the dermomyotome. The dermomyotome subsequently develops into two structures, an outer dermatome cell layer that contributes to the skin and an inner myotome cell layer which differentiates into myotomal muscles from the outset of axial musculature (Rugh, 1968).

A. Dermal Lineage

In birds, head dermis differentiates from neural crest cells (Lelièvre and Le Douarin, 1975; Couly *et al.*, 1993), whereas lateral and ventral dermis of the trunk originates from the somatopleure and dorsal dermis derives from the dermatome or upper part of the somite (Mauger, 1972). Brill *et al.* (1995) have demonstrated that cells of the epithelial dermatome need neural tube-derived cues to develop into the mesenchyme constituting the dorsal dermis, a process which may require neurotrophin-3 as the neural-tube derived signal in the early stages. Yet, although many works are currently being devoted to dermis development in birds, virtually nothing is known about this process in mammals.

As early as 24 hr after grafting of 9 dpc mouse somites, implanted cells first develop into dermatome. After 2 days, dermatome loses its epithelial structure and gives rise to scattered mouse cells beneath the chick ectoderm. The location of these cells overlaps the grafted zone and is restricted to the dorsal and lateral parts of the host embryo (Fig. 3). Regardless of the level of origin of the implanted somites, they give rise to dermis in the chick. When somitic dermomyotomes of 10-dpc mouse embryos unilaterally replace the dorsal halves of somites from 18- to 20-somite chick embryos, the dermomyotomes from which the graft is cut out comprise several cell types: dermal precursor cells, myoblasts in their proliferative phase, and myotubes. Macroscopic examination shows no alterations in the bilateral symmetry of recipient embryos. Grafted structures appear to be normal in size compared to those of the unoperated side. Transplantation does not disrupt subsequent somite development, and mouse dermomyotome participates in chick host somitogenesis. Conventional histology shows mouse cells in dermis in a zone overlapping the dorsolateral part of the grafted area.

B. Sclerotomal Lineage

After de-epithelialization of somites, medial sclerotomal cells migrate and proliferate toward the notochord to form the anlagen of vertebral bodies and intervertebral disks, whereas lateral sclerotomal cells form the pedicles and laminae of neural arches and ribs (Christ and Wilting, 1992). The notochord and the neural tube are involved in the polarization of somites that leads to the de-epithelialization of their ventral halves. The influence of the notochord has been analyzed mainly in avian embryos (Watterson *et al.*, 1954; Avery *et al.*, 1956; Cooper, 1965) in which both

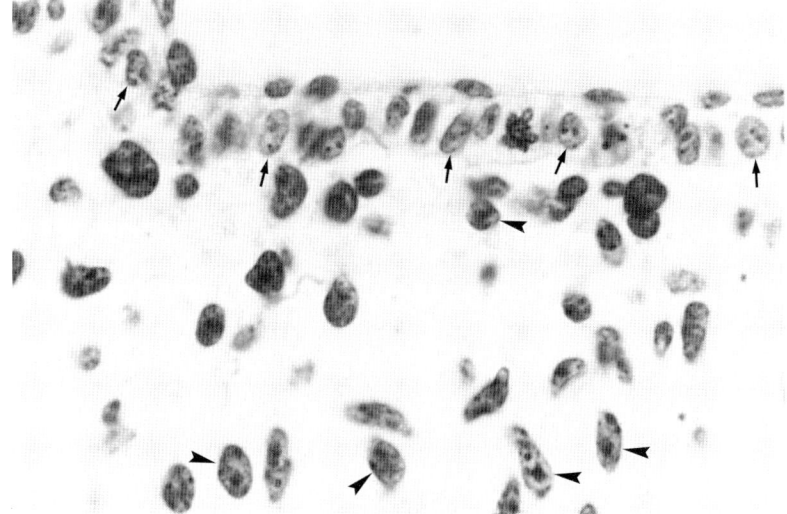

Figure 3 Transverse section of a 4.5-day-old chick host embryo implanted at the 18-somite stage with 9-dpc mouse somites. Dermatome gives rise to dispersed mouse cells (arrowheads) located beneath the chick ectoderm (arrows). Feulgen–Rossenbeck staining (magnification: ×900).

notochord and floor plate can induce epithelial somite cells to become Pax1-positive sclerotome. Pax1 is a DNA-binding transcriptional activator expressed in the developing sclerotome, facial mesenchyme, limb buds, and thymus during embryogenesis (Deutsch et al., 1988; Ebensperger et al., 1995). In the mouse, the availability of notochord mutants is conducive to a genetic approach to analyzing the function of notochord in skeletal development. *Danforth's short-tail* is a skeletal mutant in which the notochord is affected (Dunn et al., 1940). Abnormalities of the vertebral column result from structural changes in the notochord, particularly in the tail region of 10-dpc mouse embryos. *Undulated* mutation affects the development of the mouse skeleton and produces a point mutation in the paired box of Pax1. On the whole, these data clearly show that a combination of notochord effects on sclerotomal induction exists in mouse and chick embryos.

Very shortly after surgery, somite ventral cells in mouse–chick chimera deepithelialize and develop into sclerotomes that grow symmetrically with those of the unoperated side (Fig. 4a), although they sometimes appear to be more ventrally located. No visible signs of death are noted. Implanted mouse cells survive and thrive, and considerable mitosis occurs, as evidenced by Feulgen–Rossenbeck staining. At graft level, after 4 days of postoperative incubation, mouse sclerotomal cells participate in host vertebrae (Fig. 4b,c) and sometimes give rise to additional vertebral pieces. When the graft level includes the thoracic region, mouse cells are observed in host ribs, and an additional rib can sometimes be observed.

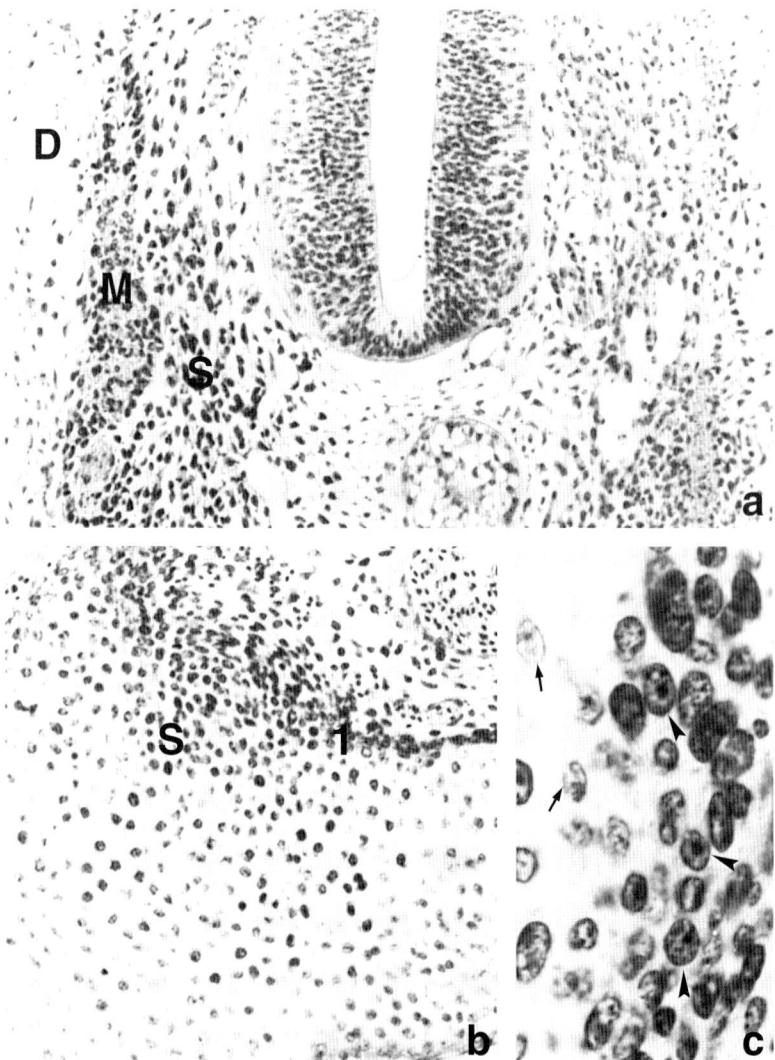

Figure 4 (a) Transverse section of a 3-day-old chick embryo implanted at the 18-somite stage, at brachial level, with 9-dpc mouse somites. Grafted mouse somites insert well and develop on the right side of the chick host. On the operated side, dermatome (D), myotome (M), and sclerotome (S) are composed of graft-derived cells. (Magnification: ×240). (b) Transverse section of a 6-day-old chick host embryo grafted at the 18-somite stage, at brachial level, with 9-dpc mouse somites. Mouse sclerotomal-derived cells (S) participate in the formation of chick host vertebra (magnification: ×240). (c) High-power view of zone 1 in b showing mouse nuclei (arrowheads) in host vertebra which are distinct from chick nuclei (arrows) (magnification: ×960).

After a mouse dermomyotome graft into the last 5-somite level of chick embryo at the 18–20 somite stage, no implanted cells participate in the sclerotome. In later development, the vertebrae appear to be entirely composed of chick cells.

C. Myogenic Lineage

In the vertebrate embryo, myogenic precursor cells of the body (excluding head muscles) derive from somites (Chevallier *et al.*, 1977; Christ *et al.*, 1977). In somite manipulation experiments in the chick embryo, particular attention has been given to the specification of the myogenic lineage in the somite. The molecular mechanisms underlying myogenic specification in the embryonic somite have been increasingly documented, and many factors now appear to be involved in myogenic determination. Studies of the spatiotemporal appearance of myogenic regulatory factor (MRF) genes indicate that they constitute the earliest known markers of myotome precursor cells in the somite. In the mouse, the first myogenic factor detected is Myf5 at 8.5 dpc (Ott *et al.*, 1991; Buckingham, 1992), which appears in the dorsomedial quadrant of the somite prior to formation of the dermomyotome. A few hours later, myogenin is clearly detectable in the myotome, at the same time as the initial appearance of -cardiac actin. MRF4 is also detected at this time, but transiently (Bober *et al.*, 1991; Hinterberger *et al.*, 1991). MyoD expression occurs in the myotome 2 days after initiation of muscle differentiation (Sassoon *et al.*, 1989). However, the control of myogenic specification by MRFs alone remains unclear because mice carrying null mutation in MRFs Myf5 or MyoD apparently have normal skeletal muscles. Mice lacking a functional MyoD gene are viable and fertile and exhibit no morphological and physiological abnormalities in skeletal muscles (Rudnicki *et al.*, 1992). Similarly, newborn mice lacking a functional Myf5 gene display no defects in musculature (Braun *et al.*, 1992; Tajbakhsh and Buckingham, 1994), whereas newborn mice deficient for the two myogenic factors (Myf5 and MyoD) appear to be devoid of all skeletal muscles (Rudnicki *et al.*, 1993). These results suggest that these different factors could be jointly involved in the determination of myoblasts. However, other findings provide evidence that myogenin, another member of the basic helix–loop–helix (bHLH) gene family, is essential for muscle development. Myogenin gene disruption results in perinatal lethality because of muscle deficiency (Hasty *et al.*, 1993; Nabeshima *et al.*, 1993). Thus, myogenin seems to be required for skeletal muscle development *in vivo*, a function in sharp contrast with that of Myf5 and MyoD.

Once a cell has been determined to be of myogenic lineage, numerous myofibrillar proteins are known to be coordinately induced during skeletal myogenic differentiation. MRFs are believed to play an important role in the transcriptional regulation of most specific genes coding for skeletal muscle proteins (Emerson, 1990; Olson, 1992; Ordahl, 1992), including contractile proteins whose distinctive expression pattern characterizes muscle development. In the myotome, differences occur in the timing and extent of encoding gene expression of contractile proteins (Buckingham and Tajbakhsh, 1993). The first myosin heavy chain transcripts are detectable only a day after cardiac actin (Lyons *et al.*, 1990). Desmin, another muscle cytoskeletal protein whose gene belongs to the family of intermediate filament proteins (Lazarides and Hubbard, 1976; Small and Sobieszek, 1977; Geisler and Weber, 1982), is expressed before skeletal muscle actin and myosin heavy chain (Babai *et al.*, 1990; Mayo *et al.*, 1992). This component is the major subunit of intermediate filament expressed by skeletal, cardiac, and smooth muscle cells in both embryonic and adult tissues. Its expression has been previously studied in chick developing somites (Borman and Yorde, 1994). Borman *et al.* (1994) examined whether each somite has an inherent capacity to support myogenesis evidenced by desmin accumulation. They found that the ability to activate desmin expression is not the same for all somites observed. To consider how desmin expression is regulated *in vivo* and trace the embryonic fate of myogenic cells, Li *et al.* (1993) obtained stable lines of transgenic mice harboring reporter genes coding for *Escherichia coli* LacZ linked to the 1-kb DNA 5´ regulatory sequence of the desmin gene. Whole-mount staining for LacZ activity revealed high levels of expression of this transgene by 9 dpc in somites (Li *et al.*, 1993). The rostrocaudal progression of transgene expression was parallel to the pattern of somite maturation and concordant with the time-course for expression of endogenous desmin gene. LacZ-positive cells were observed in the limb bud from 11 dpc.

In mammals, the main line of evidence for the participation of somites in limb myogenesis has come from descriptive studies. In histology, Milaire (1976) observed somitic cells migrating from the ventrolateral edge of the dermomyotome to the dorsoproximal territory of the hind limb in 28-somite stage mouse embryos. Similar findings have been reported for the forelimb region of 22-somite stage mouse embryo (Milaire, 1987). In experimental studies, Kieny *et al.* (1987) grafted mouse somites into the chick embryo but failed to demonstrate that mouse cells participate in the formation of host brachial musculature. In explants combining tritiated thymidine-labeled somites with unlabeled limb buds, Agnish and Kochhar (1977) showed that labeled cells helped to form cartilage and skeletal muscles in the proximal region of the limb. Beddington and Martin (1989) transplanted somites from a Tg(Act-LacZ)-1 transgenic mouse line into a wild-type mouse, a procedure ensuring that transgenic labeling is not diluted in the embryo. After 48 hr in culture, the limb vasculature of the grafted embryos was stained positive for LacZ, in agreement with results in the chick showing that limb blood vessels derived from somites (Wilting *et al.*, 1995). Beddington and Martin (1989) also reported a small population of labeled cells around the vascular plexus that they tentatively identified as myogenic cells. Sze *et al.* (1995), using donor somites arising from the Des 1-nLacZ mouse (Li *et al.*, 1993) in which transgene is specifically expressed in the skeletal muscle lineage, demonstrated, though indirectly, that myogenic cells stained blue for LacZ activity are finally identified in the mouse host limb bud. This study strongly suggests that somitic cells reach the limb bud as the limb elongates during development. In our mouse–chick somite chimera, the use

7. Mouse–Chick Chimera: Studying Somite Development

of the desmin LacZ transgenic line has provided precise analysis of the development of muscular somitic derivatives.

1. Epaxial Myogenesis in *in Ovo* Implanted Mouse Somites

In stable lines of transgenic mice harboring a LacZ reporter gene linked to region 75 to 115 of the human desmin gene, Li *et al.* (1993) observed transgene activity from 8 dpc in some particular neurectoderm regions and in the conductive system of the heart rudiment. By 8.5 dpc, these authors found transgene expression in the rostralmost somites, and by 9 dpc within almost all somites. In somites, staining was restricted to myotomes, allowing mapping of the cells activating desmin transcription and serving as a marker of the embryonic development of skeletal muscles. In our study, when somites from the 15- to 18-somite level of chick hosts were unilaterally replaced by the last 5 somites of 9-dpc desmin LacZ transgenic mice, 2 days postsurgery, *in toto* examination and histological analysis showed that the grafted somite level was detectable by X-gal blue staining (Fig. 5a,b)

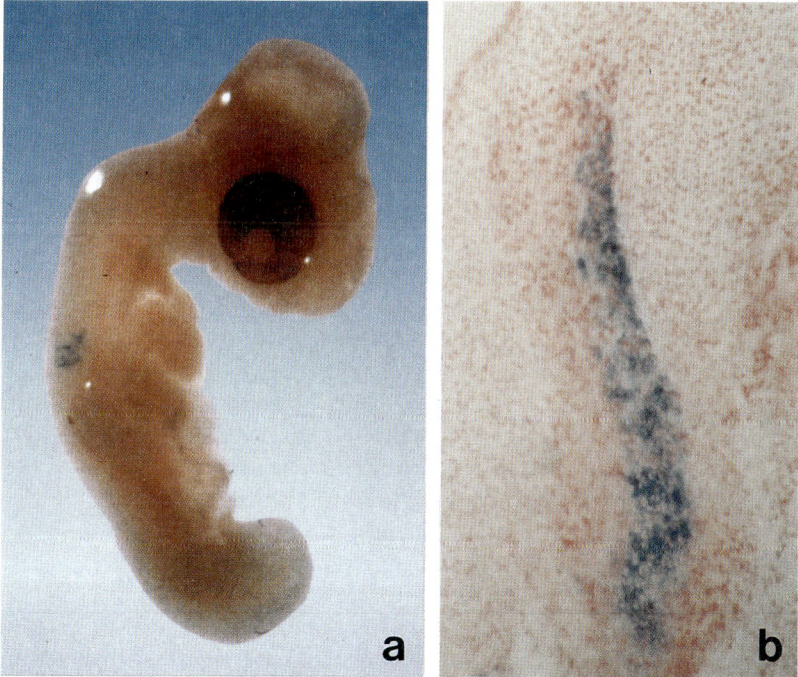

Figure 5 Expression of transgene desmin in a chick embryo examined 2 days after implantation of the brachial somites of a 9-dpc desmin transgenic mouse at the 18–20 somite level. (a) The graft site is identified by X-gal blue staining. (b) A transverse section shows evidence of transgene activity in grafted myotomes (magnification: ×240). (Fontaine-Pérus *et al.*, 1995.)

corresponding to the nuclear location of the reporter gene. During subsequent development, the paravertebral muscles located at the graft site expressed LacZ transgene and exhibited blue staining. No LacZ-positive cells were observed in the migration zone between the somite myotome and the limb bud. At day 6 postsurgery, reactive cells were readily detectable in paraxial muscles. At this age, myoblasts have fused and formed primary myotubes, accumulating fibrillar desmin and embryonic fast myosin that is well organized into striated structures. Using the same experimental procedure, anteriormost somites of 8.5 dpc transgenic mouse were implanted into the chick host. Although all grafted somites did not express the transgene at the time of surgery, *in toto* examination of grafted embryo 1 day later, after X-gal treatment, clearly showed blue staining confined to myotomal cells at the graft site. These cells subsequently differentiated into myotubes accumulating myofibrillar proteins. When grafted dermomyotomes arose from desmin LacZ transgenic mice, implanted myotomal cells were discernible by the blue staining of their nuclei, as at the time of their implantation.

A further systematic investigation was carried out on the myogenic development of somite-derived cells by considering the timing of desmin, fast, slow, and neonatal myosin heavy chain (MHC) isoform expression in the normal mouse and mouse–chick chimeras. Desmin is among the first muscle-specific proteins detected in the mammalian embryo, appearing in somitic myotomes from 9 dpc onward (Mayo *et al.*, 1992). This protein is expressed in replicating myoblasts and continues to be accumulated in high amounts as muscles differentiate (Kaufman and Foster, 1988). In 10-dpc somitic myotomes, desmin was detected in more cells than was fast MHC (F19 antibody), whereas slow MHC (S27 antibody) revealed a small population of cells within the myotomes. At this age, the first neonatal MHC-positive cells (NN6 antibody) were also detectable. At 14 dpc, nearly all primary myotubes constitutive of the paraxial muscles reacted with NN6 antibody and were also labeled by F19 antibody and anti-slow MHC S27. Isomyosins stained with these three antibodies were coexpressed in the same myotubes. At 14 dpc, the paraxial muscles of the mouse were mainly composed of primary myotubes, and few, if any, secondary myotubes were present. Primary myotubes are large in diameter, whereas secondary myotubes are small in diameter and surround each primary myotube at the time of their formation (Harris *et al.*, 1989; Duxson, 1992). Secondary myotubes react with NN6 and F19 antibodies, but not with S27 antibody. In mouse–chick chimeric embryos, during the 2 days following surgery, immunostaining of sections from the graft level treated with NN6 antibody specifically revealed the myotomes (Fig. 6a) derived from grafted mouse somites. The number of mouse myotomes observed was perfectly correlated with the number of implanted somites. During subsequent development, they took part in several paraxial muscles. The extent of their participation was slightly variable, but generally involved a length of at least three vertebral segments, as the graft-derived axial muscles were located at the level of three successive dorsal root ganglia of the chick host (Fig. 6b). Simultaneous staining with NN6 antibody and EB165 antibody (which recognizes a fast MHC isoform in the chick) showed that the ex-

7. Mouse–Chick Chimera: Studying Somite Development

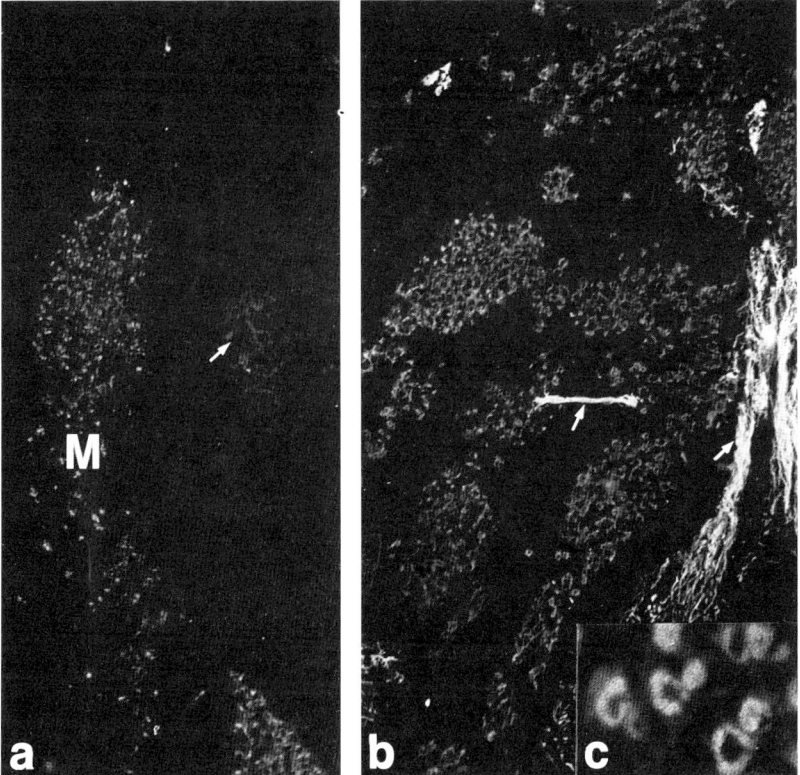

Figure 6 (a) Transverse section of a 3-day-old chick embryo implanted at the 18-somite stage, at brachial level, with 9-dpc mouse somites. Mouse myotomal cells (M) are specifically visualized after treatment with NN6 monoclonal antibody. At the graft site, chick neural crest constitutes the dorsal root ganglion (arrow), and chick motor axons, evidenced by a neurofilament monoclonal antibody, have developed (magnification: ×310). (b) In an 8-day-old chick host, examination of the graft reveals several groups of NN6 mouse cells; antineurofilament antibody identifies chick host nerves (arrows) (magnification: ×140). (c) NN6 antibody reveals clusters of primary and secondary myotubes (magnification: ×960).

perimental side was devoid of chick muscle cells at the midgraft level, whereas mixed NN6 (mouse)- and EB165 (chick)-positive cells were observed at the extremities of the graft. From 5 days postsurgery, paraxial muscles derived from implanted somites began to exhibit secondary myotubes surrounding primary ones (Fig. 6c). The criteria used to define myotube types in the graft were based on the fact that secondary myotubes, at the time of their appearance, are smaller in diameter than primary myotubes and highly reactive with fast isomyosin. It is noteworthy that a few secondary myotubes were also identified in axial muscles of the nonoperated side examined 5 days postsurgery. As on the operated side, they appeared to be small in diameter and surrounded large-diameter primary myotubes.

Their appearance coincided with that of secondary myotubes observed in paraxial muscles derived from the mouse graft. The onset of secondary myotube formation in mouse chick somite chimera, a step which can be considered to mark the boundary between embryonic and fetal phases of development, largely improves the efficiency of the experimental procedure in myogenesis study.

Some specialized nuclei in developing myotubes are capable of synthesizing acetylcholine receptor (AChR). In principle, the mechanisms responsible for local accumulation of AChR involve selective transcription of AChR genes by a subset of nuclei located beneath the motor nerve terminal, although AChR can also be present in non-nerve-contacted regions. The peculiarities of the neuromuscular synaptic pattern of chick and mouse embryos are conducive to regulation of AChR expression. The formation of AChR in epaxial muscles of normal mouse and chick embryos and mouse–chick chimeras has been systematically investigated. The time-course between the appearance of AChR clusters and the ingrowth of nerves was studied by simultaneous revelation of tetramethylrhodamine isothiocyanate (TRITC) α-BGT and neurofilament (NF) antibody, and nerve terminals were visualized with synaptic vesicle (SV2) labeling.

In chick embryo, the first AChR clusters are observed in somitic myotomes from embryonic day (E)3 of incubation. One day later, they appear to be concentrated in the ventral parts of myotomes, opposite the motor roots of the spinal cord (Fig. 7a). Although AChR clusters are observed in myotomes as early as E3, none are found in contact with nerve processes until E5. From this stage, intramuscular nerve branches grow, and at E7.5 most AChR clusters have become associated with nerve terminals. Thus, AChR clusters form and concentrate in developing epaxial muscles of chick embryo before being invaded by ingrowing axons. They subsequently appear to be colocalized with nerve terminals [as already shown by Dahm and Landmesser (1991) in developing hypaxial muscles].

In mouse embryo, the development of synaptogenesis has been examined from the time motor axons leave the spinal cord (9 dpc). When the first ventral roots emerge from the neural tube at 9 dpc, no AChR clusters are observed in somitic myotomes. At 12 dpc, developing motor nerves begin to invade somites. Even though nerve processes are growing inside the whole muscle primordia, no AChR clusters are apparent. They are first observed in developing muscle masses from 14 dpc (Fig.7b), mainly in close proximity to nerve inputs. At this time, AChR clusters appear to be quite small (7.43 ± 0.61 µm). At 15 dpc, they become enlarged in size (11.59 ± 1.32 µm), and almost all are in direct contact with nerves. These results indicate that synaptogenesis in mouse embryo is characterized first by nerve invasion of paraxial muscle primordia and then by the expression of AChR clusters in myotubes near axons. The latter observation correlates perfectly with that of Noakes *et al.* (1993) who, from 14 dpc, described anti-NF-stained nerve processes adjacent to forming AChR clusters in mouse intercostal and diaphragm muscles.

In mouse–chick chimeric embryo, the timing of AChR cluster formation and

7. Mouse–Chick Chimera: Studying Somite Development 285

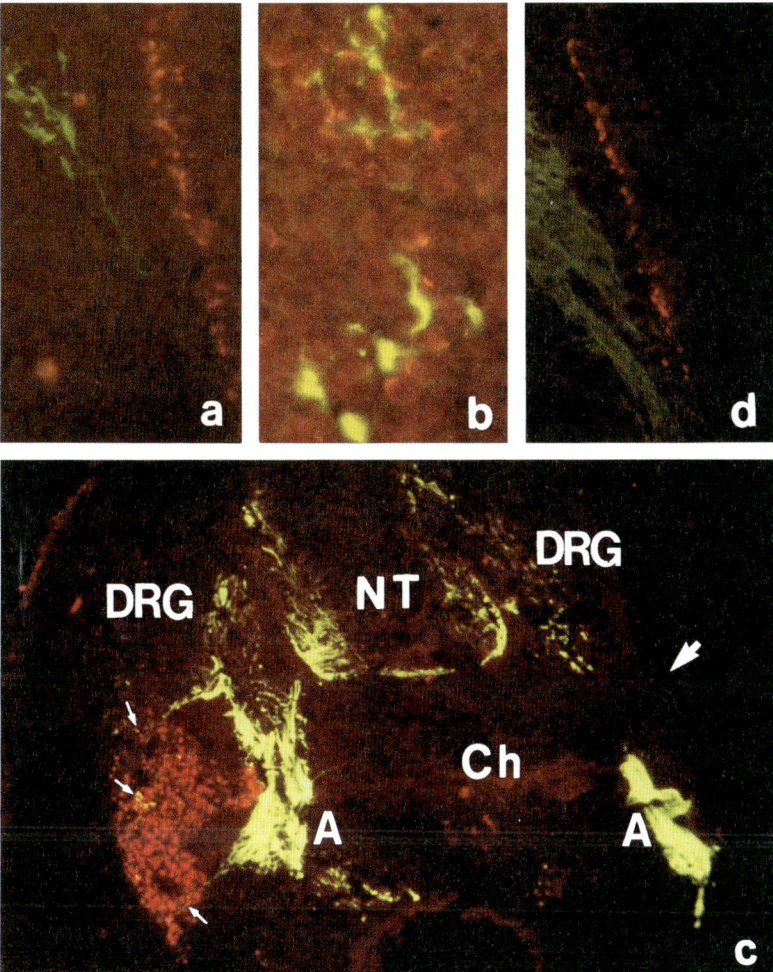

Figure 7 Simultaneous detection of nerve processes by an antineurofilament antibody revealed with a fluorescein-labeled antibody and AChR clusters, with rhodamine α-BGT. (a) In 4-day-old chick embryo, AChR clusters are vizualized along the myotome near the ventral root of spinal cord (magnification: ×160). (b) In 14-dpc mouse embryo, AChR clusters colocalized with nerves (magnification: ×240). (c, d) Transverse sections of an E4.5 chick host embryo implanted with 9-dpc mouse somites at the 1.5 day at the 4-posteriormost-somite level. (c) Simultaneous detection of nerve processes (by NF antibody revealed by a fluorescein goat anti-mouse antibody) and mouse muscle cells (by NN6 antibody revealed by rhodamine goat anti-rabbit antibody). On the operated side, NN6-reactive mouse myotomal cells are invaded by chick nerves (arrows) originating from the spinal axons of the host; on the nonoperated side, no nerves are observed in chick host myotome (arrow) (magnification: ×80). (A, axons; Ch, notochord; DRG, dorsal root ganglion; NT, neural tube.) (d) On the nonoperated side (serial section as in a), chick myotome exhibits numerous AChR clusters (revealed by rhodamine α-BGT) facing the ventral roots of the host spinal cord, none of which are observed on the implanted side (magnification: ×130). (Auda-Boucher et al., 1997.)

motor innervation was studied at the surgical level in chick host embryos by TRITC α-BGT, NN6, and NF antibody. SV2 antibody and TRITC α-BGT were used to assess the colocalization of nerve terminals and AChR clusters. From day 2.5 of postoperative incubation, mouse myotomal cells invaded by chick nerves originating from spinal motor nerves of the host embryo (Fig. 7c) do not express AChR clusters. On the control side, no nerve processes are observed in chick host myotome, which exhibits AChR clusters concentrated in the ventral part opposite the ventral roots of the chick spinal cord (Fig. 7d). Mouse myotomal cells seem capable of attracting chick motor axons at a time when, unlike chick host cells, they have not yet acquired the potentiality to synthesize AChR. Four days after grafting, the first AChR clusters are detected in axial muscle cells derived from implanted mouse somites. These clusters are small in size and adjacent to anti-NF-positive chick axons that migrated among NN6-positive mouse myotubes. Six days postsurgery, AChR clusters are enlarged (11.22 ± 0.88 μm) and exhibit the same size and morphology as nerve-contacted clusters examined in the normal mouse at 15 dpc (11.59 ± 1.32 μm). They are considerably longer than the AChR clusters of embryonic chick muscles, which are not more than half the size of those of the mouse. Simultaneous revelation of SV2 antibody and TRITC α-BGT demonstrated that all clusters formed in mouse graft at this time are colocalized with chick nerve endings. The difference in the timing of synaptogenesis in the chimeras recapitulates that of the normal mouse, suggesting that chick motoneurons respond differently to developing mouse and chick muscle.

The synaptic areas formed between chick motor axons and mouse myotubes develop according to the mouse pattern. Both the timing of their appearance and their morphology correlate perfectly with events in mouse synaptogenesis. These results indicate the important role played by postsynaptic membrane in controlling the first steps of AChR formation (Auda-Boucher *et al.,* 1997).

2. Hypaxial Myogenesis in *in Ovo* Implanted Mouse Somites

Heterospecific exchange of somites between chick and quail embryos has shown that skeletal musculature of the wing derives from somites, whereas cartilage, connective tissues, tendons, and smooth muscles are formed from the somatopleure. It is well known that specific somites contribute to specific limb muscle groups (Beresford, 1983; Chevallier *et al.,* 1977; Christ *et al.,* 1977) by migrating out from the lateral portion of the somite myotome (Christ *et al.,* 1978; Ordahl and Le Douarin, 1992). Regardless of their level of origin, mouse somites, when grafted into the chick at the brachial level, give rise to scattered or clustered cells in the chick wing bud from day 2 to 3 postsurgery. During further development, mouse cells may invade the biceps and intercostal muscles (Fig. 8a,b). They are also visualized in pectoralis major muscle on the operated side. The results of such experiments have been largely improved by the use of the transgenic desmin nlsLacZ mouse (Li *et al.,* 1993), which provides a highly sensitive means of study-

7. Mouse–Chick Chimera: Studying Somite Development 287

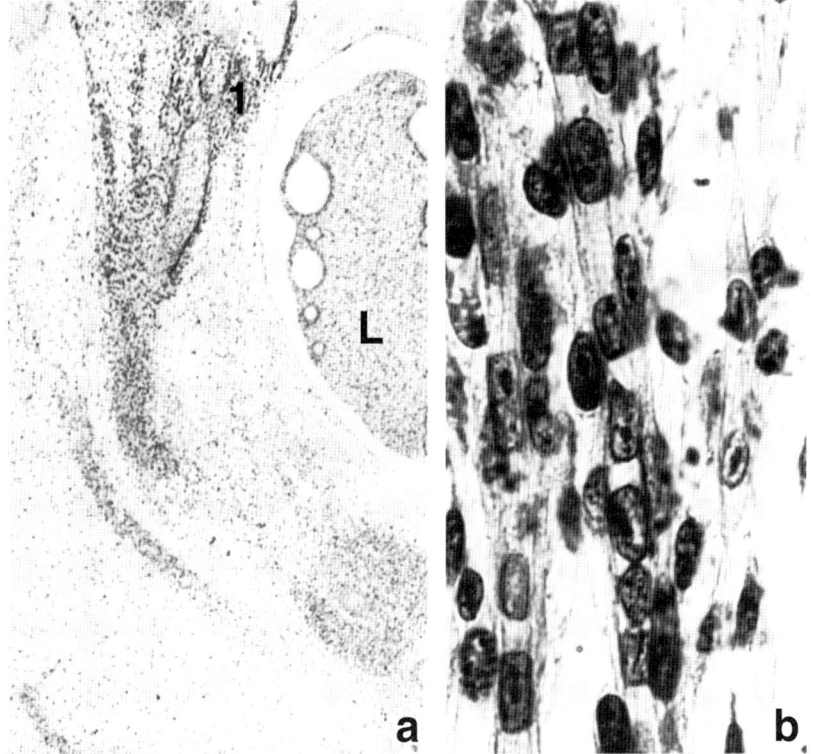

Figure 8 Transverse section of an 8-day-old chick embryo implanted at the 18-somite stage, at brachial level, with 9-dpc brachial mouse somites. (a) Low-power view of the host thoracic region, lung (L) (magnification: ×90). Graft-derived cells have infiltrated the intercostal muscles. (b) High-power view of zone 1, in a. These muscles are made up of mouse myotubes (magnification: ×960). Feulgen–Rossenbeck staining. (Fontaine-Pérus *et al.*, 1995.)

ing the behavior of myogenic cells in somites. When isolated somites arising from 9-dpc transgenic mouse embryos were grafted at the brachial level of a chick host, the limb bud muscles were stained blue. In another series of experiments, somites from the forelimb level of 18-somite chick hosts were replaced by the last six somites of 9-dpc des −/− LacZ +/+ mice (Li *et al.*, 1996, 1997). Two days postsurgery, *in toto* examination and histological analysis showed that the host limb developing on the operated side exhibited ventral and dorsal premuscle masses stained in blue (Fig. 9a,b).

Thus, the developmental potentialities of the different components of the grafted somites in the chick have been demonstrated. Orthotopic transplantation of 9-dpc somitic cells results in the participation of mouse cells in host dermis, skeleton, and muscle on the graft site. The use of transgenic mouse lines has made it possible to follow the fate of somitic myogenic derivative *in ovo* (Fontaine-Pérus *et al.*,

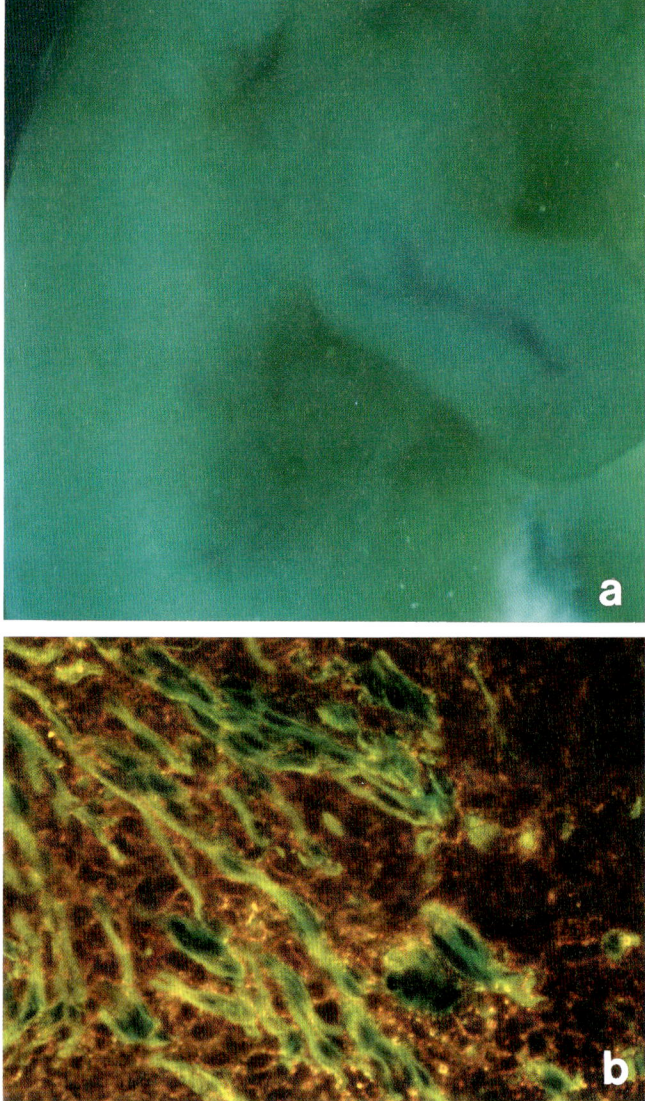

Figure 9 (a) Examination of a 4-day-old chick embryo implanted at the 18-somite stage at the brachial level with brachial somites of a 9-dpc, Des −/− LacZ +/+ transgenic mouse. The dorsal and ventral muscle masses of right limb bud are identified by X-gal blue staining (magnification: ×20). (b) In these muscles, all the myotubes exhibit blue nuclei and have accumulated myosin (magnification: ×800).

1995). Previous experiments had totally failed to demonstrate these potentialities. Kieny *et al.* (1987), after grafting mouse somitic mesoderm or limb bud premuscular mass into chick hosts, found that mouse cells were unable to participate in any skeletal musculature of the chick host. They concluded that mammalian myogenic cells are unable to differentiate in an avian matrical environment and interact with avian cells. On the contrary, it appears that mouse somitic cells are not only capable of differentiation in the chick host but can also migrate into the wing bud. Mouse graft-derived cells penetrate into some appendicular muscles adjacent to the graft level. In the chicken, cell migration from the lateral edge of the somites to the wing takes place from the 22-somite stage (Chevallier *et al.*, 1978). Mouse cells acquire migratory properties when grafted in appropriate locations and use the same migratory pathways as those of host somitic cells in reaching the host limb. These observations support the notion that the limb, in both mammals and avians, is formed by cells migrating from somitic myotomes.

IV. Epaxial Myogenic Lineage into Neural Tube Chimera

During embryogenesis, paraxial mesoderm segmentation gives rise to pairs of somites flanking the neural tube and the notochord. Cells from the newly formed somites become committed to specific fates in response to instructive signals from surrounding cell types. In birds, various studies have demonstrated that neural tube, and in some cases notochord, is critical for myotome development (Teillet and Le Douarin, 1983; Christ *et al.*, 1992; Rong *et al.*, 1992; Bober *et al.*, 1994; Pownall *et al.*, 1996; Spence *et al.*, 1996). The use of different culture systems has provided similar findings (Packard and Jacobson, 1976; Kenny-Mobbs and Thorogood, 1987; Vivarelli and Cossu, 1986; Buffinger and Stockdale, 1994; Münsterberg and Lassar, 1995; Stern and Hauschka, 1995; Gamel *et al.*, 1996). Finally, *in vitro* and *in ovo* studies have demonstrated that the notochord via Sonic hedgehog and the dorsal neural tube via Wnt-1 can induce myotomal differentiation as assayed by MyoD and myosin expression (Münsterberg *et al.*, 1995; Stern *et al.*, 1995; Pownall *et al.*, 1996).

In comparison with the chick, far less is known about how somites and the neural tube interact in mammalian embryos to determine myotome formation. One of the earliest studies in the mouse demonstrated that the myogenic potential of somitic cells isolated from 9-dpc mouse embryos was considerably enhanced when they were cultured in the presence of the neural tube (Vivarelli and Cossu, 1986). More recently, Cossu *et al.* (1996) have provided evidence that activation of different myogenic pathways is induced by mouse neural tube. The possibility of obtaining mouse mutants offers a powerful means for studying cellular interactions during development. For example, the key role of the dorsal neural tube is illustrated by the mouse *open brain* mutation, which disrupts normal development of the dorsal neural tube and prevents the formation of epaxial muscles (Spörle *et al.*, 1996). The effects of signals emanating from the mouse neural tube on myogenesis have been evaluated in the nervous system of mouse–chick chimera. Fontaine-Pérus *et al.* (1997) have analyzed MyoD gene expression in chick myotomes forming at the level of *in ovo* implanted mouse neural tube. In our study, mouse–chick chimeras were prepared by transplanting fragments of neural primordium from 8- to 8.5- and 9-dpc mouse embryos into 1.5- and 2-day-old chick embryos at different axial levels. MyoD gene expression in chick myotomes forming at the level of mouse neural tube was analyzed to determine whether mouse neural tissue in the chimeric system is capable of contributing to the signaling processes required for myogenic gene expression in somites. Accordingly, mouse neural transplants were performed at the level of segmental plate mesoderm to ascertain whether chick somites adherent to mouse tube can form and differentiate.

The demonstration that various components of the spinal cord derived from mouse neural implant could differentiate in the chick seems most essential. It has been clearly observed that marginal, mantle, and ependymal layers develop in the first 2 days following surgery, with the ependymal layer exhibiting considerable mitosis. Subsequently, the mantle layer becomes the thickest one. Among the neuroblasts composing it, motoneurons and their motor nerves develop ventrally, undergoing differentiation, as expressed by a neurofilament marker (NF, 68 kDa), and recognizing Islet-1 antibody. The main changes in size, shape, and number occur in graft-derived motoneurons between days 2 and 6 after implantation of mouse neural tube. During this period, their number decreases considerably (68% die), whereas the size of the surviving motoneurons increases (7 mm versus 12 mm in nuclear diameter). This period corresponds to the normal phase of motoneuron cell death observed at day 13–14 in *in situ* mouse development (Flanagan, 1969; Nornes and Carry, 1978; Lance-Jones, 1982). The spatial distribution of motoneurons within the motor columns does not change radically beyond day 6 postsurgery. With respect to the timing of cell death in *in ovo* developing mouse spinal cord, the peak for motoneuron cells occurs in E7–8 chick host embryo, that is, 5–6 days postgrafting, corresponding to an equivalent age of 13.5–14.5 dpc. Lance-Jones (1982) described the maximal rate of cell death in mouse spinal cord as 13.5–14.5 dpc. These data suggest that neurogenesis can occur in mammalian cells transplanted into a chick embryonic environment and that mouse neural tube *in ovo* maintains its native developmental pace.

During early development, the homeobox-containing gene Msx1 is transcribed in the mesoderm and ectoderm of the primitive streak. Up to the neurectodermal derivatives, Msx1 gene expression is restricted to the neuroepithelium that will form the dorsal part of the neural tube and the brain (see Davidson, 1995). The expression of Msx1 early in the differentiation of organs suggests that it plays a fundamental role in development. In experiments in which the prosencephalon–mesencephalon of the chick host was replaced by its mouse counterpart, dorsalization of the graft was attested by Msx gene expression. Interestingly, the chimeras assumed a rodentlike shape, with a snout and forehead bulges. Taken together, these results suggest several possible explanations. In neurogenesis, it may be admitted that the key information is provided early to ensure future programs. Thus, cells participating in the development of prosencephalon-derived structures would be committed to developmental fates by early prepatterning cues that are

7. Mouse–Chick Chimera: Studying Somite Development 291

not overcome by environmental influences. This supports the idea of Nieuwkoop (1952) on the patterning of the *Xenopus laevis* nervous system, in which prosencephalon is considered as a baseline fate. Moreover, whether environmental signals outside the nervous system are required or not for its development, these experiments, like those of Itasaki *et al.* (1996), provide direct support for the notion that fundamental processes are conserved in the patterning of spinal cord (in both mouse and chick species). The mouse neural tube graft also enters into the development of the chick host peripheral nervous system through migration of neural crest cells associated with mouse neuroepithelium. Depending on the graft level, mouse crest cells participate in the formation of various derivatives such as head components, sensory ganglia, orthosympathetic ganglionic chain, nerves, and neuroendocrine glands (Fontaine-Pérus *et al.*, 1997). Throughout the study period, chick host vertebra organizes around the mouse spinal cord by differentiation of sclerotomal somitic cells at the graft site (Fig. 10).

The demonstration that mouse neural tube develops well in a chick environment has made it possible to assess whether mouse neural tube is able to send signals conducive to chick myogenesis. Microsurgical removal of the neural tube at the level of unsegmented paraxial mesoderm ensures that somites developing postsurgically are deprived of any neural tube and neural crest influence. Despite the fact that MyoD genes are activated in somites that develop without any contact

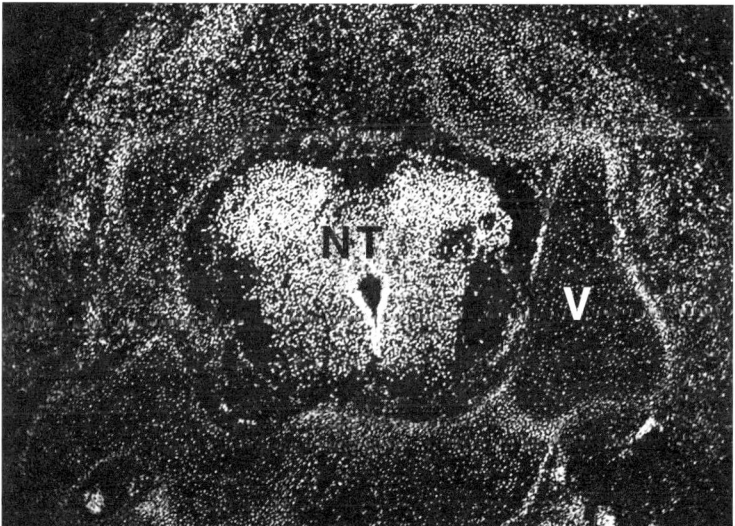

Figure 10 Transverse section of a mouse–chick chimera grafted at the 18-somite stage, at the 13- to 18-somite level, with 9-dpc mouse caudal neural tube. Examination 8 days postsurgery showed that chick host vertebra (V) organized around the mouse neural tube (NT), as revealed by Hoechst staining (magnification: ×240).

with the neural tube (Pownall *et al.*, 1996), further expression of the myogenic regulatory gene reveals that development of paraxial muscle does not proceed normally. In the 24 hr following neural tube ablation, a decrease in the level of MyoD expression is noted in the mediodorsal somitic zone, resulting in impairment of paraxial muscle development at later stages. Pownall *et al.* (1996) used microsurgery, tissue grafting, and *in situ* hybridization techniques to demonstrate that notochord provides the primary signals regulating the activation of myogenic regulatory genes when somites form from segmental plate mesoderm. These results indicate that the neural tube is required shortly after the notochord for stabilization of myogenic gene expression in newly formed somites.

In chimeras with neural tube implanted at the segmental plate level, myogenesis has been studied in chick somite differentiation at the implantation site. In all operated embryos examined between 1 and 9 days after grafting, levels of MyoD transcripts were equivalent all along the paraxial zone (Fig. 11a,b). At the surgery level, MyoD transcripts were similar in myotomes of chimeras and control embryos. The extent of the MyoD-positive zone was greater in mouse–chick chimeras than in denervated chick embryos. The presence of mouse neural tube allowed restoration of a normal level of MyoD transcription in the grafted zone (Fig. 11c,d). Although these experiments do not consider how mouse neural tube can participate in the mechanism required to maintain bHLH gene expression, they demonstrate that the grafted neural tube acts in the same manner as the chick host neural tube. Regardless of whether combinatorial signals from the neural tube, floor plate, and notochord allow an induction of myogenic gene expression in somites (Münsterberg and Lassar, 1995), mouse tissue is able to send the same signals as that of the chick.

V. What Is the Benefit of the Somite–Neural Tube Chimera?

It is now well accepted that axial structures (neural tube and notochord) play a positive role in the process leading to differentiation of epaxial skeletal muscle. Several lines of evidence indicate that Sonic hedgehog mediates the ventralizing action of notochord and floor plate on somites (Fan and Tessier-Lavigne, 1994; Johnson *et al.*, 1994; Chiang *et al.*, 1996). Conversely, dermomyotome differentiation is directed by signals derived from the surface ectoderm and dorsal part of the neural tube. Denetclaw *et al.* (1997) have recently demonstrated that the medial compartment of the dermomyotome contains the progenitors of the epaxial muscle. Members of the Wnt family may constitute the dorsal neural tube signal required to stabilize the myogenic program in the mediodorsal part of the somite (Stern *et al.*, 1995; Münsterberg *et al.*, 1995). According to Marcelle *et al.* (1997), Wnt-11 may allow muscle progenitors in the medial dermomyotome to involute into the myotome. Wnt-11 has been used by these authors to analyze the molecular interactions responsible for myotome formation. It is speculated that the absence

7. Mouse–Chick Chimera: Studying Somite Development

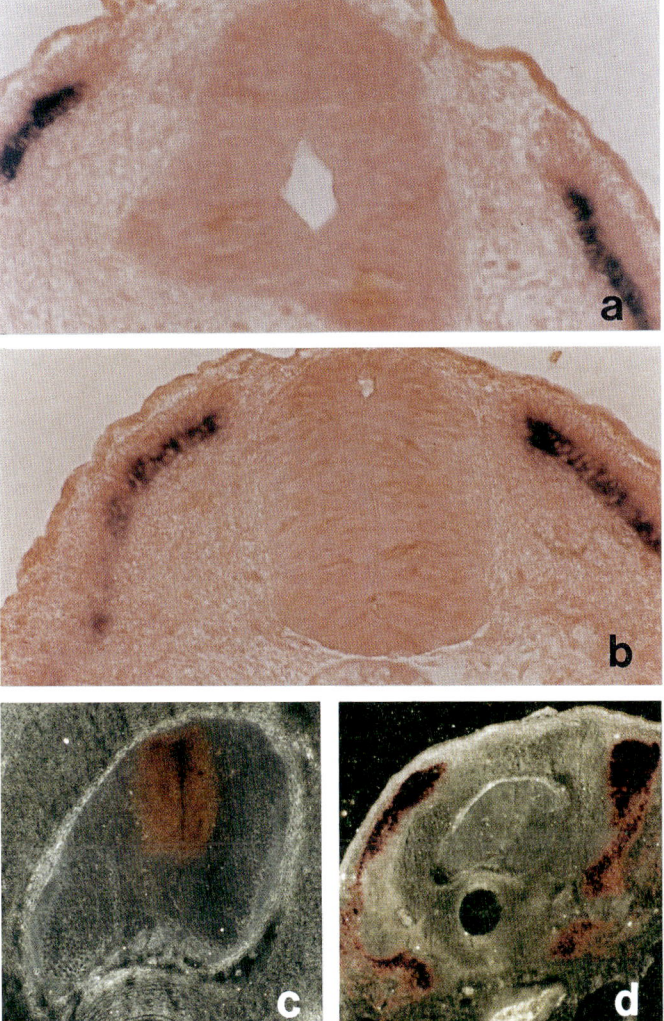

Figure 11 (a,b) Transverse sections of a mouse–chick chimera grafted at the 15-somite stage, at the segmental plate level with 9-dpc mouse neural tube and examined the day following surgery. (a) At the graft level and (b) at the chick host level, MyoD is equivalently expressed (magnification: ×240). (c) *In situ* hybridization with mouse Pax-3 probe in a mouse–chick chimera grafted at the 18-somite stage at the segmental plate level with 9-dpc mouse neural tube and examined 3 days postsurgery, showing that Pax-3 is expressed in the graft (magnification: ×180). (d) Expression of MyoD (serial section as in c) in chick somites developed at the graft level (magnification: ×60).

of Wnt-11, for instance after neural tube ablation, may be responsible for myotomal disorganization and failure of muscle progenitors to ingress into the myotome. Specific knockout may allow identification of the molecular event primitively activating myotome formation.

What is the benefit for this research field of the mouse somite–neural tube double graft in the chick? First of all, this experimental procedure may facilitate more detailed analysis of the intricate molecular combinations allowing myotome formation. It has previously been shown (Fontaine-Pérus *et al.*, 1995, 1997; Auda-Boucher *et al.*, 1997; Houzelstein *et al.*, 1999) that chick embryo constitutes a privileged environment for access to the developmental potentials of mammalian cells. This has been demonstrated in the case of somite-derived cells, neural tube cells, and neural crest derivatives. Given the importance of dorsal neural tube–somite signals, the grafting of these two structures would appear to be particularly relevant. The first step was to demonstrate the feasibility of the experimental system. For this purpose, the expression of MyoD gene in grafted somites and Pax-3 gene activation in grafted neural tube were controlled. The characterization of MyoD gene in *in ovo* grafted mouse somites has been discussed in Section III,C. Concerning Pax genes, their discovery in the mouse has been invaluable to an understanding of mouse developmental processes. Pax genes represent a family of genes that encode nuclear transcription factors characterized by the presence of a paired domain and sometimes by a paired type homeobox. In particular, the Pax-3 gene is highly patterned in the developing nervous system of early mouse embryo (Goulding *et al.*, 1991). Pax-3 transcripts are first detected in 8.5-dpc mouse embryos where they are restricted to the dorsal part of the neuroepithelium and the adjacent segmental dermomyotome. As Pax-3 expression during early neurogenesis is limited to mitotic cells of the ventricular zone of the developing spinal cord and distinct regions in the hindbrain, midbrain, and diencephalon, it can be used to trace the early steps of neural tube development (see Fig. 11c).

In chimeras in which somite and neural tube were implanted at the last 5-somite level, macroscopic examination on the first day after surgery revealed a perfect insertion of the grafted structures. Pax-3 mouse transcripts and MyoD mouse transcripts were visualized *in toto* only in grafted dorsal neural tube and somites, respectively. In fact, previous studies of somite and neural tube mouse–chick chimera had clearly predicted the development of both of these structures in the graft. The major challenge for this experiment will be study the behavior of the axial and/or paraxial structures isolated from different specific knockouts.

VI. Concluding Remarks

In chimeras, the analysis of cell lineage depends on the ability to distinguish between cells of different origins. Markers are required to trace the developmental fate of cells and recognize them at all ages in chimeric embryos. In mammals, genetic manipulations appear to be valuable means of elucidating the normal role of genes and identifying altered cells. Furthermore, the environment of the chimera allows these cells to survive and display their phenotype. In chimeras, the identification of all the descendents of genetically modified cells has thus far provided a reliable and sensitive means of measuring alterations in embryogenesis. In conjunction with an *in situ* marker, this may help in determining the cell type and the nature of the functions affected by the particular mutant. Extremely sensitive labeling is possible with transgenic mouse lines since cells that integrate foreign DNA, such as a reporter gene encoding for *E. coli* β-galactosidase, are detectable by simple histochemical revelation, which can be used to differentiate grafted cells from host cells. At present, the major means of exploring this genetic tool is by *in vitro* experiments in which mutant-induced tissue is cultured with wild tissue. However, the culture system is efficient in tissue development for only limited periods, and the media used may affect tissue outcome. *In vivo* micromanipulation remains the most powerful means of studying the fate of cells, their origins, and the cell–cell interactions that govern development. Because mouse embryo implantation is particularly unsuitable for *in vivo* manipulation, we chose to use the chick embryo as the site for developing mouse embryonic cells. Studies in such chimeras provide considerable information on lineage and the differentiation of mouse embryonic cells since *in ovo* grafted mouse cells are accessible throughout chick host embryogenesis. Moreover, the results presented in this chapter are indicative of the information to be obtained through genetic manipulations in conjunction with embryonic tissue transplantation. It has been clearly demonstrated that *in ovo* transplantation provides a suitable environment for the development of mammalian cells and that the information supplied by this environment is capable of promoting mouse cell differentiation. The replacement of chick somitic cells and neural tube by their mouse counterparts helps, respectively, to specify the development of somite components and identify effects on the somitic lineage. It would now appear easier to determine whether similar environmental cues to those described in avian embryos are required to promote mammalian somite development. For that purpose, *in ovo* engrafting of tissue from knockout mice appears to be an excellent technique. Moreover, the combined somite–neural tube graft offers additional advantages because information on the nature of the different factors (axial and/or paraxial) determining cell fate within the somite is more easily accessible.

Acknowledgments

The author is grateful to Gwenola Auda-Boucher and Yvonnick Cheraud for their collaboration during this research, Martine Molina for typing the manuscript, Félix Crossin for contributing photographic materials, Thierry Rouaud and Maryvonne Zampieri for their technical assistance, and Dr. James Gray for reading the manuscript. This work was supported by the University of Nantes, the Centre National de la Recherche Scientifique, and the Association Française contre les Myopathies.

References

Agnish, N. D., and Kochhar, M. (1977). The role of somites in the growth and early development of mouse limb buds. *Dev. Biol.* **56**, 174–183.
Auda-Boucher, G., and Fontaine Pérus, J. (1994). Differentiation potentialities of distinct myogenic cell precursors in avian embryos. *Dev. Dyn.* **201**, 95–107.
Auda-Boucher, G., Jarno, V., Fournier-Thibault, C., Bulter-Browne, G., and Fontaine-Pérus, J. (1997). Acetycholine receptor formation in mouse–chick chimera. *Exp. Cell Res.* **236**, 29–42.
Avery, G., Chow, M. and Holtzer, H. (1956). An experimental analysis of the development of the spinal column V. Reactivity of chick somites. *J. Exp. Zool.* **132**, 409–425.
Babai, F., Musevi-Aghdam, J., Schurch, W., Royal, A., and Gabiani, G. (1990). Coexpression of α-sarcomeric actin, α-smooth muscle actin and desmin during myogenesis in rat and mouse embryos. I. Skeletal muscle. *Differentiation* **44**, 132–142.
Beddington, R. S. P., and Martin, P. (1989). An *in situ* transgenic enzyme marker to monitor migration of cells in the mid-gestation mouse embryo: Somite contribution to the limb. *Mol. Biol. Med.* **6**, 263–274.
Beresford, B. (1983). Brachial muscles in the chick embryo: The fate of individual somites. *J. Embryol. Exp. Morphol.* **77**, 99–116.
Bober, E., Lyons, G. E., Braun, T., Cossu, G., Buckingham, M., and Arnold, H. H. (1991). The muscle regulatory gene, myf-6 has a biphasic pattern of expression during early mouse development. *J. Cell Biol.* **133**, 1255–1265.
Bober, E., Brand-Saberi, B., Ebensperger, C., Wilting, J., Balling, R., Paterson, B. M., Arnold, H. H., and Christ B. (1994). Initial steps of myogenesis in somites are independent of influence from axial structures. *Development* **120**, 3073–3082.
Borman, W. H., and Yorde, D. E. (1994). Analysis of chick somite myogenesis by *in situ* confocal microscopy of desmin expression. *J. Histochem. Cytochem.* **42**, 265–272.
Borman, W. H., Urlakis, K. J., and Yorde, D. E. (1994). Analysis of the *in vivo* myogenesis status of chick somites by desmin expression *in vitro*. *Dev. Dyn.* **199**, 268–279.
Braun, T., Rudnicki, M., Arnold, H., and Jaenisch, R. (1992). Targeted inactivation of the muscle regulatory gene myf-5 results in abnormal rib development and prenatal death. *Cell* **71**, 369–382.
Brill, G., Kahane, N., Carmeli, C., Von Schack D., Barde, Y. A., and Kalcheim, C. (1995). Epithelial–mesenchymal conversion of dermatome progenitors requires neural tube-derived signals: Characterization of the role of Neurotrophin-3. *Development* **121**, 2583–2594.
Buckingham, M. (1992). Making muscle in mammals. *Trends Genet.* **8**, 144–149.
Buckingham, M., and Tajbakhsh, S. (1993). Expression of myogenic factors in the mouse: myf5, the first member of the MyoD gene family to be transcribed during skeletal myogenesis. *C.R. Acad. Sci. Life Sci.* **316**, 1040–1046.
Buffinger, N., and Stockdale, F. E. (1994). Myogenic specification in somites induction by axial structures. *Development* **120**, 1443–1452.
Chiang, C., Litingtung, Y., Lee, E., Young, K. E., Corden, J. L., Westphal, H., and Beach, P. (1996). Cyclopia and defective axial patterning mice lacking sonic hedgehog gene function. *Nature* **383**, 407–413.
Chevallier, A., Kieny, M., and Mauger, A. (1977). Limb-somite relationship: Origin of the limb musculature. *J. Embryol. Exp. Morphol.* **41**, 245–258.
Chevallier, A., Kieny, M., and Mauger, A. (1978). Limb-somite relationship: Effect of removal of somitic mesoderm on the wing musculature. *J. Embryol. Exp. Morphol.* **43**, 263–278.
Christ, B., and Wilting, J. (1992). From somites to vertebral column. *Ann. Anat.* **174**, 23–32.
Christ, B., Jacob, H., and Jacob, M. (1977). Experimental analysis of the origin of the wing musculature in avian embryos. *Anat. Embryol.* **150**, 171–186.
Christ, B., Jacob, H., and Jacob, M. (1978). On the formation of the myotomes in avian embryos. An experimental and scanning electron microscope study. *Experientia* **34**, 514–516.

7. Mouse–Chick Chimera: Studying Somite Development

Christ, B., Brand-Saberi, B., Grim, M., and Wilting, J. (1992). Local signalling in dermomyotomal cell type specification. *Anat. Embryol.* **186**, 505–510.

Cossu, G., Kelly, R., Tajbakhsh, S., Di Donna, S., Vivarelli, E., and Buckingham, M. (1996). Activation of different myogenic pathways: Myf-5 is induced by the neural tube and Myo-D by the dorsal ectoderm in mouse paraxial mesoderm. *Development* **122**, 429–438.

Cooper, G. W. (1965). Induction of somite chondrogenesis by cartilage and notochord: A correlation between inductive activity and specific stages of cytodifferentiation. *Dev. Biol.* **12**, 185–212.

Couly, G. F., Coltey, P. M., and Le Douarin, N. M. (1993). The triple origin of skull in higher vertebrates —A study in quail–chick chimeras. *Development* **117**, 409–429.

Dahm, L., and Landmesser, L. (1991). The regulation of synaptogenesis during normal development and following activity blockade. *J. Neurosci.* **11**, 238–255.

Davidson, D. (1995). The function and evolution of Msx genes-pointers and paradoxes. *Trends Genet.* **11**, 405–411.

Denetclaw, W., Christ, B., and Ordahl, C. (1997). Location and growth of epaxial myotome precursor cells. *Development* **124**, 1601–1610.

Deutsch, U., Dressler, G. R., and Gruss, P. (1988). Pax-1, a member of a paired box homologous murine gene family, is expressed in segmented structures during development. *Cell* **53**, 617–625.

Dunn, L. C., Glueksohn-Schoenheimer, S., and Bryson, V. (1940). A new mutation in the mouse affecting spinal column and urogenital system. *J. Hered.* **31**, 343–348.

Duxson, M. J. (1992). The relationship of nerve to myoblasts and newly-formed secondary myotubes in the 4th lumbrical muscle of the rat foetus. *J. Neurocytol.* **21**, 574–588.

Ebensperger, C., Wilting J., Brand-Saberi, B., Mizutani, Y., Christ, B., Balling, R., and Koseki, H. (1995). Pax-1, a regulator of sclerotome development is induced by notochord and floor plate signals in avian embryos. *Anat. Embryol.* **191**, 297–310.

Emerson, C. P. (1990). Myogenesis and developmental control genes. *Curr. Opin. Gen. Dev.* **2**, 1065–1075.

Fan, C., and Tessier-Lavigne, M. (1994). Patterning of mammalian somites by surface ectoderm and notochord: Evidence for sclerotome induction by a hedgehog homologue. *Cell* **79**, 1175–1186.

Flanagan, A. E. H. (1969). Differentiation and degeneration in the motor column of foetal mouse. *J. Morphol.* **129**, 281–305.

Feulgen, R., and Rossenbeck, H. (1924). Mikroskopisch-chemicher nachweiss einer nucleinsäure und die darauf beruhende elektive färbung von zellkernen in mikroskopischen präparaten. *Hoppe-Seyler's Z. Physiol. Chem.* **135**, 203–252.

Fontaine-Pérus, J., Chanconie, M., and Le Douarin, N. M. (1985). Embryonic origin of substance P-containing neurons in cranial and spinal sensory ganglia. *Dev. Biol.* **107**, 227–238.

Fontaine-Pérus, J., Jarno, V., Fournier Le Ray, C., Li, Z., and Paulin, D. (1995). Mouse–chick chimera: A new model to study the *in ovo* developmental potentialities of mammalian somites. *Development* **121**, 1705–1718.

Fontaine-Pérus, J., Halgand, P., Chéraud, Y., Rouaud, T., Velasco, M. E., Cifuentes Diaz, C., and Rieger, F. (1997). Mouse–chick chimera: A developmental model of murine neurogenic cells. *Development* **124**, 3025–3036.

Gamel, A.J., Brand-Saberi, B., and Christ, B. (1996). Halves of epithelial somites and segmental plate show distinct muscle differentiation behavior *in vitro* compared to entire somites and segmental plate. *Dev. Biol.* **172**, 625–639.

Geisler, N., and Weber, K. (1982). The amino acid sequence of chicken muscle desmin provides a common structural model for intermediate filament proteins. *EMBO J.* **1**, 1649–1656.

Goulding, M. D., Chalepakis, G., Deutsch, U., Erselius, J., and Gruss, P. (1991). Pax-3, a novel murine DNA binding protein expressed during early neurogenesis. *EMBO J.* **10**, 1135–1147.

Harris, A. J., Fitzsimons, R. B., and McEwans, J. C. (1989). Myonuclear birthdates distinguish the origins of primary and secondary myotubes in embryonic mammalian skeletal muscles. *Development* **107**, 751–769.

Hasty, P., Bradley, A., Morris, J. H., Edmonson, D. G., Venuti, J. M., Olson, E. N., and Klein, W. H.

(1993). Muscle deficiency and neonatal death in mice with a targeted mutation in the myogenin gene. *Nature* **364**, 501–506.

Hinterberger, J. T., Sassoon, D. A., Rhodes, S. J., and Konieczny, S. F. (1991). Expression of the muscle regulatory factor MRF4-during somite and skeletal myofiber development. *Dev. Biol.* **147**, 144–156.

Houzelstein, D., Auda-Boucher, G., Chéraud, Y., Rouaud, T., Blanc, I., Tajbakhsh, S., Buckingham, M. E., Fontaine-Pérus, J., and Robert, B. (1999). The homeobox gene Msx1 is expressed in a subset of somites, and in muscle progenitor cells migrating into the forelimb. *Development* **126**, 2689–2701.

Itasaki, N., Sharpe, J., Morrison, A., and Krumlauf, R. (1996). Reprogramming Hox expression in the vertebrate hindbrain: Influence of paraxial mesoderm and rhombomere transposition. *Neuron* **16**, 487–500.

Johnson, R. L., Laufer, E., Riddle, R. D., and Tabin, C. J. (1994). Ectopic expression of sonic hedgehog alters dorsal ventral patterning of the somites. *Cell* **79**, 1165–1174.

Kaufman, S. J., and Foster, R. F. (1988). Replicating myoblasts express a muscle-specific phenotype. *Proc. Natl. Acad. Sci. U.S.A.* **85**, 9606–9610.

Kenny-Mobbs, T., and Thorogood, P. (1987). Autonomy of differentiation in avian brachial somites and the influence of adjacent tissues. *Development* **100**, 449–462.

Kieny, M., Chevallier, A., and Pautou, M. P. (1987). Attempt to produce a chick/mouse heteroclass musculature. *Roux's Arch. Dev. Biol.* **196**, 321–327.

Lance-Jones, C. (1982). Motoneuron cell death in the developing lumbar spinal cord of the mouse. *Dev. Brain Res.* **4**, 473–479.

Lazarides, E., and Hubbard, B. D. (1976). Immunological characterization of the subunit of the 100 A filaments from muscle cells. *Proc. Natl. Acad. Sci. U.S.A.* **73**, 4344–4348.

Le Douarin, N. (1969). Particularités du noyau interphasique chez la caille japonaise (*Coturnix coturnix japonica*). Utilisation de ces particularités comme "marquage biologique" dans les recherches sur les interactions tissulaires et les migrations cellulaires au cours de l'ontogenèse. *Bull. Biol. Fr. Belg.* **103**, 435–452.

Le Douarin, N. (1973). A feulgen-positive nucleolus. *Exp. Cell Res.* **77**, 459–468.

Lee, K. H., and Sze, L.Y. (1993). Role of the brachial somites in the development of appendicular musculature in rat embryos. *Dev. Dyn.* **198**, 86–93.

Lelièvre, C., and Le Douarin, N. M. (1975). Mesenchymal derivatives of the neural crest: analysis of chimaeric quail and chick embryos. *J. Embryol. Exp. Morphol.* **34**, 125–154.

Li, Z., Marchand, P., Babinet, C., and Paulin, D. (1993). Desmin sequence elements regulating skeletal muscle-specific expression in transgenic mice. *Development* **117**, 947–959.

Li, Z., Colucci-Guyon, E., Pincon-Raymond, M., Mericskay, M., Pournin, S., Paulin, D., and Babinet, C. (1996). Cardiovascular lesions and skeletal myopathy in mice lacking desmin. *Dev. Biol.* **175**, 362–366.

Li, Z., Mericskay, M. Agbulut, O., Butler-Browne, G., Carlsson, L., Thornell, L. E., Babinet, C., and Paulin, D. (1997). Desmin is essential for the tensile strength and integrity of myofibrils but not for myogenic commitment, differentiation, and fusion of skeletal muscle. *J. Cell Biol.* **139**(1), 129–144.

Lyons, G. E., Ontell, M., Cox, R., Sassoon, D., and Buckingham, M. (1990). The expression of myosin genes in developing skeletal muscle in the mouse embryo. *J. Cell. Biol.* **111**, 1465–1476.

Mauger, A. (1972). Rôle du mésoderme somitique dans le développement du plumage dorsal chez l'embryon de poulet. I. Origine, capacités de régulation et détermination du mésoderme plumigène. *J. Embryol. Exp. Morphol.* **28**, 313–341.

Marcelle, C., Stark, M. R., and Bronner-Fraser, M. (1997). Coordinate actions of BMPs, Wnts, Shh and Noggin mediate patterning of the dorsal somite. *Development* **124**, 3955–3963.

Mayo, M., Brinbas, M., Santos, V., Shum, L., and Slavkin, H. C. (1992). Desmin expression during early mouse tongue morphogenesis. *J. Dev. Biol.* **36**, 255–263.

Milaire, J. (1976). Contribution cellulaire des somites à la genèse des bourgeons de membre postérieurs chez la souris. *Arch. Biol.* **87,** 315–343.
Milaire, J. (1987). Early structural changes associated with the onset of fore-limb development in the mouse. *Arch. Biol.* **98,** 67–98.
Münsterberg, A. E., and Lassar, A. B. (1995). Combinatorial signals from the neural tube, floor plate and notochord induce myogenic bHLH gene expression in the somite. *Development* **121,** 651–660.
Münsterberg, A. E., Kitajewski, J., Bumcrot, D. A., McMahon, A., and Lassar, A. B. (1995). Combinatorial signaling by sonic hedgehog and Wnt family members induces myogenic bHLH gene expression in the somite. *Genes Dev.* **9,** 2911–2922.
Nabeshima, Y., Hanaoka, K., Hayasaka, M., Esumi, E., Li, S., Nonaka, I. and Nabeshima, Y. (1993). Myogenein gene disruption results in prenatal lethality because of severe defect. *Nature* **364,** 532–535.
Nieuwkoop, P. D. (1952). Activation and organisation of the central nervous system in amphibians. *J. Exp. Zool.* **120,** 1–108.
Noakes, P. G., Phillips, W. D., Hanley, T. A., Sanes, J. R., and Merlie J. P. (1993). 43K protein and acetylcholine receptors colocalize during the initial stages of neuromuscular synapse formation *in vivo. Dev. Biol.* **155,** 275–280.
Nornes, W. O., and Carry, M. (1978). Neurogenesis in spinal cord of mouse: An autoradiographic analysis. *Brain Res.* **159,** 1–16.
Olson, E. N. (1992). Interplay between proliferation and differentiation within the myogenic lineage. *Dev. Biol.* **154,** 261–272.
Ordahl, C. P. (1992). Developmental regulation of sarcomeric gene expression. *In* "Current Topics in Development Biology" (E. Bearer, Ed.), Vol. 26, pp. 145–168. Academic Press, New York.
Ordahl, C. P., and Le Douarin, N. M. (1992). Two myogenic lineages within the developing somite. *Development* **111,** 339–353.
Ott, M., Bober, E., Lyons, G., Arnold, H., and Buckingham, M. (1991). Early expression of the myogenic regulatory gene, myf-5, in precursor cells of skeletal muscle in the mouse embryo. *Development* **111,** 1097–1107.
Packard, D., and Jacobson, A. (1976). The influence of axial structures on chick somite formation. *Dev. Biol.* **53,** 36–48.
Pownall, M. E., Strunk, K. E., and Emerson, C. P. (1996). Notochord signals control the transcriptional cascade of myogenic bHLH genes in somites of quail embryos. *Development* **122,** 1475–1488.
Rong, P., Teillet, M. A., Ziller, C., and Le Douarin, N. M. (1992). The neural tube/notochord complex is necessary for vertebral but not limb and body wall striated muscle differentiation. *Development* **115,** 657–672.
Rudnicki, M. A., Braun, T., Hinuma, S., and Jaenisch, R. (1992). Inactivation of MyoD in mice leads to up-regulation of the myogenic HLH gene myogenin and results in apparently normal muscle development. *Cell* **71,** 383–390.
Rudnicki, M. A., Schnegelsberg, P.N., Stead, R.H., Braun, T., Arnold, H.H., and Jaenish, R. (1993). MyoD or Myf5 is required for the formation of skeletal muscle. *Cell* **15,** 1531–1539.
Rugh, R. (1968). "The Mouse Its Reproduction and Development." Oxford Univ. Press: Oxford Science Publications, London.
Sanes, J. R., Rubinstein, J. L., and Nicolas, J. F. (1986). Use of a recombinant retrovirus to study post-implantation cell lineage in mouse embryos. *EMBO J.* **5,** 3133–3142.
Sassoon, D., Lyons, G., Wright, W. E., Lin, V., Lassar, A., Weintraub, H., and Buckingham, M. (1989). Expression of two myogenic regulatory factors myogenin and MyoD during mouse embryogenesis. *Nature* **341,** 303–307.
Serbedzija, G. N., Fraser, S. E., and Bronner-Fraser M. (1990). Pathways of trunk neural crest cell migration in the mouse embryo as revealed by vital dye labelling. *Development* **108,** 605–612.

Serbedzija, G. N., Bronner-Fraser, M., and Fraser, S. E. (1994). Developmental potential of trunk neural crest in the mouse. *Development* **120,** 1709–1718.

Small, J. V., and Sobieszek, A. (1977). Studies on the function and composition of the 10 nm (100-A) filaments of vertebrate smooth muscle. *J. Cell Sci.* **23,** 243–268.

Spence, M., Yip, J., and Erickson, C. (1996). The dorsal neural tube organizes the dermomyotome and induces axial myocytes in the avian embryo. *Development* **122,** 231–241.

Spörle, R., Günther, T., Struwe, M., and Schughart, K. (1996). Severe defects in the formation of epaxial musculature in open brain (obp) mutant mouse embryos. *Development* **122,** 79–86.

Stern, H. M., and Hauschka, S. D. (1995). Neural tube and notochord promote *in vitro* myogenesis in single somite explants. *Dev. Biol.* **167,** 87–103.

Stern, H. M., Brown, A. M. C., and Hauschka, S. D. (1995). Myogenesis in paraxial mesoderm: Preferential induction by dorsal neural tube and by cells expressing Wnt-1. *Development* **121,** 3675–3686.

Sze, L. Y., Lee, K. K. H., Webb, S. E., Li, Z., and Paulin, D. (1995). Migration of myogenic cells from the somites to the fore-limb buds of developing mouse embryos. *Dev. Dyn.* **203,** 324–336.

Tajbakhsh, S., and Buckingham, M. (1994). Mouse limb muscle is determined in the absence of the earliest myogenic factor Myf5. *Proc. Natl. Acad. Sci. U.S.A.* **91,** 747–751.

Tam, P. P. L. (1981). The control of somitogenesis in mouse embryos. *J. Embryol. Exp. Morphol* **65,** 103–128.

Tam, P. P. L., and Beddington, R. S. P. (1986). The metameric organization of the presomitic mesoderm and somite specification in the mouse embryo. *In* "Somites in Developing Embryos." (R. Bellairs, A. D. Ede, and J. W. Lash, Eds.), pp.17–36. Plenum, New York.

Tam, P. P. L., and Tan, S. S. (1992). The somitic potential of cells in the primitive streak and the tail bud of the organogenesis stage mouse embryo. *Development* **115,** 703–715.

Tam, P. P. L., Parameswaran, M., Kinder, S. J., Weinberger, R. P. (1997). The allocation of epiblast cells to the embryonic heart and other mesodermal lineages: The role of ingression and tissue movement during gastrulation. *Development* **124**(9), 1631–1643.

Teillet, M. A., and Le Douarin, N. M. (1983). Consequences of neural tube and notochord excision on the development of the peripheral nervous system in the chick embryo. *Dev. Biol.* **98,** 192–211.

Trainor, P. A., and Tam, P. P. L. (1995). Cranial paraxial mesoderm and neural crest cells of the mouse embryo: Co-distribution in the craniofacial mesenchyme but distinct segregation in branchial arches. *Development* **121,** 2569–2582.

Vivarelli, E., and Cossu, G. (1986). Neural control of early myogenic differentiation in cultures of mouse somites. *Dev. Biol.* **117,** 319–325.

Watterson, R. L., Fowler, J., and Fowler, B. J. (1954). The role of the neural tube and notochord in development of the axial skeleton of the chick. *Am. J. Anat.* **95,** 337–399.

Wilting, J., Brand-Saberi, B., Huang, R., Zhi, Q., Köntges, G., Ordahl, C. P., and Christ, B. (1995). Angiogenic potential of the avian somite. *Dev. Dyn.* **202,** 165–171.

Wilting, J., Eichmann, A., and Christ, B. (1997). Expression of the avian VEGF receptor homologues quek 1 and quek 2 in blood—vascular and lymphatic endothelial and nonendothelial cells during quail embryonic development. *Cell Tissue Res.* **288**(2), 207–224.

8
Transcriptional Regulation during Somitogenesis

Dennis Summerbell and Peter W. J. Rigby[1]
Division of Eukaryotic Molecular Genetics
MRC National Institute for Medical Research
The Ridgeway, Mill Hill
London NW7 1AA, England

I. Introduction
II. Regulation of *Hox* Gene Expression
III. Control of Myogenic Regulatory Factor Gene Expression
IV. Roles of Non-MRF Transcription Factors in Muscle Development
V. The Identification of Novel Somite-Specific Transcripts
  References

I. Introduction

We are interested in the biochemical mechanisms involved in the regulation of transcription during the development of the mouse embryo. Specifically we would like to understand how the extra- and intercellular signals that pattern the embryo regulate the nuclear transcription machinery such that recipient cells express the genes appropriate to their position or fate within the embryo. We have focused our attention on the process of somitogenesis for several reasons. First, each pair of somites has a distinct segmental identity, most obviously revealed by the morphology of the vertebra derived from it, and somitogenesis thus provides an excellent system in which to try to understand how positional identity along the rostrocaudal axis is specified, in a fashion that is quite independent of decisions regarding cell fate. Conversely, as each somite compartmentalizes to give rise to the three lineages that are the precursors to bone and cartilage, skeletal muscle, and skin, well-defined decisions are made about cell identity and somitogenesis thus provides a fine model in which to study the mechanisms by which pluripotential progenitor cells are directed toward a particular differentiation pathway. Finally, the anatomy of the mid-gestation mouse embryo means that it is relatively easy to define, with considerable precision, patterns of gene expression. One can thus compare the expression of the endogenous gene, as revealed by *in situ*

[1] Author to whom correspondence should be addressed.

hybridization, with the expression of wild-type or mutant transgenes, as revealed by histochemical staining to detect the product of an appropriate reporter gene.

Our general approach has been to take genes known to act at a fairly high level in a particular genetic pathway and then to try to understand how their transcription is controlled using the power of transgenic mouse technology. In each case our ultimate objective is to identify (or know the name) of each of the transcription factors that control the chosen regulatory gene so that we can then study how the activities of these factors are modulated, and thus build up a picture of the signal transduction pathways, and signals, involved in the developmental decision. Segmental identity in the vertebrate embryo is controlled, at least in part, by the four clusters of *Hox* genes (reviewed by Krumlauf, 1994), and we have therefore chosen to analyze in detail the regulation of one such gene, *Hoxb-4*. One of the earliest transcriptional responses in the epithelial somite is the activation of the cascade of basic helix–loop–helix (bHLH) transcription factors that together regulate the commitment of skeletal myoblasts and their subsequent differentiation (reviewed by Cossu *et al.*, 1996), and we have also studied the regulation of two of these genes, *myf5* and *myogenin*. Although we have considerable knowledge of genes that function during somitogenesis, it is clear that many more such genes remain unknown, and we have therefore sought to develop techniques that will allow the discovery of novel genes which act during defined aspects of the process.

II. Regulation of *Hox* Gene Expression

In the mouse and other mammals there are four clusters of *Hox* genes, in each of which there is a direct relationship between the position of a gene within a cluster and both the timing of its expression and its final expression domain, a phenomenon known as colinearity which appears to have arisen very early in animal evolution. Genes at the 5′ ends of clusters are activated late and expressed in posterior domains, while genes at the 3′ ends of clusters are activated early and expressed in anterior domains. The paralogous genes across the four clusters, for example, *Hoxa-4, Hoxb-4, Hoxc-4,* and *Hoxd-4,* have similar but not identical anterior boundaries of expression (reviewed by Maconochie *et al.*, 1996). Thus each pair of somites expresses a particular combination of *Hox* genes, and it was therefore proposed by Kessel and Gruss (1991) that segmental identity was determined by the *Hox* code (for a detailed discussion of this point, see Chap. 18 by Burke, this volume). An analogous code acts in the segmentation of the hindbrain into rhombomeres (Hunt *et al.*, 1991). Experiments in which *Hox* gene function is inactivated by targeted mutagenesis, or in which a given *Hox* gene is ectopically expressed in more anterior somites, indicate that the concept of the *Hox* code is generally valid (Krumlauf, 1994). Given this, it follows that the establishment of the correct anterior boundary of expression is crucial, and we have sought to understand how this is achieved.

Our approach has been to construct transgenes in which the expression of the *Escherichia coli* β-galactosidase reporter gene is controlled by sequences from around and within the *Hoxb-4* gene and then to assay the functionality of these transgenes in F_0 mouse embryos generated by pronuclear injection. In this way it is possible to analyze a large number of constructs in a relatively short time. Constructs that show interesting expression patterns are then reinjected in order to generate transgenic lines which can then be used to derive more detailed pictures of the expression pattern, particularly with regard to developmental timing. In our initial experiments we showed that 7.4 kb of DNA, beginning 926 bp upstream of the 5′-most transcriptional start site and including the entire gene and 3′ flanking DNA, was sufficient to accurately recapitulate the temporal and spatial expression pattern of the endogenous gene (Whiting *et al.,* 1991). That this can be achieved is not true of all *Hox* genes. It has also been done for *Hoxa-7* (Puschel *et al.,* 1991), but in other cases, for example, *Hoxb-7* and *Hoxc-8,* it has not been possible even when quite large segments of DNA have been used (Vogels *et al.,* 1993; Bradshaw *et al.,* 1996). That it can be done at all shows that the correct temporal and spatial expression pattern can be achieved outside the context of the cluster, although it remains possible that correct levels of expression absolutely require that a gene is located within the appropriate cluster. Further analysis showed that the expression of *Hoxb-4* is controlled by two spatially specific enhancers. One of these, located in the 3′ flanking DNA and called region A, sets the hindbrain boundary between rhombomeres six and seven, while the other, located within the gene and called region C, controls most of the rest of the pattern and sets the mesodermal boundary between somites six and seven (Whiting *et al.,* 1991). In the absence of these enhancers the "promoter" of the gene, the 5′ flanking DNA, is silent in the embryo although it is constitutively active in transient transfection experiments that have been used to map a number of positively and negatively acting elements (Gutman *et al.,* 1994). This is in marked contrast to *Hoxa-7,* for which the opposite logic prevails; the promoter is ubiquitously active and it is restrained by control elements located within the gene (Puschel *et al.,* 1991). Perhaps there is no common theme to *Hox* gene regulation in vertebrates?

The two *Hoxb-4* enhancers have been studied in considerable detail. As originally defined these enhancers were quite large, 3 kb in the case of region A and 1.4 kb in the case of region C. The question therefore arose as to how best to analyze such a segment of DNA when the only available assay was the construction of transgenic mice; as the issue is how are the boundaries set, no cell culture system is going to be informative. Moreover, the size of the regions in question, and the logistics of the transgenic assay, preclude a random mutagenesis approach by, for example, linker scanning. In other systems, perhaps most notably in work on the globin gene clusters (Tuan *et al.,* 1985; Forrester *et al.,* 1987; Grosveld *et al.,* 1987), DNase hypersensitive site analysis has been particularly useful, but we were unable to obtain useful data from this approach (J. D. Gilthorpe and P. W. J. Rigby, unpublished data, 1992), primarily because we were unable to obtain sufficient numbers of homogeneous cells from the embryo. However, we were extremely successful in using DNA sequence comparisons and gained much from our analyses of the *Hox* genes of the Japanese puffer fish, *Fugu rubripes* (Aparicio et al., 1995). *Fugu* has essentially the same gene complement as man or mouse, but its genome is some eight times smaller (Brenner et al., 1993). One reason for this is that many introns in *Fugu* are markedly smaller than their mammalian counterparts; we reasoned that if the intron in the *Fugu Hoxb-4* gene was 10 times smaller than that in the mouse then our analysis would be considerably simplified. In this case the *Fugu* intron is slightly larger, presumably because the presence within the intron of a considerable number of transcriptional control elements sets a minimum size beyond which it cannot be compressed. Nonetheless, comparison of the sequences of the *Fugu* and mouse introns identified a single block of extreme conservation; when this block was deleted from the mouse region C enhancer all mesodermal expression was lost (Aparicio et al., 1995) (see Fig. 1).

Having identified a region of 92 bp which contains one or more elements necessary for the somitic expression of *Hoxb-4*, we turned to the strategy that we successfully employed in our analysis of the gene's promoter (Gutman et al., 1994). Overlapping oligonucleotides covering the region of conservation were used in electrophoretic mobility shift assays (EMSAs) that identified, in the first iteration, a pair of proteins which bind, in a noncompetitive fashion, to two overlapping sites. These proteins are YY1 and NF-Y; mutation of the YY1 site leads to a marked decrease in mesodermal expression relative to neural tube expression, but mutation of the NF-Y site abolishes all mesodermal expression (J. D. Gilthorpe, M. Vandromme, and P. W. J. Rigby, unpublished data, 1999). It is of considerable interest that these same two proteins also bind to the essential *cis*-acting element in the promoter (Gutman et al., 1994, in which paper NF-Y is referred to as HoxTF), and to an additional site in the right-hand half of the intron. NF-Y is unusual among eukaryotic transcription factors in that it is a heterotrimer and all three subunits are required for DNA binding (reviewed by Maity and de Crombrugghe, 1998); it has been shown to interact with several histone acetyltransferases, including p300, P/CAF, and hGCN5 (Currie, 1998; Li et al., 1998). YY1 is involved in the regulation of a large number of genes and can act as both an activator and a repressor of transcription (reviewed by Shi et al., 1997). It binds to the histone deactetylase RPD3 (Yang et al., 1996), and, of particular importance in the present context, its DNA-binding zinc fingers are very closely related to those of the *Drosophila* polycomb-group gene *pleiohomeotic* (Brown et al., 1998). The involvement of NF-Y and YY1 raises interesting questions about the role of histone acetylation/deacetylation in the regulation of *Hox* gene expression.

We have continued to apply this general approach and are in the process of characterizing five additional proteins that are putative regulators of *Hoxb-4* transcription (J. D. Gilthorpe, T. Brend, and P. W. J. Rigby, unpublished data, 1999). We think that it is likely that we will soon know the names of all of the transcription factors which bind to the region C enhancer that controls expression in the somites. An inherent limitation of the approach that we have used is that it can only

8. Transcriptional Regulation during Somitogenesis

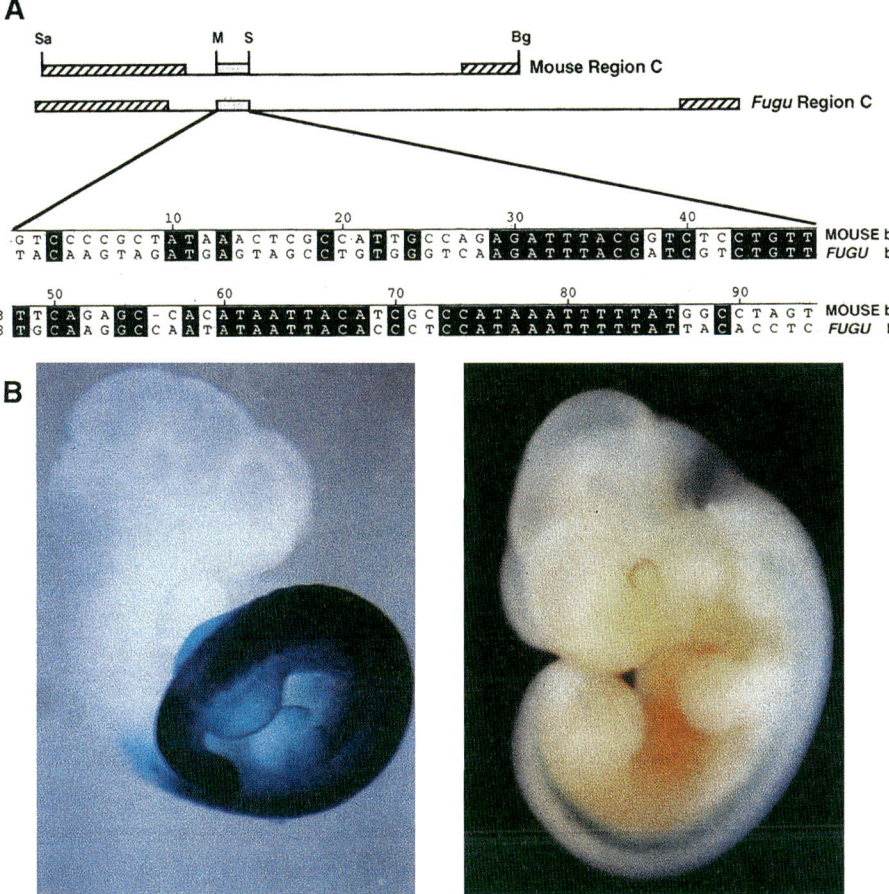

Figure 1 Identification of the critical regulatory elements in *Hoxb-4* region C by sequence homology to *Fugu* (Aparicio *et al.*, 1995). (A) Sequence comparison of mouse and *Fugu* region C identified a 92-bp sequence with 60% homology. (B) Pattern of expression in a transgenic mouse at 10.5 dpc using region C driving a lacZ reporter construct. (C) Pattern of expression in a transgenic mouse at 11.5 dpc using a similar construct in which the 92-bp fragment has been internally deleted.

identify proteins which bind to the DNA *in vitro* with an affinity that allows their detection under a limited set of EMSA conditions. Important proteins that do not bind directly to the DNA may well be found by interaction-trap approaches such as the yeast two-hybrid system, but proteins that bind to the DNA weakly, or that only bind to chromatin, may be very difficult to identify by experiments of this general sort. Nonetheless, we are of the view that the knowledge acquired by this approach has put us in a position where we can now ask mechanistic questions about *Hox* gene regulation.

III. Control of Myogenic Regulatory Factor Gene Expression

The four known myogenic regulatory factors (MRFs), myf5, myogenin, mrf4, and myoD, are members of the bHLH family of transcription factors (for reviews, see Arnold and Braun, 1996; Molkentin and Olson, 1996; Yun and Wold, 1996). Each of these proteins binds, as an obligate heterodimer with one of the products of the ubiquitously expressed *E2A* gene (Murre *et al.,* 1989; Brennan and Olson, 1990; Lassar *et al.,* 1991), to a consensus sequence called the E-box (CANNTG). Such E-boxes are found in the promoters and/or enhancers of many muscle-specific genes, for example, those encoding muscle-specific isoenzymes and proteins of the contractile apparatus, where they mediate the transcriptional response to the MRFs. Each MRF has a distinct pattern of expression in the developing mouse embryo, and genetic analyses strongly suggest that each gene has a distinct function.

The precursors to skeletal muscle arise at a number of distinct sites in the mouse embryo. The first cells to express an MRF are located at the dorsomedial lip (Fig. 2) of the dermomyotome of the somites (Ott *et al.,* 1991). The exact timing varies depending on the identity of the somite (see Smith *et al.,* 1994) but a good generalization is that by middle and later stages it comes on in the most recently formed somite. The timing of MRF expression at the other extreme of the dermomyotome, the ventrolateral edge, is even more complicated, again in general being dependent on the craniocaudal level along the embryo. At early stages ventrolateral expression lags by 9–10 somites, but by the time that the sacral somites are forming it is almost simultaneous with dorsomedial expression. As judged by *in situ* hybridization, there is no MRF expression in the presomitic mesoderm, although it is entirely possible that some MRFs are expressed at a very low level prior to somite formation (Cossu *et al.,* 1996). Grafting experiments have shown that in the chick the migratory myoblasts of the limb are derived only from the lateral half of the somite, whereas the medial half gives rise to trunk muscles (Ordahl and Le Douarin, 1992; Christ and Ordahl, 1995). It is important to remember that somitogenesis in the mouse is not the same as that in the chick, particularly during the early stages; for example, epithelial somites are transient in the mouse, whereas they persist for some considerable time in the chick. Although neither grafting nor lineage-tracing data are available for the mouse, the general view would be that the dorsomedial cells give rise to the epaxial extensor muscles of the back, while the ventrolateral cells give rise to the hypaxial muscles of the lateral and ventral body wall and, at the appropriate levels, to the migratory myoblasts destined to form the appendicular muscles of the limbs. Myoblasts also migrate from the somites, via the hypoglossal cord, to form the tongue muscles. Facial muscles do not derive from somites but from the cranial paraxial mesoderm, which is not overtly segmented. It is thus the case that a single tissue type, skeletal muscle, is derived from cells which arise at quite distinct locations within the embryo. It seems highly unlikely that the signaling environments will be the same in

8. Transcriptional Regulation during Somitogenesis 307

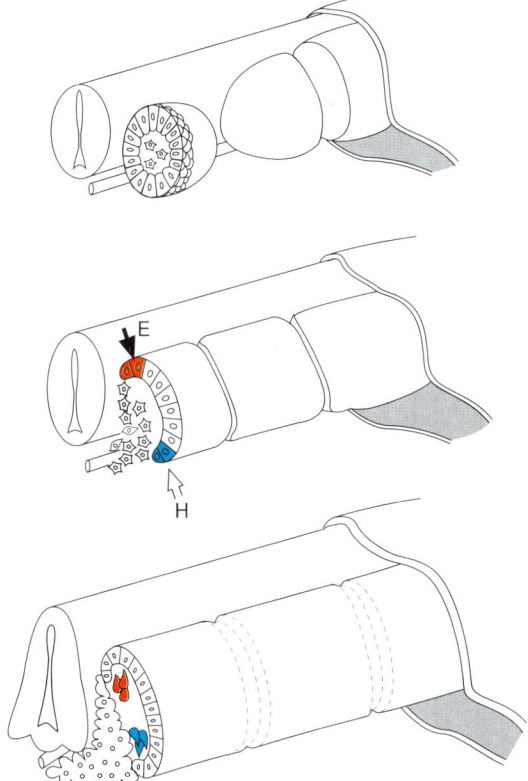

Figure 2 Myogenic lineages in the somites. Diagram of somites at successive stages of development showing the likely origin of epaxial (dorsomedial) and hypaxial (ventrolateral) components.

each of these locations, and one might therefore predict that there will be marked differences in the transcriptional circuitry which operates in each type of muscle progenitor.

In mouse the first MRF gene to be expressed is *myf5*, the mRNA appearing at 8.0 days *post coitum*—(*dpc*) in the epaxial precursors of the most cranial newly formed somites (Ott *et al.*, 1991). Expression in the hypaxial precursors begins in somite IX or X of the 21-somite embryo. *Myogenin* mRNA appears in the myotomes of successive somites some 12 hr after the expression of *myf5* (Sassoon *et al.*, 1989), which is to be expected as *myogenin* is probably a direct transcriptional target of myf5 (see below). *MyoD* mRNA does not appear until 10.5 dpc (Sassoon *et al.*, 1989), and the relationship between its activation in the epaxial and hypaxial precursors is complicated and dependent on the somitic level. The early phase of mrf4 expression, between 9.5 and 10.5 dpc, has not been precisely defined, and we therefore do not know the relationship between the expression of this gene and the various precursor populations (Bober *et al.*, 1991; Hinterberger *et al.*, 1991).

All four of the MRF genes have been knocked out using homologous recombination in embryonic stem cells (Rudnicki *et al.*, 1992, 1993; Tajbakhsh and Buckingham, 1994). The data derived from such experiments are consistent with a view in which *myf5* is the determination gene for skeletal muscle, while *myogenin* is required for overt cytodifferentiation. The role of *myoD* in the specification and/or differentiation of embryonic skeletal muscle has been the subject of a vigorous, and often confused, debate. It is often supposed that the medial precursors express myf5 while the lateral precursors express myoD, and that this difference can somehow be extrapolated to the existance of two "lineages". This view received support from experiments in which both *myf5* alleles were targeted, one with *lacZ*, and one with a selectable marker, and the resultant embryonic stem (ES) cells were differentiated *in vitro* (Braun and Arnold, 1996); it was concluded that there is a "myf5 lineage" and a "myoD lineage." However, the summation of the data available on the line of mice in which *lacZ* has been knocked in to the *myf5* locus indicates that the vast majority of muscle cells express *myf5* (Tajbakhsh *et al.*, 1996a, b; 1997), and there can be absolutely no doubt that the ventolateral precursors express *myf5*, they just do so slightly later than the epaxial precursors. Moreover, Tajbakhsh *et al.* (1997) have shown that *myoD* expression is dependent on myf5, which is not what one would expect if there were indeed two lineages. We do not know the physiological relevance of the *in vitro* differentiation of ES cells. From what we know of the situation in the embryo, there is no reason to suppose that myoD plays a determinative role, although it may well function at later stages of limb muscle development (Hughes *et al.*, 1993). The major role of myoD is likely to be in the regeneration of muscles in the adult (Megeney *et al.*, 1996). All three knockouts of *mrf4* are likely to have affected the transcription of *myf5* (Olson *et al.*, 1996), and a clear definition of the role of this gene will not be possible until a mutation that affects only the biochemical activity of mrf4 has been produced.

Our understanding of the roles of the various MRFs in muscle development would be considerably advanced if we knew how their expression was controlled. We, and others, have studied the regulation of *myogenin* transcription in some detail (Cheng *et al.*, 1993; Yee and Rigby, 1993). All of the *cis*-acting sequences required for the correct spatial and temporal regulation of *myogenin* expression in the embryo are contained within the 133 bp upstream of the transcription start site (Fig. 3). Further upstream sequences are required for full levels of transcription, and they may well function in adult responses, for example, to denervation. As far as we can tell, the promoter–proximal sequences that we have defined function in all committed myoblasts. Mutational analyses have identifed three necessary binding sites within this minimal control region. One of these is an E-box that is absolutely required, indicating that there is a bHLH protein immediately upstream in the regulatory cascade of *myogenin*. It is extremely likely that this protein is

8. Transcriptional Regulation during Somitogenesis 309

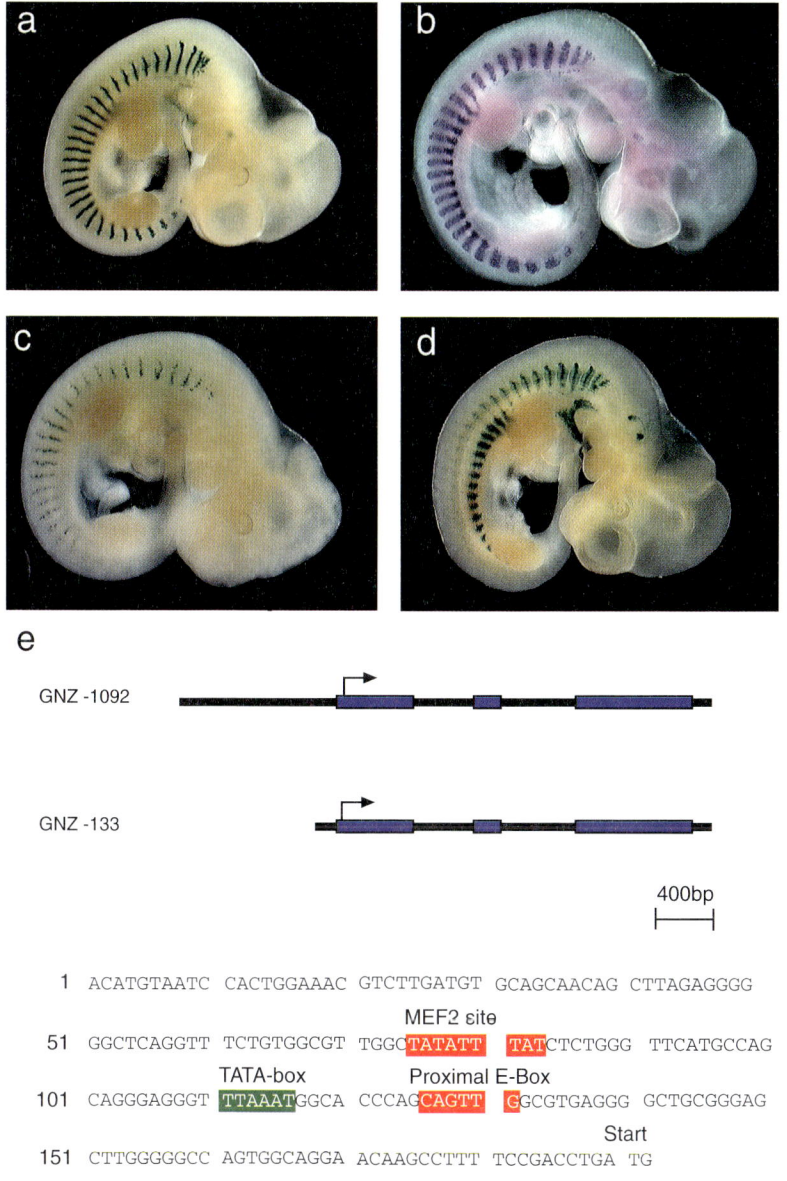

```
  1 ACATGTAATC CACTGGAAAC GTCTTGATGT GCAGCAACAG CTTAGAGGGG
                              MEF2 site
 51 GGCTCAGGTT TCTGTGGCGT TGGCTATATT TATCTCTGGG TTCATGCCAG
              TATA-box          Proximal E-Box
101 CAGGGAGGGT TTAAATGGCA CCCAGCAGTT GGCGTGAGGG GCTGCGGGAG
                                               Start
151 CTTGGGGCC AGTGGCAGGA ACAAGCCTTT TCCGACCTGA TG
```

Figure 3 The regulation of myogenin transcription. (a) Expression pattern of a lacZ reporter under the control of 1092 bp upstream of the transcriptional start site at 10.5 dpc. The pattern accurately recapitulates the normal spatial and temporal pattern. (b) Normal expression pattern using *in situ* hybridization. (c) Truncation of the 1092-bp construct to 133 bp results in the same pattern but at a reduced level of intensity. (d) Mutation of the MEF2 binding site in the 1092-bp context results in the loss of expression in anatomical subcomponents of the pattern. (e) Nucleotide sequence of the 133-bp promoter region showing the identified binding sites.

myf5. It is the only candidate known to be expressed in the right place at the right time; *myogenin* expression is delayed in the myf5 knockout, and cell culture experiments indicate that myf5 can directly activate *myogenin* transcription (Braun *et al.*, 1992). If this E-box is mutated in the context of a larger construct containing some 1 kb of upstream DNA then one sees that expression is apparently normal in trunk muscle but delayed in the limb. This result provided one of the first molecular correlates of the distinct origin of limb muscles established by Ordahl and Le Douarin (1992). The *myogenin* minimal control region also contains a binding site for the MEF2 proteins that have been shown to play important roles in the development of both skeletal and cardiac muscle (Edmondson *et al.*, 1994). Mutation of the MEF2 binding site dramatically delays expression in epaxial precursors but has little effect on hypaxial expression, a result which showed that the regulatory circuits in these two classes of trunk muscle precursors were likely to be different. It has recently been shown that MEF3 also plays a role in the regulation of myogenin transcription in particular sets of muscles (Spitz *et al.*, 1998; P. R. Ashby and P. W. J. Rigby, unpublished data, 1998). One can thus conclude that despite the fact that the *myogenin* minimal control region is very compact, it is nonetheless regulated in different ways in embryologically distinct subsets of skeletal muscles.

We have begun to analyze the transcriptional regulation of *myf5* (Fig. 4), which is located on mouse chromosome 10, some 6.5 kb downstream of the gene encoding mrf4. We have used the same transgenic approach that we employed in our studies of *Hoxb-4* and *myogenin,* but in addition to the use of conventional plasmid-based constructs we have also employed bacterial artificial chromosomes (BACs), modified using the procedures developed by Yang *et al.* (1997). These studies, together with work done using the introduction of yeast artificial chromosomes (YACs) into ES cells and the subsequent generation of chimeric embryos (Zweigerdt *et al.*, 1997), reveal that the elements required for correct myf5 transcription are spread over 100 kb of DNA. Regions in or just upstream of the promoter similar to those able to drive normal expression of *myogenin* produce reporter expression only in the branchial arches (Patapoutian *et al.*, 1993). We have shown that *myf5* is controlled by a series of quite distinct enhancers, each of which acts only in muscle precursors of a particular embryological origin, for example, epaxial, hypaxial, and facial (D. Summerbell, P. R. Ashby, O. Coutelle, D. Cox, S.-P. Yee, and P. W. J. Rigby, unpublished data, 1995). We interpret these results as being entirely consistent with the thesis advanced above that the various classes of muscle precursors arise in distinct signaling environments. Each combination of signals will require interpretation by a particular constellation of transcription factors, and the structure of each enhancer must reflect this. We believe that such a mode of transcriptional control would be highly appropriate for a determination gene that will act on a number of different cell populations that are in receipt of distinct inductive cues to produce the same myogenic end point. We also note that

8. Transcriptional Regulation during Somitogenesis

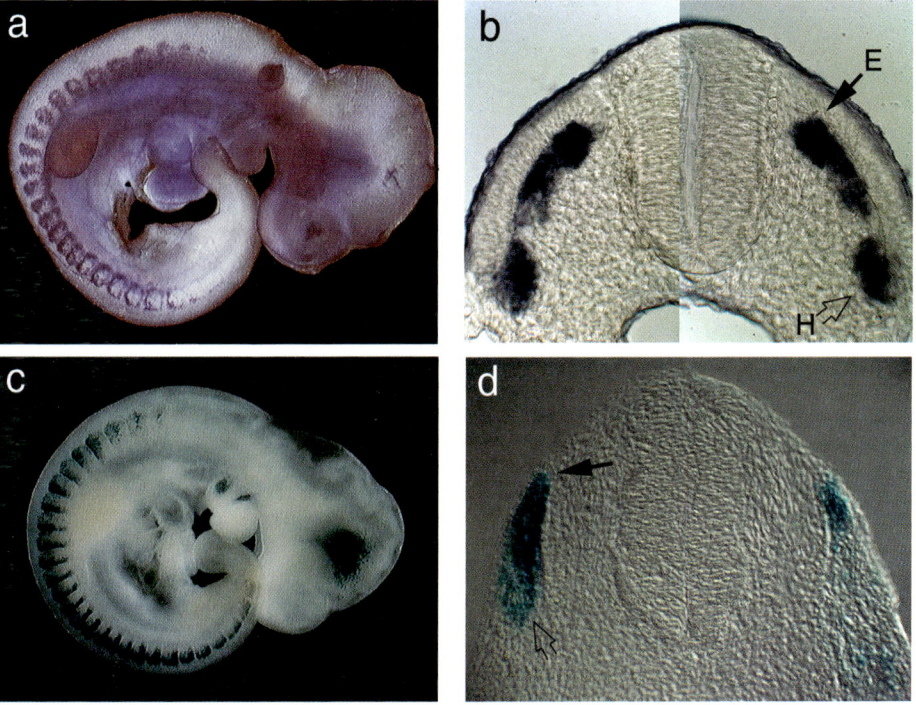

Figure 4 The regulation of myf5 transcription. (a) Normal expression pattern at 9.75 dpc (28 somites). (b) Transverse section through the thoracic region showing epaxial and hypaxial regions. (c) Expression pattern of the myf5/lacZ reporter construct using 14 kb of DNA in the immediate vicinity of the *myf5* gene. (c) Transverse section through the thoracic region with equivalent areas arrowed.

the transcriptional regulation of *myoD* is quite different from that of myf5. While two enhancers have been defined, one at -20 kb (Faerman *et al.*, 1995) and one at -6 kb (Tapscott *et al.*, 1992), it would appear from the published data that each operates in most, if not all, classes of muscle. There is certainly no evidence for the type of control that we have characterized for *myf5*, and this might well be a reflection of the fact that *myoD* transcription in the embryo does not have to respond in the same way to inductive signals.

IV. Roles of Non-MRF Transcription Factors in Muscle Development

While it is clear that members of the bHLH family play crucial roles in skeletal muscle specification and differentiation, they are by no means the only transcription factors involved in the process. The *Pax* gene family encodes proteins with paired-type homeodomains that act in a wide variety of developmental processes (for review, see Mansouri *et al.*, 1996; Dahl *et al.*, 1997). It has been known for some time that in *Splotch* mice, which carry a mutation in *Pax-3*, the limb musculature is absent, and that this is due to a failure of the myoblasts to migrate (Bober *et al.*, 1994; Goulding *et al.*, 1994; Daston *et al.*, 1996). More recent studies have shown that this mutation also affects the trunk musculature, and that in embryos doubly mutant for *myf5* and *Pax-3* all muscles of the body are absent although the head musculature is normal (Tajbakhsh *et al.*, 1997).

The two known *Mox* genes encode divergent homeodomain proteins that are expressed during somitogenesis. *Mox-1* is expressed in the dermomyotome while *Mox-2* is expressed in the myoblasts migrating into the limb, and also in a small group of distal limb mesenchymal cells which are probably of lateral plate origin (Candia and Wright, 1996; Mankoo *et al.*, 1999). In mice in which the *Mox-2* gene has been inactivated by homologous recombination there is a marked reduction in the muscle mass of the limbs, and specific muscles are absent from the fore but not the hind limbs, although there are no defects in the body musculature (Mankoo *et al.*, 1999). Crossing a *myogenin-nLacZ* reporter transgene into the null mutant background allowed the ready visualization of muscle development. It is clear that a reduced number of cells migrate into the limbs and that there are alterations in the pattern of muscle development which account for the defects observed in newborn animals. *In situ* hybridization studies using markers which identify migratory myoblasts indicate that migration is normal but the expression of *Pax-3* in the migrating cells is markedly reduced, suggesting that Mox-2 may play a role in the maintenance of *Pax-3* expression. In the mutant limbs *myoD* expression is normal but *myf5* expression is absent. It thus appears that the regulation of *myoD* expression occurs by distinct mechanisms in the trunk and the limbs and that Mox2 plays a role in myoD-independent pathways which are crucial for the normal development of the limb musculature. It is highly likely that Mox2 functions in the limb myoblasts upstream of *Pax-3* and *myf5*, but it is also possible that the protein could be involved in some aspect of connective tissue differentiation which then affects muscle development. Whatever the mechanism it is clear that the Mox homeoproteins are important new players in the process of skeletal muscle differentiation, and one suspects that there are other important transcriptional regulators of the process to be discovered.

V. Identification of Novel Somite-Specific Transcripts

In addition to the transcriptional regulators discussed above, we know of a considerable number of other genes that exhibit precise and provocative patterns of expression in the somites, the ligands and receptors of the delta-notch system being a good example (McGrew and Pourquié, 1998). Nonetheless, it would be

8. Transcriptional Regulation during Somitogenesis

highly desirable to undertake systematic searches for genes having novel expression patterns in the somite. One would certainly like to have more candidate genes for the morphogenetic processes that occur, for example the movements of muscle precursors involved in the formation of the myotome, and it would be extremely useful to have markers that further subdivided each of the three compartments of the somite.

There are many ways of searching for such genes but we have focused our attention on the use of differential display (Liang and Pardee, 1992). This PCR-based procedure generates "fingerprints" of the mRNA populations present in a sample and should, in theory, be capable of providing detailed comparisons of mRNA populations even in circumstances in which the amounts of mRNA that can be obtained are limiting. However, the original procedure gives an unacceptably high false positive rate that precludes its application to the study of developmental processes such as somitogenesis. We have modified this methodology to allow the comparison of mRNA populations in defined embryonic structures such as somites, in such a way that one could go directly from a differential display band to an *in situ* hybridization probe and thus test in the context of the three-dimensional structure of the embryo the validity or otherwise of the differential display procedure. In order to develop the requisite methods we have exploited the germ layer specific cDNA libraries constructed by Harrison *et al.* (1995), which derive from dissected endoderm, ectoderm, and mesoderm of gastrulation stage embryos. Using these libraries, we could show that the use of a secondary screening procedure involving Southern blot hybridization of the entire libraries was highly effective at identifying truly differential bands, and that the recovery of full length cDNA clones corresponding to such bands then allowed *in situ* analysis which could indeed lead to the characterization of transcription units with the predicted spatial expression pattern (Gupta *et al.,* 1998). We have since optimized the methodology such that the false positive rate is reduced to acceptable levels, and the lengths of the DNA species from the differential display bands are of sufficient size that they can be used directly for the preparation of *in situ* hybridization probes. Figure 5 shows an example of this procedure applied to a comparison of the mRNA populations of the presomitic mesoderm and somites II through VII. It is apparent that differentially expressed transcripts are readily identified, and we are now undertaking such comparisons on a systematic basis. The quantity of mRNA required for the execution of this procedure is extremely small, and it is our view that the limiting factor is now the manual skill of the dissector. Comparisons of one pair of somites with its neighbours are certainly possible, and it may well be feasible to compare subdomains within somites, for example, anterior and posterior, or medial and lateral halves. Work of this sort should lead to the identification of new genes that play important roles in somitogenesis and to more precise definitions of the cell populations present in somites at the various stages of their development and differentiation.

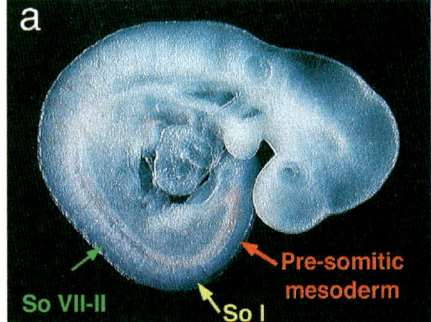

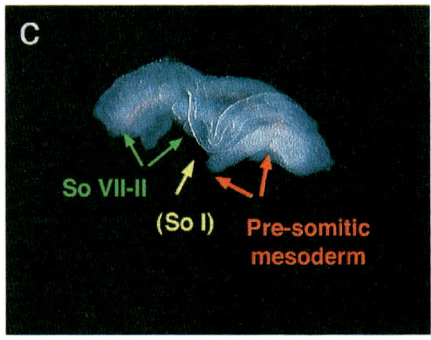

Figure 5 Differential display. (a–c) Dissection of presomitic mesoderm and somites II–VII from a 9.5-dpc embryo (22-somite stage) in which dispase (Boehringer) was used to facilitate removal of the ectoderm. In (c) the most recently formed somite (I) has first been discarded and the intact presomitic mesoderm is visible (red arrows) in the process of being detached from the neural tube and the lateral plate mesoderm. In this example six somites or a whole presomitic mesoderm have been pooled for RNA extraction. More typically now we would use single somites or subfractions of the presomitic mesoderm. (d) Typical autoradiograph of differential display analysis with obvious common and differentially expressed bands.

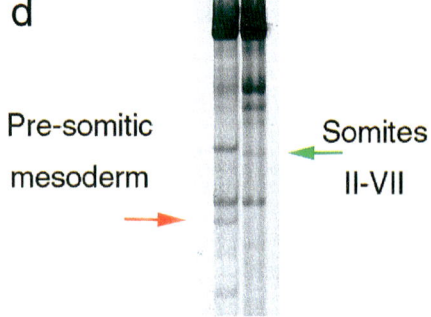

8. Transcriptional Regulation during Somitogenesis   315

Acknowledgments

We are grateful to Peter Ashby, Jon Gilthorpe, and Raj Gupta for providing us with illustrations and to all members of the Rigby laboratory for helpful discussions. Our work is paid for by the U.K. Medical Research Council.

References

Aparicio, S., Morrison, A., Gould, A., Gilthorpe, J., Chaudhuri, C., Rigby, P., Krumlauf, R., and Brenner, S. (1995). Detecting conserved regulatory elements with the model genome of the Japanese puffer fish, *Fugu rubripes*. *Proc. Natl. Acad. Sci., U.S.A.* **92**, 1684–1688.

Arnold, H. H., and Braun, T. (1996). Targeted inactivation of myogenic factor genes reveals their role during mouse myogenesis: A review. *Int. J. Dev. Biol.* **40**, 345–353.

Bober, E., Lyons, G. E., Braun, T., Cossu, G., Buckingham, M., and Arnold, H. H. (1991). The muscle regulatory gene, *Myf-6*, has a biphasic pattern of expression during early mouse development. *J. Cell Biol.* **113**, 1255–1265.

Bober, E., Franz, T., Arnold, H.-H., Gruss, P., and Tremblay, P. (1994). Pax-3 is required for the development of limb muscles: A possible role for the migration of dermomyotomal muscle progenitor cells. *Development* **120**, 603–612.

Bradshaw, M. S., Shashikant, C. S., Belting, H.-G., Bollekens, J. A., and Ruddle, F. H. (1996). A long-range regulatory element of Hox-c8 identified by using the pClasper vector. *Proc. Natl. Acad. Sci. U.S.A.* **93**, 2426–2430.

Braun, T., and Arnold, H.-H. (1996). Myf-5 and myoD genes are activated in distinct mesenchymal stem cells and determine different skeletal muscle cell lineages. *EMBO J.* **15**, 310–318.

Braun, T., Bober, E., and Arnold, H. H. (1992). Inhibition of muscle differentiation by the adenovirus E1a protein: Repression of the transcriptional activating function of the HLH protein Myf-5. *Genes Dev.* **6**, 888–902.

Brennan, T. J., and Olson, E. N. (1990). Myogenin resides in the nucleus and acquires high affinity for a conserved enhancer element on heterodimerization. *Genes Dev.* **4**, 582–595.

Brenner, S., Elgar, G., Sandford, R., Macrae, A., Venkatesh, B., and Aparicio, S. (1993). Characterization of the pufferfish (*Fugu*) genome as a compact model vertebrate genome. *Nature* **366**, 265–268.

Brown, J. L., Mucci, D., Whiteley, M., Dirksen, M. L., and Kassis, J. A. (1998). The *Drosophila* Polycomb group gene pleiohomeotic encodes a DNA binding protein with homology to the transcription factor YY1. *Mol. Cell* **1**, 1057–1064.

Candia, A. F., and Wright, C. V. (1996). Differential localization of Mox-1 and Mox-2 proteins indicates distinct roles during development. *Int. J. Dev. Biol.* **40**, 1179–1184.

Cheng, T. C., Wallace, M. C., Merlie, J. P., and Olson, E. N. (1993). Separable regulatory elements governing myogenin transcription in mouse embryogenesis. *Science* **261**, 215–218.

Christ, B., and Ordahl, C. P. (1995). Early stages of chick somite development. *Anat. Embryol.* **191**, 381–396.

Cossu, G., Tajbakhsh, S., and Buckingham, M. (1996). How is myogenesis initiated in the embryo? *Trends Genet.* **12**, 218–223.

Currie, R. A. (1998). NF-Y is associated with the histone acetyltransferases GCN5 and P/CAF. *J. Biol. Chem.* **273**, 1430–1434.

Dahl, E., Koseki, H., and Balling, R. (1997). Pax genes and organogenesis. *BioEssays* **19**, 755–765.

Daston, G., Lamar, E., Olivier, M., and Goulding, M. (1996). Pax-3 is necessary for migration but not differentiation of limb muscle precursors in the mouse. *Development* **122**, 1017–1027.

Edmondson, D. G., Lyons, G. E., Martin, J. F., and Olson, E. N. (1994). Mef2 gene expression marks the cardiac and skeletal muscle lineages during mouse embryogenesis. *Development* **120,** 1251–1263.

Faerman, A., Goldhammer, D. J., Puzis, R., Emerson, C. P., Jr., and Shani, M. (1995). The distal human MyoD enhancer sequences direct unique muscle-specific patterns of lacZ expression during mouse development. *Dev. Biol.* **171,** 27–38.

Forrester, W. C., Takegawa, S., Papayannopoulou, T., Stamatoyannopoulos, G., and Groudine, M. (1987). Evidence for a locus activation region: The formation of developmentally stable hypersensitive sites in globin-expressing hybrids. *Nucleic Acids Res.* **15,** 10159–10177.

Goulding, M., Lumsden, A., and Paquette, A. J. (1994). Regulation of Pax-3 expression in the dermomyotome and its role in muscle development. *Development* **120,** 957–971.

Grosveld, F., van Assendelft, G. B., Greaves, D. R., and Kollias, G. (1987). Position-independent, high-level expression of the human beta-globin gene in transgenic mice. *Cell* **51,** 975–985.

Gupta, R., Thomas, P., Beddington, R. S. P. and Rigby, P. W. J. (1998). Isolation of developmentally regulated genes by differential display screening of cDNA Libraries. *Nucleic Acids Res.* **26,** 4538–4539.

Gutman, A., Gilthorpe, J., and Rigby, P. W. J. (1994). Multiple positive and negative regulatory elements in the promoter of the mouse homeobox gene Hoxb-4. *Mol. Cell. Biol.* **14,** 8143–8154.

Harrison, S. M., Dunwoodie, S. L., Arkell, R. M., Lehrach, H., and Beddington, R. S. (1995). Isolation of novel tissue-specific genes from cDNA libraries representing the individual tissue constituents of the gastrulating mouse embryo. *Development* **121,** 2479–2489.

Hinterberger, T. J., Sassoon, D. A., Rhodes, S. J., and Konieczny, S. F. (1991). Expression of the muscle regulatory factor MRF4 during somite and skeletal myofiber development. *Dev. Biol.* **147,** 144–156.

Hughes, S. M., Taylor, J. M., Tapscott, S. J., Gurley, C. M., Carter, W. J., and Peterson, C. A. (1993). Selective accumulation of MyoD and myogenin mRNAs in fast and slow adult skeletal muscle is controlled by innervation and hormones. *Development* **118,** 1137–1147.

Hunt, P., Gulisano, M., Cook, M., Sham, M-H., Faiella, A., Wilkinson, D., Boncinelli, E., and Krumlauf, R. (1991). A distinct Hox code for the branchial region of the vertebrate head. *Nature* **353,** 861–864.

Kessel, M., and Gruss, P. (1991). Homeotic transformations of murine vertebrae and concomitant alteration of Hox codes induced by retinoic acid. *Cell* **67,** 89–104.

Krumlauf, R. (1994). Hox genes in vertebrate development. *Cell* **78,** 191–201.

Lassar, A. B., Davis, R. L., Wright, W. E., Kadesch, T., Murre, C., Voronova, A., Baltimore, D., and Weintraub, H. (1991). Functional activity of myogenic HLH proteins requires heterooligomerization with E12/E47-like proteins *in vivo*. *Cell* **66,** 305–315.

Li, Q., Herrler, M., Landsberger, N., Kaludov, N., Ogryzko, V. V., Nakatani, Y., and Wolffe, A. P. (1998). *Xenopus* NF-Y pre-sets chromatin to potentiate p300 and acetylation-responsive transcription from the *Xenopus* hsp70 promoter *in vivo*. *EMBO J.* **17,** 6300–6315.

Liang, P., and Pardee, A. B. (1992). Differential display of eukaryotic messenger RNA by means of the polymerase chain reaction. *Science* **257,** 967–971.

McGrew, M. J., and Pourquié, O. (1998). Somitogenesis: Segmenting a vertebrate. *Curr. Opin. Genet. Dev.* **8,** 487–493.

Maconochie, M., Nonchev, S., Morrison, A., and Krumlauf, R. (1996). Paralogous Hox genes: Function and regulation. *Annu. Rev. Genet.* **30,** 529–556.

Maity, S. N., and de Crombrugghe, B. (1998). Role of the CCAAT-binding protein CBF/NF-Y in transcription. *Trends Biochem. Sci.* **23,** 174–178.

Mankoo, B. S., Collins, N. S., Ashby, P., Grigorieva, E., Pevny, L. H., Candia, A., Wright, C., Rigby, P. W. J., and Pachnis, V. (1999). Mox2, a critical component of the genetic hierarchy controlling limb muscle development. *Nature* **400,** 69–73.

Mansouri, A., Hallonet, M., and Gruss, P. (1996). Pax genes and their roles in cell differentiation and development. *Curr. Opin. Cell Biol.* **8,** 851–857.

8. Transcriptional Regulation during Somitogenesis

Megeney, L. A., Kablar, B., Garrett, K., Anderson, J. E., and Rudnicki, M. A. (1996). MyoD is required for myogenic stem cell function in adult skeletal muscle. *Genes Dev.* **10**, 1173–1183.

Molkentin, J. D., and Olson, E. N. (1996). Defining the regulatory networks for muscle development. *Curr. Opin. Genet. Dev.* **6**, 445–453.

Murre, C., McCaw, P. S., Vaessin, H., Caudy, M., Jan, L. Y., Jan, Y. N., Cabrera, C. V., Buskin, J. N., Hauschka, S. D., Lassar, A. B., Weintraub, H., and Baltimore, D. (1989). Interactions between heterologous helix-loop-helix proteins generate complexes that bind specifically to a common DNA sequence. *Cell* **58**, 537–544.

Olson, E. N., Arnold, H.-H., Rigby, P. W. J., and Wold, B. (1996). Know your neighbors: Three phenotypes in null mutants of the myogenic bHLH Gene MRF4. *Cell* **85**, 1–4.

Ordahl, C. P., and Le Douarin, N. M. (1992). Two myogenic lineages within the developing somite. *Development* **114**, 339–353.

Ott, M. O., Bober, E., Lyons, G., Arnold, H., and Buckingham, M. (1991). Early expression of the myogenic regulatory gene, *myf-5*, in precursor cells of skeletal muscle in the mouse embryo. *Development* **111**, 1097–1107.

Patapoutian, A., Miner, J. H., Lyons, G. E., and Wold, B. (1993). Isolated sequences from the linked Myf-5 and MRF4 genes drive distinct patterns of muscle-specific expression in transgenic mice. *Development* **118**, 61–69.

Puschel, A. W., Balling, R., and Gruss, P. (1991). Separate elements cause lineage restriction and specify boundaries of Hox-1.1 expression. *Development* **112**, 279–287.

Rudnicki, M. A., Braun, T., Hinuma, S., and Jaenisch, R. (1992). Inactivation of MyoD in mice leads to up-regulation of the myogenic HLH gene *Myf-5* and results in apparently normal muscle development. *Cell* **71**, 383–390.

Rudnicki, M. A., Schnegelsberg, P. N., Stead, R. H., Braun, T., Arnold, H. H., and Jaenisch, R. (1993). MyoD or Myf-5 is required for the formation of skeletal muscle. *Cell* **75**, 1351–1359.

Sassoon, D., Lyons, G., Wright, W. E., Lin, V., Lassar, A., Weintraub, H., and Buckingham, M. (1989). Expression of two myogenic regulatory factors myogenin and MyoD1 during mouse embryogenesis. *Nature* **341**, 303–307.

Shi, Y., Lee, J. S., and Galvin, K. M. (1997). Everything you have ever wanted to know about Yin Yang 1.... *Biochim. Biophys. Acta* **1332**, 49.

Smith, T. H., Kachinsky, A. M., and Miller, J. B. (1994). Somite subdomains, muscle cell origins, and the four muscle regulatory factor proteins. *J. Cell Biol.* **127**, 95–105.

Spitz, F., Demignon, J., Porteu, A., Kahn, A., Concordet, J. P., Daegelen, D., and Maire, P. (1998). Expression of myogenin during embryogenesis is controlled by Six/sine oculis homeoproteins through a conserved MEF3 binding site. *Proc. Natl. Acad. Sci. U.S.A.* **95**, 14220–14225.

Tajbakhsh, S., and Buckingham, M. E. (1994). Mouse limb muscle is determined in the absence of the earliest myogenic factor myf-5. *Proc. Natl. Acad. Sci. U.S.A.* **91**, 747–751.

Tajbakhsh, S., Bober, E., Babinet, C., Pournin, S., Arnold, H., and Buckingham, M. (1996a). Gene targeting the myf-5 locus with nlacZ reveals expression of this myogenic factor in mature skeletal muscle fibres as well as early embryonic muscle. *Dev. Dyn.* **206**, 291–300.

Tajbakhsh, S., Rocancourt, D., and Buckingham, M. (1996b). Muscle progenitor cells failing to respond to positional cues adopt nonmyogenic fates in Myf-5 null mice. *Nature* **384**, 266–270.

Tajbakhsh, S., Rocancourt, D., Cossu, G., and Buckingham, M. (1997). Redefining the genetic hierarchies controlling skeletal myogenesis: Pax-3 and Myf-5 act upstream of MyoD. *Cell* **89**, 127–138.

Tapscott, S. J., Lassar, A. B., and Weintraub, H. (1992). A novel myoblast enhancer element mediates MyoD transcription. *Mol. Cell Biol.* **12**, 4994–5003.

Tuan, D., Solomon, W., Li, Q., and London, I. M. (1985). The beta-like-globin gene domain in human erythroid cells. *Proc. Natl. Acad. Sci. U.S.A.* **82**, 6384–6388.

Vogels, R., Charite, J., de Graaff, W., and Deschamps, J. (1993). Proximal cis-acting elements cooperate to set Hox-b7 (Hox-2.3) expression boundaries in transgenic mice. *Development* **118**, 71–82.

Whiting, J., Marshall, H., Cook, M., Krumlauf, R., Rigby, P. W. J., Stott, D., and Allemann, R. K.

(1991). Multiple spatially specific enhancers are required to reconstruct the pattern of Hox-2.6 gene expression. *Genes Dev.* **5,** 2048–2059.

Yang, W. M., Inouye, C., Zeng, Y., Bearss, D., and Seto, E. (1996). Transcriptional repression by YY1 is mediated by interaction with a mammalian homolog of the yeast global regulator RPD3. *Proc. Natl. Acad. Sci. U.S.A.* **93,** 12845–12850.

Yang, X. W., Model, P., and Heintz, N. (1997). Homologous recombination based modification in *Escherichia coli* and germline transmission in transgenic mice of a bacterial artificial chromosome. *Nat. Biotechnol.* **15,** 859–865.

Yee, S. P., and Rigby, P. W. J. (1993). The regulation of myogenin gene expression during the embryonic development of the mouse. *Genes Dev.* **7,** 1277–1289.

Yun, K. S., and Wold, B. (1996). Skeletal-muscle determination and differentiation—Story of a core regulatory network and its context. *Curr. Opin. Cell Biol.* **8,** 877–889.

Zweigerdt, R., Braun, T., and Arnold, H.-H. (1997). Faithful expression of the *Myf-5* gene during mouse myogenesis requires distant control regions: A transgene approach using yeast artificial chromosomes. *Dev. Biol.* **192,** 172–180.

9
Determination and Morphogenesis in Myogenic Progenitor Cells: An Experimental Embryological Approach

Charles P. Ordahl, Brian A. Williams, and Wilfred Denetclaw
Department of Anatomy
Cardiovascular Research Institute
University of California, San Francisco
San Francisco, California

I. Introduction
   A. The Interplay of Fate and Determination in the Generation of Tissue Anlage during Embryonic Development
   B. A Molecular Model for Cellular Memory
   C. Embryonic Models for Myogenic Determination
II. The Diversity and Location of Muscle Precursor Cells in the Vertebrate Embryo
   A. Early Ideas about the Location of Muscle Precursor Cells in Vertebrate Embryos
   B. The Quail–Chick Era of Embryonic Fate Mapping
   C. Lineage Diversity in Muscle Precursor Cells
   D. Molecular Marker Expression in Epaxial Myotome Precursor Cells
   E. Location of Myotome Precursor Cells within the Somite
III. When Do Myotome Precursor Cells Become Determined?
   A. Signal Sources That Influence Somite Development and Myogenic Specification
   B. The Notochord Challenge Assay
   C. Progressive Increase in Myogenic Potential in Epaxial Muscle Precursor Cells
IV. Gene Expression and Determination in Migratory Limb Muscle Precursor Cells
   A. Pax3 Gene Expression Marks Mesoderm Cells in the Early Stages of Myogenic Development
   B. Early Experiments on Limb Muscle Determination
V. A Cellular and Molecular Framework for Understanding Myogenic Tissue Development
   A. Morphogenetic Movements of Myogenic Precursor Cells
   B. Growth Dynamics in the Early Embryonic Myotome
   C. Myogenic Organizers and Progenitor Stem Cells
   Dedication
   Acknowledgments
VI. Appendix: Key Word Guide
   References

I. Introduction[1]

This chapter reviews historical and recent uses of experimental embryology to analyze early stages of vertebrate embryonic muscle development in support of a model in which pluripotent embryonic stem cells give rise to a unique class progenitor stem cell lineages that constitute the embryonic tissue primordia (or anlage). As their name implies, progenitor stem cells expand through both symmetric and asymmetric mitotic division to produce the differentiated cell types within developing tissues and also the adult stem cell lineages necessary for growth and repair of the tissue throughout life. Progenitor stem cells have the capacity to undergo tissue or organ morphogenesis; the embryonic process that underlies the ultimate integration of form and function in biological systems. The underlying mechanisms that govern these early, decisive events in embryogenesis have been the subject of speculation for centuries, and continuing interest in these fundamental problems is evident in the extraordinary progress in understanding the molecular and genetic aspects of insect and mammalian development. A particularly noteworthy recent advance in vertebrate myogenesis is a molecular model that has emerged to explain myogenic determination; the embryonic process by which the broad developmental potential of embryonic stem cells (pluripotentiality) is restricted as unipotent myogenic progenitor cells develop, expand, and form muscle tissues. This chapter will review recent and historical experiments indicating that the molecular events underlying myogenic determination occur within discrete regions of the somite and that they may be coupled to the morphogenetic movements of muscle precursor cells during somitogenesis.

A. The Interplay of Fate and Determination in the Generation of Tissue Anlage during Embryonic Development

Because all of the tissues and organs of the adult body are ultimately derived from a single cell, the zygote, the question arises as to how the many daughters of this first cell sort themselves out during development to form the tissue primordia. The construction of a conceptual framework for understanding this process has been the major preoccupation of developmental biologists throughout the twentieth century (Spemann 1938; Hamburger 1988). Wilhelm Roux began to experimentally investigate this over a century ago by killing one of two blastomeres of an amphibian embryo with a hot needle. When a half-tadpole developed from the ma-

---

[1] A guide to key words is contained in the appendix at the end of this chapter.

nipulated embryo, he concluded that development was *determinative* and that subcellular determinants were partitioned during each cleavage event, establishing lineages for the right and left sides of the body and their respective component parts (Roux, 1897). Driesch (1891), by contrast, found that the separated blastomeres of echinoderm and ascidian embryos could each develop into a complete separate organism, indicating that developmental potential becomes partitioned later in development. He proposed the alternative hypothesis of *regulative* development, in which multipotential embryonic cells are sensitive to extrinsic forces that direct their development by inducting and/or restricting their developmental potential to its final end point or fate.

As maps of embryonic cell fate and developmental potential emerged it became clear that, as a general rule, developmental potential is more widespread than fate, particularly at earlier stages of embryonic development (for an extensive and authoritative account of this material, see Hamburger 1988). At later stages of development discrete subpopulations of cells segregate into the anlage or primordia of adult tissues and organs, and these cells undergo phenotypic restriction in which developmental potential decreases until ultimately it becomes equal to fate. These primordia build tissues and organs through exquisitely complex cellular movements and molecular interactions between multiple cell types, each often having a distinct lineage history and potential for forming one or another tissue component. The exact timing of determination of the cells within these primordia is not well established. Once acquired, however, the determined status of adult tissues is highly stable. For example, transplantation of tissues and organs from one region of the body to another disrupts neither the phenotypic characteristics of the cells nor their morphological organization. Since this is true for both the fetus and the adult, it can be assumed that determination occurs sometime between gastrulation, when and where pluripotent embryonic stem cells are found, and the fetal stage of development, from which only determined stem cell lines can be isolated (Fig. 1C). This focuses attention on the intervening period of development, specifically the period of organogenesis, when the primary organ rudiments first form during embryogenesis.

Early experimental embryologists were mainly concerned with determination with respect to morphogenetic movements and tissue development ("dynamic determination") in contrast to histological or cell-type determination ("material determination") (Spemann, 1938). Inasmuch as their primary interest was in the distribution of developmental potential (or potency) during development, they recognized that the determined state could only be recognized through experimental manipulation that would disrupt the behavior of undetermined cells or tissues but not that of their determined counterparts. Determined cells can be characterized as having several basic attributes (see Fig. 1A): first, the positive attribute of acquired competence for one or more differentiated end points and second, the negative attribute of loss of potency for alternative differentiated end points. In

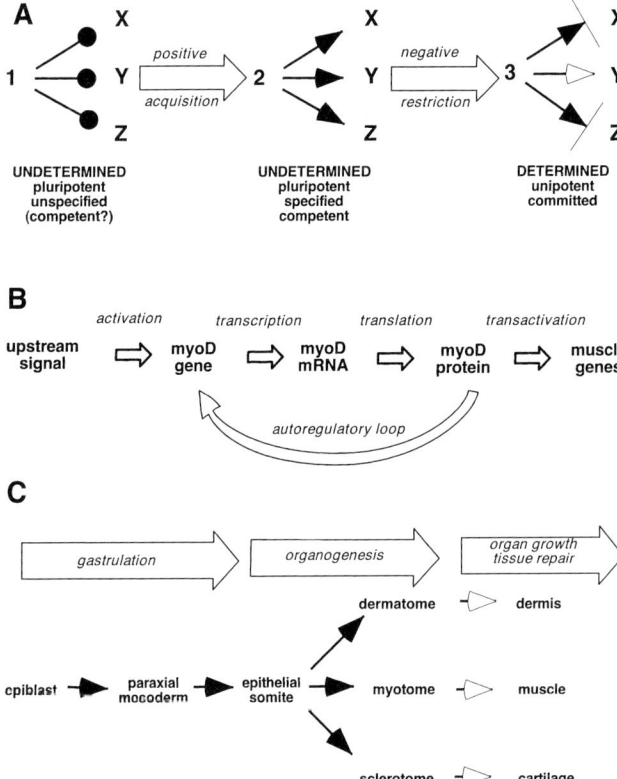

Figure 1 Cell determination models. (A) Hypothetical transitions in the developmental potential of an embryonic tissue primordium whose potential fate encompasses differentiated tissue types X, Y, and Z. In state 1, the cells are unable to express any differentiated product (lines ending in filled circles). Some or all of the cells in state 1 may undergo a positive change to state 2 in which they acquire competence to form all three differentiated tissues (solid black arrows). This state corresponds approximately to the "specified" state as defined by Slack (1983). The transition from state 2 to 3 corresponds to the negative component of determination, in which competence to form alternative differentiated tissues (X and Z) is lost but the competence to form tissue Y is retained. State 3 embryonic tissues contain both differentiated cells and determined progenitor stem cells (see C). (B) Molecular model for cellular determination and memory. Activation of myoD gene transcription leads to production of myoD protein that in turn activates the entire muscle gene expression program (positive aspect of determination). The process of cellular memory is explained by the autoregulatory nature of the myoD protein whereby transcriptional activity of the myoD gene is maintained. These two basic attributes of the model remain essentially unchanged despite substantial increase in the complexity of this model as a result of recent advances (see Chap. 12 by Arnold and Braun and Chap. 15 by Tajbakhsh and Buckingham, this volume). (C) Relating the phases of embryonic development to the search for myogenic progenitor cells. Epiblast cells gastrulate forming the paraxial mesoderm that segments into somites. In classic embryological texts the somite is considered to be the organ rudiment giving rise to three adult tissue types through three embryonic intermediates: the dermatome, myotome, and sclerotome (using these terms in their broadest sense; see footnote 3).

9. Determination and Morphogenesis in Myogenic Progenitor Cells

combination, these two attributes result in restriction of developmental potency to a single cell-type product, but early workers recognized that they might be separately acquired (Vogt, 1927, as related in Spemann, 1938). A third attribute, autonomy, is the ability of embryonic tissue to execute its positive attribute independently, in the absence of instructive influences from extrinsic sources.

Primary tissue culture, which was also established about the same time (Hamburger 1988), in which organs or tissues are dissociated and the cells cultured in dishes, shows that cells from tissues of the body have a property of determination that is best described as cellular memory, or the ability to maintain cellular identity under a variety of circumstances. One form of cellular memory is entirely cell autonomous and does not require cell division. For example, differentiated (postmitotic) cardiac myocytes in primary culture initially lose many of their phenotypic characteristics and begin to resemble fibroblast cells but, after extended periods, re-express the molecular and physiological attributes of differentiated heart cells, even re-establishing many aspects of their former *in vivo* morphology (Eppenberger *et al.*, 1995). Another form of memory is dependent on tissue-specific or determined stem cells which can mitotically expand in culture to give rise to differentiated progeny that resemble cells within the tissue of origin. Skeletal muscle tissues, for example, contain satellite cells that are undifferentiated cells capable of replacing or repairing damaged or lost portions of the tissue (Bischoff, 1994). Differentiated skeletal muscle cells are typically multinucleated and rarely survive transfer into primary culture, but satellite cells do survive and expand through asymmetric mitotic divisions to give rise to two types of cells: (1) myoblasts that directly differentiate into myocytes or undergo limited symmetric divisions to give daughters that form differentiated myocytes and (2) additional myogenic stem cells that can be repeatedly passaged in culture (and even immortalized) without loss of the capacity to form myoblasts (Konigsberg, 1979; Yaffe, 1968, 1969; Bonner and Hauschka, 1974; Hauschka 1974a,b; Yablonka-Reuveni, 1995). In adult skeletal muscle, therefore, phenotypic memory resides in the capacity of satellite stem cells to expand and generate progeny that will faithfully differentiate as skeletal muscle. A third type of phenotypic memory might best be termed tissue or morphogenetic memory because it represents the ability of differentiated cells and/or stem cells to re-establish functional tissue that is properly organized and integrated with other tissues and cells. Satellite cells may have lost some aspects of morphogenetic memory because, after reintroduction into adults, cultured satellite cells do not readily contribute to muscle tissue (Gussoni *et al.*, 1992). Similar experiments in embryonic limb buds show that cultured myogenic cell lines can be incorporated into nascent muscle tissue but show limited mitotic expansion (Antin *et al.*, 1991). Embryonic tissue primordia, on the other hand, retain an astonishing level of morphogenetic memory and growth capacity even when grown outside the embryo (Holtfreter, 1944, 1945; Holtfreter and Hamburger, 1955).

B. A Molecular Model for Cellular Memory

A molecular approach to the study of myogenic determination began to emerge when it was shown that transfection of genomic DNA from muscle cells could induce multipotential 10T1/2 cells to form myogenic stem cell lines (Konieczny *et al.,* 1986). Use of a cDNA transfection strategy, similar to that employed to isolate oncogenes, resulted in the isolation of a cDNA clone dubbed "myoD-1" that could transform fibroblast cells into myogenic stem cells (Davis *et al.,* 1987). MyoD was the first cloned representative of a family of four related MDF (myoD family) transcriptional regulatory factors (Davis *et al.,* 1987; Braun *et al.,* 1989; Rhodes and Konieczny, 1989; Wright *et al.,* 1989). The basic helix–loop–helix (bHLH) domains that these proteins share in common were found to be the essential regions responsible for each having the ability to stably convert a variety of nonmuscle cell types into myogenic stem cells in culture (Weintraub, 1993). The quality of muscle differentiation in the colonies formed by the daughters of MDF-transformed stem cells is extraordinarily high, equivalent to that of cultures of primary myogenic stem cells and much higher than is typical of highly passaged muscle stem cell lines (Lin *et al.,* 1994). Within the differentiated muscle colonies that result from MDF transfection in cell culture reside myogenic stem cells whose memory for the myogenic phenotype is stably inherited after an indefinite number of rounds of mitotic division. This suggests that MDFs can take over and reconfigure the entire molecular repertoire of virtually any cell and cause it to become "trans-determined" into another cell lineage regardless of its embryonic origins (Weintraub, 1993).

Harold Weintraub and co-workers proposed a model (Fig. 1B) to explain how expression of a single molecule, myoD protein, could transform a non-muscle cell to establish and heritably maintain the myogenic phenotype (Weintraub, 1993). The positive aspect of determination—that is, elaboration of a specific phenotype, muscle—was proposed to occur through myoD protein-mediated activation of the promoters of muscle genes and cross-activation of other MDF genes. There is extensive biochemical support for this aspect of the model inasmuch as the molecular details of these binding interactions at relevant sites of most, and possibly all, muscle gene promoters resulting in their trans-activation is well established (see Chap. 12 by Arnold and Braun, this volume). Autonomy and memory were hypothesized to occur through an autoregulatory loop whereby, once expressed, myoD would autoactivate its own transcriptional promoter, maintaining both its own expression and that of downstream targets. Results with transgenic analysis have led to modification of the autoregulatory aspect of the model (see Chap. 12 by Arnold and Braun, this volume). The negative aspect of determination, that is, repression of potential for alternative cell types, has been addressed in general molecular terms (Caplan and Ordahl, 1978) but not in context of recent advances.

9. Determination and Morphogenesis in Myogenic Progenitor Cells

C. Embryonic Models for Myogenic Determination

Developmental biologists quickly responded to place the myoD model within the context of embryonic development. The time and place of the onset of MDF expression was shown to occur in the somite (Fig. 1C); the embryonic progenitor organ known to give rise to skeletal muscle (de la Brousse and Emerson, 1990; Ott et al., 1991). That information, coupled with results showing that different somite regions give rise to muscle in different regions of the body (Ordahl and Le Douarin, 1992), suggested that the relationship between myogenic determination, as embryologically defined, and the expression of genes implicated in myogenic determination, as defined by molecular methods, could be elucidated through classic experimental embryological methods. At a minimum, the application of classic constraints could serve to test our molecular notions of cell determination. Its maximum value, however, is its potential to reveal new aspects of development that might yield fundamental insights. Toward such ends, this chapter reviews classic and modern experimental embryological experiments that address the following: first, the precise location of precursor cells within the developing embryo that give rise to muscle primordia and myogenic stem cells; second, the changing molecular expression pattern of those precursor cells; and, third, the timing of determination in those cells. Finally, a framework is proposed for an embryo-tissue-based model for understanding myogenic determination and the development of myogenic progenitor cell lineages.

II. The Diversity and Location of Muscle Precursor Cells in the Vertebrate Embryo

A. Early Ideas about the Location of Muscle Precursor Cells in Vertebrate Embryos

Although the source of skeletal muscle in the vertebrate body (excluding the head[2]) is now known to be the somites (Fig. 2), this has been a subject of controversy for over a century. Fischel (1895) deduced from histological analysis of somite development that the somites were a major source of developing muscle. He accurately described the myotome, a layer of nascent muscle interposed between

---

[2] Skeletal muscles of the head arise from head mesoderm, which includes a covertly segmented paraxial component and the unsegmented prechordal mesoderm, a small region of mesoderm located at the cranial tip of the notochord (Noden 1983, 1988; Couly et al., 1992). Myogenic development in head mesoderm has been reviewed by Hacker and Guthrie (1998) and is not further considered in this review.

the sclerotome and dermomyotome.[3] Fischel also proposed that the dorsomedial lip of the dermomyotome was the site of generation of the myotome and he also noted that mesenchyme cells appeared to be exiting the lateral margin of the dermomyotome (Fig. 3). Williams (1910) was perhaps the first to propose that all of the muscle of the body was derived from somites, although he believed that muscle expanded into the body and developing limbs through growth of the myotome itself rather than expansion and migration of the dermomyotome cells. This erroneous idea continues to be propagated in some basic textbooks (Sadler, 1995), even though myotome cells were shown to be postmitotic 30 years ago (Langman and Nelson, 1968). The idea that somites were the *sole* source of body muscle was disfavored during most of the twentieth century because limb and body wall muscle could form after experimental ablation and extirpation of the adjacent somites (Byrnes, 1898; Lewis, 1910; Detwiler, 1918, 1929). This led to the notion that precursor cells that formed the muscles of the developing limb buds (Fig. 2) were born within the lateral plate mesoderm and therefore represented a separate lineage from the somitic precursors to the muscles of the back (see, e.g., p. 235 in Hamilton, 1952). Despite the occasional publication of findings to the contrary (Hazelton, 1970; Bellairs, 1963) this view prevailed into the 1970s (Searls, 1965,

---

[3]The original names for the three major types of embryonic tissues derived from the somite are sclerotome (cartilage), myotome (muscle), and dermatome (dermal connective tissue). During the twentieth century, the latter two terms have become highly heteronymous. For example, in medical terminology "myotome" refers to motor innervation derived from specific spinal cord levels (e.g., the intrinsic muscles of the hand are innervated by the first thoracic myotome), and, correspondingly, "dermatome" refers to the area of skin innervated by a specific spinal cord level (e.g., the upper surface of the shoulder is innervated by the fourth cervical dermatome). Embryologists also use the term "myotome" heteronymously; on the one hand, as a general term referring to all of the muscle derived from the somite (i.e., the muscles of the body proper and the tongue) and, on the other hand, as the specific layer of embryonic muscle tissue that forms under the dorsal epithelium of the somite. Recent work indicates that there are two sources of myotome from the dermomyotome. The first myotome fibers to form are generated from the medial portion of the dermomyotome and are the beginnings of the epaxial muscle. Subsequently, at thoracic and abdominal levels, the lateral portions of the dermomyotome begin to generate hypaxial muscle. We refer to these two nascent muscle groups as the "epaxial myotome" and "hypaxial myotome," respectively.

Once the sclerotome begins to form, through de-epithelialization of the ventral portion of the somite, we refer to the dorsal epithelium as the "dermomyotome" (also spelled dermamyotome), a term that suitably reflects the dual fate of the cells within this epithelium. Early embryologists also used the term "dermatome" for this epithelium after the first myotome fibers form, and this usage continues to be widespread today. Such a designation carries the unfortunate implication that this epithelium is destined to give rise to only dermis even though it is now clear that muscle progenitor cells continue to be generated from the edges (lips) of this epithelium. The sheet region of this epithclium (the central region not including the lips) may also give rise to muscle later in development (W. Denetclaw, unpublished observations, 1997), but the relative quantitative contribution of these cells to the epaxial muscle has not yet been rigorously analyzed. In this chapter (as well as elsewhere; Christ and Ordahl, 1995), therefore, we prefer to use the term "dermomyotome" to designate this epithelium in order to avoid the potentially misleading implication that its cells are destined to give rise only to dermis.

9. Determination and Morphogenesis in Myogenic Progenitor Cells

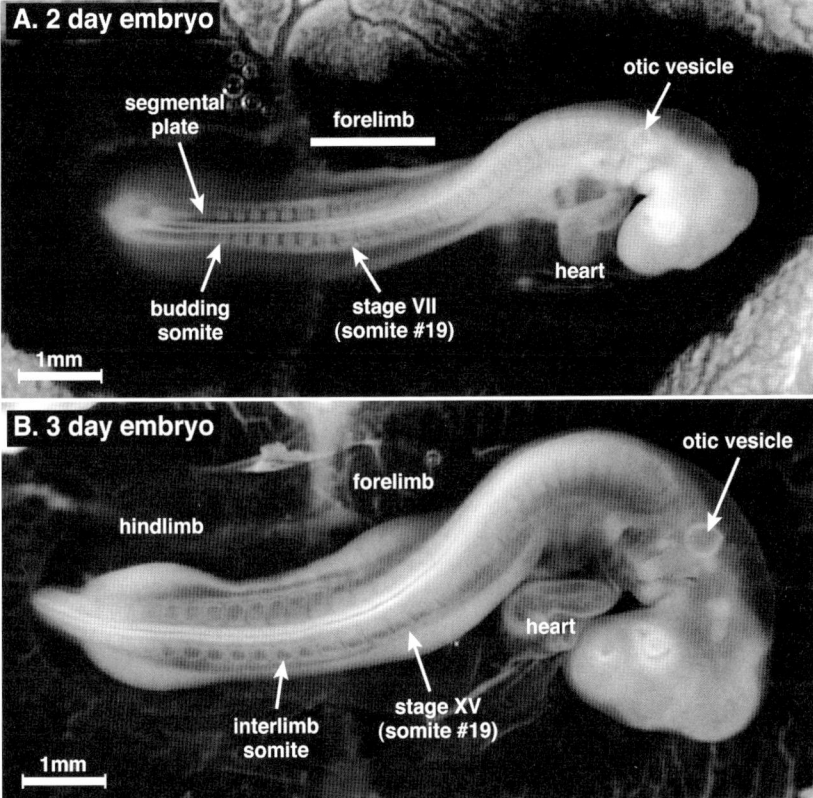

Figure 2 Chicken embryos with somites. Chick embryos at 2(A) and 3(B) days of embryonic development are shown at the same magnification. The nineteenth somite of both embryos is indicated to illustrate the disproportionate growth of the somite in the mediolateral dimension as compared to the craniocaudal dimension. Several embryonic organ rudiments are indicated.

1967; Searls and Janner, 1969; Caplan and Koutroupas, 1973; Straus and Rawles, 1953; Shellswell and Wolpert, 1977).

B. The Quail-Chick Era of Embryonic Fate Mapping

The principal difficulty in unequivocally ascertaining the origins of embryonic precursor cells was the lack of a rigorous and reliable method for the construction of embryonic fate maps, which are diagrammatic representations of the location within the early embryo of precursor cells for adult tissues (Ordahl, 1999). The partially reliable fate maps that were available by the mid-twentieth century were based on vital dye staining and carbon marking, valuable techniques that were,

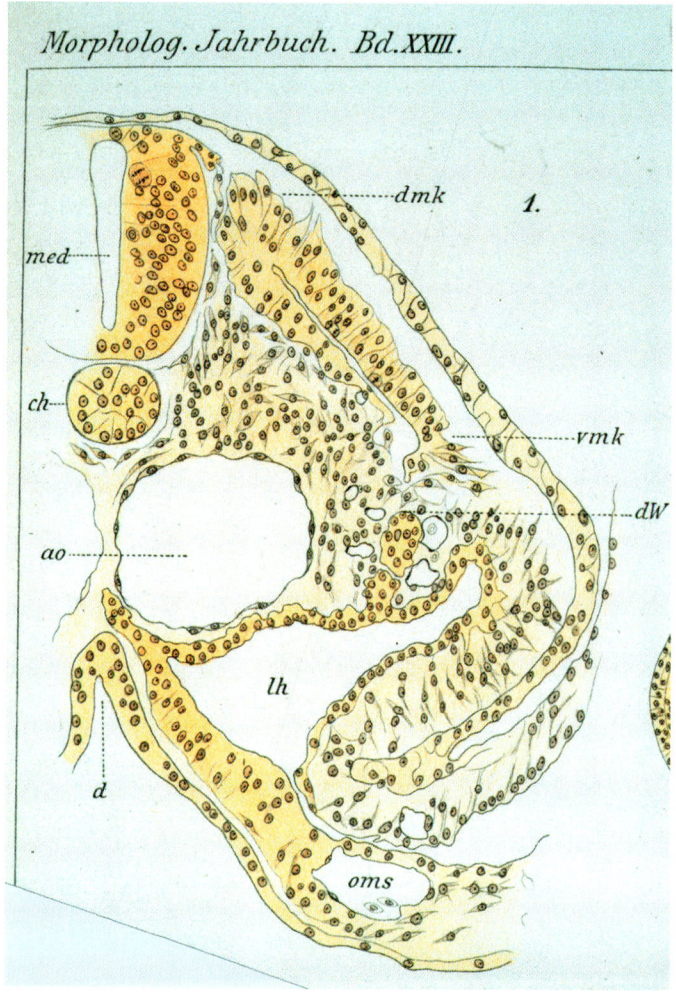

Figure 3 Fischel drawing of developing somite. An early illustration of somite activities postulated to underly the development of the epaxial myotome (dmk, medial lip of dermomyotome) and the limb muscle (vmk, migratory cells leaving the lateral lip of dermomyotome). This photographic copy of the original illustration (published in 1895) was generously provided by Bodo Christ.

nonetheless, inadequate to follow the fate of embryonic cells that migrate over long distances. Although DNA radiolabeling coupled with embryonic tissue transplantation offered an improved method for labeling embryonic tissues for fate mapping (Weston, 1963; Hazelton, 1970), detection difficulties coupled with the reduction in labeling during cell division rendered this approach impractical for general use in the creation of long-term fate maps.

This situation dramatically changed in the 1970s with the introduction, by

Nicole Le Douarin, of a general method for fate mapping vertebrate development. She had found that the nuclei of the Japanese quail (*Coturnix coturnix japonica*) contained an unusual nucleolus that stained densely and distinctively magenta with the Feulgen reagent (Le Douarin, 1973) whereas the stained nuclei of chickens, by comparison, were rather a uniform and pale pink (see, e.g., Fig. 7). With this knowledge, she developed a general strategy for fate mapping that involves transplantation of embryonic tissue from donor quail embryos into host chick embryos. The transplanted quail cells can move and develop freely in their new host environment because immunologic rejection of the donor-derived quail tissue does not appear until days or weeks after hatching (Ohki *et al.*, 1988). The quail cells retain their nuclear characteristics indefinitely, so that after a subsequent development period of days or even weeks the quail–chick chimeras can be histologically analyzed and the location of the quail-derived cells and tissues immediately recognized by virtue of their distinctive nucleolar marker. Over the past 25 years, this approach has illuminated the far corners of vertebrate cell lineage, particularly lineages that undergo extensive cellular migration such as myeloid and neural crest cells (for a compendium of examples, see Le Douarin *et al.*, 1990). More importantly, the results of fate mapping experiments in mammalian embryos (which are much more arduous to perform and limited in feasibility) have usually supported the results obtained through quail–chick chimeras (see Chap. 1 by Tam *et al.*, this volume). Thus, the detailed accuracy of modern vertebrate fate maps is largely established through analysis of avian development.

The power and elegance of the Le Douarin technique was quickly perceived by other experimental embryologists, and early application of her methods to the study of the paraxial mesoderm led to the seminal discovery that some somite cells must also undergo extensive migration because quail–chick replacement of brachial-level somites of the early embryo resulted in the extensive labeling of fetal forelimb skeletal muscle with quail nuclei (Christ *et al.*, 1974; Chevallier *et al.*, 1977; Christ *et al.*, 1977). These studies, coupled with electron microscopic analysis (Christ *et al.*, 1977, 1978), finally confirmed the conclusions of Fischel (1895) that cells at the lateral margin of the brachial dermomyotomes are migratory and are responsible for forming the muscle tissue within the adjacent limb buds. Continuing analysis of the somite fate by these methods has led to an extensive understanding of the role of somites at all axial levels to the development of muscle and other tissues (reviewed in Christ and Ordahl, 1995; and in Chap. 9 by Brand-Saberi and Christ, this volume).

C. Lineage Diversity in Muscle Precursor Cells

1. Early Segregation of Skeletal Muscle Lineages during Development

Lineage-specific attributes of skeletal muscle cells cultured from specific regions of embryos were noted in the 1950s by Holtzer and Detwiler (Holtzer *et al.*, 1957),

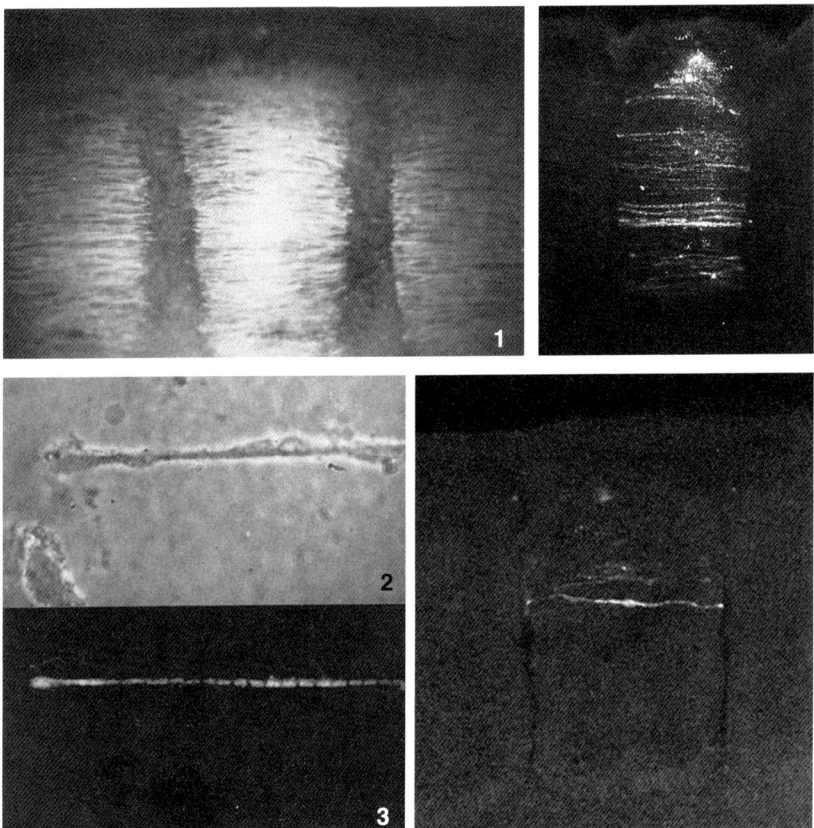

Figure 4 Individual myotome cells visualized by fluorescence. The panels on the left-hand side numbered 1, 2, and 3 are photocopies of the first published images of fluorescent localization of molecular marker expression in early embryonic myogenesis. Howard Holtzer and co-workers employed fluorescent antibody to myosin (Holtzer et al., 1957). Panel 1 shows multiple myotome fibers that span the craniocaudal width of the somite. After dissociation single fibers can be seen to be mononucleate under phase-contrast microscopy (panel 2). The characteristically precise myosin banding pattern can be seen in this cell through the fluorescent image emitted by the antimyosin antibody (panel 3). The two panels on the right-hand side show myotome fibers imaged by confocal microscopy after injection of their precursor cells within the dermomyotome with fluorescent diI. The upper panel shows many such fibers spanning an individual somite at a similar magnification as in panel 1. The lower panel shows a single, labeled myotome fiber with the central thickening characteristic of the presence of its single nucleus and with its attachment at the cranial and caudal borders of the somite. Several more lightly fluorescent fibers can be seen located medial to the highly stained fiber.

who observed that early myotome cells differed from their adult counterparts in that the former are mononucleated (Fig. 4) while the latter are highly multinucleated. Culture of myogenic precursor cells from specific regions of the embryo also shows qualitative and quantitative differences that are attributable to the early diversification of muscle lineages during amniote development (Konigsberg, 1979; Yaffe, 1968, 1969; Bonner and Hauschka, 1974; Hauschka, 1974a,b; Stockdale, 1992; Cossu *et al.*, 1996). The quail–chick surgical chimera method has proved itself to be valuable as the investigation of myogenic lineage diversification has been pushed ever earlier in development. Using this method, the first half-somite transplantation experiment confirmed that the dorsal half of the newly formed somite (the portion that gives rise to the epithelial dermomyotome; see Fig. 5) contains precursor cells for both dermis and muscle (Christ *et al.*, 1978). Experiments in which either the cranial or caudal halves of chick somites are replaced by those of quail indicated that both the cranial and caudal portions of the dermomyotome contribute to muscle formation but that profound craniocaudal polarity exists in the formation of the sclerotome (See Chap. 9 by Brand-Saberi and Christ and Chap. 10 by Dockter, this volume) and in the formation of the spinal nerves (see Chap. 7 by Bronner-Fraser, this volume). Subdividing the chick somite along its cranial–caudal axis and replacing either the medial and lateral halves with those of quail revealed a major division in the muscle progenitors of the body (Ordahl and Le Douarin, 1992). Cells within the medial half of wing-level somites are destined to give rise to back muscle, while cells within the lateral half are destined to give rise to muscles of the limb. Taken together with dye-labeling results showing that the medial and lateral portions of the somite have separate origins within the primitive streak and node (Selleck and Stern, 1991), these findings led to the conclusion that the early distribution of muscle precursor cells prefigured a major organizational aspect of the vertebrate body, namely, that of the epaxial and hypaxial domains (Ordahl and Le Douarin, 1992; Ordahl, 1993; Ordahl and Williams, 1998).

2. The Epaxial and Hypaxial Domains of the Vertebrate Body

Back and limb muscles respectively reside within two major regions of the vertebrate body, referred to as the epaxial and hypaxial domains (Fig. 6). Epaxial and hypaxial refer to the location of structures with respect to the vertebral column or body axis in the basic vertebrate body plan (Hyman, 1922; Romer, 1949). Epaxial (Greek *epi*, upon + Latin *axis*, axis) muscles lie on both sides of the vertebral spines immediately dorsal to the plane of the transverse processes. The fibers of the epaxial muscles are oriented in a craniocaudal direction, and these muscles act as extensors of the vertebral column. All of the remaining muscles of the body are hypaxial (Greek *hypo*, beneath) muscles including the limb muscles, the intercostal (rib) muscles, and the muscles of the abdominal wall. Innervation provides a simple method for distinguishing between epaxial and hypaxial muscles: All epaxial muscles are innervated by the dorsal rami of spinal nerves, whereas all hypaxial muscles are innervated via ventral rami of spinal nerves (Gray, 1989). The muscles of the epaxial and hypaxial domains of the body at the forelimb level are, respectively, derived from precursor cells located in the medial and lateral regions of the somite (Fig. 5).

Is the relationship between the somite medial–lateral organization and the

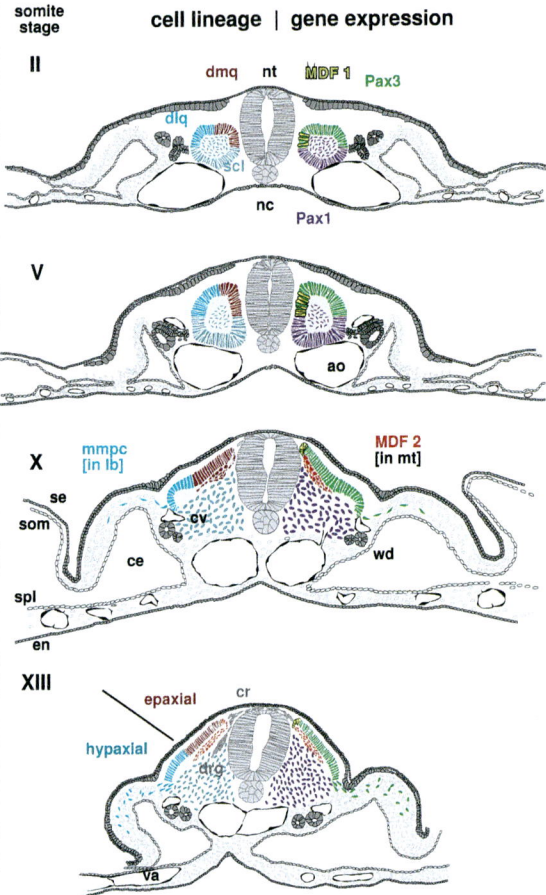

Figure 5 Cell lineage and gene expression in somite development. Illustrations of somites in transverse section at different stages of development redrawn from figures in Lillie's Development of the Chick (Hamilton, 1952).[4] The precision of the original drawings, and the inclusion of precise information regarding the axial level of somite depicted and its position in the somite ladder, permit the reliable representation of somite stages. Panel A shows brachial (forelimb) level somites at somite stages II, V, X, and XIII (derived from Lillie's Figures 118, 117, 114, and 115, respectively). Thoracic somite stages I, IV, and XIII (derived from Lillie's Figures 112, 113, and 116, respectively) show the more rapid development of somites at this axial level. Compare, for example, the extent of sclerotome development in stage V brachial versus stage IV thoracolumbar somites. Abbreviations for embryonic tissues: ao, dorsal aorta; ce, coelom; cr, neural crest; cv, cardinal vein; dlq, dorsolateral quadrant of the somite; dmq, dorsomedial quadrant of the somite; drg, dorsal root ganglion; en, endoderm; lb, limb bud; mmpc, migratory muscle progenitor cells; nc, notochord; nt, neural tube; scl, sclerotome; se, skin ectoderm; soc, somitocoel; som, somatic mesoderm; spl, splanchnic mesoderm; va, vitelline artery; wd, Wolffian duct. In each panel, the coloring on the left side follows cell lineage and fate while that on the right side illustrates gene expression. Cell lineage/fate color coding is indexed at the earliest somite stage shown as follows: aquamarine (blue) = dorsolateral quadrant of the somite (dlq); dark red (brown) = dorsomedial quadrant of the somite (dmq); slate blue = ventral half of somite epithelium and somitocoel cells. Note migration of dlq cells into the adjacent limb bud in panel A. Gene expression color coding (right side of each panel) is schematized from *in situ* hybridization data showing cells/tissues expressing the mRNAs of the following: green = Pax3 gene; yellow = MDF phase 1 (cells expressing one or more MDF mRNAs prior to overt differentiation); bright red = MDF phase 2 (differentiated myocytes expressing one or more MDF mRNAs and muscle differentiation genes, such as myosin); purple = Pax 1 gene. The boundaries for gene expression and lineage represented here are shown for purposes of illustration only and, while believed by the authors to be approximately accurate, have not been cross-referenced in any precise manner to specific results or measurements.

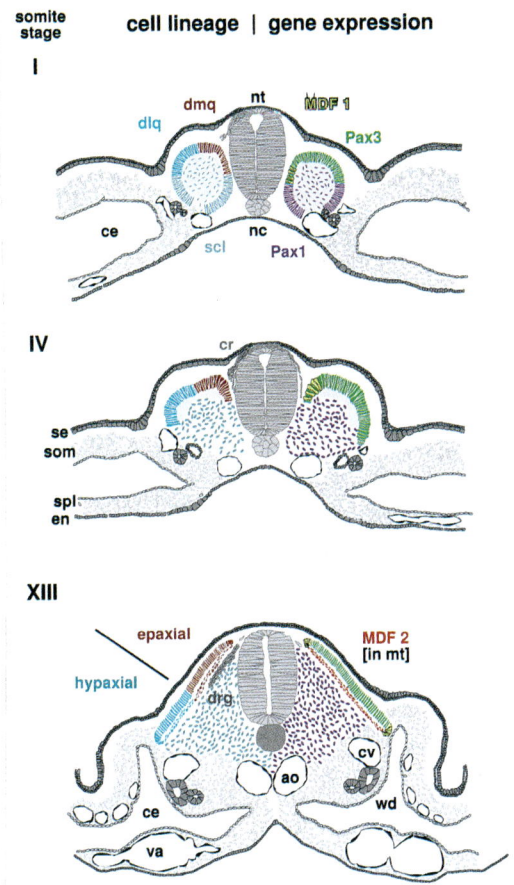

[4] *Lillie's Development of the Chick* was a combination text and laboratory manual for "beginners in embryology" that was authored by Dr. Frank R. Lillie, a noted American experimental embryologist. The third edition, which was used as the source for the illustrations in Fig. 5, was commissioned as a complete revision and condensation of the original multivolume work and whose author, Dr. Howard L. Hamilton, worked under the supervision of Dr. B. H. Willier, a former student of Dr. Lillie's who was appointed to act as as Editorial Advisor (Hamilton 1952). Even the third edition has been out of print for half a century, so copies of any edition are now extremely rare. The third edition photographically reproduced many of the hand-drawn scientific illustrations from the first edition (1908), and it is clear even from those relatively low-resolution reproductions that the original drawings must have an extraordinarily high level of precision in illustrating the relative dimensions and placement of embryonic structural elements (a precision also evident in Fig. 3). Owing to that accuracy, and the detailed information in the figure legends regarding the axial position of each somite section, it was possible to establish the somite stage for each illustration with a high degree of confidence. The cartoon treatment in Fig. 5 is intended to minimize details of cell structure while reproducing the precision of the originals in terms of the relative size and position of the structural elements of the somite and surrounding structures. In cases where the original drawing represented only one half of the embryo, its mirror image is also reproduced here to provide bilateral symmetry. The preface to the first edition (reprinted in Hamilton, 1952) indicates that most of the original drawings were made by Dr. Lillie himself, and that many were also made by Mr. Keni Toda. Two other artists, Mr. K. Hayashi and Mr. Willard C. Green, are also listed, and for the latter there is a specific credit for the original drawing of the stage II brachial somite reproduced in Fig. 5A. The preface to the first edition advised, "Some of the figures may be studied with advantage for points not described in the text."

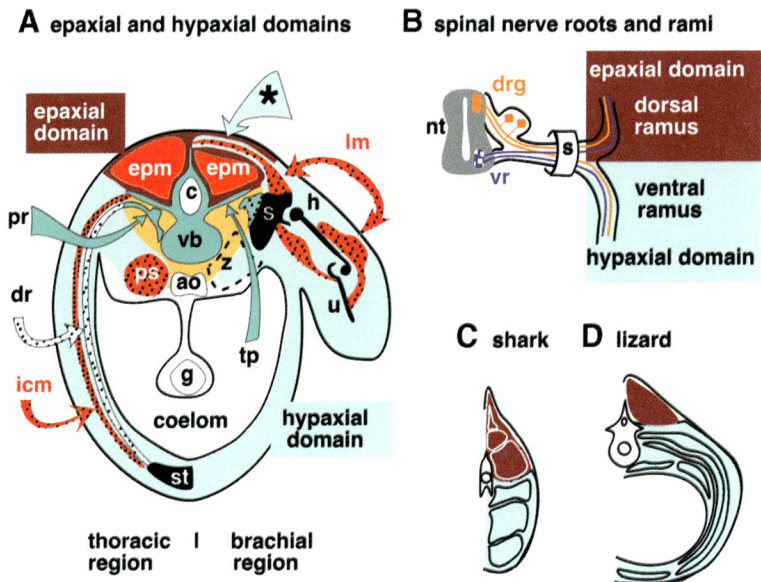

Figure 6 Epaxial–hypaxial domain structure of the adult body. (A) Schematized cross section through the body of an adult bird (or mammal) showing the distribution of somite-derived structures in the epaxial and hypaxial domains. The epaxial and hypaxial domains are colored dark red (brown) and aquamarine (blue), respectively, as in Fig. 5. A third, as yet only partially characterized domain (tan coloring) has been tentatively defined as the "ventral epaxial domain" for operational considerations (Williams and Ordahl 1997). Note that the ribs comprise both proximal and distal segments that derive from the medial and lateral halves of the somite, respectively (see Figs. 5 and 8). Structures derived from the lateral half of the somite are stippled while those derived from the medial half are not. Epaxial dermis is somite-derived while hypaxial dermis is not. Bones derived from lateral plate mesoderm are shown in black. The asterisk shows how the hypaxial domain "invades" the epaxial domain in regions where limb muscles attach to connective tissue over the vertebral spines. These muscles also carry their innervation from the ventral rami of the spinal nerves (see B). a, Aorta; c, vertebral canal; dr, distal rib segment; epm, epaxial muscle; g, gut; h, humerus; icm, intercostal muscle; lm, limb muscles (rhomboid, triceps long head, and brachialis illustrated); pr, proximal rib segment; ps, psoas muscle; s, scapula; st, sternum; tp, transverse process; u, ulna; vb, vertebral body; z, zone of mixing (dashed enclosure) observed in brachial half somite transplant experiments (Ordahl and Le Douarin, 1992). (B) The distibution of the sensory and motor components of the spinal nerves as they course through the ventral and dorsal rami to the epaxial and hypaxial regions of the body. Sensory nerve cell bodies (gold) in the dorsal root ganglion (drg) send processes into the dorsal neural tube (nt) and the spinal nerve (s). Axons from motor neuron cell bodies in the ventral region of the neural tube leave the neural tube through the ventral root (vr) to reach the spinal nerve, where motor and sensory fibers are mixed. The dorsal and ventral rami are the two main terminal branches of the spinal nerves that provide both sensory and motor innervation to the epaxial and hypaxial domains, respectively. (C and D) The epaxial and hypaxial domain structure of other vertebrates as adapted from Romer (1949).

epaxial–hypaxial domains also true for other regions of the body? Medial and lateral half-somite replacement experiments were performed at the thoracic level to test this hypothesis. The results (Fig. 7) indicate that the proximal segments of the ribs are derived from the medial halves of thoracic somites which also give

9. Determination and Morphogenesis in Myogenic Progenitor Cells 335

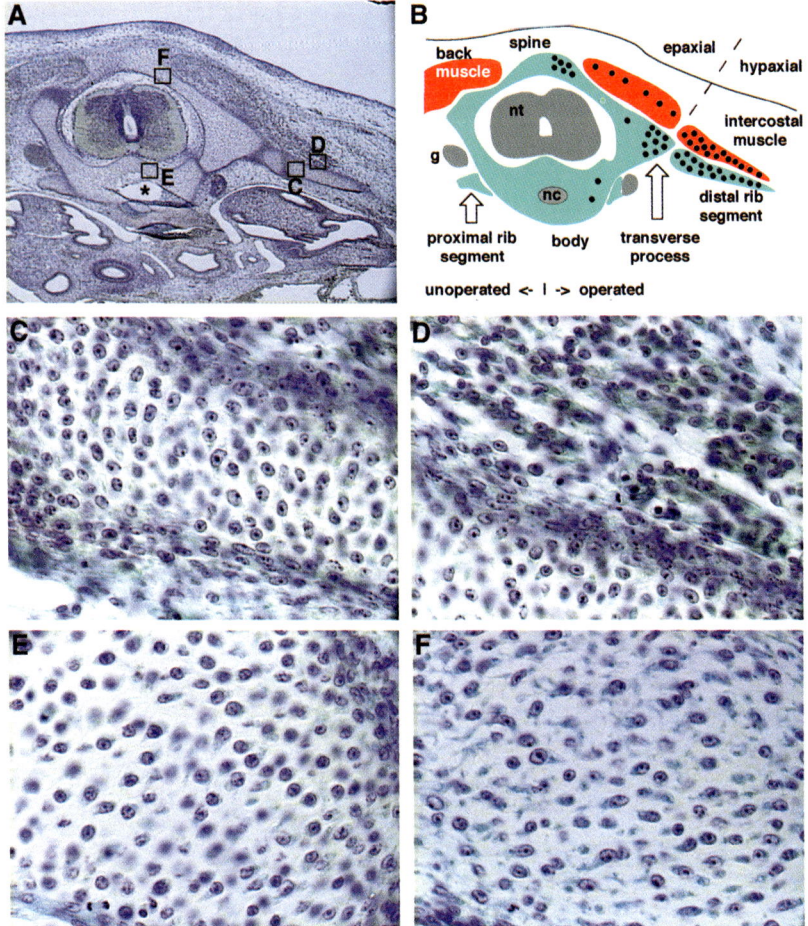

Figure 7 Lateral half-somite derivatives at the thoracic level. Experiments that entail replacement of the lateral half of a thoracic somite of a chick embryo with that of a quail at day 2 of embryoic development result in a pattern of tissue labeling at embryonic day 9 that is consistent with the example shown here. (A) A low magnification, Feulgen-stained, histological section through the labeled region after reincubation for 1 week, to the fetal stage of development. Boxes in A show regions shown at higher magnification in C–F. Quail nuclei under a microscope appear to have nucleoli that are brightly magenta but which in photographs appears as a nuclear density that is absent from chick nuclei. Quail nuclei are abundant in the distal portion of the rib (C) and intercostal muscle (D) but are absent, or present at much lower levels, in the vertebral body (E) and vertebral spine (F). (B) A cartoon tracing of the essential features in panel A with the presence of quail nuclei indicated by black circles.

rise to the body of the vertebra (for more detailed discussion of these issues, see Chap. 9 by Brand-Saberi and Christ, Chap. 8 by Monsoro-Burq and LeDouarin, and Chap. 10 by Dockter, this volume). In contrast, the lateral portion of the rib is composed, predominantly, of cells derived from the somite lateral half. Similar results have been independently obtained in the laboratory of Olivier Pourquié (personal communication). Recently, the rib primordia have been proposed to be derived from the dermomyotomal portion of the somite (Kato and Aoyama, 1998). This would place the distal rib precursor within the somite dorsolateral quadrant (dlq), a position at some variance with earlier work (see Chap. 10 by Dockter and Chap. 9 by Brand-Saberi and Christ, this volume). Nevertheless, these results all tend to support the general proposition that cells within the medial somite half are destined to form epaxial stuctures while those of the lateral half are destined to give rise to hypaxial structures (see also Figs. 5B and 6).

Other work also lends support to the general proposition that the medial and lateral portions of the somite have different fates. In the early 1950s, Holtzer and Detwiler showed that ablation of the two medial thirds of amphibian somites resulted in selective loss of back as compared to body wall muscle (Detwiler and Holtzer, 1956, and references therein). Taking another approach, Gonzalo-Sanz showed that a sheet of foil placed between the medial and lateral halves of chick epithelial somites came to lie between the epaxial and hypaxial muscles after further development (Gonzalo-Sanz, 1972). Dye labeling experiments from our laboratory also support the idea that myogenic precursor cells are segregated into medial and lateral regions of thoracic somites in a manner that prefigures the formation of the epaxial and hypaxial myotome. Finally, recombinant mosaic analysis provides single-cell resolution to the idea that the initial segregation of precursors to the epaxial and hypaxial myotome occur very early in development, probably prior to gastrulation (see Chap. 17 by Eloy-Trinquet *et al.*, this volume). Taken together, these experiments all support the notion that the precursors to epaxial and hypaxial muscle have separate origins within the dorsomedial and dorsolateral quadrants of the somite, respectively.

3. Other Considerations Regarding Somites and the Epaxial/Hypaxial Organization of the Body

There is an interesting difference in the distribution of somite cells after half-somite transplantation at the brachial versus thoracolumbar levels. At brachial levels, the cells derived from the medial and lateral somite halves, while somewhat intermingled close to the body axis, had a distinctive nonoverlapping distribution on either side of a sharp boundary between the limb and body (Ordahl and Le Douarin, 1992). In contrast, after half-somite transplantation at the thoracic or abdominal levels there was always significant mixing of cells from the medial half-somite into the hypaxial domain and vice versa (Fig. 7). Assuming that such mixing reflects imprecision in surgically replacing half-somites, why was a sharp epaxial/hypaxial boundary observed in the half-somite experiments conducted with wing-level somites but not with thoracic-level somites? Kieny and co-workers (Mauger and Kieny, 1980) showed that cells within the limb premuscle masses would remigrate to the limb after back transplantation to the paraxial region, suggesting that medial and lateral half-somite cells may become programmed

9. Determination and Morphogenesis in Myogenic Progenitor Cells   337

to partition themselves within the epaxial and hypaxial domains. If such programming occurs it must be acquired sometime after somite formation because transplantation of a quail medial half-somite into the lateral half-somite position resulted in a substantial fraction of quail cells populating wing muscle (Ordahl and Le Douarin, 1992). A second contributing mechanism may emerge from the observations of Kathryn Tosney and co-workers (K. Tosney, personal communication) that wing-level somites undergo apoptosis along their lateral borders at later stages of development and this could include cells from the lateral and medial halves of the somite which might come to be located on the "wrong" side of that boundary in the brachial region of the body. These hypothetical mechanisms for establishing and regulating domain identity at the wing level are less evident at thoracic levels and may therefore explain the more extensive mixing of surgically replaced medial and lateral half-somite derivatives at these axial levels.

The contribution of the medial half-somite to the vertebral body and proximal rib segment provides a subtle inconsistency in the general proposition that medial portions of the somite distribute solely to epaxial structures. Being ventral to the transverse process, the vertebral body and proximal rib segment technically lie outside the epaxial domain and mixing of medial and lateral half-somite cells was observed in this region (Ordahl and Le Douarin, 1992). Moreover the psoas major, a bona fide hypaxial muscle, lies ventral to the vertebral bodies and is populated by cells from the lateral half of the somite (C. P. Ordahl, unpublished observations, 1991). Because of such potential ambiguities, one of us (B. A. Williams) coined the working expression "ventral epaxial domain" to refer to this region and, in particular, to describe the results of surgical experimentation (Williams and Ordahl, 1997).

Boney structures in the limb hypaxial domains are predominantly derived from the somatic mesoderm (except for a small portion of the scapula, which is somite-derived; see Chap. 10 by Dockter, this volume), as is the sternum, to which some ribs attach distally. The syrinx bones in the avian neck are also derived from the somatic mesoderm (Noden, 1988). Thus, the boney contributions of the lateral half of the somite to the hypaxial domain appear to be limited to the distal rib segments. Kato and Aoyama (1998) recently showed that the cartilage cells in the distal rib are derived from the dermomyotome, rather than the lateral sclerotome. Those results, in combination with the half-somite transplants cited above, lead to the conclusion that the distal rib segment is derived from the lateral half of the dermomyotome; the same region that contains the precursor cells for the hypaxial myotome, the primordia of the intercostal muscles.

D. Molecular Marker Expression in Epaxial Myotome Precursor Cells

Molecular detection techniques greatly expand the ability to identify subpopulations of embryonic cells through the embryonic expression patterns of specific marker genes. As indicated above, the first experiments using molecular markers to study the initial phases of embryonic myogenesis were performed over 40 years ago by Holtzer and Detwiler (Holtzer et al., 1957) who used myosin antibodies to localize nascent myocytes within the somites of chick embryos (Fig. 4). That study defined many of the essential features of the early myotome, not the least of which is the observation that early myotome fibers are mononucleated cells that span the cranial to caudal boundaries of the somite and have organized, cross-striated myofibrils that are identical to those found in multinucleated muscle cells of the adult. Thirty years later, Christ and co-workers using antibodies to desmin to localize the nascent differentiated myotomal cells as they first emerge from the dermomyotome (Kaehn et al., 1988), elegantly showed how myotome cells progressively arise in a somite-age-dependent fashion. A somite staging system was developed to allow more systematic evaluation and comparison of the development of somites in general, and in particular with respect to the specification of myogenic precursor cells (Ordahl, 1993).

In using the somite staging scheme (see Figs. 2, 5, and 8), somite developmental age is reflected by roman numerals, such that the most newly formed somite is designated as stage I and the next oldest (in a cranial direction) as stage II and so on. The first myotome cells could be detected as desmin-positive cells in wing-level somites that had reached somite stage VII (Kaehn et al., 1988) (e.g., somite number 19 in an embryo having formed a total of 25 somites) (see Figs. 2A, 5A, and 8). Subsequently, however, the rate at which desmin-positive cells first appear in individual somites (reflecting the onset of myotome cell formation) differs at different axial levels (Borman et al., 1994; Borman and Yorde, 1994). Thus, myotome development in cervical somites is delayed initially and then progresses very rapidly. This is true for many, if not all aspects of somite development. Compare for example the degree of sclerotome development in a stage V wing-level somite to that of a stage IV thoracic-level somite (Fig. 5A, B). Thus, staging information is most useful for comparing the development of somites from the same axial level. Owing to axial differences in the rate of somite differentiation, caution must be observed when comparing somites developing at different axial levels. Nevertheless, the results of Kaehn et al. (1988) showed that the first myotome cells appear at stage VII in the craniomedial corner of forelimb level somites thus delimiting the beginning of the differentiation phase of myotome development in these somites (see Fig. 8).

The advent of nucleic acid probes and *in situ* hybridization has opened the somite up to more detailed molecular analysis of the predifferentiation phase of the development of myogenic precursor cells in the somite. As mentioned in Section I, shortly after the discovery of the MDF family of bHLH genes, two groups used *in situ* hybridization to determine which MDF gene is first activated in the newly formed somites of avian and mammalian embryos (de la Brousse and Emerson, 1990; Ott et al., 1991). In the avian (quail) embryo, the initially expressed MDF is the myoD-1 homologue (de la Brousse and Emerson, 1990), whereas *myf5* is the first MDF expressed in the mouse embryo (Ott *et al.,* 1991). In both cases, notably, the first MDF-mRNA positive cells were found to be located in the dorsomedial quadrant of the newly formed somite.[5]

The onset of MDF expression in the medial portion of the somite precedes the onset of epaxial myotome formation by two- to five-somite stages (see Fig. 6). Here we designate as "MDF phase 1" this earliest phase of MDF expression that precedes the onset of expression of differentiated muscle marker genes (e.g., desmin and myosin). MDF expression in differentiated myocytes is here referred to as "MDF phase 2." In a simple view, phase 1 MDFs are involved in the steps leading up to differentiation, whereas during phase 2 the expressed MDFs effect and maintain the differentiated state. Phase 2 MDF expression can be highly complex (Cornelison and Wold, 1997) in ways that reflect either diversity in developmental lineage and/or timing of onset of differentiation within the diverse myogenic precursor cell populations (see also below). We will not otherwise consider it here because it occurs shortly before or after the myocytes have become postmitotic and, more to the point, occurs after myogenic determination, the main subject of this chapter.

E. Location of Myotome Precursor Cells within the Somite

Despite the wealth of detailed embryological and gene expression information described above, the precise location of myotome precursor cells within the somite has proved to be controversial throughout the twentieth century and even up to the present. Williams (1910) was the first to suggest that myotome cells are generated from the cells located in the dorsomedial lip of the dermomyotome based on the location of the earliest myotome cells in histological sections. Williams' view was generally accepted until the 1960s when Langman and Nelson (1968), using tritiated thymidine to follow cell division, concluded that myotome precursor cells arise simultaneously from all areas of the dermomyotome, a view that was shared by earlier histological observers of somites (Remak, 1855; His, 1888; Bardeen, 1900; reviewed in Langman and Nelson, 1968).

This view began to change as more modern techniques were applied to the study of myotome formation. As outlined above, Christ and co-workers (Kaehn *et al.,* 1988), began analyzing early myotome formation using an antibody to desmin,

---

[5] Only the issue of timing of first expression is considered here. There are many, far more complex issues regarding the timing and position of subsequent MDF gene activation events (see Chap. 12 by Arnold and Braun, Chap. 15 by Tajbakhsh and Buckingham, and Chap. 14 by Borycki and Emerson, this volume). For example, in chick embryos, the myf5 mRNA can be detected in this same region shortly after the onset of myoD expression (Williams and Ordahl, 1994). Recent analyses of somite development show increasing precision in the localization of MDF and other gene expression (Pownall and Emerson, 1992; Lin-Jones and Hauschka, 1996; Dietrich, *et al.,* 1998; Hacker and Guthrie, 1998).

the earliest protein expressed by postmitotic, differentiating myocytes (Lin et al., 1994). In serial sections of stained embryos, desmin-positive myocytes were first detected in stage VII somites where they formed a small right triangle immediately subjacent to the craniomedial corner of the dermomyotome. This desmin-stained triangle expanded progressively in increasingly older somites until its medial side spanned the entire cranial–caudal dimension of the somite (see Fig. 8 below). The right triangular shape of the early myotome suggested to Christ and co-workers that the hypotenuse represented the tips of myocytes growing from the cranial to caudal edge of the somite (Kaehn et al., 1988) and, therefore, that the

Figure 8 Somite staging with gene expression. Gene expression during early myogenic development in the epaxial (A) and hypaxial (C) domains. (B) A schematic dorsal view of an HH 14 chick embryo showing the early stages somite development and the timing and disposition of precursor cells within the dorsomedial quadrant of the somite that give rise to the earliest differentiated muscle cells within the epaxial domain (left somite string) and those residing in the dorsolateral quadrant of the somite that similarly give rise to muscle in the hypaxial domain (right somite string). (A) Schematic cross sections of somites at selected stages of somite development (roman numerals) showing the location of selected gene expression and signal sources affecting somite development. MDF gene expression is represented as occurring in two phases based on the anatomy of the somite and the coexpression of differentiated muscle gene markers. Phase I MDF expression occurs in cells in the dorsomedial quadrant of the early somites and in the medial lip of the dermomyotome of more mature somites. Phase II MDF expression is seen in cells of the myotome which, unlike phase I MDF cells, coexpress differentiated muscle genes such as desmin and myosin. The onset of phase I MDF expression occurs about 12 hr before the onset of differentiation and phase II MDF expression. After the first differentiated myocytes appear at somite stage VII, phase I and phase II MDF expression occur simultaneously but are restricted to the dorsomedial lip of the dermomyotome and the myotome, respectively. Signals hypothesized to be responsible for the changes in gene expression are as follows: vs, ventral axial signals; ds, dorsal axial signals; ls, lateral signals. The sources and identities of signals are discussed in detail elsewhere in this volume (see Chap. 5, Volume 48 by Borycki and Emerson). The dashed line delineates the DMQ region of the somite analyzed in notochord challenge experiments. The results of those experiments is indicated to the left as the percentage of dmq grafts that gave either muscle tissue or myoclusters at each of the somite stages indicated. The method for analysis of the notochord challenge experiments is summarized in Fig. 11. (B) Schematic dorsal view of somites showing gene expression in the myogenic precursors of the epaxial myotome (left-hand somite string) and progenitors of hypaxial muscle of the limb (right-hand somite string). After stage II, in the left hand somite string, only the medial lip (MDF phase I, yellow) and the myotome (MDF phase II, red) are shown. The right-hand somite string shows gene expression in descendants of the dorsolateral lateral quadrant of the segmental plate and newly formed somite only. (C) Gene expression in precursor cells from the dorsolateral quadrant of the somite after they have migrated into the developing limb bud. The dorsal view of dermomyotome development (B, right-hand somite string) shows Pax3 gene expression in the lateral dermomyotome only (can also be seen in cross sections in A). Lateral migration of cells, induced by lateral signaling (ls, see A), begins to occur from the lateral margin of the dermomyotome at about somite stage VII and continues for at least 24 hr. Somite cells within the limb bud coalesce into dorsal and ventral muscle masses and reside there for at least 24 hr before activating MDF (hypothetical MDF phase I), followed shortly by activation of differentiated muscle markers (hypothetical MDF phase II) at about HH 24. Note that at late stages of somitogenesis at the forelimb level (e.g., HH 21) the medial lip region of the dermomyotome persists while the main epithelium has disintegrated.

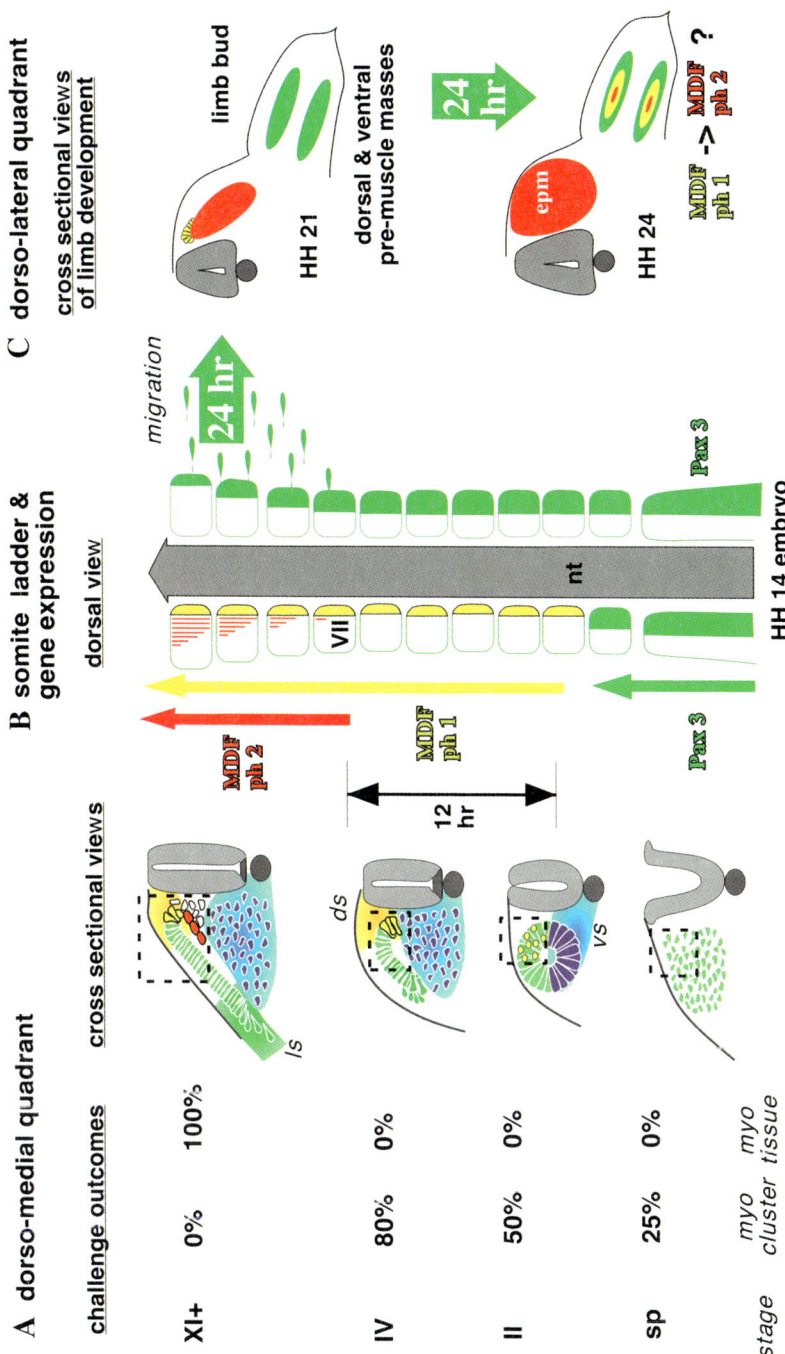

youngest (shortest) myotome cells were located at the extreme lateral edge of the myotome. Although this model provided the first mechanistic explanation for myotome formation it raised new questions. For example, if myotome cells are born in a medial-to-lateral sequence along the cranial border of the dermomyotome, then the MDF-positive cells along the medial lip must not be myogenic precursors. If the MDF-positive embryonic cells are not muscle precursor cells, then what is their function in development?

To more precisely analyze the precursor–product relationship between the dermomyotome and myotome we turned to vital dye staining, a technique that has undergone modernization through the use of fluorescent dyes (Serbedzija et al., 1989). One of us (W. Denetclaw) microinjected the carbocyanine dye, DiI, into predetermined regions of the dermomyotome in a grid pattern (see Fig. 9) (Denetclaw et al., 1997). After 16–24 hr of further incubation the somite was subjected to confocal microscopy to determine the distribution of fluorescent dye label within the dermomyotome and subjacent myotome. Because the long axis of dermomyotome cells is oriented perpendicular to the elongated myotome fibers, the label retained by these two cell types could be readily distinguished in confocal scans as punctate and elongate, respectively (see Fig. 4). Using the grid-injection approach, the frequency with which myotome fiber labeling resulted from injections at each specific site was tabulated and indicated the location of myotome precursor cells within the dermomyotome. Consistent with earlier observations (Kaehn et al., 1988), the highest incidence of myotome precursor cell labeling occured at the craniomedial corner of the dermomyotome (sites B and C in Fig. 9A), but, rather than being restricted there, myotome precursor cells were distributed along the entire medial lip of the dermomyotome (Fig. 9E). Therefore, the MDF phase 1 region of the dermomytome (the medial lip) coincides with the region where the direct precursors to myotome cells reside.

The dye-injection analysis of myotome development supported a myotome fiber birth date sequence that is the opposite of that predicted in the model of Kaehn et al. (1988) (Fig. 9D,E) and is based on two lines of evidence (Denetclaw et al., 1997). First, after sequentially injecting the same location in the medial lip of the dermomyotome with dyes that fluoresced at different wavelengths, myotome fibers labeled by the first injection were always found to be located lateral to fibers labeled by the second dye. Thus, in the early myotome, fibers located more laterally were born earlier than fibers located more medially. A second observation corroborated this fiber birth date conclusion. When analyzed a short time (7 hr) after labeling of the medial lip of the dermomyotome, myotome fiber distributions were organized as right triangles with the longest fibers located laterally in the myotome (see Denetclaw et al., 1997). Using the same line of reasoning as Kaehn et al. (1988), we conclude that the hypotenuse of this right triangle represents the growing tips of myotome fibers as they extend toward the cranial and/or caudal boundaries of the somite. Since the shorter (most recently born) fibers in such triangles are always located medially and longer (oldest) fibers located more laterally,

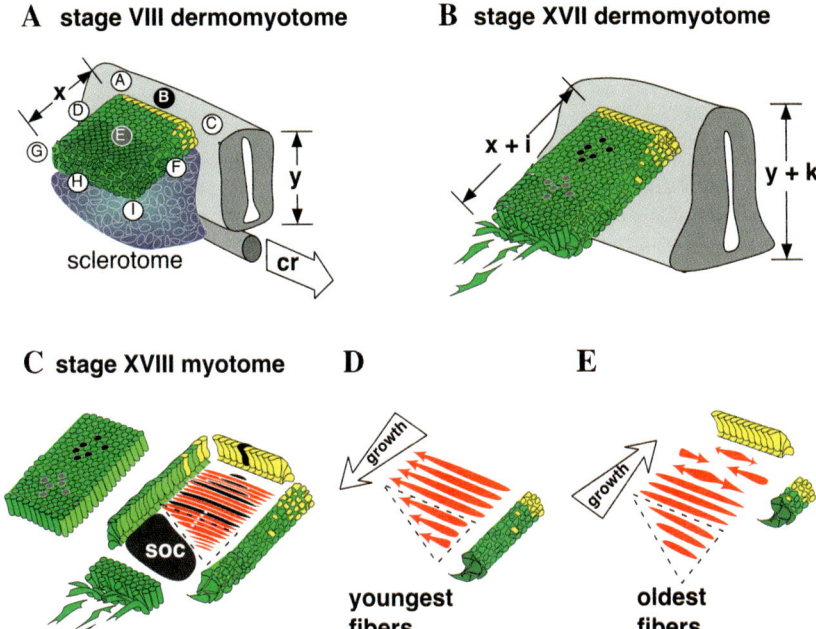

Figure 9 Myotome development models. (A) A three-dimensionalized view of a stage VIII somite dermomyotome showing sites of fluorescent labeling used to identify locations of myotome precursor cells. Sites B and E are darkly and lightly shaded to mark the origin of similarly labeled cells in panels B and C. The double-headed arrow labeled x represents the medial–lateral width of the somite dermomyotome, while that labeled y represents the dorsoventral height of the neural tube (gray). Arrow labeled cr points in cranial direction. (B) The same dermomyotome epithelium as in A, except after approximately 16 hr of development (to somite stage XVII). The position of darkly and lightly labeled cells in the dermomytote reflect the growth of the dermomyotome as indicated by the double-headed arrow marked x + i where i represents the incremental increase in dermomyotome height in the mediolateral dimension. Note the corresponding increase k in the dorsoventral dimension of the neural tube. (C) Same dermomyotome as in B except that specific regions have been displaced to reveal the underlying myotome (red) and somitocoel (soc). At this stage of development the lateralmost region of the myotome has a triangular shape (indicated by dashed triangle). The position of labeled myotome cells from a site B injection (black fibers) is indicated. No myotome progeny are observed 16 hr after labeling at site E. (D) Model for myotome development proposed by Kaehn et al., (1988) in which myotome cells (red) are born along the cranial lip of the dermomyotome epithelium and then elongate in a caudal direction to contact the caudal lip of the dermomyotome (arrowheads on myotome cells indicate direction of growth). In this model, new myotome cells are born at the most lateral extremity of the myotome as reflected in the lateral triangle interpreted to be myotome cells that had not yet completed elongation to the caudal lip. Growth (arrow) in this model was thought to procede in a medial to lateral direction. (E) Denetclaw et al. (1997) model for myotome development based on precursor product labeling of dermomyotome cells. The most active sites of myotome cell generation are along the medial lip and medialmost portion of the cranial lip of the dermomytote. Cells can be born at any point along these lips and then each cell elongates in a cranial, caudal or both cranial and caudal direction to span the distance between the cranial and caudal lips of the dermomyotome. In this model, the lateralmost portion of the myotome contains the oldest myotome fibers. The triangular shape of the myotome in this region may be explained by the presence of the somitocoel (see text). The growth of the myotome is medial-directed.

this further supports the lateral to medial birth order of the early myotome fibers and indicates that the epaxial myotome grows in a dorsomedial direction. What then causes the lateral triangle shape of the myotome that first was noted by Kaehn *et al.* (1988) and is evident in many published images of somites stained with muscle protein antibodies or *in situ* hybridization probes? We hypothesize that the hypotenuse of this triangle is defined by contact between the growing tips of myotome cells with the somitocoel (see Fig. 9C), with the boundary of the latter being displaced caudolaterally as a result (see Fig. 9C). Early myotome fibers do not cross under or over the somitocoel (see, e.g., Williams and Ordahl, 1994), consistent with the notion that the very first epaxial myotome fibers which form along the craniomedial edge of the somite are prevented from fully elongating due to the presence of the somitocoel.

III. When Do Myotome Precursor Cells Become Determined?

A. Signal Sources That Influence Somite Development and Myogenic Specification

The newly formed somite quickly shows regional specializations that are believed to be dependent on signals emanating from three regional sources (see Fig. 8) consistent with a regulative model for development as outlined in the introduction. Dorsal–axial signals (from the dorsal region of the neural tube and surface ectoderm overlying the somite) are generally thought to emit signals that are responsible for muscle specification (possibly in conjunction with, or independent of, muscle inducing signals from the notochord; see Chap. 14 by Emerson and Borycki, this volume, for detailed discussion). Lateral signals (from the lateral plate mesoderm) appear to transiently suppress MDF activation within the lateral region of the dermomyotome epithelium (Pourquié *et al.,* 1996) but later stimulate the migration of these cells to the adjacent limb bud (Bladt *et al.,* 1995). Last, signals from ventromedial sources (notochord and floor plate of the neural tube) appear to induce "ventralization," an epithelial–mesenchymal response in the ventral region of the somite to form the sclerotome, which then goes on to form the cartilage models of the vertebrae (see Chap. 10 by Dockter, this volume, for a detailed review of this field). *In vivo* experiments demonstrated that a supernumerary notochord placed in the groove between the segmental plate and neural tube could ventralize the epaxial domain of the embryo, completely eliminating epaxial muscle and replacing it with ectopic cartilage (van Straaten *et al.,* 1985, 1988, 1989; Brand-Saberi *et al.,* 1993; Pourquié *et al.,* 1993). Those results indicated that, if challenged early enough by signals from the notochord, the cells within the dorsomedial portion of the segmental plate can be influenced to become cartilage rather than muscle under the influence of the notochord.

B. The Notochord Challenge Assay

Myogenic competence in embryonic tissues exists well prior to the onset of detectable muscle markers (Bonner and Hauschka 1974; Christ *et al.*, 1974; Chevallier *et al.*, 1977; Christ *et al.*, 1977; George-Weinstein *et al.*, 1996, 1998), including MDFs (Tajbakhsh and Buckingham, 1994). At what point during development do such tissues that pass the myogenic competency test also acquire the negative aspect of determination, that is, the exclusion of potential to differentiate into cell types other than myogenic? This can be experimentally assessed by challenging the ability of embryonic tissues or cells to progress along alternative differentiation pathways. Classically, this has involved translocating the tissue to a new environment (embryonic or culture) where novel influences might promote differentiation along such an alternative path(s). One problematic aspect of such assays is the difficulty in knowing how many novel influences impact the transplanted tissue, and, perhaps more importantly, it is unclear whether those influences are meaningful in context of the normal development of the tissue being evaluated. Aoyama and Asamoto (1988) took a new approach to the analysis of embryonic determination by rotating the somite *in situ,* thereby minimally changing the environment but dramatically changing the exposure of specific regions of the somite to external signals (Fig. 10). Their results indicated that the cells in the newly formed somite (somite stages I and II) are undetermined and, regardless of their normal fate, can be redirected by extrinsic signals to develop into either muscle or cartilage in the normal, expected position. Rotation of a stage III somite, on the other hand, resulted in the formation of a "ventral" myotome (Fig. 10). This result indicated that muscle precursor cells present in the dorsal region of the stage III somite could no longer switch to a cartilage differentiation pathway despite their now ventral position. This altered morphology, with mesenchyme dorsally rather than ventrally located, also suggested that anatomical integrity of the somite as an embryonic organ anlage might also influence the developmental plasticity of its constituent cells.

To allow more flexibility in the analysis of myogenic specification and determination in different regions of the somite, one of us (B. A. Williams) developed a novel experimental design; instead of translocating the embryonic tissue to a new environment, or rotating an entire somite, Williams surgically transplanted defined fragments of quail paraxial mesoderm into the early somite environment of a chick host embryo. That host environment was further modified, however, to amplify one local signaling source known to influence the specification of somitic cells. Signaling changes were effected through juxtaposition of the test tissue and

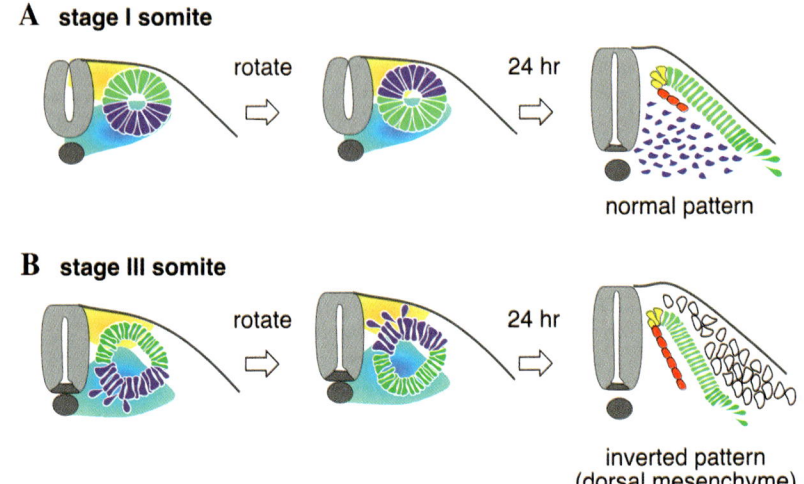

Figure 10 The somite rotation experiment of Aoyama and Asamoto (1988). This simple yet elegant experiment has come to be regarded as a classic in the analysis of somite specification. (A) Cartoon showing the results of rotating a stage I somite to invert its dorsoventral axis. After 24 hr a normal pattern of myotome and sclerotome is observed consistent with the notion that the Pax1-expressing presclerotome region became reprogrammed to form muscle and that the Pax3-expressing premyogenic region became reprogrammed to form sclerotome. In contrast, rotation of a stage II somite (B) results in an inverted pattern, with sclerotome-like mesenchyme dorsal to a relatively normal-appearing myotome and dermomyotome. The conclusion is that dorsoventral patterning is plastic in the somite until about stage III when it becomes fixed. Color coding is as in previous figures.

known signaling source tissues within the host environment, or provided through additional implants, or both. In this way, it was reasoned, that context was embryologically meaningful in terms of (1) position within the embryo, (2) developmental history, and (3) signaling environment.

The experimental logic is outlined in Fig. 11a. During specification, paraxial mesoderm cells are competent to differentiate into either muscle or cartilage. Such cells have been defined as "specified" to indicate that they are plastic, that is, competent for both phenotypic end points and therefore determined for neither (Slack, 1983). The experiment by Aoyama and Asamoto (1988) indicates that this specified state persists for a period of time after somite epithelialization but that at some point somite cells become resistant to that signaling and thus phenotypically restricted. Phenotypically restricted somite cells are refractile to changes in their signaling environment and go on to make their fated cell types. This signal-nonresponsivity, therefore, is the criterion for cell type determination in this assay.

9. Determination and Morphogenesis in Myogenic Progenitor Cells 347

A *notochord challenge: experimental logic*

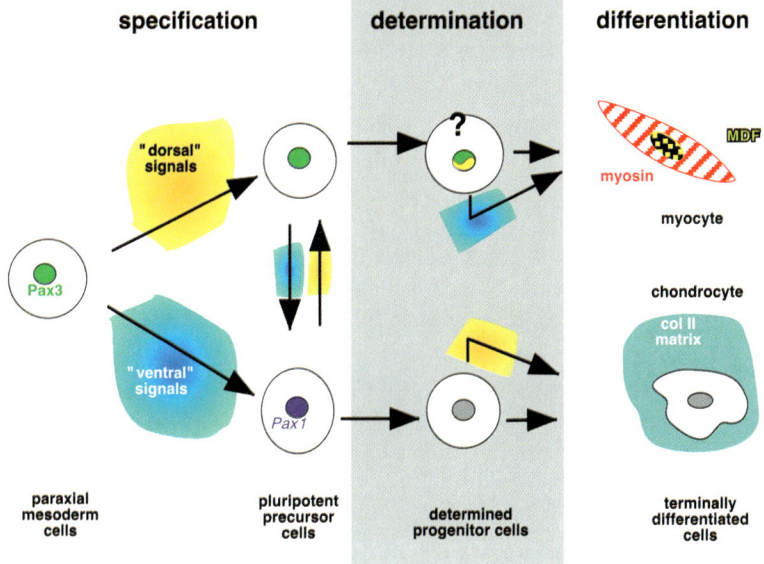

B *notochord challenge: experimental execution*

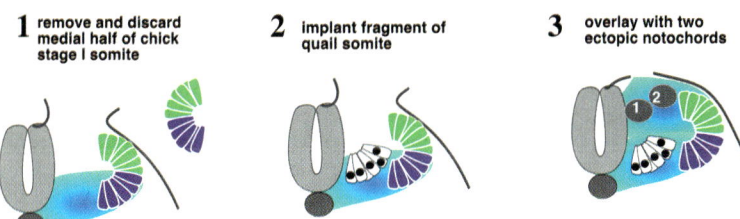

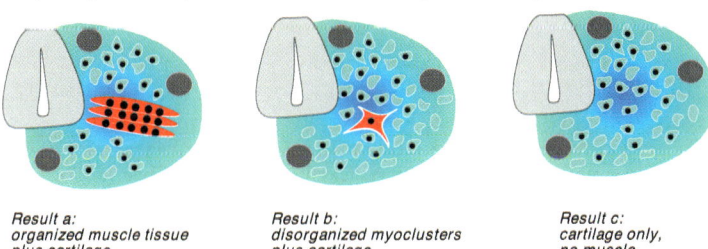

C. Progressive Increase in Myogenic Potential in Epaxial Muscle Precursor Cells

In practice (see Fig. 11B) the experiment involves transplanting quail premyogenic tissue, derived from the dorsal portion of the somite, to a ventral somite position adjacent to the notochord of a chick host embryo. Two additional notochords are then implanted to surround the test fragment with "ventral signals" arising from the host and donor notochords. As discussed above, and elsewhere (Chap. 10 by Dockter, this volume), it is well established that signals from the notochord are responsible for inducing sclerotome, which then goes on to form the axial cartilage. Moreover, a supernumerary notochord placed adjacent to the premyogenic region of the segmental plate blocks myogenesis and promotes chondrogenesis in its place (Pourquié *et al.*, 1993). Thus, ectopic signals from a supernumerary notochord can induce cells fated to become myotome to proceed down an alternative differentiation pathway, to chondrogenesis. In this environment, presumptive myogenic precursor cells that are only specified (undetermined) should differentiate as cartilage (result c in Fig. 11B). On the other hand, presumptive myogenic precursor cells that have already become determined are expected to differentiate as muscle (results a and b in Fig. 11B). By comparing the timing of appearance of such determined cells with the onset of expression of MDFs, it should be possible to establish whether MDF expression coincides with myogenic determination *in vivo*.

This experimental design was executed for the somite dorsomedial quadrant (DMQ), the region of the somite known to contain the precursor cells for the epaxial myotome (see above). Differentiated myocytes were detected in 25% of grafts from the DMQ of the segmental plate, and this percentage increases with somite age, reaching 100% by somite stage XI (Fig. 8). This progressive increase in "determined" muscle precursor cells is consistent with (but not conclusive proof of; see George-Weinstein *et al.*, 1998) the notion that they are induced by extrinsic

---

Figure 11 Notochord challenge model for assessing determination of embryonic muscle precursors. (A) Experimental logic. Early paraxial mesoderm cells respond to dorsal or ventral signals to be induced to develop toward myogenic or chondrogenic development (upper and lower pathways, respectively). If the signal environment is changed early during specification both groups of cells have the capacity to switch developmental fates and differentiate into the alternative tissue type. At some point, however, the fate of premyogenic cells becomes determined, and they continue to develop into muscle despite changing their signaling environment to expose them to ventral signals. This loss of signal responsivity is taken as the marker for determination in this assay. (B) Experimental design. Four steps in performing a notochord challenge transplantation experiment are illustrated. Steps 1–3 involve removal of the medial half of a chick host somite, implantation of the dorsomedial quadrant of a stage II quail somite, and implantation of two chick notochords dorsal to the implanted quail somite fragment. Analysis of the results, in step 4, illustrates the three types of results obtained when the somite dorsomedial quadrant is analyzed. A breakdown of these results for different stages of somite development is given in Fig. 7.

9. Determination and Morphogenesis in Myogenic Progenitor Cells

influences. The short developmental interval between the earliest appearance of such determined cells in the segmental plate precedes the overt expression of the first MDF at somite stage II (100–200 min), and the increasing frequency and size of myoclusters in later staged somites tend to support the conclusion that MDF phase I gene expression and myogenic determination might be directly related. In other words, the segmental plate may contain isolated MDF-positive cells that are undetectable by conventional *in situ* hybridization techniques but which, nevertheless, show up in the notochord challenge assay as individual myocytes within a myocluster. On the other hand, the low frequency of detection of determined cells from the segmental plate (25%) may underestimate the frequency with which such cells arise in the segmental plate because of the dependence of the development of these cells on the proximity to spinal nerve. There was a 100% coincidence between myocluster and spinal nerve location consistent with the conclusion that the former depends on the latter. Myocluster differentiation can be detected within hours of initiating the notochord challenge assay (Williams and Ordahl, 1997), suggesting that the frequency with which myoclusters are observed at the earliest stages may be related to the accessibility of such clusters to spinal nerve later. This tends to undermine the notion outlined above that myocluster frequency may be related to MDF expression in a semiquantitative manner. Thus, while the notochord challenge assay more clearly defines the timing of myogenic determination in the somite, the issue of whether determination precedes or follows the onset of MDF phase I remains equivocal.

The notochord challenge analysis of the somite DMQ also revealed unexpected and potentially important information about the onset of appearance of myogenic progenitor cells possessing different types of myogenic memory. Two observations suggested that myogenic stem cells with increasing mitotic capacity progressively appear in a somite-stage-dependent fashion. First, although the sizes of transplanted somite DMQ fragments were similar, the mass of myogenic tissue obtained using later-staged DMQ fragments was far greater than that from early-staged DMQs. Second, DNA labeling showed that myogenic precursor cells within the DMQ had passed through the S phase of the cell cycle prior to differentiating in the presence of the notochord challenge (Williams and Ordahl, 1997). A third observation suggests that these stem cells also had morphogenetic capacity characteristic of progenitor stem cells because myogenic cells from early staged somites formed "myoclusters" that consisted of 50 or fewer disorganized, mononucleated myocytes while later somites gave rise to multinucleated myofibers that were highly organized into parallel fiber bundles. The organizational characteristics of these muscle tissue masses resemble that of muscle formed at later stages of development and are consistent with the birth of these myogenic progenitor cells occurring later than the onset of MDF phase I and possibly coincident with the timing of MDF phase II. It should be pointed out, however, that the exact location of these myogenic progenitor cells is not yet evident from the notochord challenge experiment (see below).

Finally, and surprisingly, the notochord challenge analysis of the somite DMQ did not indicate that more widespread phenotypic restriction, the negative attribute of cell determination (Fig. 1), accompanied the onset of myogenic determination. Regardless of the somite stage from which DMQ fragments were taken, each gave rise to abundant cartilage under the influence of the notochord challenge environment. Does this reflect chondrogenic potential within the dermomyotome epithelium, or does cartilage potential in DMQ grafts result from a small number of adherent cells from other regions of the somite, for example the more ventral sclerotome? Histological examination of dissected DMQs from early stage somites (e.g., stage IV) showed little evidence of contamination (Williams and Ordahl, 1997) consistent with the hypothesis that, at least at this early stage of somite development, the DMQ epithelium contains both determined myogenic precursor cells and undetermined cells. At later somite stages (XI–XIII), however, contamination by a small number of mesenchyme cells (represented by white cells in Fig. 6) cannot be ruled out so the question of whether the epithelial cells within the somite dermomyotome remain pluripotent (with respect to muscle and cartilage forming capacity) remains open. Conclusive information regarding the source of chondrogenic potential within the somite DMQ, and where within the dermomyotome "determined" cells reside is currently unavailable. Furthermore, as discussed above, recent experiments indicate that the dermomyotomes of thoracic somites have a chondrogenic fate (Kato and Aoyama, 1998).

IV. Gene Expression and Determination in Migratory Limb Muscle Precursor Cells

Because epaxial and hypaxial muscles are derived, respectively, from the medial and lateral halves of the somite it is clear that MDF expression in the early somite is restricted to cells that give rise to only epaxial muscle. By contrast, MDF expression in forelimb muscle precursor cells does not begin until 48 hr after MDF expression first occurs in the medial halves of forelimb-level somites (Fig. 8). This raises two different, but related questions. First, does expression of other genes mark such cells prior to MDF expression, ideally during their migration from somite to limb bud? Second, when do limb muscle precursor cells become determined?

A. Pax3 Gene Expression Marks Mesoderm Cells in the Early Stages of Myogenic Development

A marker for migratory limb muscle precursor cells has potential, particularly in the case of mammalian embryos where the origin of limb muscle in mammals remains somewhat controversial. In fact, experiments in which marked somites were transplanted into cultured mouse embryos failed to fully replicate the results of quail–chick transplantation in that somite cells were not present in limb muscle (Beddington and Martin, 1989; Sze and Paulin, 1995), leading to a measure of skepticism that such migration might even occur in mammalian embryos. In related experiments, Tajbakhsh and Buckingham (1994) elegantly showed that mouse limb buds contained muscle precursor cells prior to expression of myf5. Evidence that limb muscle was derived from migratory cells from the mouse somite finally came via a circuitous route that was triggered by the careful analysis of the development of a naturally occurring mouse mutant, Splotch. The Splotch mouse had been studied for over 40 years before its mutation was discovered to be in the gene encoding Pax3 (Epstein *et al.*, 1991), a transcription factor implicated in development of the central nervous system (Goulding *et al.*, 1991). Homozygous Splotch embryos die during development while heterozygotes are viable with splotch coat coloring, hence the name. It took careful examination of the embryos and fetuses of matings between heterozygote Splotch parents to reveal that the limb muscle primordia were absent in homozygous splotch littermates, whereas the development of back and body wall muscles was relatively normal (Franz *et al.*, 1993). This seminal link between the Pax3 gene and limb muscle development led several groups to analyze Pax3 expression in somite and limb muscle development (Bober *et al.*, 1994; Goulding *et al.*, 1994; Williams and Ordahl, 1994).

Pax3 expression in the paraxial mesoderm precedes that of MDFs (see Figs. 5 and 8). In the newly formed somite, Pax3 expression is restricted to the dorsal epithelium, the future dermomyotome. Several hours later in development, after the stage IV somite has undergone dorsoventral patterning, three domains within the dermomyotome can be defined with respect to the level of Pax3 gene expression (Williams and Ordahl, 1994). First, the lateral half of the dermomyotome, which contains cells destined to migrate to the limb buds, expresses high levels of Pax3 mRNA (dark green). Second, the medial half of the dermomyotome, which contains cells destined to form the muscles and dermis of the epaxial domain, expresses Pax3 at relatively lower levels at these early stages of somite development (light green). The third domain comprises the MDF-positive cells at the extreme medial edge of the dermomyotome (yellow) that have extinguished Pax3 gene expression (Williams and Ordahl, 1994). Because the myoD-1-positive cells in the dermomyotome medial lip are the direct precursors of the early epaxial myotome (see above), and because these cells were expressing Pax3 earlier, it can be concluded that Pax3 gene expression marks all myogenic precursor cells prior to the onset of MDF expression. In older somites Pax3 expression is restricted to the dermomyotome and is excluded from the myotome as well as the medial lip.

In older embryos the dermomyotome epithelium disintegrates, releasing mesenchyme cells that express Pax3 at relatively high levels (Williams and Ordahl, 1994). In wing-level somites, these cells are derived from the medial half of the dermomyotome and therefore can be considered to have upregulated the Pax3 gene concomitant with their mesenchymalization in a manner that could be comparable to that of the lateral half-dermomyotome cells that migrated earlier to the limb. The fate(s) of these newly mesenchymalized cells is incompletely understood. Some are known to eventually migrate dorsally, to the area immediately subjacent to the skin ectoderm (the true "dermatome") and will form the dermis of the skin in the epaxial domain (see also Fig. 6). Others may also migrate in the opposite direction, into the developing myotomal musculature, to contribute to the future growth of these muscle masses during fetal development. These cells may also contribute to the satellite cells present in adult muscle.

In the limb bud somites, Pax3-expressing cells leave the lateral margin of the dermomyotome and migrate into the limb (Figs. 5A, 8C). Initially, the Pax3-positive cells appear to be widely dispersed in the limb bud, but within a few hours these cells coalesce into two distinct groups, the dorsal and ventral premuscle masses, that, respectively, prefigure the future extensor and flexor compartments of the forelimb. Notably, these cells are still not detectably expressing MDFs at this time (Fig. 8C). Two days after emigration from the somite (HH 24)* myogenic precursor cells in the limb bud initiate MDF expression within the broader domain of Pax3 expressing cells. As myogenic differentiation proceeds in the developing limb bud the Pax3 *in situ* hybrization signal progressively disappears consistent with the notion that Pax3 gene expression is extinguished concommitant with the onset of differentiation. Because myoblasts and myocytes are intermingled in the limb bud there is no anatomical separation between the Pax3 and MDF expression domains as there is in the somite dermomyotome and myotome.

Additional markers have shown up for somite-derived myogenic precursor cells although none are expressed through the full range of Pax3 gene expression in these cells. The related gene Pax7 is expressed in the somite dermomyotome and migratory cells (Jostes *et al.,* 1991) as are the bHLH genes Sim 1 and 2 (Fan and Tessier-Lavigne, 1994; Pourquié *et al.,* 1996) and the ladybird gene (Dietrich *et al.,* 1997, 1998). In addition, the cMet receptor is expressed in migratory muscle precursor cells, and genetic knockout of the cMet receptor blocks migration of limb muscle precursor cells (Bladt *et al.,* 1995) in a fashion similar to that in the Pax3$^{-/-}$ Splotch mouse. The lack of migration of precursor cells in these two cases raises the interesting question of whether these mutations affect muscle specification per se or migration, with secondary effects on those muscles that are migration-dependent. An elegant test of this question was designed by Goulding and co-workers, who showed that muscle precursor cells from the lateral halves of brachial somites from Splotch embryos could differentiate into muscle after transplantation into chick limb buds (Daston *et al.,* 1996). Recently, Pax3 has been implicated as a potential myogenic specification gene: forced expression of Pax3 via viral infection induces muscle formation in tissues that normally do not give rise to muscle (Maroto *et al.,* 1997), and Splotch/myf 5 double mutants show severely reduced trunk and limb muscle (Tajbakhsh *et al.,* 1997). Thus, the precise role of Pax3 in these early phases of myogenic development remains a subject of contro-

*Hamburger Hamilton Staging (Hamilton, 1952).

versy and speculation. Finally, although Pax3 gene expression marks somite-derived myogenic precursor cells, it is not a marker for myogenic precursor cells within the head mesoderm (Hacker and Guthrie, 1998).

B. Early Experiments on Limb Muscle Determination

The results of the notochord challenge experiments outlined above pertain only to the precursors of the epaxial myotome because they were restricted to the somite DMQ. What is the timing of determination of cells within the somite dorsolateral quadrant (DLQ), the region of the somite containing the hypaxial muscle precursor cells? Kieny and co-workers prepared chimeric chick embryos into which quail somites had been transplanted in order to "program" the developing wing buds with quail muscle precursor cells. When the quail cells from such chimeric limb buds were back-transplanted to the somite region some cells remigrated to the limb bud and, after doing so, differentiated exclusively into muscle (Mauger and Kieny, 1980), indicating that these cells were determined both for migration and for their ultimate cell type. Wachtler and Christ performed similar experiments which indicated that somite-derived cells in the limb buds of HH 19 embryos were already determined to form muscle (Wachtler et al., 1981).

However, when the chick and quail cells within such chimeric limb buds were removed, dispersed by enzymatic digestion, and then repacked into limb ectoderm jackets that were then allowed to develop further, a surprising result ensued (Kieny et al., 1981) (see Fig. 12). While the vast majority of quail cells formed muscle, as expected, a very small percentage of cells were found within cartilage tissue (see Fig. 12F). The question of whether such cells were simply undifferentiated muscle precursor cells trapped within cartilage tissue or were authentically "trans differentiated" cartilage cells was not directly addressed, but the quail cells were morphologically indistinguishable from their chicken chondrocyte neighbors. Given these results, it will be interesting to determine the timing of determination in these precursors using the notochord challenge assay.

V. A Cellular and Molecular Framework for Understanding Myogenic Tissue Development

A model that fully explains and integrates molecular and cellular events of myogenic specification, determination and myotome formation continues to be elusive. The experimental embryonic strategies reviewed in this chapter provide information regarding the cell and tissue dynamics related to early myogenic determination and differentiation. This information allows certain conclusions to be made but, more importantly, raises some questions that might lead to a deeper integration between cellular and molecular models of myogenic development.

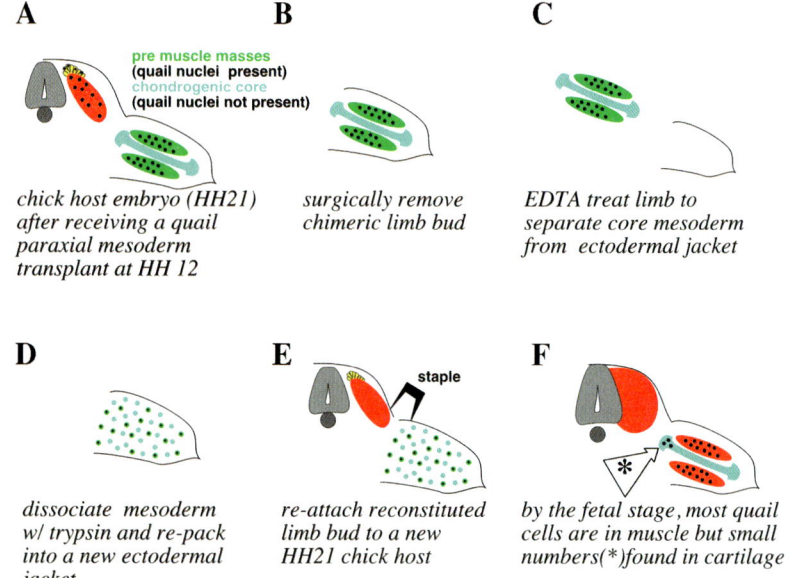

Figure 12 Are limb muscle precursor cells myogenically determined? Diagrammatic representation of a classic experiment performed by Madeleine Kieny and co-workers to analyze the stability of the determined state in limb muscle precursor cells (Kieny *et al.*, 1981). (A) The distribution of quail nuclei (magenta) in a HH 21 chick embryo after having received a quail segmental plate transplant at stage 12. After removal of the chimeric wing bud (B), the mesoderm and ectoderm are separated (C), and the mesoderm component is dissociated, repacked into a new ectodermal jacket (D), and reattached to a new HH 21 chick host (E). After further development (F), quail nuclei are predominantly found in muscle cells but a small number are also found in cartilage (magenta arrow).

A. Morphogenetic Movements of Myogenic Precursor Cells

Using the information reviewed above it is possible to begin to reconstruct the developmental history of cells that are precursors to the myoblast cells which form the early epaxial myotome fibers and myogenic progenitor cells which will give rise to differentiated muscle at some later time in the epaxial domain (see Fig. 13). These cells have been shown to be located in the medial half of the segmental plate and the newly formed somite (Ordahl and Le Douarin, 1992) and in the dorsomedial quadrant of later somites (Denetclaw *et al.*, 1997; Williams and Ordahl, 1997). Precursor cells for the medial half of the segmental plate have been shown to reside within Hensen's node (Selleck and Stern, 1991; and see Chap. 1 by Tam *et al.*, this volume), the ridge of epiblast epithelium where cells are undergoing an epitheliomesenchymal transition to form, among other things, the paraxial mesoderm subjacent to the epiblast and ectoderm epithelia. This is the first epitheliomesenchymal transition of early epaxial myotome precursor cells (α in Fig. 13A)

9. Determination and Morphogenesis in Myogenic Progenitor Cells 355

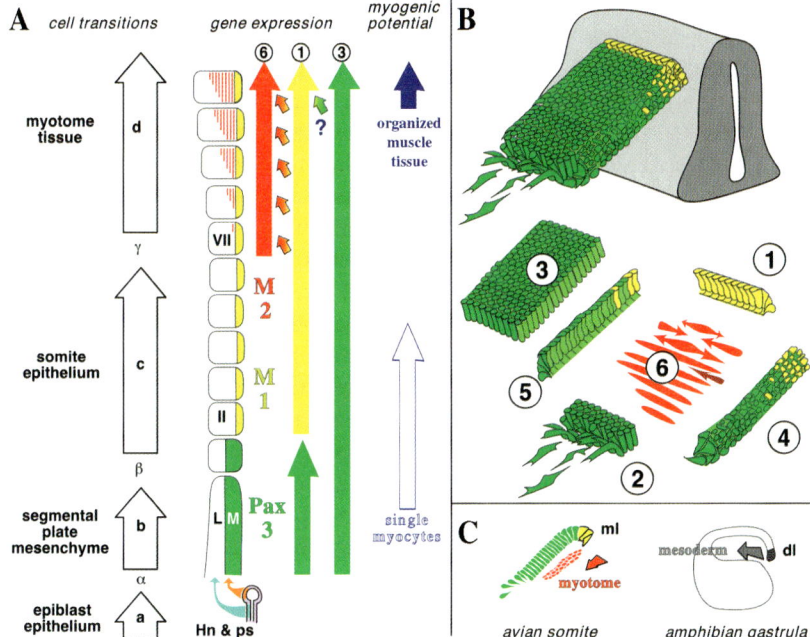

Figure 13 Summary of development of the myotome and myogenic precursor cells. (A) The relationship between transitions in cell organization, gene expression, and the birth of myogenic progenitor cells. Three key cellular transitions are indicated: α, the epithelial–mesenchymal transition of epiblast cells in Hensen's node (state a) entering the segmental plate region of the paraxial mesoderm (state b); β, the mesenchymal–epithelial transition as cells in segmental plate epithelialize to form somites (state c); and γ, the epithelial–myocyte transition as cells within the dermomyotome epithelium (state c) translocate to the myotome layer and differentiate in an organized myocyte array. (B) An expanded schematic view of the dermomyotome and myotome of a stage XVII somite depicting six subregions of the dermomyotome and myotome: (1) the medial lip of the dermomyotome containing MDF-expressing cells and the most active site of myotome cell production in the dermomyotome; (2) the lateral margin where migratory cells leave the dermomyotome to populate the muscle beds destined to form within the adjacent limb bud; (3) the central sheet region of the dermomyotome epithelium that may carry stem cell and/or precursors which contribute to later waves of myogenesis; (4) the cranial lip of the dermomyotome, the medial portion of which gives rise to myotome cells at a reduced rate as comparison to the medial lip; (5) the caudal lip of the dermomyotome; and (6) the myotome, which contains both fully mature myocytes that span between the recurved edges of the cranial and caudal lips of the dermomyotome and immature myocytes that are in the process of elongating toward one or both of those lips. A seventh population has not yet been precisely defined but are the cells that translocate between the dermomyotome and myotome layers. (C) Comparison between the medial lip (ml) of the avian somite and the dorsal lip (dl) of the amphibian embryo blastopore illustrating similarities in the cellular transitions involved.

as they move from the epithelial organization of the epiblast (a) to the mesenchymal organization of the segmental plate (b). The second is a mesenchymal–epithelial transition (β) as epithelial somites form at the tip of the segmental plate. And finally, a third de-epithelialization transition (γ) occurs as the myotome precursor cells within the medial lip region translocate to the subjacent myotome layer.[6] The specification of sclerotome, dermatome, and the migratory limb muscle precursor cells also involve epithelial–mesenchymal γ transitions.

Each of these transitions is associated with changes in developmental potential and gene expression. Transition α involves the localized transcription of a number of early developmental genes and the allocation of epiblast cells to the embryonic primordial tissue (for an authoritative review, see Smith and Schoenwolf, 1998) including the paraxial mesoderm which then activates the gene encoding Pax3 (Williams and Ordahl, 1994) and the timed waves of expression of the gene c-Hairy (see Chap. 4 by Pourquié, this volume). Shortly after somite budding, transition β, cells within the somite dorsomedial quadrant repress Pax3 and activate MDF phase I expression. Transition γ is accompanied by tissue-primordia-specific changes in gene expression. In the cells fated to become early epaxial myotome cells, the onset of expression of Pax3, MDF phase 1, and MDF phase 2, respectively, occur shortly after each of the cellular transformations noted above (see Fig. 13B). In contrast, however, Pax3 expression is continuous both before and after the epitheliomesenchymal transition preceding the migration of limb muscle precursor cells, even though a similar MDF phase 1 and MDF phase 2 could be hypothesized to occur much later in development when somite-derived mesenchyme cells reside within the limb premuscle masses (see Fig. 8). Thus, the degree to which such cellular transformations might be coupled to changes in gene expression remains a subject of intense scrutiny (see, e.g., Chap. 4 by Pourquié and Chap. 6 by Rawls *et al.*, this volume).

As outlined in Section I, early embryologists recognized that the morphogenetic movements of cells during the epitheliomesenchymal transitions of gastrulation are coupled to the fundamental determinative events in development. It is attractive to speculate that the subsequent transitions in cellular organization noted above are similarly interlinked with the changes in gene expression and developmental potential in such a manner as to be codeterminative events in the generation and organization of the early myotome primordia. It is perhaps no coincidence, therefore, that the cellular transformations which occur at the dorsal lip of the amphibian blastopore (Keller, 1985; Keller *et al.*, 1985; Moore *et al.*, 1995; Minsuk and Keller, 1996; Winklbauer and Keller, 1996) resemble those occurring as myotome cells are born at the dorsomedial lip of the dermomyotome of chick embryos (Fig. 13C). In both instances, cells emigrate from the recurved edge of an epithelium or organized cell sheet. Soon after emigrating, mesenchyme cells in

---

[6] This last transition may not be a true epithelial-mesenchymal transition because the early myotome has some epithelial characteristics, one being a basement membrane continuous with that of the dermomyotome (Christ and Ordahl, 1995; also see Chap. 9 by Brand-Saberi and Christ, this volume).

9. Determination and Morphogenesis in Myogenic Progenitor Cells 357

the midline of the early amphibian gastrula undergo a process known as "convergent extension" to form and extend the notochord (Keller *et al.*, 1985). Similarly, newborn myotome cells emerging underneath the medial lip of the avian embryo dermomytome undergo a bipolar convergent extension to form a densely packed myotome layer beneath the dermomyotome. The fact that newly born myotome cells have just undergone a developmental transition (pre-muscle to muscle) is further consistent with the developmental transition occurring at gastrulation during which cells begin the process of specification for a variety of embryonic and adult cell lineages. It is interesting to speculate that the physical constraints placed on the epithelial cells within the dermomyotome medial lip and dorsal lip of the blastopore may play some role to affect or constrain the developmental transitions occurring at those sites.

B. Growth Dynamics in the Early Embryonic Myotome

The third cellular transition (γ in Fig. 13A) is also coupled to the process of growth occurring within the epaxial domain of the early embryo. Superficial examination of a chick or mouse embryo (at embryonic days 3 and 9.5, respectively) reveals that the dermomyotomes of older somites are rectangular in shape (i.e., elongated in their mediolateral dimension) as compared to more recently formed (caudal) somites, which are smaller and essentially square in outline (see Fig. 2). The mediolateral expansion correlates with the deposition of myotomal myocytes in a lateral-to-medial direction and indicates that the growth of the somite dermomyotome is coupled with the formation of the early myotome. This growth is further coupled as a tangential vector relative to the dorsoventral vector of axial growth (Fig. 9) that is, at least in part, a result of dorsal expansion of the neural tube due to cell division (Fowler and Watterson, 1953; Gasser, 1979). Together these two growth processes account for a substantial fraction of the cell mass that fills the epaxial domain of the developing embryo (see Fig. 2).

What might be the driving force for the mediolateral expansion of the dermomyotome? A definitive answer is unknown but multiple forces may be at work. Possible contributors are the nascent myotome cells themselves that could exert an active role through a convergent extension mechanism (Keller *et al.*, 1985). The bipolar elongation of newly born myotome fibers (Figs. 4, 9, and 13) is a highly stereotaxic example of convergent extension that could act to expand growth in a medial direction by mechanisms that are essentially identical to those by which convergent extension has been proposed to drive the axial elongation of the notochord (Moore *et al.*, 1995). A second potentially important component is channeling of dermomyotome expansion from mitotic activity in a mediolateral direction through dynamic instability along its medial margin (and the medial portion of its cranial margin) through depletion of myoblastic cells from this region via translocation to the myotome layer with concomitant disruption of cell–cell interactions and possibly disruption of cell–extracellular matrix interactions. The relative stability of the cranial and caudal borders, possibly a mechanical consequence of their close juxtaposition to adjacent somites, would therefore direct expansion in the medial direction toward the unstable border.

Is it not yet possible to evaluate the contributions of these two potential mechanisms for the medial-ward growth of the somite, but it may soon be possible to glean information in this regard from genetic knockout experiments involving MDF and other genes. For example, homozygous Myf5 knockout mice are reported to be smaller than their littermates (Braun et al., 1992), consistent with the notion that retardation of early epaxial, and later, hypaxial myotome formation leads an overall diminution in body size. On the other hand, images of early mouse embryos bearing homozygous knockouts of the *myf5* gene suggest at least some medial–lateral expansion of the dermomyotomes between 8.5 and 10 days of gestation (see Chap. 15 by Tajbakhsh and Buckingham, this volume). Quantitative morphometric analysis of dermomyotome expansion in these and other mutants may therefore be useful in understanding of the cellular mechanics of early embryonic growth and morphogenesis of the somite.

C. Myogenic Organizers and Progenitor Stem Cells

As outlined in Section I of this chapter, although early embryologists distinguished between the cell differentiation and morphogenetic aspects of differentiation they were primarily concerned with the harmonious development of whole organisms and organ systems. Spemann used the term "organizer" to describe the ability of embryonic tissue fragments to re-create the "wholeness" or totality of the integrated structure and function of biological tissues and organs (Spemann 1938; see Smith and Schoenwolf, 1998, for a recent discussion of this). In the challenge assays described previously, somite DMQ fragments from earlier stages gave rise to muscle consisting of a few isolated mononucleate cells with myosin-positive but unorganized cytoplasm (sp, II, IV; see Fig. 8). In contrast, DMQs from more mature somites (stages XI–XIII) consistently showed the ability to generate substantial amounts of well-organized muscle tissue. The transition from unorganized cells to organized tissue in the challenge assay may reflect the emergence of a minimalistic or quantal aspect of the broader organizer activity indicated above and one that may indicate a shift of myogenic progenitor cell lineages within the DMQ.

Where within the somite do these two different myogenic progenitor cell lineages reside? At least six anatomic regions of the dermomyotome and myotome can be identified and surgically isolated from a stage XVIII wing-level somite (Fig. 13C). The medial lip (number 1 in Fig. 13B) is the coincident site of direct precursors to myotome cells and localized MDF(phase 1)-positive cells, consistent with the notion that these two are one and the same. It is not yet known, however, whether these cells are (1) myogenic progenitor cells possessing self-renewing (stem cell) capacity or (2) myoblast cells destined to differentiate as muscle but lacking self-renewing capacity. Relatively few new daughter cells could be generated during the 500-min interval between the first detectable appearance of MDF in the somite DMQ (stage II) and the first appearance of differentiated myocytes (stage VII), but this lip continues to produce myotome cells for extended periods. Therefore, the cells that translocate into the myotome layer must be replenished either by stem cell activity within the lip itself or from cells in the immediately adjacent dermomyotome epithelium (region numbers 1 and 3, respectively, in Fig. 13C). Early myotome cells are mononucleate and have a simple bipolar organization parallel to the embryonic axis and attached to the recurved cranial and caudal lips of the dermomyotome (see Figs. 4, 9C, and 13B). Should the morphogenesis of these early myotome cells depend on the organization of the dermomyotome, it is not difficult to conclude that notochord signals which disrupt those signals also disrupt the morphogenetic organization of those individual cells in the notochord challenge assays described previously.

The epaxial muscle cells that develop later differ from the early myotome cells in several important ways: first, these fibers are long, spanning multiple axial segments, and highly multinucleated; second, they are hierarchically organized in fiber bundles; and, third, a substantial fraction are oriented at an angle to, rather than parallel with, the body axis. Myogenic progenitor cells that emerge in late stage somite DMQs after the notochord challenge give rise to muscle tissue with these basic characteristics. Interestingly, the observation that muscle tissue in the notochord challenge was always oriented at an angle to the body axis (Williams and Ordahl, 1997) might also reflect characteristics intrinsic to the late appearing cells. The identity of the cells responsible for the organizational component of epaxial myogenesis can only be speculated upon at this time. Experiments with limb muscle development suggest that the connective tissue component of these muscles is responsible for its morphogenesis (Chevallier and Kieny, 1982). The embryonic origin of muscle connective tissue in the epaxial domain is unknown, and it is therefore possible, at least theoretically, that this later wave of progenitor cells has the potential to generate both cell types. On the other hand, there is no reason to rule out the birth of a second progenitor cell population within the somite; one that generates myogenic connective tissue. The precise location of these cells within the dermomyotome remains to be determined.

Nevertheless, the ability of these cells for extensive mitotic expansion, tissue organization and orientation suggests that they can be distinguished as a class of cells within the embryo that directly produce the cells which make up the tissue and organ rudiments and their descendants. The use of the term progenitor stem cells to designate such cells is in accordance with nomenclature developed elsewhere in this book (see Chap. 17 by Eloy-Trinquet *et al.,* this volume) and helps to distinguish them from other types of cells. Myogenic progenitor cells, therefore, are the type of progenitor stem cells that give rise to muscle tissue during embryonic development. A central characteristic of stem cells is the capacity for self-renewal, a process by which some daughters retain the mitotic and differentiative capacity of the mother stem cell (see Chap. 17 by Eloy-Trinquet *et al.,* this volume). *Embryonic stem cells* are stem cell lines derived from very early embryos and which are multipotential, retaining the ability to give rise to multiple cell types (see, e.g., Anderson, 1994; Galimi *et al.,* 1994; Bronner-Fraser, 1995; Gage *et al.,* 1995; Golden and Cepko, 1996; Ikeshima-Kataoka *et al.,* 1997). At the other end of the spectrum, and as discussed in Section I, adult tissues also contain stem cells (such as satellite cells) that are unipotential and capable of replacing one cell type that has been lost or injured within an organ or tissue. Progenitor stem cells develop within the primary organ rudiments of the developing embryo and differ from embryonic stem cells in that their developmental potential is restricted to one (or a very few) differentiated end point cell types (i.e., determined stem cell lines). Progenitor stem cells are further distinguished from their adult counterparts in that they have the morphogenetic ability to interact with other embryonic cells and tissues to build tissue architecture *de novo.* The progenitor stem cells of skeletal muscle, or myogenic progenitor cells, themselves probably represent multiple, distinct embryonic cell lineages such as those giving rise to epaxial and hypaxial muscle (Ordahl and Le Douarin, 1992; Ordahl, 1993; Ordahl and Williams, 1998) and those giving rise to the slow and fast muscle lineages (Stockdale, 1992). It remains to be determined what lineage relationships do or do not exist between the multiple myogenic progenitor populations that arise during embryonic development.

Dedication

This chapter is dedicated to the memory of Harold Weintraub, one of the principal authors of the myoD model of myogenic determination and memory, who died prematurely and tragically on March 28, 1995.

Acknowledgments

The authors would like to thank Bodo Christ for contributing the photographic copy of Fischel's somite drawings shown in Fig. 2, Howard Holtzer for agreeing to the republication of his original fluorescent antibody experiments shown in Fig. 4, Olivier Pourquié and Kathryn Tosney for communicating unpublished information, and Rebecca Greene for expert technical assistance in the thoracolumbar half-somite transplant experiments. The experimental embryological work in this review was supported by grants to C.P.O. from The Muscular Dystrophy Association of America and the National Institutes of Health (AR59693).

VI. Appendix: Key Word Guide

With malice toward none and apologies to all, we tender this brief guide to terms as they are used in this chapter.

developmental potential: The ability of embryonic tissues to form organized structures, tissues, and organs. The Spemann–Mangold "organizer" (dorsal lip of amphibian blastopore) could direct the assembly of an essentially complete embryo (Spemann and Mangold, 1924) and had the "harmonious" properties of the "whole" organism (as discussed further in Spemann, 1938). Holtfeter, a student of Spemann (and a U.S. resident for many decades), showed that this developmental potential was parsed out during development giving rise to tissues with partial or restricted potential (Holtfreter, 1944, 1945; Holtfreter and Hamburger, 1955). The key issues then, as now, surround the mechanisms that direct such diminution of developmental potential. Pluripotentiality, multipotentiality, and unipotentiality therefore refer to perceived stages in this parsing process.

embryonic stem cells: A general term for embryonic cells that are multipotential stem cells, able to regenerate themselves (as in all stem cells), and able to give rise to daughters that differentiate into multiple cell types.

experimental embryology: The analysis of embryonic events, phenomena, and concepts through experimental means whose roots can be traced to the late nineteenth century.

memory: A recent (and clearly anthropomorphic) term to describe the quality of cells and tissues to reproduce, regenerate, or grow specific biological structures. The term "cellular memory" refers to the varied abilities of cells to express a specific cellular phenotype (see text for examples).

morphogenesis: The emergence of form and ultimately function from "undifferentiated" embryonic primordial tissues. This involves the morphogenetic movements of embryonic cells upon one another and within the extracellular matrix as well as changes in cell shape.

myogenic determination: Developmental potential from the viewpoint of muscle tissue; the embryonic process by which pluripotent embryonic cells produce daughters that produce only muscle tissue.

precursor cell: A neutral term for cell lineage or fate in development that carries no implication regarding developmental potential of either precursor or product.

progenitor stem cells: A proposed general term for cells within the embryonic tissue primordia (anlage) that, through stem cell activity, produce daughters with two unique and essential properties: first, *unipotentiality* with respect to a given tissue or organ and, second, *morphogenetic competence* to organize that tissue or organ. In the case of muscle primordia, the term *myogenic progenitor cell* signifies this dual capacity as well as distinguishing it from a partially synonymous term "myogenic stem cell" that is often used heteronymously for myoblastic stem cells such as satellite cells. This terminology is also consistent with the use of progenitor cell as defined in Chap. 17 by Eloy-Trinquet *et al.,* this volume.

stem cells: A general term for cells capable of undergoing "asymmetric" mitoses producing daughter pools that contain two cell types: blast cells (e.g., myoblast cells), which give rise to differentiated progeny (with or without subsequent cell division), and additional stem cells. Stem cells may be multipotential or unipotential. Myoblastic cells isolated from adult muscle (satellite cells) or from fetal muscle tissue (fetal myoblasts) represent unipotential (myoblastic) stem cells that can even be immortalized as such.

References

Anderson, D. J. (1994). "Stem cells and transcription factors in the development of the mammalian neural crest." *FASEB J.* **8,** 707–717.
Antin, P., Karp, G., and Ordahl, C. (1991). Transgene expression in QM cells. *Dev. Biol.* **43,** 122–129.
Aoyama, H., and Asamoto, K. (1988). Determination of somite cells: Independence of cell differentiation and morphogenesis. *Development* **104,** 15–28.
Bardeen, C. R. (1900). The development of the musculature of the body wall in the pig. *Bull. Johns Hopkins Hosp.* **9,** 367–399.
Beddington, R., and Martin, P. (1989). An *in situ* transgenic enzyme marker to monitor migration of

cells in the mid-gestation mouse embryo: Somite contribution to the early forelimb bud. *Mol. Biol. Med.* **6,** 263–274.

Bellairs, R. (1963). The development of the somites in the chick embryo. *J. Embryol. Exp. Morphol.* **11,** 691–714.

Bischoff, R. (1994). The satellite cell and muscle regeneration. In "Myology" (A. G. Engle and C. Franzini-Armstrong, Eds.), pp. 97–118. McGraw-Hill, New York.

Bladt, F., Riethmacher, D., Isenmann, S., Aguzzi, A., and Birchmeier, C. (1995). Essential role for the c-met receptor in the migration of myogenic precursor cells into the limb bud. *Nature* **376,** 768–771.

Bober, E., Franz, T., Arnold, H.-H., Gruss, P., and Tremblay, P. (1994). Pax-3 is required for the development of limb muscles: A possible role for the migration of dermomyotomal muscle progenitor cells. *Development* **120,** 603–612.

Bonner, P., and Hauschka, S. (1974). Clonal analysis of vertebrate myogenesis I: Early developmental events in the chick limb. *Dev. Biol.* **37,** 317–328.

Borman, W. H., and Yorde, D. E. (1994). Analysis of chick somite myogenesis by *in situ* confocal microscopy of desmin expression. *J. Histochem. Cytochem.* **42,** 265–272.

Borman, W. H., Urlakis, K. J., Jr., and Yorde, D. E. (1994). Analysis of the *in vivo* myogenic status of chick somites by desmin expression *in vitro*. *Dev. Dyn.* **199,** 268–279.

Brand-Saberi, B., Ebensperger, C., Wilting, J., Balling, R., and Christ, B. (1993). The ventralizing effect of the notochord on somite differentiation in chick embryos. *Anat. Embryol.* **188,** 239–245.

Braun, T., Buschhausen-Denker, G., Bober, E., Tannich, E., and Arnold, H.-H. (1989). A novel human muscle factor related to but distinct from MyoD1 induces myogenic conversion in 10T1/2 fibroblasts. *Embo J.* **8,** 701–709.

Braun, T., Rudnicki, M. A., Arnold, H.-H., and Jeanisch, R. (1992). Targeted inactivation of the muscle regulatory gene *myf-5* results in abnormal rib development and perinatal death. *Cell* **71,** 369–382.

Bronner-Fraser, M. (1995). Origins and developmental potential of the neural crest. *Exp. Cell Res.* **218,** 405–417.

Byrnes, E. F. (1898). Experimental studies on the development of limb-muscles in Amphibia. *J. Morphol.* **14,** 105–140.

Caplan, A. I., and Koutroupas, J. (1973). The control of muscle and cartilage development in the chick limb: The role of differential vascularization. *J. Embryol. Exp. Morphol.* **29,** 571–583.

Caplan, A. I., and Ordahl, C. P. (1978). Irreversible gene repression model for control of development. *Science* **201,** 120–130.

Chevallier, A., Kieny, M., and Mauger, A. (1977). Limb–somite relationship: Origin of the limb musculature. *J. Embryol. Exp. Morphol.* **41,** 245–258.

Chevallier, A., and Kieny, M. (1982). On the role of the connective tissue in the patterning of the chick limb musculature. *Wilhelm Roux's Arch. Dev. Biol.* **191,** 277–280.

Christ, B., and Ordahl, C. P. (1995). Early stages of chick somite development. *Anat. Embryol.* **191,** 381–396.

Christ, B., Jacob, H., and Jacob, M. (1974). Uber den ursprung der flugelmuskulature. *Experientia* **30,** 1446–1449.

Christ, B., Jacob, H. J., and Jacob, M. (1977). Experimental analysis of the origin of the wing musculature in avian embryos. *Anat. Embryol.* **150,** 171–186.

Christ, B., Jacob, H., and Jacob, M. (1978). On the formation of the myotomes in avian embryos. An experimental and scanning electron microscope study. *Experientia* **34,** 514–516.

Cornelison, D. D. W., and Wold, B. J. (1997). Single-cell analysis of regulatory gene expression in quiescent and activated mouse skeletal muscle satellite cells. *Dev. Biol.* **191,** 270–280.

Cossu, G., Kelly, R., Tajbakhsh, S., Di Donna, S., Vivarelli, E., and Buckingham, M. (1996). Activation of different myogenic pathways: myf-5 is induced by the neural tube and MyoD by the dorsal ectoderm in mouse paraxial mesoderm. *Development* **122,** 429–437.

9. Determination and Morphogenesis in Myogenic Progenitor Cells

Couly, G. F., Coltey, P. M., and Le Douarin, N. M. (1992). The developmental fate of cephalic mesoderm in quail–chick chimeras. *Development* **114,** 1–15.
Daston, G., Lamar, E., Olivier, M., and Goulding, M. (1996). Pax-3 is necessary for migration but not differentiation of limb muscle precursors in the mouse. *Development* **122,** 1017–1027.
Davis, R., Weintraub, H., and Lassar, A. (1987). Expression of a single transfected cDNA converts fibroblasts to myoblasts. *Cell* **51,** 987–1000.
de la Brousse, C. F., and Emerson, C. P. (1990). Localized expression of a myogenic regulatory gene, *qmf1*, in the somite dermatome of avian embryos. *Genes Dev.* **4,** 567–581.
Denetclaw, W. J., Christ, B., and Ordahl, C. P. (1997). Location and growth of epaxial myotome precursor cells. *Development* **124,** 1601–1610.
Detwiler, S. R. (1918). Experiments on the development of the shoulder-girdle and the anterior limb of *Ablystoma punctatum*. *J. Exp. Zool.* **25,** 499–538.
Detwiler, S. R. (1929). Transplantation of anterior-limb mesoderm from *Amblystoma* embryos in the slit-blastopore stage. *J. Exp. Zool.* **52,** 315–324.
Detwiler, S. R., and Holtzer, H. (1956). The developmental dependence of the vertebral column upon the spinal cord in the urodele. *J. Exp. Zool.* **132,** 299–310.
Dietrich, S., Schubert, F. R., and Lumsden, A. (1997). Control of dorsoventral pattern in the chick paraxial mesoderm. *Development* **124,** 3895–3908.
Dietrich, S., Schubert, F. R., Healy, C., Sharpe, P. T., and Lumsden, A. (1998). Specification of the hypaxial musculature. *Development* **125,** 2235–2249.
Driesch, H. (1891). Entwicklungsmechanisch studien I. Der wert der beiden ersten furchungszellen in der Echinodermenentwicklung. Experimentelle erzeugung von teil- und doppelbildungen. *Z. Wiss. Zool. Abt. 3* **53,** 160–183.
Eppenberger, H. M., Eppenberger-Eberhardt, M., and Hertig, C. (1995). Cytoskeletal rearrangements in adult rat cardiomyocytes in culture. *Ann. N.Y. Acad. Sci.* **752,** 128–130.
Epstein, D. J., Vekemans, M., and Gros, P. (1991). Splotch (Sp2H), a mutation affecting development of the mouse neural tube, shows a deletion within the paired homeodomain of Pax-3. *Cell* **67,** 767–774.
Fan, C. M., and Tessier-Lavigne, M. (1994). Patterning of mammalian somites by surface ectoderm and notochord: Evidence for sclerotome induction by a hedgehog homolog. *Cell* **79,** 1175–1186.
Fischel, A. (1895). Zur entwicklung der ventralen rumpf- und extremitatenmuskulatur der Vogel und Saugetiere. *Morphol. Jb.* **23,** 544–561.
Fowler, I., and Watterson, R. L. (1953). The role of the neural tube in development of the axial skeleton of the chick. *Anat. Rec.* **117,** 555–556.
Franz, T., Kothary, R., Surani, M. A. H., Halata, Z., and Grim, M. (1993). The Splotch mutation interferes with muscle development in the limbs. *Anat. Embryol.* **187,** 153–160.
Gage, F. H., Ray, J., and Fisher, L. J. (1995). Isolation, characterization, and use of stem cells from the CNS. *Annu. Rev. Neurosci.* **18,** 159–192.
Galimi, F., Bagnara, G. P., Bonsi, L., Cottone, E., Follenzi, A., Simeone, A., and Comoglio, P. M. (1994). Hepatocyte growth factor induces proliferation and differentiation of multipotent and erythroid hemopoietic progenitors. *J. Cell Biol.* **127,** 1743–1754.
Gasser, R. F. (1979). Evidence that sclerotomal cells do not migrate medially during normal embryonic development of the rat. *Am. J. Anat.* **154,** 509–524.
George-Weinstein, M., Gerhart, J., Reed, R., Flynn, J., Mattiacci, M., Callihan, B., Foti, G., Miehl, C., Lash, J. W., and Weintraub, H. (1996). Skeletal myogenesis: The preferred pathway of chick embryo epiblast cells *in vitro*. *Dev. Biol.* **173,** 728–741.
George-Weinstein, M., Gerhart, J., Mattiacci, M., Simak, E., Blitz, J., Reed, R., and Knudsen, K. (1998). The roles of stably committed and uncommitted cells in establishing tissues of the somite. *Ann. N.Y. Acad. Sci.* **842,** 16–27.
Golden, J. A., and Cepko, C. L. (1996). Clones in the chick diencephalon contain multiple cell types and siblings are widely dispersed. *Development* **122,** 65–78.

Gonzalo-Sanz, L. (1972). Wechselwirkung zwischen crista neuralis und somiten. *Acta Anat.* **81**, 396–408.
Goulding, M. D., Chalepakis, G., Deutsch, U., Erselius, J. R., and Gruss, P. (1991). Pax-3, a novel murine DNA binding protein expressed during early neurogenesis. *EMBO J.* **10**, 1135–1147.
Goulding, M., Lumsden, A., and Paquette, A. J. (1994). Regulation of Pax-3 expression in the dermomyotome and its role in muscle development. *Development* **120**, 957–971.
Gray, H. (1989). "Gray's Anatomy." Churchill Livingstone, Edinburgh.
Gussoni, E., Pavlath, G. K., Lanctot, A. M., Sharma, K. R., Miller, R. G., Steinman, L., and Blau, H. M. (1992). Normal dystrophin transcripts detected in Duchenne muscular dystrophy patients after myoblast transplantation. *Nature* **356**, 435–438.
Hacker, A., and Guthrie, S. (1998). A distinct developmental programme for the cranial paraxial mesoderm in the chick embryo. *Development* **125**, 3461–3472.
Hamburger, V. (1988). "The Heritage of Experimental Embryology: Hans Spemann and the Organizer." Oxford Univ. Press, Oxford and New York.
Hamilton, H. L. (1952). "Lillie's Development of the Chick: An Introduction to Embryology." Holt, New York.
Hauschka, S. D. (1974a). Clonal analysis of vertebrate myogenesis. 3. Developmental changes in the muscle-colony-forming cells of the human fetal limb. *Dev. Biol.* **37**, 345–368.
Hauschka, S. D. (1974b). Clonal analysis of vertebrate myogenesis. II. Environmental influences upon human muscle differentiation. *Dev. Biol.* **37**, 329–344.
Hazelton, R. D. (1970). A radioautographic analysis of the migration and fate of cells derived from the occipital somites in the chick embryo with specific reference to the development of the hypoglossal musculature. *J. Embryol. Exp. Morphol.* **24**, 455–465.
His, W. (1888). Untersuchungen uber die erste anlage des wirbeltierleibes. Die erste entwicklung des Huhnchens im Ei. *Leipzig* **16**.
Holtfreter, J. (1944). Neural differentiation of ectoderm through exposure to saline solution. *J. Exp. Zool.* **95**, 307–340.
Holtfreter, J. (1945). Neuralization and epidermization of gastrula ectoderm. *J. Exp. Zool.* **98**, 161–209.
Holtfreter, J., and Hamburger, V. (1955). Amphibians. *In* "Analysis of Development" (B. H. Willier, P. Weiss, and V. Hamburger, Eds.), pp. 230–296. Saunders, Philadelphia and London.
Holtzer, H., Marshall, J. M., and Finck, H. (1957). An analysis of myogenesis by the use of fluorescent antimyosin. *J. Biophys. Biochem. Cytol.* **3**, 705–729.
Hyman, L. H. (1922). "A Laboratory Manual for Comparative Vertebrate Anatomy." Univ. of Chicago Press, Chicago, Illinois.
Ikeshima-Kataoka, H., Skeath, J. B., Nabeshima, Y., Doe, C. Q., and Matsuzaki, F. (1997). Miranda directs Prospero to a daughter cell during *Drosophila* asymmetric divisions. *Nature* **390**, 625–629.
Jostes, B., Walther, C., and Gruss, P. (1991). The murine paired box gene, *Pax 7*, is expressed specifically during the development of the nervous and muscular system. *Mech. Dev.* **33**, 27–38.
Kaehn, K., Jacob, H., Christ, B., Hinrichsen, K., and Poelmann, R. (1988). The onset of myotome formation in the chick. *Anat. Embryol.* **177**, 191–201.
Kato, N., and Aoyama, H. (1998). Dermomyotomal origin of the ribs as revealed by extirpation and transplantation experiments in chick and quail embryos. *Development* **125**, 3437–3443.
Keller, R. E. (1985). The cellular basis of amphibian gastrulation. *Dev. Biol.* **2**, 241–327.
Keller, R. E., Danilchik, M., Gimlich, R., and Shih, J. (1985). The function and mechanism of convergent extension during gastrulation of *Xenopus laevis*. *J. Embryol. Exp. Morphol.* **89**(Suppl.), 185–209.
Kieny, M., Pautou, M., and Chevallier, A. (1981). On the stability of the myogenic cell line in avian limb bud development. *Arch. Anat. Microsc.* **70**, 81–90.
Konieczny, S., Baldwin, A., and Emerson, C. (1986). Myogenic determination and differentiation of 10T1/2 cell lineages: Evidence for a simple genetic regulatory system. *In* "Molecular Biology

9. Determination and Morphogenesis in Myogenic Progenitor Cells

of Muscle Development, UCLA Symposium on Molecular and Cellular Biology" (C. Emerson, D. Fischman, B. Nadal-Ginard, and M. A. Q. Siddiqui, Eds.), New Series, Vol. 29, pp. 21–34. Alan R. Liss, New York.

Konigsberg, I. R. (1979). Skeletal myoblasts in culture. *In* "Cell Culture" (W. B. Jacoby and I. H. Pastan, Eds.), Vol. 58, pp. 511–527. Academic Press, San Diego.

Langman, J., and Nelson, G. R. (1968). A radioautographic study of the development of the somite in the chick embryo. *J. Embryol. Exp. Morphol.* **19,** 217–226.

Le Douarin, N. (1973). A Feulgen-positive nucleolus. *Exp. Cell Res.* **77,** 459–468.

Le Douarin, N., Dieterlen-Lievre, F., and Smith, J. (1990). "The Avian Model in Developmental Biology: From Organism to Genes," p. 319. Editions du CNRS, Paris.

Lewis, W. (1910). The relation of the myotomes to the ventro-lateral musculature and to the anterior limbs in *Amblystoma. Anat. Rec.* **4,** 183–190.

Lin, Z., Lu, M. H., Schultheiss, T., Choi, J., Holtzer, S., DiLullo, C., Fischman, D. A., and Holtzer, H. (1994). Sequential appearance of muscle-specific proteins in myoblasts as a function of time after cell division: Evidence for a conserved myoblast differentiation program in skeletal muscle. *Cell Motil. Cytoskeleton* **29,** 1–19.

Lin-Jones, J., and Hauschka, S. D. (1996). Myogenic determination factor expression in the developing avian limb bud: An RT-PCR analysis. *Dev. Biol.* **174,** 407–422.

Maroto, M., Reshef, R., Munsterberg, A. E., Koester, S., Goulding, M., and Lassar, A. B. (1997). Ectopic Pax3 activates *MyoD* and *Myf-5* expression in embryonic mesoderm and neural tissue. *Cell* **89,** 139–148.

Mauger, A., and Kieny, M. (1980). Migratory and ontogenetic capacities of muscle cells in bird embryos. *Wilhelm Roux's Arch. Dev. Biol.* **189,** 123–134.

Minsuk, S. B., and Keller, R. E. (1996). Dorsal mesoderm has a dual origin and forms by a novel mechanism in *Hymenochirus,* a relative of *Xenopus. Dev. Biol.* **174,** 92–103.

Moore, S. W., Keller, R. E., and Koehl, M. A. (1995). The dorsal involuting marginal zone stiffens anisotropically during its convergent extension in the gastrula of *Xenopus laevis. Development* **121,** 3131–3140.

Noden, D. M. (1983). The embryonic origins of avian cephalic and cervical muscle and associated connective tissues. *Amer. J. Anat.* **168,** 257–276.

Noden, D. M. (1988). Interactions and fates of avian craniofacial mesenchyme. *Development* **103** (Suppl.), 121–140.

Ohki, H., Martin, C., Coltey, M., and Le Douarin, N. M. (1988). Implants of quail thymic epithelium generate permanent tolerance in embryonically constructed quail/chick chimeras. *Development* **104,** 619–630.

Ordahl, C. P. (1993). Myogenic lineages within the developing somite. *In* "Molecular Basis of Morphogenesis" (M. Bernfield, Ed.), Wiley, New York.

Ordahl, C. P. (1999). Fate maps. *In* "Encyclopedia of Molecular Biology," Wiley, New York.

Ordahl, C., and Le Douarin, N. (1992). Two myogenic lineages within the developing somite. *Development* **114,** 339–353.

Ordahl, C. P., and Williams, B. A. (1998). Knowing chops from chuck: Roasting myoD redundancy. *BioEssays* **20,** 357–362.

Ott, M., Bober, E., Lyons, G., Arnold, H., and Buckingham, M. (1991). Early expression of the myogenic regulatory gene, *myf-5,* in precursor cells of skeletal muscle in the mouse embryo. *Development* **111,** 1097–1107.

Pourquié, O., Coltey, M., Teillet, M.-A., Ordahl, C., and Le Douarin, N. (1993). Control of dorsoventral patterning of somitic derivatives by notochord and floor plate. *Proc. Nat. Acad. Sci., U.S.A.* **90,** 5242–5246.

Pourquié, O., Fan, C.-M., Coltey, M., Hirsinger, E., Watanabe, Y., Breant, C., Francis-West, P., Brickell, P., Tessier-Lavigne, M., and Le Douarin, N. M. (1996). Lateral and axial signals involved in avian somite patterning: A role for BMP4. *Cell* **84,** 461–471.

Pownall, M. E., and Emerson, C. P. (1992). Sequential activation of three myogenic regulatory genes during somite morphogenesis in quail embryos. *Dev. Biol.* **151,** 67–79.
Remak, R. (1855). *Froriep's N. Notiz* **55,** 57.
Rhodes, S. J., and Konieczny, S. F. (1989). Identification of MRF4: A new member of the muscle regulatory factor gene family. *Genes Dev.* **3,** 2050–2061.
Romer, A. S. (1949). "The Vertebrate Body." Saunders, Philadelphia.
Roux, W. (1897). "Programm und Forschungsmethoden der Entwicklungsmechanik der Organismen." W. Engelmann, Leipzig.
Sadler, T. W. (1995). "Langman's Medical Embryology." Williams & Wilkins, Baltimore.
Searls, R. L. (1965). An autoradiographic study of the uptake of S35-sulfate during the differentiation of limb bud cartilage. *Dev. Biol.* **11,** 155–168.
Searls, R. L. (1967). The role of cell migration in the development of the embryonic chick limb bud. *J. Exp. Zool.* **166,** 39–45.
Searls, R. L., and Janner, M. Y. (1969). The stabilization of cartilage properties in the cartilage-forming mesenchyme of the embryonic chick limb. *J. Exp. Zool.* **170,** 365–376.
Selleck, M., and Stern, C. (1991). Fate mapping and cell lineage analysis of Hensen's node in the chick embryo. *Development* **112,** 615–626.
Serbedzija, G. N., Bronner-Fraser, M., and Fraser, S. E. (1989). A vital dye analysis of the timing and pathways of avian trunk neural crest cell migration. *Development* **106,** 809–816.
Shellswell, G. B., and Wolpert, L. (1977). The pattern of muscle and tendon development in the chick wing. *In* "Vertebrate Limb and Somite Morphogenesis" (D. A. Ede, J. R. Hinchliffe, and M. Balls, Eds.), pp. 71–86. Cambridge University Press, Cambridge.
Slack, J. M. W. (1983). "From egg to embryo: Determinative events in early development." Cambridge University Press, New York.
Smith, J. L., and Schoenwolf, G. C. (1998). Getting organized: New insights into the organizer of higher vertebrates. *Curr. Top. Dev. Biol.* **40,** 79–110.
Spemann, H. (1938). "Embryonic Development and Induction," 1st Ed. Yale Univ. Press, New Haven, Connecticut.
Spemann, H., and Mangold, H. (1924). "Induction of Embryonic Primordia by Implantation of Organizers from a Different Species," Foundations of Experimental Embryology (J. M. Oppenheimer, Ed.), pp. 144–184. Hafner, New York.
Stockdale, F. E. (1992). Myogenic cell lineages. *Dev. Biol.* **154,** 284–298.
Straus, W. L., and Rawles, M. E. (1953). An experimental study of the origin of the trunk musculature and ribs in the chick. *Am. J. Anat.* **92,** 471–509.
Sze, L. Y., and Paulin, D. (1995). Migration of myogenic cells from the somites to the fore-limb buds of developing mouse embryos. *Dev. Dyn.* **203,** 324–336.
Tajbakhsh, S., and Buckingham, M. E. (1994). Mouse limb muscle is determined in the absence of the earliest myogenic factor myf-5. *Proc. Natl. Acad. Sci. U.S.A.* **91,** 747–751.
Tajbakhsh, S., Rocancourt, D., Cossu, G., and Buckingham, M. (1997). Redefining the genetic hierarchies controlling skeletal myogenesis: *Pax-3* and *Myf-5* act upstream of *MyoD*. *Cell* **89,** 127–138.
van Straaten, H., Thors, F., Wiertz-Hoessels, L., Hekking, J., and Drukker, J. (1985). Effect of a notochordal implant on the early morphogenesis of the neural tube and neuroblasts: Histochemical and histological results. *Dev. Biol.* **110,** 247–254.
van Straaten, H., Hekking, J., Wiertz-Hoessels, E., Thors, F., and Drukker, J. (1988). Effect of notochord on the differentiation of a floor plate area in the neural tube of the chick embryo. *Anat. Embryol.* **177,** 317–324.
van Straaten, H., Hekking, J., Beursgens, J., Terwindt-Rouwenhorst, E., and Drukker, J. (1989). Effect of the notochord on proliferation and differentiation in the neural tube of the chick embryo. *Development* **107,** 793–803.
Wachtler, F., Christ, B., and Jacob, H. (1981). On the determination of mesodermal tissues in the avian embryonic wing bud. *Anat. Embryol.* **161,** 283–289.

9. Determination and Morphogenesis in Myogenic Progenitor Cells

Weintraub, H. (1993). The myoD family and myogenesis: redundancy, networks and thresholds. *Cell* **75,** 1241–1244.

Weston, J. A. (1963). A radioautographic analysis of the migration and localization of the trunk neural crest cells in the chick. *Dev. Biol.* **6,** 279–310.

Williams, B. A., and Ordahl, C. P. (1994). *Pax-3* expression in segmental mesoderm marks early stages in myogenic cell specification. *Development* **120,** 785–796.

Williams, B. A., and Ordahl, C. P. (1997). Emergence of determined myotome precursor cells in the somite. *Development* **124,** 4983–4997.

Williams, L. (1910). The somites of the chick. *Am. J. Anat.* **11,** 55–100.

Winklbauer, R., and Keller, R. E. (1996). Fibronectin, mesoderm migration, and gastrulation in *Xenopus*. *Dev. Biol.* **177,** 413–426.

Wright, W., Sassoon, D., and Lin, V. (1989). Myogenin, a factor regulating myogenesis, has a domain homologous to MyoD. *Cell* **56,** 607–618.

Yablonka-Reuveni, Z. (1995). Development and postnatal regulation of adult myoblasts. *Microsc. Res. Techn.* **30,** 366–380.

Yaffe, D. (1968). Retention of differentiation potentialities during prolonged cultivation of myogenic cells. *Proc. Natl. Acad. Sci. U.S.A.* **61,** 477–483.

Yaffe, D. (1969). Cellular aspects of muscle differentiation in vitro. *Curr. Top. Dev. Biol.* **4,** 37–77.

Index

Please note: Index includes entries for volumes 47 and 48. Volume numbers are in **bold.**

A

Acetylcholine receptor clusters, in somitic myotomes, **48**:284, 286
Adaxial cells
  elongation, **47**:268–269
  Myo-D expression, **47**:253–254
  radial migration across myotome, **48**:234
Amphibians
  anterior–posterior regulatory wave, **47**:235
  cell signaling and cell decisions, **47**:233–234
  dermatome and sclerotome, **47**:226–230
  presomitic mesoderm, movements, **47**:185–198
  resegmentation, **47**:230–231
  somitomeres, **47**:234–235
Amphibian system
  patterning of somitic mesoderm by adjacent tissues, **47**:231–232
  segmentation, role of morphomechanical molecules, **47**:235–239
Amphioxus, mesodermal cell fate, **48**:156
Anatomical development, sclerotome, **48**:79–83
Ancestral cells
  clonal analysis, **47**:39–42
  myotome segment, contribution to other segments, **47**:56–57
  primary expansion period, **47**:43–44
  regionalization and coherence, **47**:65
Angioblasts, somite-derived, **48**:33–34
Anterior–posterior wave, regulating morphogenic properties, **47**:235
Antibodies, F19 and S27, detection in myotome cells, **48**:282
Anurans
  fate maps, **47**:189–190
  resegmentation, **47**:230–231
  somite and myotome formation, **47**:220–226
Apoptosis, *see also* Cell death
  in signaling molecule mutants, **48**:245

somitic cells, **48**:175
wing-level somites, **48**:337
Archenteron roof, in urodeles, **47**:194
Axial elements, development, **48**:39
Axial identity, lower vertebrates, **47**:125–126
Axial levels, somites at, differences between, **48**:235–237
Axial structures
  induction of somite chondrogenesis, **48**:87–90, 95
  ventral, promoting role in myogenesis, **48**:69–70
Axial tissues, control of epaxial somite myogenesis, **48**:175–176

B

Basic helix–loop–helix
  encoded by *Paraxis*, **47**:98
  myogenic, gene expression, **48**:132–134
  role in somitogenesis, **47**:135–140
  as sclerotome markers, **48**:102–103
  skeletal muscle, **48**:130
bHLH, *see* Basic helix–loop–helix
Bilaterality
  presomitic mesoderm, **47**:65
  short clones along embryonic axis, **47**:57
Bilateralization, preceded by mediolateral regionalization, **47**:61
Birefringence, reduced, in mutant zebrafish embryos, **47**:265–267
Blastopore, *Xenopus laevis,* **47**:199–200
  postgastrula, **47**:207–209
BMP4
  effect on dorsal mesenchyme and cartilage, **48**:60–65
  expressed in dorsal neural tube, **48**:153–154
  and *Msx1*, dorsal domain, **48**:66, 68–69
  and Noggin, regulation of somite domain boundaries, **48**:200–203
  Noggin protein effect, **48**:109

369

BMP4 (*continued*)
  positive regulator of hypaxial genes, **48**:208–209
  produced by lateral mesoderm, **48**:243
  role in vertebral chondrogenesis, **48**:68
Bmp4, expression in roof plate, **48**:51–52
Bombina, somite and myotome formation, **47**:220–222
Bone morphogenetic proteins
  during early dorsoventral patterning, **48**:52
  role in somite dorsoventral patterning, **48**:113–114
  and sclerotome development, **48**:110
  signaling and transduction in embryos, **48**:201–203
  signal transduction mechanisms, **48**:200–201
Borders
  intersomitic, **47**:110–111
  metameric
    added sequentially, **47**:108
    in segmental organization, **47**:110
Boundaries
  anterior/posterior
    intrasomitic, **47**:137
    Notch activation along, **47**:146–147
  dorsal/ventral, Notch signaling along, **47**:141–142
  formation and maintenance, molecular basis, **47**:122–125
  Hox, and AP patterning, **47**:169, 172–173
  mediolateral, somite domains, **48**:200–203
  notochordal–somitic mesodermal, **47**:202, 204
  segmental, in lamprey, **48**:35
Brachyury
  mutant mouse, vertebral development in, **48**:84
  mutation, **47**:9–10
Breeding temperature, effect on somite formation, **47**:83–84
Bufo bufo, somite and myotome formation, **47**:222–223

C

N-Cadherin
  expression in somitic mesoderm, **47**:237
  loss, resulting in abnormal somite shape, **47**:22
  in myogenic cell distribution in limb bud, **48**:28
  in relation to myogenic competence, **48**:31
  role in myoblast migration, **48**:152
Cartilage
  discernible differentiation, **48**:82–83
  dorsal, formation in subectodermal position, **48**:51–54
  formation
    grafting experiments, **48**:87–88
    neural tube effect, **48**:90, 92
    inducing activity, **48**:94
  sclerotome-derived, early development, **48**:78–79
  subectodermal, prevention of formation, **48**:54–57
  superficial vertebral, differentiation, **48**:60–65
Caudal primary, growth cone, **47**:270
Cell adhesion molecules, role in *Xenopus* somite morphogenesis, **47**:237–238
Cell cycle model, somite formation, **47**:93, 115–117
Cell death, *see also* Apoptosis
  in sclerotome, **48**:82
Cell decisions, in making intersomitic furrow, **47**:232–235
Cell derivatives, dermomyotomal, **48**:234–235
Cell determination, within tissue primordia, **48**:321, 323
Cell intercalation, in *Xenopus* prospective somitic mesoderm, **47**:200–202, 208
Cell–matrix interactions, role in neural crest migration, **47**:292
Cell signaling, in amphibian somitogenesis, **47**:233–234
Ceratophrys ornata, zones in gastrocoel roof, **47**:191
c-hairy-1
  discrete phases of expression, **47**:120–122
  expressed in cyclic waves, **47**:139
  oscillations of expression, **47**:126
  smooth wave of expression, **47**:119–120
  transcription rounds, **47**:21
  variable expression domain, **47**:117–118
Chick embryo
  AChR clusters, **48**:284
  2-day-old host, preparation for creation of chimeras, **48**:271–273
  explants of segmental plate and somites, **48**:177–178
  research on somites, **48**:6–9
  sclerotome development, **48**:79–83
  tail bud, somitogenic potential, **47**:14–15

Chondrogenesis
  somites
    early inducer searches, **48**:97–98
    in vitro experiments, **48**:92–95
    in vivo experiments, **48**:85–92
    roles of *Shh* and *Noggin*, **48**:109–110
    subectodermal, inhibition by sonic hedgehog, **48**:57–60
  vertebral
    BMP4 role, **48**:68
    molecular pathways leading to, **48**:57–65
    role of axial organs, **48**:88
Chorioallantoic membrane, grafting experiments, **48**:85–88
Ci transcription factor, complex with Fu and Cos2, **48**:192
Clock and wavefront model, somite formation, **47**:92–93, 112–113
Clonal analysis, basic logic of, **47**:39–45
Clones
  bilateral, comparison of left and right contributions, **47**:62–63
  contributing to several segments of dermomyotome, **47**:47, 52
  LaacZ labeling system, **47**:35, 38–39
  localization in embryonic anatomical structures, **47**:54–55
  pools, vertical propagation, **47**:55–56
Cloning, positional, zebrafish, **47**:250–251
c-met
  cross-talk with scatter factor, **48**:8
  signaling pathway, **48**:259
cMet receptor, expressed in migratory muscle precursor cells, **48**:352
Coherent organization
  along dermomyotome axes, **47**:64–65
  primitive streak, **47**:73
Colinearity, *Hox* genes, **47**:166–167, 175–176
Compartmentalization, sclerotome, and resegmentation, **47**:85–89
Competence
  myogenic, N-cadherin in relation to, **48**:31
  paraxial mesoderm cells, **48**:346, 348
Conservation
  evolutionary, metamerism, **48**:2–5
  genes in segmental organization, **47**:108–110
  segmentation mechanisms between insect and vertebrate, **47**:256–258
Contractile proteins, MRF role, **48**:279–280
Craniofacial muscles, origins, **48**:167–168
Cuboidal cells, lateral lamella of, **47**:227–228

Cyclic AMP, somitic levels, **48**:106
Cytoskeleton, amphibian, **47**:238

D

Delta-D
  overexpression in zebrafish, **47**:144
  tissue expression, **47**:21
Delta genes
  differential expression in presomitic mesoderm, **47**:23–24
  expression in relation to somite boundary, **47**:115
  role in
    formation of segmental boundaries, **47**:122–123
    somitogenesis, **47**:143–145
Derepression, and Notch pathway, **48**:245–246
Dermatome
  controlling extent of migration, **47**:137
  fate map, **47**:162
  Xenopus, **47**:226–228
Dermis
  lineage, into somite mouse–chick chimera, **48**:276
  somite-derived, **48**:32–33
Dermomyotome
  development, ephrin receptor expression, **47**:285
  differentiation, **48**:292
  domains
    central, **48**:232–234
    defined according to Pax3 expression, **48**:351–352
    epaxial and hypaxial, **48**:227, 230–232
  formation, **48**:275–276
  lateral edge, **48**:8–9
  muscle lineage arising from, **48**:27–28
  mouse donor
    implantation level, **48**:273
    preparation for creation of chimeras, **48**:271
  muscle progenitor cells originating from, **48**:252
  nonmuscle derivatives, **48**:234–235
  Paraxis expression restricted to, **48**:183
  relationship to myotome, **48**:342
  and sclerotome, cell migration between, **48**:168
  segments
    longitudinal organization, **47**:52–55

Dermomyotome (*continued*)
  mediolateral organization, **47**:61–64
  polyclonal contribution, **47**:47
Desmin
  in 10-dpc somitic myotomes, **48**:282
  expressed before skeletal muscle actin, **48**:280
  in studies of myotome formation, **48**:339–340
Determination
  and fate, in generation of tissue anlage, **48**:320–323
  limb muscle, early studies, **48**:353–354
  muscle progenitor cells
    migratory limb, **48**:350–354
    Myf5 and *MyoD* roles, **48**:250–260
  myogenic, *see* Myogenic determination
  myotome precursor cells, time frame, **48**:344–350
  sclerotome, and quail–chick grafting, **48**:111
Developmental potency
  loss of, **48**:321, 323
Developmental potential
  and epithelial–mesenchymal transitions, **48**:356
Differential affinity, epithelializing cell subdivision by, **47**:123
Differential display, in comparison of mRNA populations, **48**:313
Differentiation
  cartilage
    discernible, **48**:82–83
    superficial vertebral, **48**:60–65
  dermomyotome, **48**:292
  dorsal mesenchyme, floor plate effect, **48**:56–57
  embryonic skeletal muscle, MyoD role, **48**:308
  embryonic stem cells *in vitro*, **48**:308
  mouse neural implant-derived spinal cord, in chick, **48**:290–291
  myocluster, **48**:349
  myocytes, myogenin gene role, **48**:139–142
  somites
    governed by extrinsic factors, **48**:242–246
    and tissue interactions, **47**:219–220
  Xenopus prospective somitic mesoderm role, **47**:203–206
Differentiation pathway, somitic, nonrigidity of, **47**:195
Diffusibility, *Shh*, **48**:106–107

Diversity, muscle precursor cells in vertebrate embryo, **48**:325–344
DNA libraries, germ layer-specific, **48**:313
Dorsal cells, Fng protein expression, **47**:142
Dorsalizing molecules, candidate, **47**:260
Drosophila
  hairy/enhancer of split, vertebrate homologs, **47**:139–140
  Notch signaling in, **47**:141–143
  and vertebrates, segmentation homology, **48**:39
DSL family proteins, ligands of Notch receptors, **48**:206
Dye-injection analysis, myotome development, **48**:342, 344

E

Early studies
  limb muscle determination, **48**:353
  location of muscle precursor cells in vertebrate body, **48**:325–327
  resegmentation, **48**:13–19
  search for inducers, **48**:97–98
  somitogenesis models, **47**:89–90
E-box, MRFs binding to, **48**:306, 308, 310
Ectoderm
  BMP4-positive, **48**:68
  surface
    combinatorial signals from, **48**:209
    signals controlling somite formation, **48**:182–184
    Wnt signals, **48**:196–200
Elongation
  adaxial cells, **47**:268–269
  cellular, mechanism and regulation, **47**:218–219
Embryo
  amniote, neural crest cell migration, **47**:280
  BMP signaling and transduction in, **48**:201–203
  chick, *see* Chick embryo
  cultures, xenograft implants, **48**:270
  FGF signaling and transduction in, **48**:205–206
  lacking notochord, gastrulation in, **47**:205
  mouse, *see* Mouse embryo
  mutant, somitogenic defects in, **47**:144
  myogenic tissue development, **48**:353–360
  Notch signaling and transduction in, **48**:206–207

Index 373

Pax3(splotch)/Myf5 double mutant, **48**:236
shark, two-layered somites, **48**:35, 38
Sonic hedgehog signaling and transduction in, **48**:192–194
TGFβ signaling and transduction in, **48**:204–205
vertebrate, *see* Vertebrate embryo
Wnt signaling and transduction in, **48**:197–200
zebrafish
  mutants, **47**:265–266, 271
  wild-type, **47**:256
Embryology, experimental, fate maps from, **47**:161–165
Embryonic axis, somitic precursor cell sequential allocation to, **47**:12
Embryonic models, for myogenic determination, **48**:325
En1, identification of central dermomyotome domain, **48**:232–234
Endoderm
  prospective, suprablastoporal and subblastoporal, **47**:186
  urodeles, **47**:190–191
Endodermalizing pathway, amphibian, **47**:197
Endoplasmic reticulum, PS1 protein role, **47**:146
Enhancers
  Hoxb-4 expression, **48**:303–304
  Myf5, **48**:310
Epaxial–hypaxial organization, vertebrate body, **48**:336–337
EphA4, hindbrain expression, **47**:124–125
Eph receptors, and ligands
  inhibitory interactions, **47**:289–291
  in neural patterning, **47**:283–287
Epiblast
  paraxial mesodermal domain, **47**:5–7
  spatial organization, **47**:74
Epiblast cells
  activated myogenic regulatory genes in, **48**:210
  developmental fate, **47**:3–4
Epithelialization
  lack of, effect on somitogenesis, **47**:136–138
  segmentation occurring in absence of, **47**:259
  somite, **47**:98
    role in sclerotome induction, **48**:113
Epithelializing cells, subdivision by differential affinity, **47**:123

Epithelial layer, amphibian, removing prospective mesoderm from, **47**:191–195
Epithelial–mesenchymal transitions
  c-met role, **48**:259
  dermomyotome as site of, **48**:235
  epaxial myotome precursor cells, **48**:354–356
  and sclerotome formation, **48**:113
Epithelial organization
  Notch–Delta signaling pathway role, **48**:206–207
  superficial cells, **47**:197
Epithelium, dermomyotome, **48**:227, 230–232
Evolution, metamerism in vertebrates, **48**:34–40
Expansion period, primary, pool of ancestral cells, **47**:43–44
Extracellular matrix
  components, distribution in sclerotome, **48**:10
  neural crest cell effects, **47**:283
  role in segmentation, **47**:236–237
  in somite chondrogenesis, **48**:98
  Wnt proteins associated with, **48**:116

F

Fate
  areal, sclerotome subdivided according to, **48**:44
  and determination, in generation of tissue anlage, **48**:320–323
  dorsal mesenchyme, **48**:47–51
  mesenchymalized cells, **48**:351–352
  mesodermal cells, in amphioxus, **48**:156
  myotome, in amphibians, **47**:195
  somitic, commitment to, **47**:15–16
Fate mapping
  embryonic, quail–chick studies, **48**:327–329
  evidence for resegmentation from, **47**:87
  experimental strategy for, **47**:3–5
  studies on mouse gastrula, **47**:8–9
Fate maps
  amphibian, variability, **47**:185–198
  anurans, **47**:189–190
  from experimental embryology, **47**:161–165
  myotome, **47**:162–163
  scapula, **47**:163–165
  sclerotome and dermatome, **47**:162
  urodeles, **47**:190–191
FGF, *see* Fibroblast growth factors

Fibers, myotome, distributions, **48**:342, 344
Fibroblast growth factor-6, expression in myogenic lineage, **48**:145–146
Fibroblast growth factors
  functional redundancy, **48**:136
  signaling and transduction in embryos, **48**:205–206
  signal transduction mechanisms, **48**:205
Fibronectin, loss, resulting in failure to form somites, **47**:22
Floor plate
  alternative source of notochord signaling molecules, **48**:179
  effect on dorsal mesenchyme differentiation, **48**:56–57
Fng protein, expression in dorsal cells, **47**:142
Founder cells
  clonal analysis, **47**:39–42
  of independent origin for segmental groups, **47**:46–47
  intermediary pool, **47**:44–45
  muscle system segments, mediolateral organization, **47**:63
FREK, myotomal cells expressing, **48**:241–242
fss-type mutants, in zebrafish, **47**:253–256, 258–259, 270–271
Fugu rubripes, Hoxb-4 gene intron, **48**:304

G

Gastrotheca, somite and myotome formation, **47**:220–222
Gastrula, mouse and chick, cells ingressing through primitive streak, **47**:8–9
Gastrulation
  in embryo lacking notochord, **47**:205
  Hox gene expression throughout, **47**:73–74
Gating mechanism, in cell cycle model of somite formation, **47**:117
Genealogical cohorts, formed by descendants of founder cells, **47**:42
Gene expression
  c-hairy-1, Morse code, **47**:120–122
  consistent with parasegmental boundary position, **47**:111
  in developing somitic cells, **48**:44–51
  hairy1, and vertebrate segmentation, **47**:93–94
  Hox, **48**:302–305
  MDF, transcriptional control, **48**:148–151
  in migratory limb muscle progenitor cells, **48**:350–353

myogenic regulatory factors
  control of, **48**:306–311
  patterns, **48**:255–256
  in somite, **48**:237–242
Pax-1, Noggin role, **48**:108–109
regional, in presomitic mesoderm, **47**:18–26
sequential, during patterning, **47**:74
temporal, myogenic bHLH, **48**:132–133
Genes, *see also* Reporter genes
  affecting vertebrate segmentation and somitogenesis, **47**:98
  disruptions, and MRF4 role in myogenesis, **48**:142–145
  fly neurogenic, vertebrate homologs, **47**:94–98
  hairylike segmentation, **47**:21
  hypaxial, BMP4 as positive regulator, **48**:208–209
  lacking MDF binding sites, muscle-specificity, **48**:148
  mutating, somite phenotype produced by, **47**:9–11
  myogenic factor, targeted inactivation, **48**:133–134, 136–145
  myogenic regulatory factor, role in somite myogenesis, **48**:170–172
  with novel expression patterns in somite, **48**:312–313
  pair-rule, **47**:99, 101, 257–258
  realizator, **47**:175
  in segmental organization, conservation, **47**:108–110
  signal transduction, in Shh pathway, **48**:195–196
  target, regulated by *Myf5* and *MyoD*, **48**:211–212
Genetic analysis, zebrafish, **47**:251
Genetic markers
  in fate mapping, **47**:4
  identification of central dermomyotome domain, **48**:232–234
  for mouse–chick chimeras, **48**:274
  of sclerotome, **48**:99–103
Geometry, *Xenopus* prospective somitic mesoderm, **47**:198–200
Gli genes, role in control of myogenesis, **48**:195–196
Global patterning
  comparative data and evolutionary implications, **47**:168–173
  Hox function, **47**:173–175

Index

Hox genes and vertebrate body plan, **47**:165–168
somitic cells, **47**:155–157
Grafting
chorioallantoic membrane, **48**:85–88
ectopic
BMP-producing cells, **48**:61, 63, 65
dorsal neural tube, **48**:51–54
limb bud premuscular mass, **48**:288–289
notochord, response of somites to, **48**:176–177
quail–chick, and sclerotome determination, **48**:111
Green fluorescent protein, in zebrafish analysis, **47**:250
Growth
clonal and nonclonal, ancestral cells, **47**:42–43
dynamics, early embryonic myotome formation, **48**:357–358
Growth zone, amniote vertebrates and fishes, **47**:108

H

hairy1, gene expression, and vertebrate segmentation, **47**:93–94
her1, *hairy* homolog in zebrafish, **47**:257
Hh family proteins, lineage-specific myogenic functions in zebrafish, **48**:193–194
Histology
Xenopus somite formation, **47**:209–216
zebrafish muscle differentiation mutants, **47**:265–266
HNF-3β, mutant mice, **48**:178
Hox code, **47**:167–168
paraxial, relevance to lateral structures, **47**:172–173
Hox function
downstream targets, and models, **47**:174–175
upstream factors, **47**:173–174
Hox genes
expression throughout gastrulation, **47**:73–74
regulation, **48**:302–305
and vertebrate body plan, **47**:165–168
Hymenochirus boettgeri, migration of lateral endodermal crests, **47**:191, 193
Hypocentrum, development, **48**:39

I

Identity, somites
determination of, **47**:125–126
positional and segmental, **48**:301–302

Id proteins, heterodimerization with MDFs, **48**:146
Inducers
early studies on, **48**:97–98
sclerotome
candidate, **48**:103–106
Shh as, **48**:106–108
somite chondrogenesis, various tissues, **48**:93–95
Induction
paraxial mesoderm, timing of, **47**:25
Pax-1 expression, by notochord, **48**:11
permissive and instructive, **48**:78, 116
prospective endoderm, in involuting marginal zone, **47**:196
sclerotome
role of somite epithelialization, **48**:113
Shh role, **48**:106–108
in vivo and *in vitro* experiments, **48**:103–106
sclerotome formation, by neural tube, **48**:115–116
somite chondrogenesis
by axial structures, **48**:87–90, 95
by notochord, **48**:88, 95, 115–116
tissue, differences in, reflected in fate map variations, **47**:195–196
Information
local, in somitic mesoderm, **47**:159–160
positional
for correct pattern, **47**:156
translated by *Hox* genes, **47**:176
Ingression
stage-specific, in amphibian, **47**:191
through primitive streak, **47**:7–12
Inhibitory cues, from notochord, for trunk neural crest migration, **47**:292–293
Inhibitory feedback mechanism, Notch/Delta-mediated, **47**:233–235
Initiation
epaxial somite myogenesis, notochord signals for, **48**:176–179
sclerotome, **48**:108–109
segmentation, and somitomeres, **47**:84–85
Innervation, somitic musculature, zebrafish, **47**:269–271
Insects
long germ band
parasegments, **47**:110–111
segmentation by subdivision, **47**:108
segmental patterning, **47**:168
and vertebrates

Insects (continued)
  conservation of segmentation mechanisms, 47:256–258
  segmentation differences, 48:4–5
Integrin, role in somite morphogenesis, 47:236
Intersomitic furrow, cellular decisions on, 47:232–235
Intervertebral disks
  origin, 48:43–44
  vertebral bodies growth from, 48:11–12
Intervertebral fissure, early description, 48:15, 17
Involuting marginal zone
  amphibian annular region, 47:185–186
  in anurans and urodeles, 47:189–190
  initial differences in, 47:196–198
  prospective endoderm induction in, 47:196
  superficial layer, in urodeles, 47:193–194
  Xenopus, deep mesenchymal layer, 47:186–189

K

Knock-in mice, myogenin/Myf-5, 48:140–141
Knockouts
  Hox, 47:175
  in ovo engrafting of tissue from, 48:295
  mrf4, 48:308
  myf5, 48:358
  Myf-5 mice, 48:136–138, 143
  Notch pathway components, 47:252–253
  residual segmentation in, 47:258–259
  Shh, 48:107–108

L

LaacZ labeling system, of clones, 47:35, 38–39
LacZ, as genetic marker for mouse–chick chimeras, 48:274
LacZ, grafted embryos stained positive for, 48:280
Lateral plate, *Hox* expression, 47:173
Lbx1, somite-expressed, 48:173
Ligands
  Eph-family receptor tyrosine kinases, 47:283–287
  Eph receptors, inhibitory interactions, 47:289–291
  Notch receptors: DSL family proteins, 48:206
Limb, hypaxial domains, boney structures in, 48:337

Limb bud
  invaded by myogenic precursor cells, 48:28
  Pax3-expressing somites, 48:352
  premuscular mass, grafting, 48:288–289
Limb muscles
  determination, early experiments, 48:353
  origins, 48:168
Limb musculature, development, 48:151–153
Lineage
  cell, analysis in chimeras, 48:294–295
  diversity, in muscle precursor cells, 48:329–337
  epaxial myogenic, into neural tube mouse–chick chimera, 48:289–292
  muscle, 48:26–31
  myogenic
    FGF-6 expression in, 48:145–146
    Myf-5 role, 48:139
  permanent, in nonbilateralized embryonic structures, 47:57
  somitic, into somite mouse–chick chimera, 48:274–289
  tracing
    muscle precursors of expaxial myotome, 48:231–232
    in *Myf5* mutant mice, 48:171
    transient, presomitic mesoderm, 47:58
Longitudinal organization
  myotome in E11.5 mouse embryo, 47:45–46
  presomitic mesoderm, 47:75
  segments of dermomyotome
    clone classes, 47:52–54
    localization, 47:54–55
lunatic fringe, expression in presomitic mesoderm, 47:23–24

M

Markers
  cytological, for mouse–chick chimeras, 48:274
  genetic, *see* Genetic markers
  molecular
    expression in epaxial myotome precursor cells, 48:337–339
    of sclerotome, 48:99–103
    sclerotomal and myotomal, expressed in *paraxis* mutants, 47:136–137
MDFs, *see* Myogenic determination factors
Mediolateral intercalation behavior, 47:200–202

Index

Mediolateral organization
   dermomyotome segments, **47**:61–64
   murine muscle system, model, **47**:64–65
   myotome in E11.5 mouse embryo, **47**:46
   somite, **48**:331, 334–336
MEF2, cooperation with MDFs in myogenesis, **48**:147–148
Meinhardt model, somite formation, **47**:113–115
Memory
   cellular
      autonomous and phenotypic, **48**:323
      molecular model, **48**:324
   myogenic, **48**:349
Meristic pattern, somitomeres, **47**:17–18
Meristic prepattern, implied by regionalized gene expression, **47**:20–21
Mesenchymalized cells, fate, **48**:351–352
Mesenchyme
   dorsal
      development, **48**:60–65
      differentiation, **48**:56–57
      origin and fate, **48**:47–51
   hypoglossal chord, **48**:236
   sclerotomal, associated gene expression, **48**:100
   subectodermal, *Msx* gene expression, **48**:46–47
Mesoderm
   lateral, inhibitory signals from, **48**:184–185
   myogenic potential, **48**:155–157
   origin of skeletal muscle in vertebrate embryos, **48**:167–168
   paraxial, *see* Paraxial mesoderm
   patterns, vertebrate body plan, **47**:157–159
   presomitic, *see* Presomitic mesoderm
   prospective, removal from epithelial layer in amphibians, **47**:191–195
   somitic, *see* Somitic mesoderm
Mesoderm cells, Pax3 gene expression as marker, **48**:350–353
Mesp2 mutation
   and abnormal somitogenesis, **47**:97
   effect on sclerotome, **47**:138
Messenger RNA
   c-hairy1, **47**:93–94
   MRF genes, **48**:307
   Pax-9, **48**:100, 102
   populations, comparison with differential display, **48**:313
   Scleraxis, **48**:102

Metameres, *see also* Segments
   acquisition of functional specializations, **47**:125
   and borders
      added sequentially, **47**:108
      in segmental organization, **47**:110
      pattern established from unsegmented tissue, **47**:133–134
Metamerism, *see also* Segmentation
   associated mobility at segment boundaries, **47**:107–108
   evolutionary conservation, **48**:2–5
   evolution in vertebrates, **48**:34–40
Migration
   differentiated muscle cells, **48**:168
   mesenchymal cells of somitic origin, **48**:65–66
   muscle cells, cellular interactions in, **48**:152
   muscle progenitor cells, **48**:230–232
   radial, adaxial cells across myotome, **48**:234
   sclerotome cells to notochord, **48**:81
   trunk neural crest
      dynamic analysis, **47**:287–289
      effect of somitic rostrocaudal polarity, **47**:280–282
      PNA effect, **47**:291–292
Molecular clock, linked to vertebrate segmentation, **47**:93–94
Molecular model, for cellular memory, **48**:324
Molecular pathways, leading to chondrogenesis in vertebra, **48**:57–65
Molecules
   dorsalizing, candidate, **47**:260
   extracellular and cell surface, patterns in somites, **47**:282–283
   morphomechanical, role in amphibian segmentation, **47**:235–239
   peanut lectin-binding, effect on neural crest migration pattern, **47**:291–292
   signaling
      controlling myogenesis in somites, **48**:185, 191–207
      and interference, **48**:243–244
Morphogenesis
   organization, role of prospective somitic mesoderm, **47**:203–206
   regulatory role of anterior–posterior wave, **47**:235
   somite, role of morphomechanical molecules, **47**:235–239

Morphogenetic movements
  early embryonic myotome formation,
    48:357–358
  myogenic precursor cells, **48**:354–357
Morphomechanical molecules, role in amphibian segmentation, **47**:235–239
Morse code, *c-hairy-1* gene expression, **47**:120–122
Motoneurons
  graft-derived, **48**:290
  primary, zebrafish somitic musculature, **47**:269–271
Mouse–chick chimera
  cell markers, **48**:274
  creation of, **48**:271–273
  neural tube, epaxial myogenic lineage into, **48**:289–292
  somite, somitic lineage into, **48**:274–289
  somite–neural tube, benefits of, **48**:292–294
Mouse embryo
  AChR clusters, **48**:284
  E9.5
    epaxial *Myf5* expression, **48**:194
    neural tube/notochord from, **48**:181
    focus on signaling mechanisms, **48**:166
  myotome at E11.5
    longitudinal organization, **47**:45–46
    of muscle system, **47**:58–61
    mediolateral organization, **47**:46
    presomitic clonal organization, **47**:47, 52
    questions concerning structural formation, **47**:46–47
    segmental longitudinal organization, **47**:52–58
    segmental mediolateral organization, **47**:61–64
  temporal expression of myogenic bHLH genes, **48**:132–133
Mouse mutants, *see also* Transgenic mice
  HNF-3β, **48**:178
  Myf5, **48**:171
  Myf-5, **48**:136–138
  Myf5-nlacZ null, **48**:248, 251–252
  Myf5 null, **48**:254–255
  MyoD, **48**:138–139
  open brain, **48**:182
  Shh, **48**:107–108, 244–245
  Splotch, **48**:351–352
  undulated, **47**:164
Mox genes, expression during somitogenesis, **48**:312

MRF4
  role in myogenesis, **48**:142–145
  temporal expression
    in embryogenesis, **48**:133
    with myogenin, **48**:140
MRF4
  expression waves, **48**:239
  role in somite myogenesis, **48**:170–172
Msx
  expression in
    host mesenchymal cells, **48**:61, 63
    subectodermal mesenchyme, **48**:46–47
  function, tested in cell culture, **48**:69
M-twist, as sclerotome marker, **48**:102–103, 116–117
Muscle cells, *see also* Smooth muscle cells
  subpopulations, defined by *Myf5* and *MyoD*, **48**:246–250
Muscle differentiation
  mutants, zebrafish, **47**:264–267
  MyoD role, **48**:256–257
  and myotome, in *Xenopus*, **47**:216–217
  notochord role, **48**:31
  slow and fast, **47**:267–269
  synergistic role of growth factors, **48**:154–155
Muscle precursor cells
  developmental history, **47**:34–35
  epaxial, increase in myogenic potential in, **48**:348–350
  lineage diversity, **48**:329–337
  location in vertebrate embryo, **48**:325–327
  migration, **48**:352
  populations, **48**:26–28
Muscle progenitor cells
  aberrant accumulation in *Myf5-nlacZ* null, **48**:254–255
  determination, *Myf5* and *MyoD* roles, **48**:250–260
  migratory limb, gene expression and determination in, **48**:350–353
  migratory routes, **48**:230–232
  in *Myf5* null, **48**:248
  originating from dermomyotome, **48**:252
Muscles
  adult, role of MDFs, **48**:145–146
  development, role of non-MRF transcription factors, **48**:311–312
  epaxial, polymerization in, **48**:22
  of epaxial and hypaxial domains of vertebrate body, **48**:331, 334–336

Index

formation, mesoderm role, **48**:156–157
lineage, **48**:26–31
skeletal, *see* Skeletal muscle
Muscle system
  coherence, **47**:67
  models
    longitudinal organization, **47**:58–61
    mediolateral organization, **47**:64–65
Musculature
  limb, development, **48**:151–153
  somitic, innervation in zebrafish, **47**:269–271
Mutants
  embryo, somitogenic defects in, **47**:144
  Hox, **47**:167–168
  mouse, *see* Mouse mutants
  paraxis, expression of sclerotomal and myotomal markers, **47**:136–137
  short-tail, **48**:83–85
  undulated, affecting sclerotome, **48**:84
  zebrafish, *see* Zebrafish mutants
Mutational studies, segmentation prepattern in presomitic mesoderm, **47**:18–25
Mutations
  embryonic genes, **47**:9–11
  Mesp2, **47**:97
    effect on sclerotome, **47**:138
  Notch1, **47**:24
  paraxis, **47**:135–137
  PS1, **47**:97
  targeted, in Wnt genes, **48**:199–200
  Tbx6, **47**:9–10
Myf5
  expression in hypaxial precursors, **48**:307–308
  locus, myogenin targeted in, **48**:141
  mutant mice, abnormal somite development, **48**:136–138
  temporal expression in embryogenesis, **48**:133
  transcript accumulation along somite medial edge, **48**:231
  transcription, **48**:149–151
  transcriptional regulation, **48**:310–311
Myf5
  activation, governed by extrinsic factors, **48**:242–246
  activity upstream of *MyoD*, **48**:257, 259–260
  epaxial expression in E9.5 *Shh* mutant embryos, **48**:194
  muscle cell subpopulations defined by, **48**:246–250

and *MyoD*, expressed at different levels, **48**:256–257
role in
  muscle progenitor cell determination, **48**:250–260
  somite myogenesis, **48**:170–172
  spatiotemporal expression in somites, **48**:238–241
Myocluster, differentiation, **48**:349
Myocoel, formation in urodele, **47**:224, 226
Myocytes, differentiation, myogenin gene role, **48**:139–142
MyoD
  expression in Myf-5 mutant mice, **48**:136–138
  mutant mice, Myf-5 role in myogenesis, **48**:138–139
  restriction to muscle forming regions, **48**:156
  role in
    differentiation of embryonic skeletal muscle, **48**:308
    early events of myogenesis, **48**:132
    transcriptional regulation, **48**:311
MyoD
  activation
    governed by extrinsic factors, **48**:242–246
    in myogenic progenitor cells, **48**:209
    in somites, **48**:291–292
  expression
    in adaxial cells, **47**:253–254
    neural tube role, **48**:179–180
    Wnt protein role, **48**:198
    in *you*-type mutants, **47**:263
  gene regulation, **48**:149
  muscle cell subpopulations defined by, **48**:246–250
  and *Myf5*, expressed at different levels, **48**:256–257
  Myf5 and *Pax3* activity upstream of, **48**:257, 259–260
  role in
    muscle progenitor cell determination, **48**:250–260
    somite myogenesis, **48**:170–172
    spatiotemporal expression in somites, **48**:238–239
MyoD family genes
  and cellular memory, **48**:324
  expression phases, **48**:339
Myogenesis
  antagonistic role of BMPs, **48**:202

Myogenesis (continued)
  effect of structures surrounding somites, **48**:153–155
  epaxial, in in ovo implanted mouse somites, **48**:281–286
  hypaxial
    activation by Wnt protein signals, **48**:196–200
    control by surface ectoderm signals, **48**:182–184
    in in ovo implanted mouse somites, **48**:286–289
    signaling components, **48**:207–208
  MDFs in collaboration with other transcription factors, **48**:146–148
  mouse, targeted inactivation of myogenic factor genes, **48**:133–134, 136–145
  Pax3 role, **48**:259–260
  promoting role of ventral axial structures, **48**:69–70
  somite, see Somite myogenesis
Myogenic cells
  distribution in limb bud, N-cadherin role, **48**:28
  somitic, fate of, **48**:30
Myogenic determination
  acquired during somite formation, **48**:174–175
  embryonic models, **48**:325
  epaxial, role of axial tissue interactions, **48**:175–176
Myogenic determination factors
  gene expression, transcriptional control, **48**:148–151
  with other transcription factors in myogenesis, **48**:146–148
  role in adult muscle, **48**:145–146
Myogenic lineage
  epaxial, into neural tube mouse–chick chimera, **48**:289–292
  into somite mouse–chick chimera, **48**:279–289
Myogenic organizers, and progenitor stem cells, **48**:358–360
Myogenic potential
  of mesoderm, **48**:155–157
  progressive increase in epaxial muscle precursor cells, **48**:348–350
Myogenic precursor cells, morphogenetic movements, **48**:354–357
Myogenic regulatory factors
  control of myogenic specification, **48**:279

gene expression
  control of, **48**:306–311
  in somite, **48**:237–242
  and threshold levels, differences between, **48**:255–257
Myogenin
  temporal expression in embryogenesis, **48**:133
  transcript accumulation in myotomal cells, **48**:239, 241
  transcription activation by Myf5, **48**:310
Myogenin, role in
  myocyte differentiation, **48**:139–142
  somite myogenesis, **48**:170–172
Myoseptum, horizontal, zebrafish mutants, **47**:263–264
Myotome
  developing zebrafish, **47**:267–269
  early mouse, initiated by Myf5, **48**:251–254
  early studies on, **48**:325–327
  fate map, **47**:162–163
  heterogeneity, and MRF expression patterns in somite, **48**:237–242
  mouse embryo at E11.5
    longitudinal organization, **47**:45–46
    mediolateral organization, **47**:46
  and muscle differentiation, in Xenopus, **47**:216–217
  prospective, elongated cells as, **47**:211–212, 215
  residual segmentation, **47**:254
  structural origin, polyclone size at, **47**:47, 52
Myotome cells
  early, differing from epaxial muscle cells, **48**:359
  elongated, rotation, **47**:212, 215–217
  elongation, **47**:218–219
  localization to myotome surface, **48**:241–242
  in Myf-5 mutant mice, **48**:136–138
  newborn, **48**:356–357
  rotation as group or as individuals, **47**:217
  traction and orientation, **47**:217–218
Myotome formation
  cellular mechanisms, **47**:217–220
  early embryonic, and growth dynamics, **48**:357–358
  studies with desmin, **48**:339–340, 342, 344
Myotome precursor cells
  determination time frame, **48**:344–350
  epaxial
    epithelial–mesenchymal transitions, **48**:354–356
    molecular marker expression, **48**:337–339

Index 381

location within somite, **48**:339–340, 342, 344
Myotube types, identified in graft, **48**:283–284

N

Neural arch, development from single somite, **48**:21
Neural crest, trunk, migration, *see* Trunk neural crest migration
Neural crest cells
  Eph family effect, **47**:285
  midline, preceding migration, **47**:279
  migrating *in situ,* **47**:290
  neuroepithelium-associated, migration, **48**:291
  selective migration through sclerotome, **47**:144
  in spatial relation to notochord, **47**:292–293
Neural tube
  in cartilage formation, **48**:88, 90, 92
  dorsal
    ectopic grafting, **48**:51–54
    role in morphogenesis of somite dorsal domain, **48**:65
    induction of sclerotome formation, **48**:115–116
  mouse donor
    implantation level, **48**:273
    preparation for creation of chimeras, **48**:271
  signals, for maintenance of epaxial somite myogenesis, **48**:179–182
  Wnt signals, **48**:196–200
Neural tube chimera, mouse–chick, epaxial myogenic lineage into, **48**:289–292
NF-Y protein, *Hoxb-4* gene promoter-binding, **48**:304
Node
  Henson's, somitic precursor cells derived from, **47**:12–13
  mouse gastrula, and somite respecification, **47**:11
Noggin
  and BMP4, regulation of somite domain boundaries, **48**:200–203
  as sclerotome initiation factor, **48**:108–109
Noggin
  as mediator of sclerotome induction, **48**:115–116
  and somite chondrogenesis, **48**:109–110

Notch
  activation
    along A/P boundary, **47**:146–147
    in ventral cells, **47**:142
  genes, role during somitogenesis, **47**:143
  homologs, in zebrafish, **47**:253
  intracellular targets, **47**:145
  proteins, and signal transduction, **47**:140–141
  signaling and transduction in embryos, **48**:206–207
Notch5, expression in presomitic mesoderm, **47**:20
Notch–Delta signaling
  in control of boundary within somite, **47**:122–123
  role in epithelial organization, **48**:206–207
  during somitogenesis, **47**:96–97
Notch receptors, DSL family proteins as ligands, **48**:206
Notch receptors, redundancy in somites, **47**:24
Notch signaling pathway
  and derepression, **48**:245–246
  in *Drosophila,* **47**:141–143
  negative auto feedback regulation, **47**:233
  presomitic mesoderm, **47**:23–25
  and questions of somite formation, **47**:271
Notochord
  development, *short-tail* mutant affecting, **48**:83–85
  effect on myogenesis, **48**:153
  embryos lacking, gastrulation in, **47**:205
  encircled by sclerotome-derived cells, **47**:229, **48**:81
  graft, effect on cartilage formation, **48**:54–56
  inducing capabilities, **48**:104
  induction of
    Pax-1 expression, **48**:11
    somite chondrogenesis, **48**·88, 95, 115–116
  and positioning of neural crest cells, **47**:292–293
  in postgastrula *Xenopus,* **47**:206–209
  prospective, urodeles, **47**:193–195
  role in
    muscle differentiation, **48**:31
    somite development, **47**:231–232
  signals
    disruption of cell morphogenetic organization, **48**:359
    for initiation of epaxial somite myogenesis, **48**:176–179

Notochord (*continued*)
zebrafish
mutants, **47**:261–262
ventralizing activity, **47**:260
Notochordal cells, *Hymenochirus*, variant form of ingression, **47**:193
Notochord challenge assay, **48**:345–350

O

Orientation, myotome cells, **47**:218
Oscillations
cellular states, in presomitic mesoderm, **47**:16
c-hairy-1 expression, **47**:118–120
segment formation preceded by, **47**:113–115
somitogenic cells, **47**:125
Overexpression, BMP2/4, effect on dorsal mesenchyme, **48**:61, 63

P

Parasegments, metameric, in long germ band insects, **47**:110–111
Paraxial mesoderm
anatomy, **47**:1–3
cell competence, **48**:346, 348
cellular displacement in, **47**:13–14
epiblastic domain, **47**:5–7
Hox c-11 expression, **47**:169, 172
mediolateral patterning, **48**:131
Pax gene expression, **48**:7
segmentation, **47**:134–135
in zebrafish, **47**:252–259
timing of induction, **47**:25
Paraxis
expression in segmental plate, **48**:9–10
role in somitogenesis, **47**:135–138
Paraxis
encoding bHLH, **47**:98
expression restricted to dermomyotome, **48**:183
somite-expressed, **48**:172
Patterning
axonal, Eph receptor–ligand interactions, **47**:284
global, *see* Global patterning
mediolateral, paraxial mesoderm, **48**:131
sequential gene expression during, **47**:74
somite, *see* Somite patterning
somitic mesoderm, by adjacent tissues, **47**:231–232

spatial, gene expression during myogenesis, **48**:211
Patterns, *see also* Prepattern
extracellular and cell surface molecules in somites, **47**:282–283
meristic, somitomeres, **47**:17–18
mesodermal
and local information, **47**:159–160
vertebrate body plan, **47**:157–159
segmental, trunk neural crest migration
Eph receptor role, **47**:289–291
PNA effect, **47**:291–292
Pax-1, sclerotome positive for, **48**:277
Pax3
expression, as marker of early-stage mesoderm cells, **48**:350–353
role in myogenesis, **48**:259–260
transcript abundance in hypaxial somitic bud, **48**:232–233
Pax3
activity upstream of *MyoD*, **48**:257, 259–260
expression maintenance by surface ectoderm, **48**:184
upstream gene in somite myogenesis, **48**:173–174
Pax-3
effect on migratory muscle progenitors, **48**:151–152
expression during early neurogenesis, **48**:294
Pax family genes
expression in
paraxial mesoderm, **48**:7
sclerotome, **48**:45–46, 50
and *MyoD* gene regulation, **48**:149
role in muscle development, **48**:311–312
as sclerotome markers, **48**:99–102, 116–117
P cells, reiteratively produced by S cells, **47**:60
Peanut lectin-binding molecules, effect on neural crest migration pattern, **47**:291–292
Pelobates fuscus, somite and myotome formation, **47**:222–223
Perinotochordal space, **48**:11
Perinotochordal tissue, changes in, **48**:15
Periodicity, segmental, and gene expression, **47**:101
Persistence, stem cells throughout axiogenesis, **47**:67, 73
Phenotype
myogenic, and targeted mutations in Wnt genes, **48**:199–200
somite, produced by mutating genes, **48**:9–11

Index 383

Plasticity, developmental, and commitment to somitic fate, **47**:15–16
Platelet-derived growth factor, downstream target of Myf5, **48**:252
Pleurocentrum, development, **48**:39
Pleurodeles waltl, involuted trunk mesoderm, **47**:194
Polarity
  craniocaudal, in formation of sclerotome, **48**:331
  dorsoventral, somite, **48**:13
  rostrocaudal, somite, **47**:280–282
Polymerization, in epaxial muscle, **48**:22
Positional information based model, somite formation, **47**:90
Positive signal hypothesis, somitic mesoderm, **47**:204
Precursor cells
  muscle, *see* Muscle precursor cells
  myogenic, morphogenetic movements, **48**:354–357
  myotome, *see* Myotome precursor cells
  somitic, *see* Somitic precursor cells
  vertebra, gene expression heterogeneity, **48**:48, 50
Prepattern
  segmental plate, **47**:139
  of segmentation, in presomitic mesoderm, **47**:16–25, 133–134
Presenilin family members, role in segmentation, **47**:145–147
Presomitic mesoderm
  amphibian, variability, **47**:185–198
  anteroposterior inversion, **47**:88
  cocultured with mature sclerotome, **48**:104
  explant studies
    neural tube support of myogenesis, **48**:181
    Wnt protein role in somite myogenesis, **48**:198–199
  future somite precursor cells feeding into, **48**:226
  laterally arranged cells in, **47**:62–63
  longitudinal organization, **47**:75
  prepattern of segmentation in, **47**:133–134
  existence of, **47**:16–25
  rostral, segmentation, **47**:99
  rostrocaudal inversion, **48**:237
  segmented, **47**:2–3
  transient lineage in, **47**:58
Primitive streak
  coherent organization, **47**:73
  epiblast cell recruitment to, **47**:2
  posterior movement, **47**:132
  recruitment and ingression through, **47**:7–12
  S cells residing in, **47**:59–60
  somitic precursor cells derived from, **47**:13–14
Proneural cells, cell fate restriction, **47**:146
Propagation, vertical, clone pools, **47**:55–56
Prospective somitic mesoderm, *Xenopus*
  cell intercalation, **47**:200–202
  geometry, **47**:198–200
  role in morphogenesis and differentiation, **47**:203–206
PS1, mutation, and somitogenesis defects, **47**:97
PS1 protein, detection along with amyloid precursor protein, **47**:146
Ptc, role in control of myogenesis, **48**:195–196

Q

Quail–chick chimera
  evidence for resegmentation, **47**:87
  information on somite development, **48**:270
  limb muscle determination, **48**:353
  studies on embryonic fate mapping, **48**:327–329

R

Reaction–diffusion based model, somite formation, **47**:90
Receptor tyrosine kinases, Eph-family, in neural patterning, **47**:283–287
Recombinant protein, N-Shh, **48**:193
Recruitment
  epiblast cells, to primitive streak, **47**:2
  through primitive streak, **47**:7–12
Redundancy
  functional
    among myogenic bHLH transcription factors, **48**:140–141
    FGFs, **48**:136
  Notch receptor, in somites, **47**:24
Region A, and region C, enhancers of *Hoxb-4* expression, **48**:303–304
Regionalization
  epithelial somites, **48**:130–131
  mediolateral, preceding bilateralization, **47**:61
Reporter genes
  CAT, **47**:163
  LacZ, **48**:281–282

Reporter genes (*continued*)
  nls LacZ, **47**:35, 38–39
  S cell, recombination, **47**:60
Resegmentation
  addressed by somite grafting experiments, **48**:19, 21
  in amphibians, **47**:230–231
  early studies, **48**:14–19
  sclerotome, **47**:160
  and sclerotome compartmentalization, **47**:85–89
  vertebrae formed through, **47**:137
Retinoic acid, *Hox* expression sensitive to, **47**:174
Rhombomere, boundaries between, **47**:124
Rib defect
  and MRF4 expression, **48**:143
  in Myf-5 mutant mice, **48**:136–138
  in *Myf5* null mice, **48**:254–255
Rosette formation, urodeles, **47**:223–226
Rostrocaudal compartments, somites, defining, **47**:22–25
Rostrocaudal polarity, somites, effect on trunk neural crest migration, **47**:280–282
Rotation
  elongated myotome cells, **47**:212, 215–217
  role in segmentation and muscle differentiation, **47**:219
  stage III somite, **48**:345
  whole somites, experiments, **48**:111

S

Satellite cells, survival in culture, **48**:323
Scanning electron microscopy, *Xenopus* somite formation, **47**:209–216
Scapula, fate map, **47**:163–165
Scatter factor, cross-talk with c-met, **48**:8
S cells, in embryonic nonbilateralized structures, **47**:58–60
Scleraxis, as sclerotome marker, **48**:102–103
Sclerotome
  amphibian, **47**:228–230
  cartilages derived from, early development, **48**:78–79
  compartmentalization, and resegmentation, **47**:85–89
  cranial and caudal halves, **48**:10–11
  determination of, **48**:110–113
  development
    anatomical and morphological description, **48**:79–83
    and BMPs, **48**:110

fate map, **47**:162
  formation, and epithelial–mesenchymal transitions, **48**:113
induction
  in vivo and *in vitro* experiments, **48**:103–106
  Shh role, **48**:106–108
  initiation, **48**:108–109
  like and unlike halves, boundary generation, **47**:123–124
  lineage, into somite mouse–chick chimera, **48**:276–279
  and lineage restrictions, **47**:111
  Mesp2 mutation effect, **47**:138
  molecular markers, **48**:99–103
  Pax gene expression, **48**:45–46
  subdivided according to areal fate, **48**:44
  zebrafish, **47**:260–261
Segmental organization, gene conservation in, **47**:108–110
Segmental plate
  chick embryo, explants, **48**:178
  Paraxis expression, **48**:9–10
  prepatterning, **47**:139
  reorientation, **47**:135
  Xenopus, **47**:211
Segmentation, *see also* Metamerism; Resegmentation
  in animal kingdom, **47**:82–83
  homology in *Drosophila* and vertebrates, **48**:39
  initiation, and somitomeres, **47**:84–85
  Notch signaling role, **47**:140–147
  paraxial mesoderm, zebrafish, **47**:252–259
  prepattern in presomitic mesoderm, **47**:16–25, 133–134
  process, control of, **47**:94–98
  role of
    morphomechanical molecules, **47**:235–239
    Presenilin family members, **47**:145–147
  vertebrate
    model, **47**:99–101
    molecular clock linked to, **47**:93–94
  Xenopus laevis, **47**:206–220, 236–237
  somite, **47**:202
Segments, *see also* Metameres; Parasegments
  boundaries, in lamprey, **48**:35
  dermomyotome
    mediolateral organization, **47**:61–64
    polyclonal contribution, **47**:47, 52
  mesodermal, definition, **47**:82–89
  myotome at E11.5

Index 385

longitudinal structure, **47**:45–46, 52–58
mediolateral organization, **47**:46
secondary structures, effect on sclerotome, **48**:12
short-tail mutant, effect on notochord development, **48**:83–85
Signaling molecules
  controlling myogenesis in somites, **48**:185, 191–207
  and interference, **48**:243–244
Signaling pathway
  c-met, **48**:259
  Notch
    presomitic mesoderm, **47**:23–25
    role in segmentation, **47**:140–147
  Notch–Delta
    in control of boundary within somite, **47**:122–123
    role in epithelial organization, **48**:206–207
    during somitogenesis, **47**:96–97
  in processes of myogenesis, **48**:210–211
  Shh, and control of myogenesis, **48**:195–196
Signaling systems, spatiotemporal differences in, **48**:244–245
Signals
  lateral mesoderm, spatial restriction of expaxial somite myogenesis, **48**:184–185
  local, associated information, **47**:156–157
  neural tube, for maintenance of epaxial somite myogenesis, **48**:179–182
  notochord, for initiation of epaxial somite myogenesis, **48**:176–179
  sources, influencing somite development, **48**:344–345
  surface ectoderm, control of hypaxial myogenesis, **48**:182–184
  from tissues adjacent to somites, effect on myogenesis, **48**:153–155
Signal transduction
  and activation of *MRF* genes, **48**:166–167
  BMP, mechanisms, **48**:200–201
  Sonic hedgehog, mechanisms, **48**:185, 191–192
  TGFβ, mechanisms, **48**:203–204
  Wnt proteins, mechanisms, **48**:196–197
Sim1, identification of central dermomyotome domain, **48**:232–234
Sim-1, somite-expressed, **48**:172
Skeletal muscle
  bHLH specific to, **48**:130
  development in MyoD mutant mice, **48**:138–139

lineages, early segregation during development, **48**:329–331
mesodermal origins in vertebrate embryos, **48**:167–168
origins in vertebrates, **48**:130–131, 226–227
phenotypic memory, **48**:323
progenitor stem cells, **48**:360
Smooth muscle cells, somite-derived, **48**:33
Somite chimera, mouse–chick, somitic lineage into, **48**:274–289
Somite chondrogenesis
  early inducer searches, **48**:97–98
  in vitro experiments, **48**:92–95
  in vivo experiments, **48**:85–92
  roles of *Shh* and *Noggin*, **48**:109–110
Somite formation
  during axis development, **47**:2–3
  in *Bombina* and *Gastrotheca*, **47**:220–222
  cell cycle model, **47**:115–117
  characteristics, **47**:83–84
  clock and wavefront model, **47**:112–113
  clock-based models, **47**:92–93
  control by surface ectoderm signals, **48**:182–184
  early somitogenesis models, **47**:89–90
  Meinhardt model, **47**:113–115
  myogenic determination acquired during, **48**:174–175
  in *Pelobates fuscus* and *Bufo bufo*, **47**:222–223
  questions related to Notch pathway, **47**:271
  reaction–diffusion and positional information based models, **47**:90
  somitogenic cluster model, **47**:90–91
  stem cell models, **47**:91–92
  urodeles: rosette formation, **47**:223–226
  in *Xenopus laevis*, **47**:206–220
Somite myogenesis
  control by signaling molecules and transduction pathways, **48**:185, 191–207
  myogenic regulatory factor gene role, **48**:170–172
  Pax3 role, **48**:173–174
  somite-expressed regulatory gene role, **48**:172–173
  tissue interactions controlling, **48**:174–185
Somite–neural tube chimera, benefits of, **48**:292, 294
Somite patterning
  amphibian, **47**:184–185, 239
  dorsoventral, **48**:70, 113–114, 154
  mediolateral, BMP4 effect, **48**:202
  zebrafish, **47**:259–261

Somites
  components, interactions between, **48**:114
  derivatives
    angioblasts, **48**:33–34
    dermis, **48**:32–33
    smooth muscle cells, **48**:33
  development, abnormal in Myf-5 mutant mice, **48**:136–138
  at different axial levels, **48**:235–237
  differentiation
    governed by extrinsic factors, **48**:242–246
    and tissue interactions, **47**:219–220
  dorsomedial quadrant, **48**:348–350, 358–359
  epaxial and hypaxial domains, mediolateral boundaries, **48**:200–203
  and epaxial–hypaxial organization of body, **48**:336–337
  epithelialization, **47**:98
    role in sclerotome induction, **48**:113
  extracellular and cell surface molecules in, patterns, **47**:282–283
  fate, commitment to, **47**:15–16
  forming, exposed to multiple signal interactions, **48**:180–181
  half-somite transplantation, **48**:331
  head, and initiation of segmentation, **47**:84–85
  identity
    determination of, **47**:125–126
    positional and segmental, **48**:301–302
  in ovo implanted mouse
    epaxial myogenesis in, **48**:281–286
    hypaxial myogenesis in, **48**:286–289
    location of myotome precursor cells, **48**:339–340, 342, 344
  and mesodermal segments, vertebrate, **47**:83
  morphogenesis, role of morphomechanical molecules, **47**:235–239
  mouse donor
    implantation level, **48**:273
    preparation for creation of chimeras, **48**:271
  MRF expression patterns, and myotome heterogeneity, **48**:237–242
  myogenic cells, fate of, **48**:30
  Notch receptor redundancy, **47**:24
  prospective, establishment, **47**:21–23
  research on, **48**:6–13
  responses to notochord, **48**:94
  tissues adjacent to, effect on myogenesis, **48**:153–155
  transcripts specific to, identification, **48**:312–313
  zebrafish, primarily myotome, **47**:248

Somitic cells
  avian scapula formed from, **47**:163–165
  developing, gene expression in, **48**:44–51
  global patterning, **47**:155–157
  lacZ-expressing, **48**:171
  postgastrula progression of behaviors, **47**:206–209
Somitic mesoderm
  amphibian, variations in formation, **47**:195–198
  local information and patterns, **47**:159–160
  origin in *Xenopus,* **47**:186–189
  patterning, by adjacent tissues, **47**:231–232
  prospective, *see* Prospective somitic mesoderm
  superficial epithelial contributions, **47**:189–195
Somitic precursor cells
  localization, **47**:3–15
  state of commitment, **47**:15–16
Somitocoel cells, mesenchymal, **48**:21
Somitogenesis
  and control of MRF gene expression, **48**:306–311
  early models, **47**:89–90
  Notch gene role during, **47**:143
  role of bHLH transcription factors, **47**:135–140
  stem cells as self-renewing pool for, **47**:11–12
  tail bud as source of new cells for, **47**:25
  vertebrate, **47**:132–135
    model, **47**:99–101
    molecular aspects, **47**:93–98
Somitogenic cluster model, somite formation, **47**:90–91
Somitomeres
  formation in amphibians, **47**:234–235
  and initiation of segmentation, **47**:84–85
  meristic pattern, **47**:17–18
Sonic hedgehog
  inhibition of subectodermal chondrogenesis, **48**:57–60
  signaling
    BMP4 effect, **48**:66
    downstream effectors, **48**:195–196
    and transduction in embryos, **48**:192–194
    signal transduction mechanisms, **48**:185, 191–192
    synergism with Wnt proteins, **48**:199, 211
    trophic effects, **48**:194–195
Sonic hedgehog
  as sclerotome inducer, **48**:106–108, 115–116
  and somite chondrogenesis, **48**:109–110

Index

Spatial restriction, epaxial somite myogenesis, **48**:184–185
Specification
  muscle, **48**:30–31
  myogenic
    control by MRFs, **48**:279
    signal sources influencing, **48**:344–345
Spemann organizer, **47**:203–206
Spinal cord
  inducing activity, **48**:94–95
  mouse neural implant-derived, differentiation in chick, **48**:290–291
Stem cell model, somite formation, **47**:91–92
Stem cells
  embryonic
    differentiation *in vitro*, **48**:308
    in Myf-5 mutant mice, **48**:137–138
    persistence throughout axiogenesis, **47**:67, 73
  progenitor, **48**:320
    and myogenic organizers, **48**:358–360
    self-renewing pool for somitogenesis, **47**:11–12
Superficial cells
  epithelial organization, **47**:197
  late additions to somitic mesoderm, **47**:195

T

Tail bud
  chick embryo, somitogenic potential, **47**:14–15
  source of new cells for somitogenesis, **47**:25
Targeted inactivation, myogenic factor genes, role in myogenesis, **48**:133–134, 136–145
Tbx6 mutation, **47**:9–10
Tenascin, in *Xenopus* somites, **47**:237
Tissue induction, differences in, reflected in fate map variations, **47**:195–196
Tissue interactions, control of somite myogenesis, **48**:174–185, 212
Tissues
  adjacent to somites, effect on myogenesis, **48**:153–155
  anlage, and interplay of fate and determination, **48**:320–323
  inducing somite chondrogenesis, **48**:93–95
  myogenic, development in embryo, **48**:353–360
  perinotochordal, changes in, **48**:15
  surrounding neural crest cells, interrelationship, **47**:282
T mouse, *see* Brachyury
Traction, in myotome cell rotation, **47**:217–218

Transcriptional control, MDF gene expression, **48**:148–151
Transcription factors
  bHLH, *see* Basic helix–loop–helix
  Ci, complex with Fu and Cos2, **48**:192
  collaboration with MDFs in myogenesis, **48**:146–148
  muscle-specific, MyoD family, **48**:131–151
  non-MRF, role in muscle development, **48**:311–312
  region C enhancer-binding, **48**:304–305
Transforming growth factor β
  effect on muscle differentiation, **48**:154–155
  signaling and transduction in embryos, **48**:204–205
  signal transduction mechanisms, **48**:203–204
Transgenic mice, *see also* Mouse mutants
  cell lines, labeling, **48**:295
  desmin LacZ, **48**:281–282
  Hox code, **47**:167–168
Transposition, vertebrate body plan, **47**:157–159
Trunk neural crest migration
  dynamic analysis, **47**:287–289
  effect of somitic rostrocaudal polarity, **47**:280–282
  role of cell–matrix interactions, **47**:292
  segmental pattern
    Eph receptor role, **47**:289–291
    peanut lectin-binding molecule effect, **47**:291–292
twist, bHLH, expression pattern, **48**:146–147

U

Urodeles
  fate maps, **47**:190–191
  prospective notochord, **47**:193–195
  resegmentation, **47**:230–231
  rosette formation, **47**:223–226

V

Ventral axial structures, promoting role in myogenesis, **48**:69–70
Ventral cells, Notch activation in, **47**:142
Vertebrae
  chondrogenesis
    BMP4 role, **48**:68
    molecular pathways leading to, **48**:57–65
    role of axial organs, **48**:88
  development, model for, **48**:65–70
  formation
    from sclerotome, **47**:229
    through resegmentation, **47**:137

Vertebrae (*continued*)
  fusions, **48**:63, 65
  origin, **48**:43–44
  patterning, *Hox* gene role, **47**:168–169
  Shh mutant mouse embryo lacking, **48**:107–108
Vertebral body
  contribution of medial half-somite, **48**:337
  growth from intervertebral disks, **48**:11–12
Vertebrate embryo
  mesodermal origin of skeletal muscle, **48**:167–168
  muscle precursor cell location, **48**:325–327
  polyclonal origin, **47**:44–45
Vertebrates
  body plan
    epaxial and hypaxial domains, **48**:331, 334–337
    Hox genes, **47**:165–168
    transposition, **47**:157–159
  evolution of metamerism, **48**:34–40
  and insects
    conservation of segmentation mechanisms, **47**:256–258
    segmentation differences, **48**:4–5
  intersomitic furrow, cellular decisions on, **47**:232–235
  segmental patterning system, **47**:168–173
  skeletal muscle origin in, **48**:130–131, 226–227
  somitogenesis, **47**:132–135
    model, **47**:99–101
    molecular aspects, **47**:93–98
Videomicroscopy
  neural crest cell movements, **47**:289
  Xenopus explants, **47**:206–209
von Ebner's fissure, sclerotome, **48**:82

W

Wnt-1, ectopic expression, **48**:114
Wnt-3a, mutation, **47**:9–10
Wnt proteins
  differential effects on *MyoD* and *Myf5* expression, **48**:242–244
  signaling and transduction in embryos, **48**:197–200
  signal transduction mechanisms, **48**:196–197

X

Xenopus
  dermatome, **47**:226–228
  embryos, post-heat shock defects, **47**:134
  presomitic mesoderm, cell mixing, **47**:20–21
  prospective somitic mesoderm
    cell intercalation, **47**:200–202
    geometry, **47**:198–200
    role in morphogenesis and differentiation, **47**:203–206
  sclerotome, **47**:228–230
    anteroposterior subdivision, **47**:88
  segmentation
    extracellular matrix role, **47**:236–237
    and somite formation, **47**:206–220
  somite morphogenesis, cell adhesion molecule role, **47**:237–238
  somitic mesoderm, origin, **47**:186–189
  Xcad-2, **47**:174

Y

you-type mutants, in zebrafish, **47**:262–264
YY1 protein, *Hoxb-4* gene promoter-binding, **48**:304

Z

Zebrafish
  Hh family proteins, lineage-specific myogenic functions, **48**:193–194
  innervation of somitic musculature, **47**:269–271
  as model organism, **47**:250–251
  muscle differentiation
    mutants, **47**:264–267
    slow and fast, **47**:267–269
  notochord role in somite myogenesis, **48**:178–179
  segmentation of paraxial mesoderm, **47**:252–259
  somite patterning, **47**:259–261
  somites, dorsal and ventral compartments, **47**:248
Zebrafish mutants
  defective in primary motoneuron development, **47**:270–271
  flh, **48**:155–156
  fss-type, **47**:253–256, 258–259, 270–271
  genetic screens, **47**:251
  muscle differentiation, **47**:264–267
  notochord, **47**:261–262
  you-type, **47**:262–264
Zone of extension, somitic mesoderm, **47**:209

Contents of Previous Volumes

Volume 42
Cumulative Subject Index,
Volumes 20 through 41

Volume 43

1. **Epigenetic Modification and Imprinting of the Mammalian Genome during Development**
   Keith E. Latham

2. **A Comparison of Hair Bundle Mechanoreceptors in Sea Anemones and Vertebrate Systems**
   Glen M. Watson and Patricia Mire

3. **Developmental of Neural Crest in *Xenopus***
   Roberto Mayor, Rodrigo Young, and Alexander Vargas

4. **Cell Determination and Transdetermination in *Drosophila* Imaginal Discs**
   Lisa Maves and Gerold Schubiger

5. **Cellular Mechanisms of Wingless/Wnt Signal Transduction**
   Herman Dierick and Amy Bejsovec

6. **Seeking Muscle Stem Cells**
   Jeffrey Boone Miller, Laura Schaefer, and Janice A. Dominov

7. **Neural Crest Diversification**
   Andrew K. Groves and Marianne Bronner-Fraser

8. **Genetic, Molecular, and Morphological Analysis of Compound Leaf Development**
   Tom Goliber, Sharon Kessler, Ju-Jiun Chen, Geeta Bharathan, and Neelima Sinha

Volume 44

1. **Green Fluorescent Protein (GFP) as a Vital Marker in Mammals**
   Masahito Ikawa, Shuichi Yamada, Tomoko Nakanishi, and Masaru Okabe

2. **Insights into Development and Genetics from Mouse Chimeras**
   John D. West

3. **Molecular Regulation of Pronephric Development**
   Thomas Carroll, John Wallingford, Dan Seufert, and Peter D. Vize

4. **Symmetry Breaking in the Zygotes of the Fucoid Algae: Controversies and Recent Progress**
   Kenneth R. Robinson, Michele Wozniak, Rongsun Pu, and Mark Messerli

5. **Reevaluating Concepts of Apical Dominance and the Control of Axillary Bud Outgrowth**
   Carolyn A. Napoli, Christine Anne Beveridge, and Kimberley Cathryn Snowden

6. **Control of Messenger RNA Stability during Development**
   Aparecida Maria Fontes, Jun-itsu Ito, and Marcelo Jacobs-Lorena

7. **EGF Receptor Signaling in *Drosophila* Oogenesis**
   Laura A. Nilson and Trudi Schupbach

Volume 45

1. **Development of the Leaf Epidermis**
   Philip W. Becraft

2. **Genes and Their Products in Sea Urchin Development**
   Giovanni Giudice

3. **The Organizer of the Gastrulating Mouse Embryo**
   Anne Camus and Patrick P. L. Tam

4. **Molecular Genetics of Gynoecium Development in *Arabidopsis***
   John L. Bowman, Stuart F. Baum, Yuval Eshed, Joanna Putterill, and John Alvarez

5. **Digging out Roots: Pattern Formation, Cell Division, and Morphogenesis in Plants**
   Ben Scheres and Renze Heidstra

Contents of Previous Volumes

Volume 46

1. **Maternal Cytoplasmic Factors for Generation of Unique Cleavage Patterns in Animal Embryos**
   Hiroki Nishida, Junji Morokuma, and Takahito Nishikata

2. **Multiple Endo-1,4-β-D-glucanase (Cellulase) Genes in *Arabidopsis***
   Elena del Campillo

3. **The Anterior Margin of the Mammalian Gastrula: Comparative and Phylogenetic Aspects of Its Role in Axis Formation and Head Induction**
   Christoph Viebahn

4. **The Other Side of the Embryo: An Appreciation of the Non-D Quadrants in Leech Embryos**
   David A. Weisblat, Françoise Z. Huang, Deborah E. Isaksen, Nai-Jia L. Liu, and Paul Chang

5. **Sperm Nuclear Activation during Fertilization**
   Shirley J. Wright

6. **Fibroblast Growth Factor Signaling Regulates Growth and Morphogenesis at Multiple Steps during Brain Development**
   Flora M. Vaccarino, Michael L. Schwartz, Rossana Raballo, Julianne Rhee, and Richard Lyn-Cook

Volume 47

1. **Early Events of Somitogenesis in Higher Vertebrates: Allocation of Precursor Cells during Gastrulation and the Organization of a Meristic Pattern in the Paraxial Mesoderm**
   Patrick P. L. Tam, Devorah Goldman, Anne Camus, and Gary C. Schoenwolf

2. **Retrospective Tracing of the Developmental Lineage of the Mouse Myotome**
   Sophie Eloy-Trinquet, Luc Mathis, and Jean-François Nicolas

3. **Segmentation of the Paraxial Mesoderm and Vertebrate Somitogenesis**
   Olivier Pourquié

4. **Segmentation: A View from the Border**
   Claudio D. Stern and Daniel Vasiliauskas

5 **Genetic Regulation of Somite Formation**
   Alan Rawls, Jeanne Wilson-Rawls, and Eric N. Olsen

6 **Hox Genes and the Global Patterning of the Somitic Mesoderm**
   Ann Campbell Burke

7 **The Origin and Morphogenesis of Amphibian Somites**
   Ray Keller

8 **Somitogenesis in Zebrafish**
   Scott A. Holley and Christiane Nüsslein-Volhard

9 **Rostrocaudal Differences within the Somites Confer Segmental Pattern to Trunk Neural Crest Migration**
   Marianne Bronner-Fraser